CAMBRIDGE
UNIVERSITY PRESS

Cambridge IGCSE™ and O Level
Additional
Mathematics

COURSEBOOK

Sue Pemberton

D1573999

CAMBRIDGE
UNIVERSITY PRESS

University Printing House, Cambridge CB2 8BS, United Kingdom

One Liberty Plaza, 20th Floor, New York, NY 10006, USA

477 Williamstown Road, Port Melbourne, VIC 3207, Australia

314–321, 3rd Floor, Plot 3, Splendor Forum, Jasola District Centre, New Delhi – 110025, India

103 Penang Road, #05–06/07, Visioncrest Commercial, Singapore 238467

Cambridge University Press is part of the University of Cambridge.

It furthers the University's mission by disseminating knowledge in the pursuit of education, learning and research at the highest international levels of excellence.

www.cambridge.org
Information on this title: www.cambridge.org/9781009293679

First published 2016
Second edition 2018
Third edition 2023

20 19 18 17 16 15 14 13 12 11 10 9 8 7 6 5 4 3 2 1

Printed in Italy by L.E.G.O S.p.A.

A catalogue record for this publication is available from the British Library

ISBN 978-1-009-29367-9 Cambridge IGCSE™ and O Level Additional Mathematics Coursebook with Cambridge Online Mathematics (2 Years' Access)

ISBN 978-1-009-34185-1 Cambridge IGCSE™ and O Level Additional Mathematics Cambridge Online Mathematics Course – Class Licence Access Card (1 Year Access)

ISBN 978-1-009-34183-7 Cambridge IGCSE™ and O Level Additional Mathematics Coursebook with Digital Version (2 Years' Access)

Additional resources for this publication at www.cambridge.org/go

ISBN 978-1-009-34802-7 Cambridge IGCSE™ and O Level Additional Mathematics Coursebook – Digital Version (2 Years' Access)

Cambridge University Press has no responsibility for the persistence or accuracy of URLs for external or third-party internet websites referred to in this publication, and does not guarantee that any content on such websites is, or will remain, accurate or appropriate. Information regarding prices, travel timetables, and other factual information given in this work is correct at the time of first printing but Cambridge University Press & Assessment does not guarantee the accuracy of such information thereafter.

....

CAMBRIDGE DEDICATED TEACHER AWARDS

2022

Teachers play an important part in shaping futures. Our Dedicated Teacher Awards recognise the hard work that teachers put in every day.

Thank you to everyone who nominated this year; we have been inspired and moved by all of your stories. Well done to all of our nominees for your dedication to learning and for inspiring the next generation of thinkers, leaders and innovators.

Congratulations to our incredible winners!

WINNER

Regional Winner Australia, New Zealand & South-East Asia	**Regional Winner** Europe	**Regional Winner** North & South America	**Regional Winner** Central & Southern Africa	**Regional Winner** Middle East & North Africa	**Regional Winner** East & South Asia
Mohd Al Khalifa Bin Mohd Affnan Keningau Vocational College, Malaysia	Dr. Mary Shiny Ponparambil Paul Little Flower English School, Italy	Noemi Falcon Zora Neale Hurston Elementary School, United States	Temitope Adewuyi Fountain Heights Secondary School, Nigeria	Uroosa Imran Beaconhouse School System KG-1 branch, Pakistan	Jeenath Akther Chittagong Grammar School, Bangladesh

For more information about our dedicated teachers and their stories, go to
dedicatedteacher.cambridge.org

CAMBRIDGE UNIVERSITY PRESS

Brighter Thinking
Better Learning

Building Brighter Futures **Together**

Endorsement statement

Endorsement indicates that a resource has passed Cambridge International's rigorous quality-assurance process and is suitable to support the delivery of a Cambridge International syllabus. However, endorsed resources are not the only suitable materials available to support teaching and learning, and are not essential to be used to achieve the qualification. Resource lists found on the Cambridge International website will include this resource and other endorsed resources.

Any example answers to questions taken from past question papers, practice questions, accompanying marks and mark schemes included in this resource have been written by the authors and are for guidance only. They do not replicate examination papers. In examinations the way marks are awarded may be different. Any references to assessment and/or assessment preparation are the publisher's interpretation of the syllabus requirements. Examiners will not use endorsed resources as a source of material for any assessment set by Cambridge International.

While the publishers have made every attempt to ensure that advice on the qualification and its assessment is accurate, the official syllabus, specimen assessment materials and any associated assessment guidance materials produced by the awarding body are the only authoritative source of information and should always be referred to for definitive guidance. Cambridge International recommends that teachers consider using a range of teaching and learning resources based on their own professional judgement of their students' needs.

Cambridge International has not paid for the production of this resource, nor does Cambridge International receive any royalties from its sale. For more information about the endorsement process, please visit www.cambridgeinternational.org/endorsed-resources

Third party websites and resources referred to in this publication have not been endorsed by Cambridge Assessment International Education.

❯ Contents

> Introduction

This highly illustrated coursebook covers the *Cambridge IGCSE™ and O Level Additional Mathematics and O Level* syllabuses (0606/4037). The course is aimed at students who are currently studying or have previously studied *Cambridge IGCSE™ Mathematics* (0580/0980) or *Cambridge O Level Mathematics* (4024) syllabuses.

Where the content in one chapter includes topics that should have already been covered in previous studies, a prerequisite knowledge section has been provided so that you can build on your prior knowledge.

'Discussion' sections have been included to provide you with the opportunity to discuss and learn new mathematical concepts with your classmates.

'Challenge' questions have been included at the end of most exercises to challenge and stretch you.

Towards the end of each chapter, there is a summary of the key concepts to help you consolidate what you have just learnt. This is followed by a 'Past paper' questions section, which contains questions taken from past papers for this syllabus.

A Practice Book is also available in the *IGCSE™ Additional Mathematics* series, which offers you further targeted practice. This book closely follows the chapters and topics of the coursebook, offering additional exercises to help you to consolidate concepts you have learnt and to assess your learning after each chapter.

> How to use this book

Throughout this book, you will notice lots of different features that will help your learning. These are explained below.

THIS SECTION WILL SHOW YOU HOW TO:

These set the scene for each chapter, help with navigation through the Coursebook and indicate the important concepts in each topic.

PRE-REQUISITE KNOWLEDGE

This feature shows how your understanding or use of a topic covered in another area of the book will help you with the concepts in this chapter.

TIP

The information in this feature will help you complete the exercises, and give you support in areas that you might find difficult.

KEY WORDS

The key vocabulary appears in a box at the start of each chapter, and is highlighted in the text when it is first introduced. You will also find definitions of these words in the Glossary at the back of this book.

ACTIVITY

Activities give you an opportunity to apply your understanding of a concept to a practical task. When activities have answers, you can find these in the digital version of the Coursebook.

WORKED EXAMPLE

These boxes show you the step-by-step process to work through an example question or problem, giving you the skills to work through questions yourself.

CLASS DISCUSSION

At certain points in the chapters you will be given opportunities to talk about your learning and understanding of the topic in a small group or with a partner.

REFLECTION

These activities ask you to think about the approach that you take to your work, and how you might improve this in the future.

Exercises

Appearing throughout the text, exercises give you a chance to check that you have understood the topic you have just read about and practice the mathematical skills you have learned. You can find the answers to these questions in the digital version of the Coursebook.

CHALLENGE QUESTIONS

These exercises will stretch your skills in the topic you have just learned. You can find the answers to these questions in the digital version of the Coursebook.

Past paper questions

Questions at the end of each chapter provide a variety of past paper questions, some of which may require use of knowledge from previous chapters. Answers to these questions can be found in the digital version of the Coursebook.

SUMMARY

There is a summary of key points at the end of each chapter.

 This icon shows you where you should complete an exercise without using your calculator.

> How to use this series

This suite of resources supports learners and teachers following the Cambridge IGCSE™ and O Level Additional Mathematics syllabuses (0606/4037). Up-to-date metacognition techniques have been incorporated throughout the resources to meet the changes in the syllabuses content and develop a complete understanding of mathematics for learners. All of the components in the series are designed to work together.

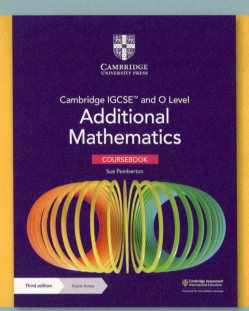

The coursebook contains sixteen chapters that together offer complete coverage of the syllabus. We have worked with NRICH to provide a variety of project activities, designed to engage learners and strengthen their problem-solving skills. Each chapter contains opportunities for formative assessment, differentiation and peer and self-assessment offering learners the support needed to make progress. Cambridge Online Mathematics is available through the digital/print bundle option or on its own without the print coursebook. Learners can review content digitally, explore worked examples and test their knowledge with quiz questions and answers. Teachers benefit from the ability to set tests and tasks with the added auto-marking functionality and a reporting dashboard to help track learner progress quickly and easily.

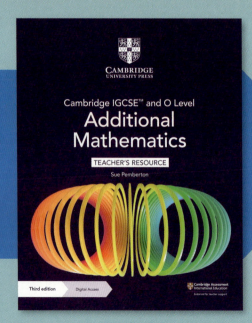

The digital teacher's resource provides extensive guidance on how to teach the course, including suggestions for differentiation, formative assessment and language support, teaching ideas and PowerPoints. The Teaching Skills Focus shows teachers how to incorporate a variety of key pedagogical techniques into teaching, including differentiation, assessment for learning, and metacognition. Answers for all components are accessible to teachers for free on the Cambridge GO platform.

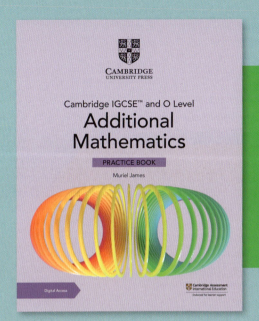

A Practice Book is available for learners that wish to have extra questions to work through. This resource which can be used in class or assigned as homework, provide a wide variety of extra maths activities and questions to help learners consolidate their learning and prepare for assessment. 'Tips' are also regularly featured to give learners extra advice and guidance on the different areas of maths they encounter. Access to the digital versions of the practice books is included, and answers can be found either here or in the back of the books.

A Worked Solutions Manual has been introduced to the series. This offers a fully worked solution, with annotated comments, to a selection of questions for teachers or learners to use as they work through the content.

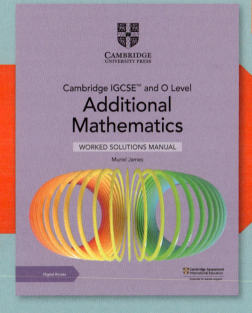

> Acknowledgements

The authors and publishers acknowledge the following sources of copyright material and are grateful for the permissions granted. While every effort has been made, it has not always been possible to identify the sources of all the material used, or to trace all copyright holders. If any omissions are brought to our notice, we will be happy to include the appropriate acknowledgements on reprinting.

Cambridge International copyright material in this publication is reproduced under licence and remains the intellectual property of Cambridge Assessment International Education. Cambridge Assessment International Education bears no responsibility for the example answers to questions taken from its past question papers which are contained in this publication. In examinations, the way marks are awarded may be different.

Thanks to the following for permission to reproduce images:

Cover Sven Krobot/EyeEm/Getty Images; *Inside* **Unit 1** Fan jianhua/Shutterstock; **Unit 2** Zhu difeng/Shutterstock; **Unit 3** Michael Dechev/Shutterstock; **Unit 4** Aitor Diago/Getty Images; **Unit 5** Peshkova/Shutterstock; **Unit 6** Ittipon/Shutterstock; **Unit 7** MirageC/Getty Images; **Unit 8** Zhu Qiu/Getty Images; **Unit 9** Munro1/Getty Images; **Unit 10** Gino Santa Maria/Shutterstock; Snake3d/Shutterstock; Keith Publicover/Shutterstock; Aleksandr Kurganov/Shutterstock; Africa Studio/Shutterstock; **Unit 11** Oscar Alvarez/Getty Images; **Unit 12** AlenKadr/Shutterstock; **Unit 13** Ben Welsh/Getty Images; **Unit 14** Nadla/Getty Images; **Unit 15** Neamov/Shutterstock; **Unit 16** Ahuli Labutin/Shutterstock

Chapter 1
Functions

PRE-REQUISITE KNOWLEDGE

Before you start…

Where it comes from	What you should be able to do	Check your skills
Cambridge IGCSE/O Level Mathematics	Find an output for a given function.	**1** If $f(x) = 5x - 1$, find $f(3)$.
Cambridge IGCSE/O Level Mathematics	Find a composite function.	**2** If $f(x) = 3x - 2$ and $g(x) = 4 - x$, find $fg(x)$.
Cambridge IGCSE/O Level Mathematics	Find the inverse of a simple function.	**3** If $f(x) = 3x + 5$, find $f^{-1}(x)$.
Cambridge IGCSE/O Level Mathematics	Sketch linear and quadratic graphs.	**4** **a** Sketch the graph of $y = 2x - 1$. **b** Sketch the graph of $y = x^2 + 1$.
Cambridge IGCSE/O Level Mathematics	Solve linear and quadratic equations.	**5** **a** Solve $5 - 3x = 8$. **b** Solve $(x + 2)^2 = 16$.

1.1 Mappings

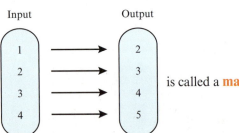

is called a **mapping diagram**.

The rule connecting the input and output values can be written algebraically as: $x \mapsto x + 1$.

This is read as 'x is mapped to $x + 1$'.

The mapping can be represented graphically by plotting values of $x + 1$ against values of x.

The diagram shows that for one input value there is just one output value.

It is called a one-one mapping.

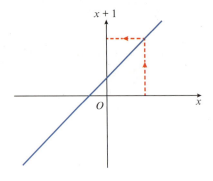

KEY WORDS

mapping diagram

function

one-one function

domain

range

composite function

modulus

absolute value

self-inverse functions

The table below shows one-one, many-one and one-many mappings.

one-one	many-one	one-many
For one input value there is just one output value.	For two input values there is one output value.	For one input value there are two output values.

Exercise 1.1

Determine whether each of these mappings is one-one, many-one or one-many.

1 $x \mapsto x + 1$ $x \in \mathbb{R}$

2 $x \mapsto x^2 + 5$ $x \in \mathbb{R}$

3 $x \mapsto x^3$ $x \in \mathbb{R}$

4 $x \mapsto 2^x$ $x \in \mathbb{R}$

5 $x \mapsto \dfrac{1}{x}$ $x \in \mathbb{R}, x > 0$

6 $x \mapsto x^2 + 1$ $x \in \mathbb{R}, x \geq 0$

7 $x \mapsto \dfrac{12}{x}$ $x \in \mathbb{R}, x > 0$

8 $x \mapsto \pm x$ $x \in \mathbb{R}, x \geq 0$

1.2 Definition of a function

A **function** is a rule that maps each x value to just one y value for a defined set of input values.

This means that mappings that are either $\begin{cases} \text{one-one} \\ \text{many-one} \end{cases}$ are called functions.

The mapping $x \mapsto x + 1$, where $x \in \mathbb{R}$, is a **one-one function**.

It can be written as $\begin{cases} f : x \mapsto x + 1 & x \in \mathbb{R} \\ f(x) = x + 1 & x \in \mathbb{R} \end{cases}$

($f : x \mapsto x + 1$ is read as 'the function f, such that x is mapped to $x + 1$')

$f(x)$ represents the output values for the function f.

So when $f(x) = x + 1$, $f(2) = 2 + 1 = 3$.

The set of input values for a function is called the **domain** of the function.

The set of output values for a function is called the **range** (or image set) of the function.

WORKED EXAMPLE 1

$f(x) = 2x - 1$ $\qquad x \in \mathbb{R}, -1 \leqslant x \leqslant 3$

a Write down the domain of the function f.

b Sketch the graph of the function f.

c Write down the range of the function f.

Answers

a The domain of f is $-1 \leqslant x \leqslant 3$.

b The graph of $y = 2x - 1$ has gradient 2 and a y-intercept of -1.

When $x = -1$, $y = 2(-1) - 1 = -3$

When $x = 3$, $y = 2(3) - 1 = 5$

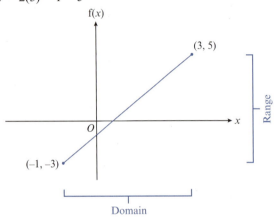

c The range of f is $-3 \leqslant f(x) \leqslant 5$.

WORKED EXAMPLE 2

The function f is defined by $f(x) = (x - 2)^2 + 3$ for $0 \leqslant x \leqslant 6$.

Sketch the graph of the function.

Find the range of f.

Answers

$f(x) = (x - 2)^2 + 3$ is a positive quadratic function, so the graph will be of the form $\bigvee$

 $(x - 2)^2 + 3$ This part of the expression is a square so it will always be $\geqslant 0$. The smallest value it can be is 0. This occurs when $x = 2$.

CONTINUED

The minimum value of the expression is $0 + 3 = 3$ and this minimum occurs when $x = 2$.

So the function $f(x) = (x - 2)^2 + 3$ will have a minimum point at the point $(2, 3)$.

When $x = 0$, $y = (0 - 2)^2 + 3 = 7$.

When $x = 6$, $y = (6 - 2)^2 + 3 = 19$.

The range of f is $3 \le f(x) \le 19$.

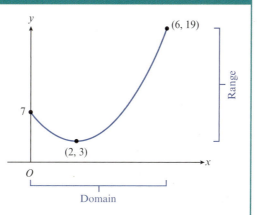

Exercise 1.2

1 Which of the mappings in **Exercise 1.1** are functions?

2 Find the range for each of these functions.

   a $f(x) = x - 5$, $-2 \le x \le 7$ b $f(x) = 3x + 2$, $0 \le x \le 5$

   c $f(x) = 7 - 2x$, $-1 \le x \le 4$ d $f(x) = x^2$, $-3 \le x \le 3$

   e $f(x) = 2^x$, $-3 \le x \le 3$ f $f(x) = \dfrac{1}{x}$, $1 \le x \le 5$

3 The function g is defined as $g(x) = x^2 + 2$ for $x \ge 0$.
   Find the range of g.

4 The function f is defined by $f(x) = x^2 - 4$ for $x \in \mathbb{R}$.
   Find the range of f.

5 The function f is defined by $f(x) = (x - 1)^2 + 5$ for $x \ge 1$.
   Find the range of f.

6 The function f is defined by $f(x) = (2x + 1)^2 - 5$ for $x \ge -\dfrac{1}{2}$.
   Find the range of f.

7 The function f is defined by $f : x \mapsto 10 - (x - 3)^2$ for $2 \le x \le 7$.
   Find the range of f.

8 The function f is defined by $f(x) = 3 + \sqrt{x - 2}$ for $x \ge 2$.
   Find the range of f.

1.3 Composite functions

Most functions that you meet are combinations of two or more functions.

For example, the function $x \mapsto 2x + 5$ is the function 'multiply by 2 and then add 5'.

It is a combination of the two functions g and f where:

$g: x \mapsto 2x$ (the function 'multiply by 2')

$f: x \mapsto x + 5$ (the function 'add 5')

So, $x \mapsto 2x + 5$ is the function described as 'first do g, then do f'.

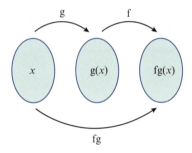

When one function is followed by another function, the resulting function is called a **composite function**.

> There are three important points to remember about composite functions
> fg only exists if the range of g is contained within the domain of f.
> In general, $fg(x) \neq gf(x)$.
> $ff(x)$ means you apply the function f twice.

WORKED EXAMPLE 3

The function f is defined by $f(x) = (x - 2)^2 - 3$ for $x > -2$.

The function g is defined by $g(x) = \dfrac{2x + 6}{x - 2}$ for $x > 2$.

Find $fg(7)$.

Answers

$fg(7)$ g acts on 7 first and $g(7) = \dfrac{2(7) + 6}{7 - 2} = 4$

$= f(4)$ f is the function 'subtract 2, square the result and then subtract 3'

$= (4 - 2)^2 - 3$

$= 1$

WORKED EXAMPLE 4

$f(x) = 2x - 1$ for $x \in \mathbb{R}$ $g(x) = x^2 + 5$ for $x \in \mathbb{R}$

Find   **a** $fg(x)$ **b** $gf(x)$ **c** $f^2(x)$.

Answers

a $fg(x)$ g acts on x first and $g(x) = x^2 + 5$

    $= f(x^2 + 5)$ f is the function 'double and subtract 1'

    $= 2(x^2 + 5) - 1$

    $= 2x^2 + 9$

b $gf(x)$ f acts on x first and $f(x) = 2x - 1$

    $= g(2x - 1)$ g is the function 'square and add 5'

    $= (2x - 1)^2 + 5$ expand brackets

    $= 4x^2 - 4x + 1 + 5$

    $= 4x^2 - 4x + 6$

c $f^2(x)$ $f^2(x)$ means $ff(x)$

    $= ff(x)$ f acts on x first and $f(x) = 2x - 1$

    $= f(2x - 1)$ f is the function 'double and subtract 1'

    $= 2(2x - 1) - 1$

    $= 4x - 3$

WORKED EXAMPLE 5

$f(x) = x^2 + 4$ for $x \in \mathbb{R}, x < 0$ $g(x) = 8 - \dfrac{x}{2}$ for $x \in \mathbb{R}, x > 4$

Find the domain and range of $fg(x)$.

Answer

Domain of $fg(x)$:

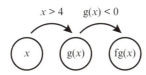

$x > 4$ and $g(x) < 0$

       $8 - \dfrac{x}{2} < 0$

         $x > 16$

Overlap of $x > 4$ and $x > 16$ is $x > 16$

Domain of $fg(x)$ is: $x \in \mathbb{R}, x > 16$

CONTINUED

Range of fg(x):

$$fg(x) = \left(8 - \frac{x}{2}\right)^2 + 4$$

This is a quadratic curve and the turning point occurs when $8 - \frac{x}{2} = 0$

$$x = 16$$

Hence turning point is (16, 4).

Range of fg(x) is: $fg(x) \in \mathbb{R},\ fg(x) > 4$

(16, 4)

Exercise 1.3

1   $f : x \mapsto 2x + 3$ for $x \in \mathbb{R}$
    $g : x \mapsto x^2 - 1$ for $x \in \mathbb{R}$
    Find fg(2).

2   $f(x) = x^2 - 1$ for $x \in \mathbb{R}$
    $g(x) = 2x + 3$ for $x \in \mathbb{R}$
    Find gf(5).

3   $f(x) = (x + 2)^2 - 1$ for $x \in \mathbb{R}$
    Find $f^2(3)$.

4   The function f is defined by $f(x) = 1 + \sqrt{x - 2}$ for $x \geq 2$.

    The function g is defined by $g(x) = \frac{10}{x} - 1$ for $x > 0$.
    Find gf(18).

5   The function f is defined by $f(x) = (x - 1)^2 + 3$ for $x > -1$.

    The function g is defined by $g(x) = \frac{2x + 4}{x - 5}$ for $x > 5$.
    Find fg(7).

6   $h : x \mapsto x + 2$ for $x > 0$
    $k : x \mapsto \sqrt{x}$ for $x > 0$
    Express each of the following in terms of h and k.

    a $x \mapsto \sqrt{x} + 2$ **b** $x \mapsto \sqrt{x + 2}$

7   The function f is defined by $f : x \mapsto 3x + 1$ for $x \in \mathbb{R}$.

    The function g is defined by $g : x \mapsto \frac{10}{2 - x}$ for $x \neq 2$.
    Solve the equation gf(x) = 5.

8   $g(x) = x^2 + 2$ for $x \in \mathbb{R}$
    $h(x) = 3x - 5$ for $x \in \mathbb{R}$
    Solve the equation gh(x) = 51.

9   $f(x) = x^2 - 3$ for $x > 0$

$g(x) = \dfrac{3}{x}$ for $x > 0$

Solve the equation $fg(x) = 13$.

10   The function f is defined by $f : x \mapsto \dfrac{3x + 5}{x - 2}$, $x \neq 2$, for $x \in \mathbb{R}$, $x \neq 2$.

The function g is defined by $g : x \mapsto \dfrac{x - 1}{2}$, for $x \in \mathbb{R}$.

Solve the equation $gf(x) = 12$.

11   $f(x) = (x + 4)^2 + 3$ for $x > 0$

$g(x) = \dfrac{10}{x}$ for $x > 0$

Solve the equation $fg(x) = 39$.

12   The function g is defined by $g(x) = x^2 - 1$ for $x \geqslant 0$.

The function h is defined by $h(x) = 2x - 7$ for $x \geqslant 0$.

Solve the equation $gh(x) = 0$.

13   The function f is defined by $f : x \mapsto x^3$ for $x \in \mathbb{R}$.

The function g is defined by $g : x \mapsto x - 1$ for $x \in \mathbb{R}$.

Express each of the following as a composite function, using only f and/or g:

a $x \mapsto (x - 1)^3$ b $x \mapsto x^3 - 1$

c $x \mapsto x - 2$ d $x \mapsto x^9$

14   $f(x) = \dfrac{x}{x + 2}$ for $x \in \mathbb{R}$, $x \neq -2$ $g(x) = \dfrac{3}{x}$ for $x \in \mathbb{R}$, $x \neq 0$

Find the domain of $fg(x)$.

15   $f(x) = x^2 - 9$ for $x \in \mathbb{R}$, $x < 0$ $g(x) = 10 - \dfrac{x}{2}$ for $x \in \mathbb{R}$, $x > 6$

Find the domain and range of $fg(x)$.

16   $f(x) = \dfrac{1}{x}$ for $x \in \mathbb{R}$, $x \neq 0$ $g(x) = \dfrac{1}{x - 1}$ for $x \in \mathbb{R}$, $x \neq 1$

Find the domain of   a $fg(x)$ b $gf(x)$

17   $f(x) = 2x - 6$ for $x \in \mathbb{R}$ $g(x) = \sqrt{x}$ for $x \in \mathbb{R}$, $x \geqslant 0$

Find the domain and range of   a $fg(x)$ b $gf(x)$

18   $f(x) = 2x^2 - x$ for $x \in \mathbb{R}$ $g(x) = \dfrac{1}{x}$ for $x \in \mathbb{R}$, $x \neq 0$

a   Find the range of   i $f(x)$ ii $g(x)$

b   Find the domain and range of   i $fg(x)$ ii $gf(x)$

19   $f(x) = 2x + 5$ for $x \in \mathbb{R}$, $x < 2$ $g(x) = (x - 3)^2$ for $x \in \mathbb{R}$, $x > 3$

a   Find the range of   i $f(x)$ ii $g(x)$

b   Find $gf(x)$.

c   Find the domain and range of $gf(x)$.

20 $f(x) = 2x - 1$ for $x \in \mathbb{R}, x > 3$ $\qquad$ $g(x) = x^2 - 2$ for $x \in \mathbb{R}, x > 7$

    a Find the range of $\quad$ **i** $f(x)$ $\quad$ **ii** $g(x)$

    b Find gf(x).

    c Find the domain and range of gf(x).

21 $f(x) = 3x + 2$ for $x \in \mathbb{R}, x > 1$ $\qquad$ $g(x) = x^2 + 1$ for $x \in \mathbb{R}, x > 8$

    a Find the range of $\quad$ **i** $f(x)$ $\quad$ **ii** $g(x)$

    b Find gf(x).

    c Find the domain and range of gf(x).

1.4 Modulus functions

The **modulus** of a number is the magnitude of the number without a sign attached.

The modulus of 4 is written $|4|$.

$|4| = 4$ and $|-4| = 4$

It is important to note that the modulus of any number (positive or negative) is always a positive number.

The modulus of a number is also called the **absolute value**.

The modulus of x, written as $|x|$, is defined as:

$$|x| = \begin{cases} x & \text{if } x > 0 \\ 0 & \text{if } x = 0 \\ -x & \text{if } x < 0 \end{cases}$$

CLASS DISCUSSION

Ali says that these are all rules for absolute values:

$$|x + y| = |x| + |y| \qquad\qquad |x - y| = |x| - |y|$$

$$|xy| = |x| \times |y| \qquad \left|\frac{x}{y}\right| = |x| \div |y| \qquad (|x|)^2 = x^2$$

Discuss each of these statements with your classmates and decide if they are:

(Always true) (Sometimes true) (Never true)

You must justify your decisions.

The statement $|x| = k$, where $k \geq 0$, means that $x = k$ or $x = -k$.

This property is used to solve equations that involve modulus functions.

So, if you are solving equations of the form $|ax + b| = k$, you solve the equations

$ax + b = k \quad$ and $\quad ax + b = -k$

If you are solving harder equations of the form $|ax + b| = cx + d$, you solve the equations

$ax + b = cx + d$ and $ax + b = -(cx + d)$.

When solving these more complicated equations, you must always check your answers to make sure that they satisfy the original equation.

> **WORKED EXAMPLE 6**
>
> Solve.
>
> **a** $|2x + 1| = 5$ **b** $|4x - 3| = x$ **c** $|x^2 - 10| = 6$ **d** $|x - 3| = 2x$
>
> **Answers**
>
> **a** $|2x + 1| = 5$
>
> $2x + 1 = 5$ or $2x + 1 = -5$
>
> $\quad 2x = 4 \qquad\qquad 2x = -6$
>
> $\quad\quad x = 2 \qquad\qquad\quad x = -3$
>
> CHECK: $|2 \times 2 + 1| = 5$ ✓ and $|2 \times -3 + 1| = 5$ ✓
>
> Solution is: $x = -3$ or 2.
>
> **b** $|4x - 3| = x$
>
> $4x - 3 = x$ or $4x - 3 = -x$
>
> $\quad 3x = 3 \qquad\qquad 5x = 3$
>
> $\quad\, x = 1 \qquad\qquad\, x = 0.6$
>
> CHECK: $|4 \times 0.6 - 3| = 0.6$ ✓ and $|4 \times 1 - 3| = 1$ ✓
>
> Solution is: $x = 0.6$ or 1.
>
> **c** $|x^2 - 10| = 6$
>
> $x^2 - 10 = 6$ or $x^2 - 10 = -6$
>
> $\quad x^2 = 16 \qquad\qquad x^2 = 4$
>
> $\quad\, x = \pm 4 \qquad\qquad\, x = \pm 2$
>
> CHECK: $|(-4)^2 - 10| = 6$ ✓, $|(-2)^2 - 10| = 6$ ✓, $|(2)^2 - 10| = 6$ ✓
>
> $\qquad$ and $|(4)^2 - 10| = 6$ ✓
>
> Solution is: $x = -4, -2, 2$ or 4.
>
> **d** $|x - 3| = 2x$
>
> $x - 3 = 2x$ or $x - 3 = -2x$
>
> $\quad x = -3 \qquad\qquad 3x = 3$
>
> $\qquad\qquad\qquad\qquad\, x = 1$
>
> CHECK: $|-3 - 3| = 2 \times -3$ ✗ and $|1 - 3| = 2 \times 1$ ✓
>
> Solution is: $x = 1$.

Exercise 1.4

1 Solve each equation for x.

a $|3x - 2| = 10$ **b** $|2x + 9| = 5$ **c** $|6 - 5x| = 2$

d $\left|\dfrac{x - 1}{4}\right| = 6$ **e** $\left|\dfrac{2x + 7}{3}\right| = 1$ **f** $\left|\dfrac{7 - 2x}{2}\right| = 4$

g $\left|\dfrac{x}{4} - 5\right| = 1$ **h** $\left|\dfrac{x + 1}{2} + \dfrac{2x}{5}\right| = 4$ **i** $|2x - 5| = x$

2 Solve each equation for x.

a $\left|\dfrac{2x - 5}{x + 3}\right| = 8$ **b** $\left|\dfrac{3x + 2}{x + 1}\right| = 2$ **c** $\left|1 + \dfrac{x + 12}{x + 4}\right| = 3$

d $|3x - 5| = x + 2$ **e** $x + |x - 5| = 8$ **f** $9 - |1 - x| = 2x$

3 Solve each equation for x.

a $|x^2 - 1| = 3$ **b** $|x^2 + 1| = 10$ **c** $|4 - x^2| = 2 - x$

d $|x^2 - 5x| = x$ **e** $|x^2 - 4| = x + 2$ **f** $|x^2 - 3| = x + 3$

g $|2x^2 + 1| = 3x$ **h** $|2x^2 - 3x| = 4 - x$ **i** $|x^2 - 7x + 6| = 6 - x$

4 Solve each pair of simultaneous equations.

a $y = x + 4$ **b** $y = x$ **c** $y = 3x$
$y = |x^2 - 16|$ $y = |3x - 2x^2|$ $y = |2x^2 - 5|$

REFLECTION

Look back at this section on solving modulus equations.

a What did you find easy?

b What did you find difficult?

c Are there any parts you need to practice more?

1.5 Graphs of $y = |\,f(x)\,|$ where $f(x)$ is linear

Consider drawing the graph of $y = |x|$.

First draw the graph of $y = x$.

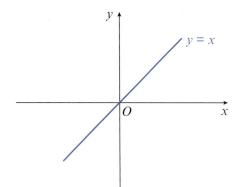

Then reflect in the x-axis the part of the line that is below the x-axis.

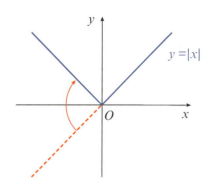

WORKED EXAMPLE 7

Sketch the graph of $y = \left| \dfrac{1}{2}x - 1 \right|$, showing the coordinates of the points where the graph intersects the axes.

Answers

First sketch the graph of $y = \dfrac{1}{2}x - 1$.

The line has gradient $\dfrac{1}{2}$ and a y-intercept of -1.

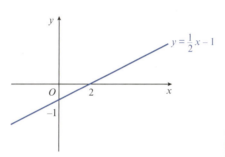

Then reflect in the x-axis the part of the line that is below the x-axis.

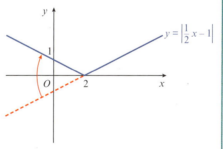

In **Worked example 5** you saw that there were two answers, $x = -3$ or $x = 2$, to the equation

$|2x + 1| = 5$.

These can also be found graphically by finding the x-coordinates of the points of intersection of the graphs of $y = |2x + 1|$ and $y = 5$ as shown.

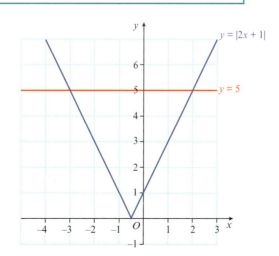

In the same worked example, you also saw that there was only one answer, $x = 1$, to the equation $|x - 3| = 2x$.

This can also be found graphically by finding the x-coordinates of the points of intersection of the graphs of $y = |x - 3|$ and $y = 2x$ as shown.

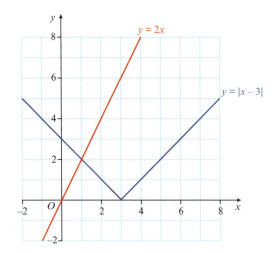

Exercise 1.5

1 Sketch the graphs of each of the following functions, showing the coordinates of the points where the graph intersects the axes.

    a $y = |x + 1|$ **b** $y = |2x - 3|$ **c** $y = |5 - x|$

    d $y = \left|\dfrac{1}{2}x + 3\right|$ **e** $y = |10 - 2x|$ **f** $y = \left|6 - \dfrac{1}{3}x\right|$

2   **a**   Complete the table of values for $y = |x - 2| + 3$.

x	−2	−1	0	1	2	3	4
y		6		4			

    b   Draw the graph of $y = |x - 2| + 3$ for $-2 \leqslant x \leqslant 4$.

3 Draw the graph of each of the following functions.

    a $y = |x| + 1$ **b** $y = |x| - 3$ **c** $y = 2 - |x|$

    d $y = |x - 3| + 1$ **e** $y = |2x + 6| - 3$

4 Given that each of these functions is defined for the domain $-3 \leqslant x \leqslant 4$, find the range of

    a $f : x \mapsto 5 - 2x$ **b** $g : x \mapsto |5 - 2x|$ **c** $h : x \mapsto 5 - |2x|$.

5 $f : x \mapsto 3 - 2x$ for $-1 \leqslant x \leqslant 4$

    $g : x \mapsto |3 - 2x|$ for $-1 \leqslant x \leqslant 4$

    $h : x \mapsto 3 - |2x|$ for $-1 \leqslant x \leqslant 4$

    Find the range of each function.

6   **a**   Sketch the graph of $y = |2x + 4|$ for $-6 < x < 2$, showing the coordinates of the points where the graph intersects the axes.

    b   On the same diagram, sketch the graph of $y = x + 5$.

    c   Solve the equation $|2x + 4| = x + 5$.

7   A function f is defined by $f(x) = |2x - 6| - 3$, for $-1 \leqslant x \leqslant 8$.

    a   Sketch the graph of $y = f(x)$.

    b   State the range of f.

    c   Solve the equation $f(x) = 2$.

8   a   Sketch the graph of $y = |3x - 4|$ for $-2 < x < 5$, showing the coordinates of the points where the graph intersects the axes.

    b   On the same diagram, sketch the graph of $y = 2x$.

    c   Solve the equation $2x = |3x - 4|$.

9   **CHALLENGE QUESTION**

    a   Sketch the graph of $f(x) = |x + 2| + |x - 2|$.

    b   Use your graph to solve the equation $|x + 2| + |x - 2| = 6$.

1.6 Inverse functions

The inverse of a function $f(x)$ is the function that undoes what $f(x)$ has done.

The inverse of the function $f(x)$ is written as $f^{-1}(x)$.

The domain of $f^{-1}(x)$ is the range of $f(x)$.

The range of $f^{-1}(x)$ is the domain of $f(x)$.

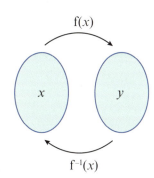

It is important to remember that not every function has an inverse.

An inverse function $f^{-1}(x)$ can exist if, and only if, the function $f(x)$ is a one-one mapping.

You should already know how to find the inverse function of some simple one-one mappings.

The steps to find the inverse of the function $f(x) = 5x - 2$ are:

Step 1: Write the function as $y = \quad \longrightarrow \quad y = 5x - 2$

Step 2: Interchange the x and y variables. $\longrightarrow \quad x = 5y - 2$

Step 3: Rearrange to make y the subject. $\longrightarrow \quad y = \dfrac{x + 2}{5}$

Therefore, $f^{-1}(x) = \dfrac{x + 2}{5}$

CLASS DISCUSSION

Discuss the function $f(x) = x^2$ for $x \in \mathbb{R}$.

Does the function f have an inverse?

Explain your answer.

How could you change the domain of f so that $f(x) = x^2$ does have an inverse?

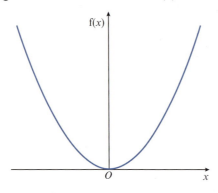

WORKED EXAMPLE 8

$f(x) = \sqrt{x + 1} - 5$ for $x \geqslant -1$

a Find an expression for $f^{-1}(x)$.

b Solve the equation $f^{-1}(x) = f(35)$.

Answers

a $f(x) = \sqrt{x + 1} - 5$ for $x \geqslant -1$

Step 1: Write the function as $y =$ $\longrightarrow$ $y = \sqrt{x + 1} - 5$

Step 2: Interchange the x and y variables. $\longrightarrow$ $x = \sqrt{y + 1} - 5$

Step 3: Rearrange to make y the subject. $\longrightarrow$ $x + 5 = \sqrt{y + 1}$

$$(x + 5)^2 = y + 1$$

$$y = (x + 5)^2 - 1$$

$f^{-1}(x) = (x + 5)^2 - 1$

b $f(35) = \sqrt{35 + 1} - 5 = 1$

$$(x + 5)^2 - 1 = 1$$

$$(x + 5)^2 = 2$$

$$x + 5 = \pm\sqrt{2}$$

$$x = -5 \pm \sqrt{2}$$

$x = -5 + \sqrt{2}$ or $x = -5 - \sqrt{2}$

The range of f is $f(x) \geqslant -5$ so the domain of f^{-1} is $x \geqslant -5$.

Hence the only solution of $f^{-1}(x) = f(35)$ is $x = -5 + \sqrt{2}$.

Exercise 1.6

1 $f(x) = (x + 5)^2 - 7$ for $x \geq -5$. Find an expression for $f^{-1}(x)$.

2 $f(x) = \dfrac{6}{x + 2}$ for $x \geq 0$. Find an expression for $f^{-1}(x)$.

3 $f(x) = (2x - 3)^2 + 1$ for $x \geq 1.5$. Find an expression for $f^{-1}(x)$.

4 $f(x) = 8 - \sqrt{x - 3}$ for $x \geq 3$. Find an expression for $f^{-1}(x)$.

5 $f : x \mapsto 5x - 3$ for $x > 0$

    $g : x \mapsto \dfrac{7}{2 - x}$ for $x \neq 2$

    Express $f^{-1}(x)$ and $g^{-1}(x)$ in terms of x.

6 $f : x \to (x + 2)^2 - 5$ for $x > -2$

    a   Find an expression for $f^{-1}(x)$.   **b**   Solve the equation $f^{-1}(x) = 3$.

7 $f(x) = (x - 4)^2 + 5$ for $x > 4$

    a   Find an expression for $f^{-1}(x)$.   **b**   Solve the equation $f^{-1}(x) = f(0)$.

8 $g(x) = \dfrac{2x + 3}{x - 1}$ for $x > 1$

    a   Find an expression for $g^{-1}(x)$.   **b**   Solve the equation $g^{-1}(x) = 5$.

9 $f(x) = \dfrac{x}{2} + 2$ for $x \in \mathbb{R}$

    $g(x) = x^2 - 2x$ for $x \in \mathbb{R}$

    a   Find $f^{-1}(x)$.   **b**   Solve $fg(x) = f^{-1}(x)$.

10 $f(x) = x^2 + 2$ for $x \in \mathbb{R}$

    $g(x) = 2x + 3$ for $x \in \mathbb{R}$

    Solve the equation $gf(x) = g^{-1}(17)$.

11 $f : x \mapsto \dfrac{2x + 8}{x - 2}$ for $x \neq 2$

    $g : x \mapsto \dfrac{x - 3}{2}$ for $x > -5$

    Solve the equation $f(x) = g^{-1}(x)$.

12 $f(x) = 3x - 24$ for $x \geq 0$. Write down the range of f^{-1}.

13 $f : x \mapsto x + 6$ for $x > 0$

    $g : x \mapsto \sqrt{x}$ for $x > 0$

    Express $x \mapsto x^2 - 6$ in terms of f and g.

14 $f : x \mapsto 3 - 2x$ for $0 \leq x \leq 5$

    $g : x \mapsto |3 - 2x|$ for $0 \leq x \leq 5$

    $h : x \mapsto 3 - |2x|$ for $0 \leq x \leq 5$

    State which of the functions f, g and h has an inverse.

15 $f(x) = x^2 + 2$ for $x \geq 0$

$g(x) = 5x - 4$ for $x \geq 0$

a Write down the domain of f^{-1}. **b** Write down the range of g^{-1}.

16 The functions f and g are defined, for $x \in \mathbb{R}$, by

$f : x \mapsto 3x - k$, where k is a positive constant

$g : x \mapsto \dfrac{5x - 14}{x + 1}$ where $x \neq -1$.

a Find expressions for f^{-1} and g^{-1}.

b Find the value of k for which $f^{-1}(5) = 6$.

c Simplify $g^{-1}g(x)$.

17 $f : x \mapsto x^3$ for $x \in \mathbb{R}$ $\qquad g : x \mapsto x - 8$ for $x \in \mathbb{R}$

Express each of the following as a composite function, using only f, g, f^{-1} and/or g^{-1}:

a $x \mapsto (x - 8)^{\frac{1}{3}}$ **b** $x \mapsto x^3 + 8$ **c** $x \mapsto x^{\frac{1}{3}} - 8$ **d** $x \mapsto (x + 8)^{\frac{1}{3}}$

1.7 The graph of a function and its inverse

In **Worked example 1** you considered the function $f(x) = 2x - 1$ for $x \in \mathbb{R}$, $-1 \leq x \leq 3$.

The domain of f was $-1 \leq x \leq 3$ and the range of f was $-3 \leq f(x) \leq 5$.

The inverse function is $f^{-1}(x) = \dfrac{x + 1}{2}$.

The domain of f^{-1} is $-3 \leq x \leq 5$ and the range of f^{-1} is $-1 \leq f^{-1}(x) \leq 3$.

Drawing f and f^{-1} on the same graph gives:

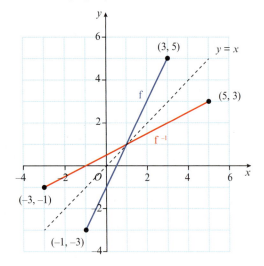

Note:

The graphs of f and f^{-1} are reflections of each other in the line $y = x$.

This is true for all one-one functions and their inverse functions.

This is because: $ff^{-1}(x) = x = f^{-1}f(x)$.

Some functions are called **self-inverse functions** because f and its inverse f^{-1} are the same.

If $f(x) = \dfrac{1}{x}$ for $x \neq 0$, then $f^{-1}(x) = \dfrac{1}{x}$ for $x \neq 0$.

So $f(x) = \dfrac{1}{x}$ for $x \neq 0$ is an example of a self-inverse function.

When a function f is self-inverse, the graph of f will be symmetrical about the line $y = x$.

WORKED EXAMPLE 9

$f(x) = (x - 2)^2, 2 \leqslant x \leqslant 5$

On the same axes, sketch the graphs of $y = f(x)$ and $y = f^{-1}(x)$, showing clearly the points where the curves meet the coordinate axes.

Answers

$y = (x - 2)^2$ This part of the expression is a square so it will always be $\geqslant 0$. The smallest value it can be is 0. This occurs when $x = 2$.

When $x = 5$, $y = 9$.

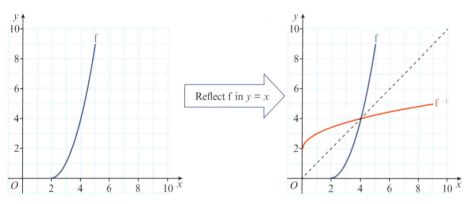

Reflect f in $y = x$

CLASS DISCUSSION

Sundeep says that the diagram shows the graph of the function $f(x) = x^x$ for $x > 0$, together with its inverse function $y = f^{-1}(x)$.

Is Sundeep correct? Explain your answer.

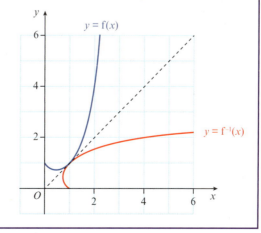

Exercise 1.7

1   On a copy of the grid, draw the graph of the inverse of the function f.

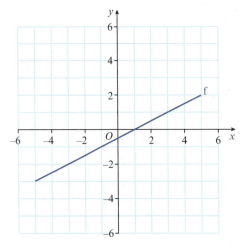

2   On a copy of the grid, draw the graph of the inverse of the function g.

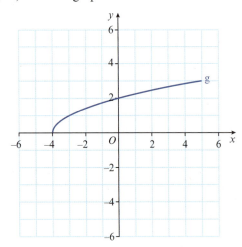

3   $f(x) = x^2 + 3, x \geqslant 0$.

On the same axes, sketch the graphs of $y = f(x)$ and $y = f^{-1}(x)$, showing the coordinates of any points where the curves meet the coordinate axes.

4   $g(x) = 2^x$ for $x \in \mathbb{R}$.

On the same axes, sketch the graphs of $y = g(x)$ and $y = g^{-1}(x)$, showing the coordinates of any points where the curves intersect the coordinate axes.

5   $g(x) = x^2 - 1$ for $x \geqslant 0$.

On the same axes, sketch the graphs of $y = g(x)$ and $y = g^{-1}(x)$, showing the coordinates of any points where the curves intersect the coordinate axes.

6   $f(x) = 4x - 2$ for $-1 \leqslant x \leqslant 3$.

On the same axes, sketch the graphs of $y = f(x)$ and $y = f^{-1}(x)$, showing the coordinates of any points where the lines intersect the coordinate axes.

7   The function f is defined by $f : x \mapsto 3 - (x + 1)^2$ for $x \geqslant -1$.

   a   Explain why f has an inverse.

   b   Find an expression for f^{-1} in terms of x.

   c   On the same axes, sketch the graphs of $y = f(x)$ and $y = f^{-1}(x)$, showing the coordinates of any points where the curves intersect the coordinate axes.

8   **CHALLENGE QUESTION**

   $f : x \mapsto \dfrac{2x + 7}{x - 2}$ for $x \neq 2$

   a   Find f^{-1} in terms of x.

   b   Explain what this implies about the symmetry of the graph of $y = f(x)$.

SUMMARY

Functions

A function is a rule that maps each x-value to just one y-value for a defined set of input values.

Mappings that are either $\begin{cases} \text{one-one} \\ \text{many-one} \end{cases}$ are called functions.

The set of input values for a function is called the **domain** of the function.

The set of output values for a function is called the **range** (or image set) of the function.

Modulus function

The modulus of x, written as $|x|$, is defined as:

$$|x| = \begin{cases} x & \text{if } x > 0 \\ 0 & \text{if } x = 0 \\ -x & \text{if } x < 0 \end{cases}$$

Composite functions

$fg(x)$ means the function g acts on x first, then f acts on the result.

$f^2(x)$ means $ff(x)$.

Inverse functions

The inverse of a function $f(x)$ is the function that undoes what $f(x)$ has done.

The inverse of the function $f(x)$ is written as $f^{-1}(x)$.

The domain of $f^{-1}(x)$ is the range of $f(x)$.

The range of $f^{-1}(x)$ is the domain of $f(x)$.

An inverse function $f^{-1}(x)$ can exist if, and only if, the function $f(x)$ is a one-one mapping.

The graphs of f and f^{-1} are reflections of each other in the line $y = x$.

Past paper questions

Worked example

The functions f and g are defined by

$$f(x) = \frac{2x}{x+1} \quad \text{for} \quad x > 0,$$

$$g(x) = \sqrt{x+1} \quad \text{for} \quad x > -1.$$

a   Find $fg(8)$.   [2]

b   Find an expression for $f^2(x)$, giving your answer in the form $\dfrac{ax}{bx+c}$, where a, b and c are integers to be found.   [3]

c   Find an expression for $g^{-1}(x)$, stating its domain and range.   [4]

Cambridge IGCSE Additional Mathematics 0606 Paper 21 Q12i,ii,iii Jun 2014

Answers

a $g(8) = \sqrt{8+1} = 3$

$fg(8) = f(3)$ substitute 3 for x in $\dfrac{2x}{x+1}$

$= \dfrac{2(3)}{3+1}$

$= 1.5$

b $f^2(x) = ff(x)$

$= f\left(\dfrac{2x}{x+1}\right)$ substitute $\dfrac{2x}{x+1}$ for x in $\dfrac{2x}{x+1}$

$= \dfrac{2\left(\dfrac{2x}{x+1}\right)}{\left(\dfrac{2x}{x+1}\right)+1}$ simplify

$= \dfrac{\dfrac{4x}{x+1}}{\dfrac{3x+1}{x+1}}$ multiply numerator and denominator by $x+1$

$= \dfrac{4x}{3x+1}$

$a = 4$, $b = 3$ and $c = 1$

c $g(x) = \sqrt{x+1}$ for $x > -1$

Step 1: Write the function as $y = $ $\longrightarrow$ $y = \sqrt{x+1}$

Step 2: Interchange the x and y variables. $\longrightarrow$ $x = \sqrt{y+1}$

Step 3: Rearrange to make y the subject. $\longrightarrow$ $x^2 = y+1$

$y = x^2 - 1$

$g^{-1}(x) = x^2 - 1$

The range of g is $g(x) > 0$, so the domain of g^{-1} is $x > 0$.

The domain of g is $x > -1$, so the range of g^{-1} is $g^{-1}(x) > -1$.

1   The functions f and g are defined for real values of x by

$f(x) = \sqrt{x-1} - 3$ for $x > 1,$

$g(x) = \dfrac{x-2}{2x-3}$ for $x > 2.$

   a   Find gf(37).   [2]

   b   Find an expression for $f^{-1}(x)$.   [2]

   c   Find an expression for $g^{-1}(x)$.   [2]

Cambridge IGCSE Additional Mathematics 0606 Paper 21 Q4 Nov 2014

2   The function f is defined by $f(x) = 2 - \sqrt{x+5}$ for $-5 \leqslant x < 0$.

   i   Write down the range of f.   [2]

   ii   Find $f^{-1}(x)$ and state its domain and range.   [4]

The function g is defined by $g(x) = \dfrac{4}{x}$ for $-5 \leqslant x < -1$.

   iii   Solve fg(x) = 0.   [3]

Cambridge IGCSE Additional Mathematics 0606 Paper 11 Q6 Jun 2016

3   i   On the axes below, sketch the graphs of $y = 2 - x$ and $y = |3 + 2x|$.   [4]

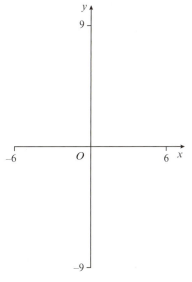

   ii   Solve $|3 + 2x| = 2 - x$.   [3]

Cambridge IGCSE Additional Mathematics 0606 Paper 12 Q4 Mar 2016

4   Diagrams **A** to **D** show four different graphs. In each case the whole graph is shown and the scales on the two axes are the same.

A

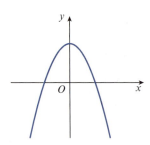

B

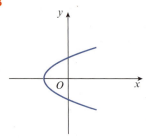

C

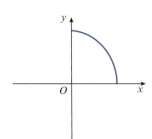

D

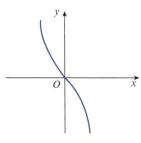

Place ticks in the boxes in the table to indicate which descriptions, if any, apply to each graph.
There may be more than one tick in any row or column of the table.   [4]

	A	**B**	**C**	**D**
Not a function				
One-one function				
A function that is its own inverse				
A function with no inverse				

Cambridge IGCSE Additional Mathematics 0606 Paper 11 Q3 June 2018

5   The function f is defined by $f(x) = \dfrac{1}{2x - 5}$ for $x > 2.5$.

i   Find an expression for $f^{-1}(x)$.   [2]

ii   State the domain of $f^{-1}(x)$.   [1]

iii Find an expression for $f^2(x)$, giving your answer in the form $\dfrac{ax+b}{cx+d}$,
         where a, b, c and d are integers to be found.   [3]

Cambridge IGCSE Additional Mathematics 0606 Paper 21 Q5 June 2018

6 **i**   Sketch the graph of $y = |5x - 3|$ on the axes below, showing the coordinates of the points where
       the graph meets the coordinate axes.

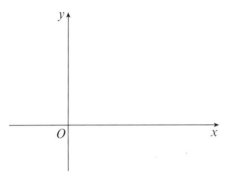

                                                                                                     [3]

   ii   Solve the equation $|5x - 3| = 2 - x$.   [3]

Cambridge IGCSE Additional Mathematics 0606 Paper 21 Q3 June 2019

7 **a**   Functions f and g are such that, for $x \in \mathbb{R}$,

$$f(x) = x^2 + 3,$$
$$g(x) = 4x - 1.$$

   i   State the range of f.   [1]

   ii   Solve $fg(x) = 4$.   [3]

   b   A function h is such that $h(x) = \dfrac{2x+1}{x-4}$ for $x \in \mathbb{R}$, $x \neq 4$.

   i   Find $h^{-1}(x)$ and state its range.   [4]

   ii   Find $h^2(x)$, giving your answer in its simplest form.   [3]

Cambridge IGCSE Additional Mathematics 0606 Paper 11 Q6 Nov 2017

8 The functions f and g are defined by

$$f(x) = \frac{x^2 - 2}{x} \text{ for } x \geq 2,$$
$$g(x) = \frac{x^2 - 1}{2} \text{ for } x \geq 0.$$

   i   State the range of g.   [1]

   ii   Explain why $fg(1)$ does not exist.   [2]

   iii Show that $gf(x) = ax^2 + b + \dfrac{c}{x^2}$, where a, b and c are constants to be found.   [3]

   iv   State the domain of gf.   [1]

   v   Show that $f^{-1}(x) = \dfrac{x + \sqrt{x^2 + 8}}{2}$.   [4]

Cambridge IGCSE Additional Mathematics 0606 Paper 22 Q11 Mar 2017

9   i   On the axes below, draw the graph of $y = |2x - 3|$.

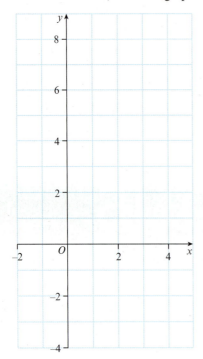

[2]

ii   Solve the equation $7 - |2x - 3| = 0$.   [3]

Cambridge IGCSE Additional Mathematics 0606 Paper 21 Q1 Nov 2019

> # Chapter 2
Simultaneous equations and quadratics

THIS SECTION WILL SHOW YOU HOW TO:

- solve simultaneous equations in two unknowns by elimination or substitution
- find the maximum and minimum values of a quadratic function
- sketch graphs of quadratic functions and find their range for a given domain
- sketch graphs of the function $y = |f(x)|$ where $f(x)$ is quadratic and solve associated equations
- determine the number of roots of a quadratic equation and the related conditions for a line to intersect, be a tangent or not intersect a given curve
- solve quadratic equations for real roots and find the solution set for quadratic inequalities.

PRE-REQUISITE KNOWLEDGE

Before you start…

Where it comes from	You should be able to…	Check your skills
Cambridge IGCSE/O Level Mathematics	Solve simultaneous equations using the elimination method.	**1** Use the elimination method to solve: **a** $4x + 3y = 1$ $2x - 3y = 14$ **b** $3x + 2y = 19$ $x + 2y = 13$
Cambridge IGCSE/O Level Mathematics	Solve simultaneous equations using the substitution method.	**2** Use the substitution method to solve: **a** $y = 3x - 10$ $x + y = -2$ **b** $x + 2y = 11$ $4y - x = -2$
Cambridge IGCSE/O Level Mathematics	Solve quadratic equations using the factorisation method.	**3** Solve by factorisation: **a** $x^2 + x - 6 = 0$ **b** $x^2 - 10x + 16 = 0$ **c** $6x^2 + 11x - 10 = 0$
Cambridge IGCSE/O Level Mathematics	Solve quadratic equations by completing the square.	**4 a** Express $2x^2 + 7x + 3$ in the form $a(x + b)^2 + c$. **b** Use your answer to **part a** to solve the equation $2x^2 + 7x + 3 = 0$.
Cambridge IGCSE/O Level Mathematics	Solve quadratic equations using the quadratic formula	**5** Solve $2x^2 - 9x + 8 = 0$. Give your answers correct to 2 decimal places.

CLASS DISCUSSION

Solve each pair of simultaneous equations.

$8x + 3y = 7$	$3x + y = 10$	$2x + 5 = 3y$
$3x + 5y = -9$	$2y = 15 - 6x$	$10 - 6y = -4x$

Discuss your answers with your classmates.
Discuss what the graphs would be like for each pair of equations.

CLASS DISCUSSION

Solve each of these quadratic equations.

$x^2 - 8x + 15 = 0$ $\qquad$ $x^2 + 4x + 4 = 0$ $\qquad$ $x^2 + 2x + 4 = 0$

Discuss your answers with your classmates. Discuss what the graphs would be like for each of the functions $y = x^2 - 8x + 15$, $y = x^2 + 4x + 4$ and $y = x^2 + 2x + 4$.

KEY WORDS

parabola

minimum point

maximum point

turning point

stationary point

completing the square

roots

discriminant

tangent

2.1 Simultaneous equations (one linear and one non-linear)

In this section you will learn how to solve simultaneous equations where one equation is linear and the second equation is not linear.

The diagram shows the graphs of $y = x + 1$ and $y = x^2 - 5$.

The coordinates of the points of intersection of the two graphs are $(-2, -1)$ and $(3, 4)$.

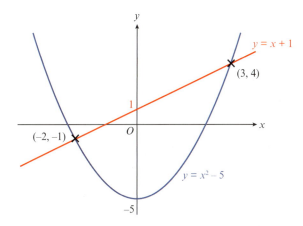

We say that $x = -2$, $y = -1$ and $x = 3$, $y = 4$ are the solutions of the simultaneous equations $y = x + 1$ and $y = x^2 - 5$.

The solutions can also be found algebraically:

$y = x + 1$ (1)
$y = x^2 - 5$ (2)

Substitute for y from (1) into (2):

$$x + 1 = x^2 - 5 \quad \text{rearrange}$$
$$x^2 - x - 6 = 0 \quad \text{factorise}$$
$$(x + 2)(x - 3) = 0$$
$$x = -2 \text{ or } x = 3$$

Substituting $x = -2$ into (1) gives $y = -2 + 1 = -1$.

Substituting $x = 3$ into (1) gives $y = 3 + 1 = 4$.

The solutions are: $x = -2$, $y = -1$ and $x = 3$, $y = 4$.

WORKED EXAMPLE 1

Solve the simultaneous equations.

$2x + 2y = 7$

$x^2 + 4y^2 = 8$

Answers

$2x + 2y = 7$ (1)

$x^2 - 4y^2 = 8$ (2)

From (1), $x = \dfrac{7 - 2y}{2}$

Substitute for x in (2):

$\left(\dfrac{7 - 2y}{2}\right)^2 - 4y^2 = 8$ expand brackets

$\dfrac{49 - 28y + 4y^2}{4} - 4y^2 = 8$ multiply both sides by 4

$49 - 28y + 4y^2 - 16y^2 = 32$ rearrange

$12y^2 + 28y - 17 = 0$ factorise

$(6y + 17)(2y - 1) = 0$

$y = -2\dfrac{5}{6}$ or $y = \dfrac{1}{2}$

Substituting $y = -2\dfrac{5}{6}$ into (1) gives $x = 6\dfrac{1}{3}$

Substituting $y = \dfrac{1}{2}$ into (1) gives $x = 3$

The solutions are: $x = 6\dfrac{1}{3}, y = -2\dfrac{5}{6}$ and $x = 3, y = \dfrac{1}{2}$

Exercise 2.1

Solve the following simultaneous equations.

1. $y = x^2$
   $y = x + 6$

2. $y = x - 6$
   $x^2 + xy = 8$

3. $y = x - 1$
   $x^2 + y^2 = 25$

4. $xy = 4$
   $y = 2x + 2$

5. $x^2 - xy = 0$
   $x + y = 1$

6. $3y = 4x - 5$
   $x^2 + 3xy = 10$

7. $2x + y = 7$
   $xy = 6$

8. $x - y = 2$
   $2x^2 - 3y^2 = 15$

9. $x + 2y = 7$
   $x^2 + y^2 = 10$

10. $y = 2x$
   $x^2 + y^2 = 3$

11. $xy = 2$
   $x + y = 3$

12. $y^2 = 4x$
   $2x + y = 4$

13. $x + 3y = 0$
   $2x^2 + 3y = 1$

14. $x + y = 4$
   $x^2 + y^2 = 10$

15. $y = 3x$
   $2y^2 - xy = 15$

16. $x - 2y = 1$
   $4y^2 - 3x^2 = 1$

17. $3 + x + xy = 0$
   $2x + 5y = 8$

18. $xy = 12$
   $(x - 1)(y + 2) = 15$

19. Calculate the coordinates of the points where the line $y = 1 - 2x$ cuts the curve $x^2 + y^2 = 2$.

20 The sum of two numbers x and y is 11.

The product of the two numbers is 21.25.

    a   Write down two equations in x and y.

    b   Solve your equations to find the possible values of x and y.

21 The sum of the areas of two squares is $818\,\text{cm}^2$.

The sum of the perimeters is $160\,\text{cm}$.

Find the lengths of the sides of the squares.

22 The line $y = 2 - 2x$ cuts the curve $3x^2 - y^2 = 3$ at the points A and B.

Find the length of the line AB.

23 The line $2x + 5y = 1$ meets the curve $x^2 + 5xy - 4y^2 + 10 = 0$ at the points A and B.

Find the coordinates of the midpoint of AB.

24 The line $y = x - 10$ intersects the curve $x^2 + y^2 + 4x + 6y - 40 = 0$ at the points A and B. Find the length of the line AB.

25 The straight line $y = 2x - 2$ intersects the curve $x^2 - y = 5$ at the points A and B.

Given that A lies below the x-axis and the point P lies on AB such that $AP : PB = 3 : 1$, find the coordinates of P.

26 The line $x - 2y = 2$ intersects the curve $x + y^2 = 10$ at two points A and B.

Find the equation of the perpendicular bisector of the line AB.

2.2 Maximum and minimum values of a quadratic function

The general equation of a quadratic function is $\text{f}(x) = ax^2 + bx + c$, where a, b and c are constants and $a \neq 0$.

The graph of the function $y = ax^2 + bx + c$ is called a **parabola**. The orientation of the parabola depends on the value of a, the coefficient of x^2.

If $a > 0$, the curve has a **minimum point** which occurs at the lowest point of the curve.

If $a < 0$, the curve has a **maximum point** which occurs at the highest point of the curve.

The maximum and minimum points are also called **turning points** or **stationary points**.

Every parabola has a line of symmetry that passes through the maximum or minimum point.

WORKED EXAMPLE 2

$f(x) = x^2 - 3x - 4 \quad x \in \mathbb{R}$

a   Find the axis crossing points for the graph of $y = f(x)$.

b   Sketch the graph of $y = f(x)$ and use the symmetry of the curve to find the coordinates of the minimum point.

c   State the range of the function $f(x)$.

Answers

a $y = x^2 - 3x - 4$

When $x = 0$, $y = -4$

When $y = 0$,

$$x^2 - 3x - 4 = 0$$
$$(x + 1)(x - 4) = 0$$
$$x = -1 \text{ or } x = 4$$

Axes crossing points are: $(0, -4)$, $(-1, 0)$ and $(4, 0)$.

b   The line of symmetry cuts the x-axis midway between -1 and 4.

So, the line of symmetry is $x = 1.5$

When $x = 1.5$, $y = 1.5^2 - 3(1.5) - 4$

$$y = -6.25$$

Minimum point $= (1.5, -6.25)$

c   The range is $f(x) \geqslant -6.25$

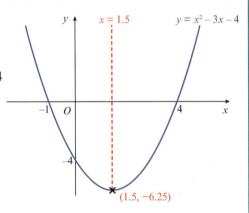

Completing the square

If you expand the expressions $(x + d)^2$ and $(x - d)^2$ you obtain the results:

$(x + d)^2 = x^2 + 2dx + d^2$ and $(x - d)^2 = x^2 - 2dx + d^2$

Rearranging these give the following important results:

$x^2 + 2dx = (x + d)^2 - d^2$

$x^2 - 2dx = (x - d)^2 - d^2$

This is known as **completing the square**.

To complete the square for $x^2 + 8x$:

$$8 \div 2 = 4$$

$$x^2 + 8x = (x + 4)^2 - 4^2$$
$$x^2 + 8x = (x + 4)^2 - 16$$

To complete the square for $x^2 + 10x - 3$:

$$10 \div 2 = 5$$

$$x^2 + 10x - 3 = (x + 5)^2 - 5^2 - 3$$
$$x^2 + 10x - 3 = (x + 5)^2 - 28$$

To complete the square for $2x^2 - 8x - 14$ you must first take a factor of 2 out of the expression:

$$2x^2 - 8x + 14 = 2[x^2 - 4x + 7]$$

$$4 \div 2 = 2$$

$$x^2 - 4x + 7 = (x - 2)^2 - 2^2 + 7$$
$$x^2 - 4x + 3 = (x - 2)^2 + 3$$

So, $2x^2 - 8x + 6 = 2[(x - 2)^2 + 3] = 2(x - 2)^2 + 6$

You can also use an algebraic method for completing the square, as shown in **Worked example 3**.

WORKED EXAMPLE 3

Express $2x^2 - 4x + 5$ in the form $p(x - q)^2 + r$, where p, q and r are constants to be found.

Answers

$2x^2 - 4x + 5 = p(x - q)^2 + r$

Expanding the brackets and simplifying gives:

$2x^2 - 4x + 5 = px^2 - 2pqx + pq^2 + r$

Comparing coefficients of x^2, coefficients of x and the constant gives:

$2 = p$ (1) $-4 = -2pq$ (2) $5 = pq^2 + r$ (3)

Substituting $p = 2$ in equation (2) gives $q = 1$.

Substituting $p = 2$ and $q = 1$ in equation (3) gives $r = 3$.

So $2x^2 - 4x + 5 = 2(x - 1)^2 + 3$.

Completing the square for a quadratic expression or function enables you to:

- write down the maximum or minimum value of the expression
- write down the coordinates of the maximum or minimum point of the function
- sketch the graph of the function
- write down the line of symmetry of the function
- state the range of the function.

In **Worked example 3** you found that:

$$2x^2 - 4x + 5 = 2(x - 1)^2 + 3$$

> This part of the expression is a square so it will always be ≥ 0. The smallest value it can be is 0. This occurs when $x = 1$.

The minimum value of the expression is $2 \times 0 + 3 = 3$ and this minimum occurs when $x = 1$.

So, the function $y = 2x^2 - 4x + 5$ will have a minimum at the point $(1, 3)$.

When $x = 0$, $y = 5$.

The graph of $y = 2x^2 - 4x + 5$ can now be sketched:

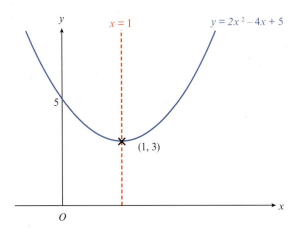

The line of symmetry is $x = 1$.

The range is $y \geq 3$.

The general rule is:

> For a quadratic function $f(x) = ax^2 + bx + c$ that is written in the form $f(x) = a(x - h)^2 + k$,
>
> **i**   if $a > 0$, the minimum point is (h, k)
>
> **ii** if $a < 0$, the maximum point is (h, k).

WORKED EXAMPLE 4

$f(x) = 2 + 8x - 2x^2 \quad x \in \mathbb{R}$

a Find the value of a, the value of b and the value of c for which $f(x) = a - b(x + c)^2$.

b Write down the coordinates of the maximum point on the curve $y = f(x)$.

c Sketch the graph of $y = f(x)$, showing the coordinates of the points where the graph intersects the x and y-axes.

d State the range of the function $f(x)$.

Answers

a $2 + 8x - 2x^2 = a - b(x + c)^2$

$2 + 8x - 2x^2 = a - b(x^2 + 2cx + c^2)$

$2 + 8x - 2x^2 = a - bx^2 - 2bcx - bc^2$

Comparing coefficients of x^2, coefficients of x and the constant gives:

$-2 = -b$ (1) $8 = -2bc$ (2) $2 = a - bc^2$ (3)

Substituting $b = 2$ in equation (2) gives $c = -2$.

Substituting $b = 2$ and $c = -2$ in equation (3) gives $a = 10$.

So, $a = 10$, $b = 2$ and $c = -2$.

b $y = 10 - 2(x - 2)^2$

> This part of the expression is a square so it will always be $\geqslant 0$. The smallest value it can be is 0. This occurs when $x = 2$.

The maximum value of the expression is $10 - 2 \times 0 = 10$ and this maximum occurs when $x = 2$. So, the function $y = 2 + 8x - 2x^2$ will have maximum at the point $(2, 10)$.

c $y = 2 + 8x - 2x^2$

When $x = 0$, $y = 2$.

When $y = 0$,

$10 - 2(x - 2)^2 = 0$

$\qquad 2(x - 2)^2 = 10$

$\qquad (x - 2)^2 = 5$

$\qquad\quad x - 2 = \pm\sqrt{5}$

$\qquad\qquad x = 2 \pm \sqrt{5}$

$\qquad\qquad x = 2 - \sqrt{5}$ or $x = 2 + \sqrt{5}$

$\qquad\qquad (x = -0.236$ or $x = 4.24$ to 3 sf$)$

Axes crossing points are: $(0, 2)$, $(2 + \sqrt{5}, 0)$ and $(2 - \sqrt{5}, 0)$.

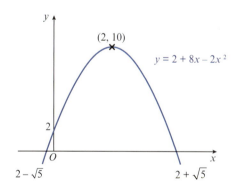

d The range is $f(x) \leqslant 10$.

Exercise 2.2

1   Use the symmetry of each quadratic function to find the maximum or minimum points.
    Sketch each graph, showing all axis crossing points.

    a $y = x^2 - 5x - 6$ **b** $y = x^2 - x - 20$ **c** $y = x^2 + 4x - 21$

    d $y = x^2 + 3x - 28$ **e** $y = x^2 + 4x + 1$ **f** $y = 15 + 2x - x^2$

2   Express each of the following in the form $(x - m)^2 + n$.

    a $x^2 - 8x$ **b** $x^2 - 10x$ **c** $x^2 - 5x$ **d** $x^2 - 3x$

    e $x^2 + 4x$ **f** $x^2 + 7x$ **g** $x^2 + 9x$ **h** $x^2 + 3x$

3   Express each of the following in the form $(x - m)^2 + n$.

    a $x^2 - 8x + 15$ **b** $x^2 - 10x - 5$ **c** $x^2 - 6x + 2$ **d** $x^2 - 3x + 4$

    e $x^2 + 6x + 5$ **f** $x^2 + 6x + 9$ **g** $x^2 + 4x - 17$ **h** $x^2 + 5x + 6$

4   Express each of the following in the form $a(x - p)^2 + q$.

    a $2x^2 - 8x + 3$ **b** $2x^2 - 12x + 1$ **c** $3x^2 - 12x + 5$ **d** $2x^2 - 3x + 2$

    e $2x^2 + 4x + 1$ **f** $2x^2 + 7x - 3$ **g** $2x^2 - 3x + 5$ **h** $3x^2 - x + 6$

5   Express each of the following in the form $m - (x - n)^2$.

    a $6x - x^2$ **b** $10x - x^2$ **c** $3x - x^2$ **d** $8x - x^2$

6   Express each of the following in the form $a - (x + b)^2$.

    a $5 - 2x - x^2$ **b** $8 - 4x - x^2$ **c** $10 - 5x - x^2$ **d** $7 + 3x - x^2$

7   Express each of the following in the form $a - p(x + q)^2$.

    a $9 - 6x - 2x^2$ **b** $1 - 4x - 2x^2$ **c** $7 + 8x - 2x^2$ **d** $2 + 5x - 3x^2$

8   **a**   Express $4x^2 + 2x + 5$ in the form $a(x + b)^2 + c$, where a, b and c are constants.

    b   Does the function $y = 4x^2 + 2x + 5$ meet the x-axis?
         Explain your answer.

9   $f(x) = 2x^2 - 8x + 1$

    a   Express $2x^2 - 8x + 1$ in the form $a(x + b)^2 + c$, where a and b are integers.

    b   Find the coordinates of the stationary point on the graph of $y = f(x)$.

10 $f(x) = x^2 - x - 5$ for $x \in \mathbb{R}$

    a   Find the minimum value of $f(x)$ and the corresponding value of x.

    b   Hence write down a suitable domain for $f(x)$ in order that $f^{-1}(x)$ exists.

11 $f(x) = 5 - 7x - 2x^2$ for $x \in \mathbb{R}$

    a   Write $f(x)$ in the form $p - 2(x - q)^2$, where p and q are constants to be found.

    b   Write down the range of the function $f(x)$.

12 $f(x) = 14 + 6x - 2x^2$ for $x \in \mathbb{R}$

    a Express $14 + 6x - 2x^2$ in the form $a + b(x + c)^2$, where a, b and c are constants.

    b Write down the coordinates of the stationary point on the graph of $y = f(x)$.

    c Sketch the graph of $y = f(x)$.

13 $f(x) = 7 + 5x - x^2$ for $0 \leqslant x \leqslant 7$

    a Express $7 + 5x - x^2$ in the form $a - (x + b)^2$, where a, and b are constants.

    b Find the coordinates of the turning point of the function $f(x)$, stating whether it is a maximum or minimum point.

    c Find the range of f.

    d State, giving a reason, whether or not f has an inverse.

14 The function f is such that $f(x) = 2x^2 - 8x + 3$.

    a Write $f(x)$ in the form $2(x + a)^2 + b$, where a and b are constants to be found.

    b Write down a suitable domain for f so that f^{-1} exists.

15 $f(x) = 4x^2 + 6x - 8$ where $x \geqslant m$

  Find the smallest value of m for which f has an inverse.

16 $f(x) = 1 + 4x - x^2$ for $x \geqslant 2$

    a Express $1 + 4x - x^2$ in the form $a - (x + b)^2$, where a and b are constants to be found.

    b Find the coordinates of the turning point of the function $f(x)$, stating whether it is a maximum or minimum point.

    c Explain why $f(x)$ has an inverse and find an expression for $f^{-1}(x)$ in terms of x.

2.3 Graphs of $y = |f(x)|$ where f(x) is quadratic

To sketch the graph of the modulus function $y = |ax^2 + bx + c|$, you must:

- first sketch the graph of $y = ax^2 + bx + c$
- reflect in the x-axis the part of the curve $y = ax^2 + bx + c$ that is below the x-axis.

WORKED EXAMPLE 5

Sketch the graph of $y = |x^2 - 2x - 3|$.

Answers

First sketch the graph of $y = x^2 - 2x - 3$.

When $x = 0$, $y = -3$.

So, the y-intercept is -3.

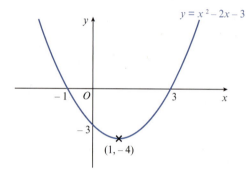

CONTINUED

When $y = 0$,

$$x^2 - 2x - 3 = 0$$
$$(x + 1)(x - 3) = 0$$
$$x = -1 \quad \text{or} \quad x = 3.$$

The x-intercepts are -1 and 3.

The x-coordinate of the minimum point $= \dfrac{-1 + 3}{2} = 1$.

The y-coordinate of the minimum point $= (1)^2 - 2(1) - 3 = -4$.

The minimum point is $(1, -4)$.

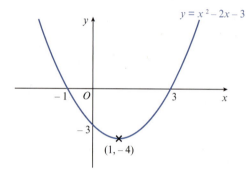

Now reflect in the x-axis the part of the curve $y = x^2 - 2x - 3$ that is below the x-axis.

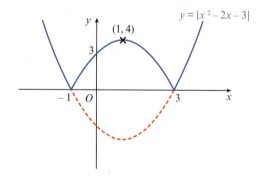

A sketch of the function $y = |x^2 + 4x - 12|$ is shown below.

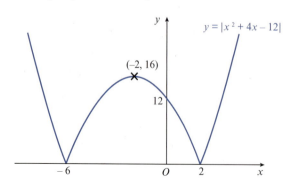

Now consider using this graph to find the number of solutions of the equation $|x^2 + 4x - 12| = k$, where $k \geqslant 0$.

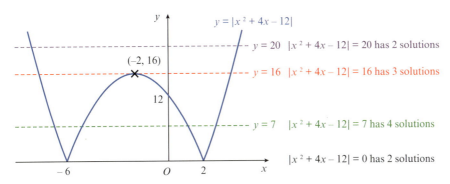

The conditions for the number of solutions of the equation $|x^2 + 4x - 12| = k$ are:

Value of k	$k = 0$	$0 < k < 16$	$k = 16$	$k > 16$
Number of solutions	2	4	3	2

Equations involving $|f(x)|$, where $f(x)$ is quadratic, can be solved algebraically:

To solve $|x^2 + 4x - 12| = 16$:

$x^2 + 4x - 12 = 16$ or $x^2 + 4x - 12 = -16$

$x^2 + 4x - 28 = 0$ or $x^2 + 4x + 4 = 0$

$$x = \frac{-4 \pm \sqrt{4^2 - 4 \times 1 \times (-28)}}{2 \times 1} \quad \text{or} \quad (x + 2)(x + 2) = 0$$

$$x = \frac{-4 \pm \sqrt{128}}{2} \qquad\qquad \text{or} \quad x = -2$$

$$x = -2 \pm 4\sqrt{2}$$

$(x = 3.66$ or $x = -7.66$ to 3 sf $)$

The exact solutions are $x = -2 - 4\sqrt{2}$ or $x = -2$ or $x = -2 + 4\sqrt{2}$.

TIP

The graph of $y = |x^2 + 4x - 12|$ is sketched here showing these three solutions.

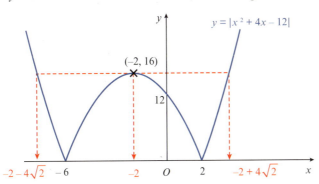

Exercise 2.3

1 Sketch the graphs of each of the following functions.

a $y = |x^2 - 4x + 3|$ **b** $y = |x^2 - 2x - 3|$ **c** $y = |x^2 - 5x + 4|$

d $y = |x^2 - 2x - 8|$ **e** $y = |2x^2 - 11x - 6|$ **f** $y = |3x^2 + 5x - 2|$

2 $f(x) = 1 - 4x - x^2$

   a Write $f(x)$ in the form $a - (x + b)^2$, where a and b are constants.

   b Sketch the graph of $y = f(x)$.

   c Sketch the graph of $y = |f(x)|$.

3 $f(x) = 2x^2 + x - 3$

   a Write $f(x)$ in the form $a(x + b)^2 + c$, where a, b and c are constants.

   b Sketch the graph of $y = |f(x)|$.

4 **a** Find the coordinates of the stationary point on the curve $y = |(x - 7)(x + 1)|$.

   b Sketch the graph of $y = |(x - 7)(x + 1)|$.

   c Find the set of values of k for which $|(x - 7)(x + 1)| = k$ has four solutions.

5 **a** Find the coordinates of the stationary point on the curve $y = |(x + 5)(x + 1)|$.

   b Find the set of values of k for which $|(x + 5)(x + 1)| = k$ has two solutions.

6 **a** Find the coordinates of the stationary point on the curve $y = |(x - 8)(x - 3)|$.

   b Find the value of k for which $|(x - 8)(x - 3)| = k$ has three solutions.

7 Solve these equations.

   a $|x^2 - 6| = 10$ **b** $|x^2 - 2| = 2$ **c** $|x^2 - 5x| = 6$

   d $|x^2 + 2x| = 24$ **e** $|x^2 - 5x + 1| = 3$ **f** $|x^2 + 3x - 1| = 3$

   g $|x^2 + 2x - 4| = 5$ **h** $|2x^2 - 3| = 2x$ **i** $|x^2 - 4x + 7| = 4$

8 **CHALLENGE QUESTION**

Solve these simultaneous equations.

   a $y = x + 1$ **b** $2y = x + 4$ **c** $y = 2x$

     $y = |x^2 - 2x - 3|$ $y = \left|\frac{1}{2}x^2 - x - 3\right|$ $y = |2x^2 - 4|$

2.4 Quadratic inequalities

You should already know how to solve linear inequalities.
Two examples are shown below.

Solve $2x - 5 < 9$ expand brackets

$\quad 2x - 10 < 9$ add 10 to both sides

$\quad\quad 2x < 19$ divide both sides by 2

$\quad\quad x < 9.5$

Solve $5 - 3x \geq 17$ subtract 5 from both sides

$\quad -3x \geq 12$ divide both sides by -3

$\quad\quad x \leq -4$

TIP

It is very important that you remember the rule that when you multiply or divide both sides of an inequality by a negative number then the inequality sign must be reversed. This is illustrated in the second of these examples, when both sides of the inequality were divided by -3.

CLASS DISCUSSION

Robert is asked to solve the inequality $\dfrac{7x + 12}{x} \geqslant 3$.

He writes: $7x + 12 \geqslant 3x$

$\qquad\qquad\quad 4x \geqslant -12$

$\qquad\qquad\; \text{So } x \geqslant -3$

Anna checks his answer using the number -4.

She writes: when $x = -4$,

$\qquad (7 \times (-4) + 12) \div (-4) = (-16) \div (-4) = 4$

$\qquad$ Hence $x = -4$ is a value of x that satisfies the original inequality

$\qquad$ So Robert's answer must be incorrect!

Discuss Robert's working out with your classmates and explain Robert's error.

Now solve the inequality $\dfrac{7x + 12}{x} \geqslant 3$ correctly.

Quadratic inequalities can be solved by sketching a graph and considering when the graph is above or below the x-axis.

WORKED EXAMPLE 6

Solve $x^2 - 3x - 4 > 0$.

Answers

Sketch the graph of $y = x^2 - 3x - 4$.

When $y = 0$, $x^2 - 3x - 4 = 0$

$\qquad\qquad (x + 1)(x - 4) = 0$

$\qquad\quad x = -1 \quad \text{or} \quad x = 4$

So, the x-axis crossing points are -1 and 4.

For $x^2 - 3x - 4 > 0$ you need to find the range of values of x for which the curve is positive (above the x-axis).

The solution is $x < -1$ and $x > 4$.

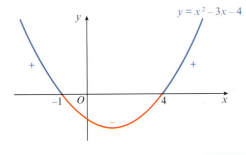

WORKED EXAMPLE 7

Solve $2x^2 \leqslant 15 - x$.

Answers

Rearranging: $2x^2 + x - 15 \leqslant 0$.

Sketch the graph of $y = 2x^2 + x - 15$.

When $y = 0$, $2x^2 + x - 15 = 0$

$$(2x - 5)(x + 3) = 0$$

$$x = 2.5 \quad \text{or} \quad x = -3$$

So, the x-axis crossing points are -3 and 2.5

For $2x^2 + x - 15 \leqslant 0$ you need to find the range of values of x for which the curve is either zero or negative (below the x-axis).

The solution is $-3 \leqslant x \leqslant 2.5$

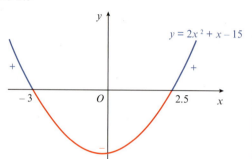

Exercise 2.4

1 Solve.

   a $(x + 3)(x - 4) > 0$ **b** $(x - 5)(x - 1) \leqslant 0$ **c** $(x - 3)(x + 7) \geqslant 0$

   d $x(x - 5) < 0$ **e** $(2x + 1)(x - 4) < 0$ **f** $(3 - x)(x + 1) \geqslant 0$

   g $(2x + 3)(x - 5) < 0$ **h** $(x - 5)^2 \geqslant 0$ **i** $(x - 3)^2 \leqslant 0$

2 Solve.

   a $x^2 + 5x - 14 < 0$ **b** $x^2 + x - 6 \geqslant 0$ **c** $x^2 - 9x + 20 \leqslant 0$

   d $x^2 + 2x - 48 > 0$ **e** $2x^2 - x - 15 \leqslant 0$ **f** $5x^2 + 9x + 4 > 0$

3 Solve.

   a $x^2 < 18 - 3x$ **b** $12x < x^2 + 35$ **c** $x(3 - 2x) \leqslant 1$

   d $x^2 + 4x < 3(x + 2)$ **e** $(x + 3)(1 - x) < x - 1$ **f** $(4x + 3)(3x - 1) < 2x(x + 3)$

4 Find the set of values of x for which

   a $x^2 - 11x + 24 < 0$ and $2x + 3 < 13$

   b $x^2 - 4x \leqslant 12$ and $4x - 3 > 1$

   c $x(2x - 1) < 1$ and $7 - 2x < 6$

   d $x^2 - 3x - 10 < 0$ and $x^2 - 10x + 21 < 0$

   e $x^2 + x - 2 > 0$ and $x^2 - 2x - 3 \geqslant 0$.

5 Solve.

a $|x^2 + 2x - 2| < 13$ b $|x^2 - 8x + 6| < 6$ c $|x^2 - 6x + 4| < 4$

6 CHALLENGE QUESTION

Find the range of values of x for which $\dfrac{4}{3x^2 - 2x - 8} < 0$.

> **REFLECTION**
>
> Look back at this exercise.
>
> a   How confident do you feel in your understanding of this section?
>
> b   What can you do to increase your level of confidence?

2.5 Roots of quadratic equations

The solutions of an equation are called the **roots** of the equation.

Consider solving the following three quadratic equations using the quadratic formula $x = \dfrac{-b \pm \sqrt{b^2 - 4ac}}{2a}$.

$x^2 + 2x - 8 = 0$	$x^2 + 6x + 9 = 0$	$x^2 + 2x + 6 = 0$
$x = \dfrac{-2 \pm \sqrt{2^2 - 4 \times 1 \times (-8)}}{2 \times 1}$	$x = \dfrac{-6 \pm \sqrt{6^2 - 4 \times 1 \times 9}}{2 \times 1}$	$x = \dfrac{-2 \pm \sqrt{2^2 - 4 \times 1 \times 6}}{2 \times 1}$
$x = \dfrac{-2 \pm \sqrt{36}}{2}$	$x = \dfrac{-6 \pm \sqrt{0}}{2}$	$x = \dfrac{-2 \pm \sqrt{-20}}{2}$
$x = 2$ or $x = -4$	$x = -3$ or $x = -3$	no solutions
2 distinct roots	**2 equal roots**	**0 roots**

The part of the quadratic formula underneath the square root sign is called the **discriminant**.

$$\text{discriminant} = b^2 - 4ac$$

The sign (positive, zero or negative) of the discriminant tells you how many roots there are for a particular quadratic equation.

$b^2 - 4ac$	Nature of roots
> 0	2 real distinct roots
$= 0$	2 real equal roots
< 0	0 real roots

There is a connection between the roots of the quadratic equation $ax^2 + bx + c = 0$ and the corresponding curve $y = ax^2 + bx + c$.

$b^2 - 4ac$	Nature of roots of $ax^2 + bx + c = 0$	Shape of curve $y = ax^2 + bx + c$
> 0	2 real distinct roots	The curve cuts the x-axis at 2 distinct points.
$= 0$	2 real equal roots	The curve touches the x-axis at 1 point.
< 0	0 real roots	The curve is entirely above or entirely below the x-axis.

WORKED EXAMPLE 8

Find the values of k for which $x^2 - 3x + 6 = k(x - 2)$ has two equal roots.

Answers

$$x^2 - 3x + 6 = k(x - 2)$$
$$x^2 - 3x + 6 - kx + 2k = 0$$
$$x^2 - (3 + k)x + 6 + 2k = 0$$

For two equal roots, $b^2 - 4ac = 0$.

$$(3 + k)^2 - 4 \times 1 \times (6 + 2k) = 0$$
$$k^2 + 6k + 9 - 24 - 8k = 0$$
$$k^2 - 2k - 15 = 0$$
$$(k + 3)(k - 5) = 0$$

So $k = -3$ or $k = 5$.

WORKED EXAMPLE 9

Find the values of k for which $x^2 + (k - 2)x + 4 = 0$ has two distinct roots.

Answers

$x^2 + (k - 2)x + 4 = 0$

For two distinct roots $b^2 - 4ac > 0$

$(k - 2^2) - 4 \times 1 \times 4 > 0$

$k^2 - 4k + 4 - 16 > 0$

$k^2 - 4k - 12 > 0$

$(k + 2)(k - 6) > 0$

Critical values are -2 and 6.

So $k < -2$ or $k > 6$.

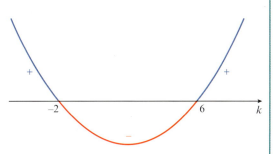

Exercise 2.5

1   State whether these equations have two distinct roots, two equal roots or no roots.

   a $x^2 + 4x + 4 = 0$ b $x^2 + 4x - 21 = 0$ c $x^2 + 9x + 1 = 0$

   d $x^2 - 3x + 15 = 0$ e $x^2 - 6x + 2 = 0$ f $4x^2 + 20x + 25 = 0$

   g $3x^2 + 2x + 7 = 0$ h $5x^2 - 2x - 9 = 0$

2   Find the values of k for which $x^2 + kx + 9 = 0$ has two equal roots.

3   Find the values of k for which $kx^2 - 4x + 8 = 0$ has two distinct roots.

4   Find the values of k for which $3x^2 + 2x + k = 0$ has no real roots.

5   Find the values of k for which $(k + 1)x^2 + kx - 2k = 0$ has two equal roots.

6   Find the values of k for which $kx^2 + 2(k + 3)x + k = 0$ has two distinct roots.

7   Find the values of k for which $3x^2 - 4x + 5 - k = 0$ has two distinct roots.

8   Find the values of k for which $4x^2 - (k - 2)x + 9 = 0$ has two equal roots.

9   Find the values of k for which $4x^2 + 4(k - 2)x + k = 0$ has two equal roots.

10   Show that the roots of the equation $x^2 + (k - 2)x - 2k = 0$ are real and distinct for all positive values of k.

11   Show that the roots of the equation $kx^2 + 5x - 2k = 0$ are real and distinct for all positive values of k ($k \neq 0$).

2.6 Intersection of a line and a curve

When considering the intersection of a straight line and a parabola, there are three possible situations.

Situation 1	Situation 2	Situation 3
2 points of intersection	1 point of intersection	0 points of intersection
The line cuts the curve at two distinct points.	The line touches the curve at one point. This means that the line is a **tangent** to the curve.	The line does not intersect the curve.

You have already learned that to find the points of intersection of the line $y = x - 6$ with the parabola $y = x^2 - 3x - 4$ you solve the two equations simultaneously.

This would give $x^2 - 3x - 4 = x - 6$

$$x^2 - 4x + 2 = 0.$$

The resulting quadratic equation can then be solved using the quadratic formula:

$$x = \frac{-b \pm \sqrt{b^2 - 4ac}}{2a}$$

The number of points of intersection will depend on the value of $b^2 - 4ac$.

The different situations are given in the table below.

$b^2 - 4ac$	Nature of roots	Line and curve
> 0	2 real distinct roots	2 distinct points of intersection
$= 0$	2 real equal roots	1 point of intersection (line is a tangent)
< 0	0 real roots	no points of intersection

The condition for a quadratic equation to have real roots is $b^2 - 4ac \geqslant 0$.

WORKED EXAMPLE 10

Find the value of k for which $y = 2x + k$ is a tangent to the curve $y = x^2 - 4x + 4$.

Answers

$$x^2 - 4x + 4 = 2x + k$$
$$x^2 - 6x + (4 - k) = 0$$

Since the line is a tangent to the curve, $b^2 - 4ac = 0$.

$$(-6^2) - 4 \times 1 \times (4 - k) = 0$$
$$36 - 16 + 4k = 0$$
$$k = -5$$

WORKED EXAMPLE 11

Find the range of values of k for which $y = x - 5$ intersects the curve $y = kx^2 - 6$ at two distinct points.

Answers

$$kx^2 - 6 = x - 5$$
$$kx^2 - x - 1 = 0$$

Since the line intersects the curve at two distinct points, $b^2 - 4ac > 0$.

$$(-1)^2 - 4 \times k \times (-1) > 0$$
$$1 + 4k > 0$$
$$k > -\frac{1}{4}$$

WORKED EXAMPLE 12

Find the values of k for which $y = kx - 3$ does not intersect the curve $y = x^2 - 2x + 1$.

Answers

$$x^2 - 2x + 1 = kx - 3$$
$$x^2 - x(2 + k) + 4 = 0$$

Since the line and curve do not intersect, $b^2 - 4ac < 0$.

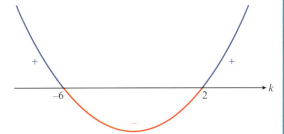

$$(2 + k)^2 - 4 \times 1 \times 4 < 0$$
$$k^2 + 4k + 4 - 16 < 0$$
$$k^2 + 4k - 12 < 0$$
$$(k + 6)(k - 2) < 0$$

Critical values are -6 and 2.

So $-6 < k < 2$.

Exercise 2.6

1   Find the values of k for which $y = kx + 1$ is a tangent to the curve $y = 2x^2 + x + 3$.

2   Find the value of k for which the x-axis is a tangent to the curve
    $y = x^2 + (3 - k)x - (4k + 3)$.

3   Find the values of the constant c for which the line $y = x + c$ is a tangent to the
    curve $y = 3x + \frac{2}{x}$.

4   Find the set of values of k for which the line $y = 3x + 1$ cuts the curve
    $y = x^2 + kx + 2$ in two distinct points.

5   The line $y = 2x + k$ is a tangent to the curve $x^2 + 2xy + 20 = 0$.

    a   Find the possible values of k.

    b   For each of these values of k, find the coordinates of the point of contact of
        the tangent with the curve.

6   Find the set of values of k for which the line $y = k - x$ cuts the curve
    $y = x^2 - 7x + 4$ in two distinct points.

7   Find the values of k for which the line $y = kx - 10$ intersects the curve
    $x^2 + y^2 = 10x$.

8   Find the set of values of m for which the line $y = mx - 5$ does not intersect
    the curve $y = x^2 - 5x + 4$.

9   The line $y = mx + 6$ is a tangent to the curve $y = x^2 - 4x + 7$.

    Find the possible values of m.

SUMMARY

Completing the square

For a quadratic function $f(x) = ax^2 + bx + c$ that is written in the form $f(x) = a(x - h)^2 + k$,

i   if $a > 0$, the minimum point is (h, k) ii   if $a < 0$, the maximum point is (h, k).

Quadratic equation ($ax^2 + bx + c = 0$) and corresponding curve ($y = ax^2 + bx + c$)

$b^2 - 4ac$	Nature of roots of $ax^2 + bx + c = 0$	Shape of curve $y = ax^2 + bx + c$
> 0	2 real distinct roots	The curve cuts the x-axis at 2 distinct points.
$= 0$	2 real equal roots	The curve touches the x-axis at 1 point.
< 0	0 real roots	The curve is entirely above or entirely below the x-axis.

CONTINUED

Intersection of a quadratic curve and a straight line

Situation 1	Situation 2	Situation 3
2 points of intersection	1 point of intersection	0 points of intersection
The line cuts the curve at two distinct points.	The line touches the curve at one point. This means that the line is a **tangent** to the curve.	The line does not intersect the curve.

Interpreting the discriminant

Solving simultaneously the equation of the curve with the equation of the line will give a quadratic equation of the form $ax^2 + bx + c = 0$. The discriminant $b^2 - 4ac$, gives information about the roots of the equation and also about the intersection of the curve with the line.

$b^2 - 4ac$	Nature of roots	Line and curve
> 0	2 real distinct roots	2 distinct points of intersection
$= 0$	2 real equal roots	1 point of intersection (line is a tangent)
< 0	no real roots	no points of intersection

The condition for a quadratic equation to have real roots is $b^2 - 4ac \geqslant 0$.

Past paper questions

Worked example

a Express $5x^2 - 14x - 3$ in the form $p(x + q)^2 + r$, where p, q and r are constants. [3]

b Sketch the graph of $y = |5x^2 - 14x - 3|$ on the axes below. Show clearly any points where your graph meets the coordinate axes. [4]

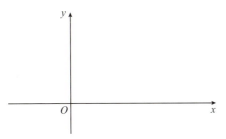

c State the set of values of k for which $|5x^2 - 14x - 3| = k$ has exactly four solutions. [2]

Cambridge IGCSE Additional Mathematics 0606 Paper 21 Q9 Jun 2018

Answers

a $\quad 5x^2 - 14x - 3 = 5\left(x^2 - \dfrac{14}{5}x - \dfrac{3}{5}\right)$

$$= 5\left[\left(x - \dfrac{7}{5}\right)^2 - \left(\dfrac{7}{5}\right)^2 - \dfrac{3}{5}\right]$$

$$= 5\left[\left(x - \dfrac{7}{5}\right)^2 - \dfrac{64}{25}\right]$$

$$= 5\left(x - \dfrac{7}{5}\right)^2 - \dfrac{64}{5}$$

b   First sketch the graph of $y = 5x^2 - 14x - 3$.

When $x = 0$, $y = -3$.

So the y-intercept is -3.

When $y = 0$,

$$5\left(x - \dfrac{7}{5}\right)^2 - \dfrac{64}{5} = 0$$

$$5\left(x - \dfrac{7}{5}\right)^2 = \dfrac{64}{5}$$

$$\left(x - \dfrac{7}{5}\right)^2 = \dfrac{64}{25}$$

$$x - \dfrac{7}{5} = \pm\dfrac{8}{5}$$

$$x = 3 \text{ or } x = -\dfrac{1}{5}$$

So, the x-intercepts are $-\dfrac{1}{5}$ and 3.

Using the answer to **part i**, the minimum point on the curve is $\left(\dfrac{7}{5}, -\dfrac{64}{5}\right)$.

Graph of $y = 5x^2 - 14x - 3$ is:

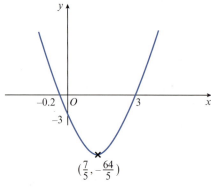

Graph of $y = |5x^2 - 14x - 3|$ is:

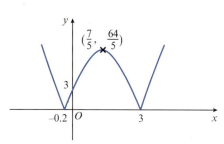

c   The values of k for which $|5x^2 - 14x - 3| = k$ has exactly four solutions are $0 < k < \dfrac{64}{5}$.

1 Find the set of values of k for which the line $y = k(4x - 3)$ does not intersect the curve $y = 4x^2 + 8x - 8$. [5]

Cambridge IGCSE Additional Mathematics 0606 Paper 11 Q4 Jun 2014

2 Find the set of values of x for which $x(x + 2) < x$. [3]

Cambridge IGCSE Additional Mathematics 0606 Paper 21 Q1 Jun 2014

3 **a** Express $2x^2 - x + 6$ in the form $p(x - q)^2 + r$, where p, q and r are constants to be found. [3]

   b Hence state the least value of $2x^2 - x + 6$ and the value of x at which this occurs. [2]

Cambridge IGCSE Additional Mathematics 0606 Paper 21 Q5 Jun 2014

4 Find the range of values of k for which the equation $kx^2 + k = 8x - 2xk$ has 2 real distinct roots. [4]

Cambridge IGCSE Additional Mathematics 0606 Paper 11 Q1 Nov 2015

5 **a** Find the set of values of x for which $4x^2 + 19x - 5 \leqslant 0$. [3]

   b **i** Express $x^2 + 8x - 9$ in the form $(x + a)^2 + b$, where a and b are integers. [2]

      ii Use your answer to **part i** to find the greatest value of $9 - 8x - x^2$ and the value of x at which this occurs. [2]

      iii Sketch the graph of $y = 9 - 8x - x^2$, indicating the coordinates of any points of intersection with the coordinate axes. [2]

Adapted from Cambridge IGCSE Additional Mathematics 0606 Paper 21 Q9 Jun 2015

6 The curve $3x^2 + xy - y^2 + 4y - 3 = 0$ and the line $y = 2(1 - x)$ intersect at the points A and B.

   i Find the coordinates of A and B. [5]

   ii Find the equation of the perpendicular bisector of the line AB, giving your answer in the form $ax + by = c$, where a, b and c are integers. [4]

Cambridge IGCSE Additional Mathematics 0606 Paper 21 Q9 Jun 2017

7 **a** Write $9x^2 - 12x + 5$ in the form $p(x - q)^2 + r$, where p, q and r are constants. [3]

   b Hence write down the coordinates of the minimum point of the curve $y = 9x^2 - 12x + 5$. [1]

Cambridge IGCSE Additional Mathematics 0606 Paper 21 Q2 Jun 2020

8 The line $y = 5x + 6$ meets the curve $xy = 8$ at the points A and B.

   a Find the coordinates of A and B. [3]

   b Find the coordinates of the point where the perpendicular bisector of the line AB meets the line $y = x$. [5]

Cambridge IGCSE Additional Mathematics 0606 Paper 11 Q6 Jun 2020

9 Solve the inequality $(x - 1)(x - 5) > 12$. [4]

Cambridge IGCSE Additional Mathematics 0606 Paper 21 Q1 Nov 2017

10 Solve the equations
$$y - x = 4$$
$$x^2 + y^2 - 8x - 4y - 16 = 0$$
[5]

Cambridge IGCSE Additional Mathematics 0606 Paper 11 Q1 June 2018

11 Find the values of k for which the line $y = kx + 3$ is a tangent to the curve $y = 2x^2 + 4x + k - 1$. [5]

Cambridge IGCSE Additional Mathematics 0606 Paper 12 Q2 Mar 2020

12 Find the values of the constant k for which the equation $kx^2 - 3(k+1)x + 25 = 0$ has equal roots. [4]

Cambridge IGCSE Additional Mathematics 0606 Paper 22 Q2 Mar 2021

13 **Do not use a calculator in this question.** The curve $xy = 11x + 5$ cuts the line $y = x + 10$ at the points A and B. The mid-point of AB is the point C. Show that the point C lies on the line $x + y = 11$. [7]

Cambridge IGCSE Additional Mathematics 0606 Paper 21 Q6 Nov 2019

14 a Show that $2x^2 + 5x - 3$ can be written in the form $a(x+b)^2 + c$, where a, b and c are constants. [3]

b Hence, write down the coordinates of the stationary point on the curve with equation $y = 2x^2 + 5x - 3$. [2]

c On the axes below, sketch the graph of $y = |2x^2 + 5x - 3|$, stating the coordinates of the intercepts with the axes.

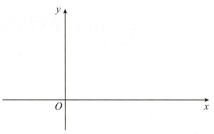

[3]

d Write down the value of k for which the equation $|2x^2 + 5x - 3| = k$ has exactly 3 distinct solutions. [1]

Cambridge IGCSE Additional Mathematics 0606 Paper 12 Q4 Mar 2021

15 i Write $x^2 - 9x + 8$ in the form $(x-p)^2 - q$, where p and q are constants. [2]

ii Hence write down the coordinates of the minimum point on the curve $y = x^2 - 9x + 8$. [1]

iii On the axes below, sketch the graph of $y = |x^2 - 9x + 8|$, showing the coordinates of the points where the curve meets the coordinate axes.

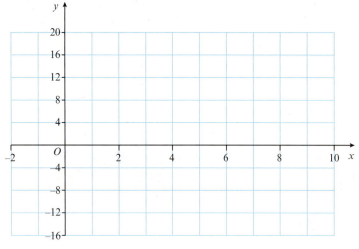

[3]

iv Write down the value of k for which $|x^2 - 9x + 8| = k$ has exactly 3 solutions. [1]

Cambridge IGCSE Additional Mathematics 0606 Paper 11 Q4 Nov 2018

Factors and polynomials

PRE-REQUISITE KNOWLEDGE

Before you start…

Where it comes from	You should be able to...		Check your skills
Cambridge IGCSE/O Level Mathematics	Expand brackets.	1	Expand and simplify: $(x^2 - 2x + 5)(3x - 2)$
Cambridge IGCSE/O Level Mathematics	Factorise quadratic expressions.	2	Factorise: $2x^2 + 5x - 12$
Cambridge IGCSE/O Level Mathematics	Perform long division on numbers and find the remainder where necessary.	3	Using long division, calculate:
		a	$4270 \div 35$
		b	$6662 \div 18$
		c	$10386 \div 27$

3.1 Adding, subtracting and multiplying polynomials

A **polynomial** is an expression of the form

$$a_n x^n + a_{n-1}x^{n-1} + a_{n-2}x^{n-2} + \ldots + a_2 x^2 + a_1 x^1 + a_0$$

where:

- x is a variable

- n is a non-negative integer

- the coefficients $a_n, a_{n-1}, a_{n-2}, \ldots, a_2, a_1, a_0$ are constants

- a_n is called the leading coefficient and $a_n \neq 0$

- a_0 is called the constant term.

The highest power of x in the polynomial is called the **degree** of the polynomial.

You already know the special names for polynomials of degree 1, 2 and 3. These are shown in the table below together with the special name for a polynomial of degree 4.

Polynomial expression	Degree	Name
$ax + b, \quad a \neq 0$	1	linear
$ax^2 + bx + c, \quad a \neq 0$	2	quadratic
$ax^3 + bx^2 + cx + d, \quad a \neq 0$	3	cubic
$ax^4 + bx^3 + cx^2 + dx + e, \quad a \neq 0$	4	quartic

The next example is a recap of how to add, subtract and multiply polynomials.

WORKED EXAMPLE 1

If $P(x) = 2x^3 - 6x^2 - 5$ and $Q(x) = x^3 + 2x - 1$, find an expression for

a $P(x) + Q(x)$ **b** $P(x) - Q(x)$ **c** $2Q(x)$ **d** $P(x)Q(x)$

Answers

a $P(x) + Q(x) = 2x^3 - 6x^2 - 5 + x^3 + 2x - 1$ collect like terms
$= 3x^3 - 6x^2 + 2x - 6$

b $P(x) - Q(x) = (2x^3 - 6x^2 - 5) - (x^3 + 2x - 1)$ remove brackets
$= 2x^3 - 6x^2 - 5 - x^3 - 2x + 1$ collect like terms
$= x^3 - 6x^2 - 2x - 4$

c $2Q(x) = 2(x^3 + 2x - 1)$
$= 2x^3 + 4x - 2$

d $P(x)Q(x) = (2x^3 - 6x^2 - 5)(x^3 + 2x - 1)$
$= 2x^3(x^3 + 2x - 1) - 6x^2(x^3 + 2x - 1) - 5(x^3 + 2x - 1)$
$= 2x^6 + 4x^4 - 2x^3 - 6x^5 - 12x^3 + 6x^2 - 5x^3 - 10x + 5$
$= 2x^6 - 6x^5 + 4x^4 - 19x^3 + 6x^2 - 10x + 5$

CLASS DISCUSSION

$P(x)$ is a polynomial of degree p and $Q(x)$ is a polynomial of degree q, where $p > q$.

Discuss with your classmates what the degree of each of the following polynomials is:

$P(x) + Q(x)$	$2P(x)$	$Q(x) + 5$
$-3Q(x)$	$P^2(x)$	$[Q(x)]^2$
$P(x)Q(x)$	$QP(x)$	$Q(x) - P(x)$

Exercise 3.1

1   If $P(x) = 3x^4 + 2x^2 - 1$ and $Q(x) = 2x^3 + x^2 + 1$, find an expression for

   a $P(x) + Q(x)$ **b** $3P(x) + Q(x)$ **c** $P(x) - 2Q(x)$ **d** $P(x)Q(x)$

2   Find the following products.

   a $(2x - 1)(4x^3 + x + 2)$ **b** $(x^3 + 2x^2 - 1)(3x + 2)$

   c $(3x^2 + 2x - 5)(x^3 + x^2 + 4)$ **d** $(x + 2)^2(3x^3 + x - 1)$

   e $(x^2 - 5x + 2)^2$ **f** $(3x - 1)^3$

3 Simplify each of the following.

   a $(2x - 3)(x + 2) + (x + 1)(x - 1)$

   b $(3x + 1)(x^2 + 5x + 2) - (x^2 - 4x + 2)(x + 3)$

   c $(2x^3 + x - 1)(x^2 + 3x - 4) - (x + 2)(x^3 - x^2 + 5x + 2)$

4 If $f(x) = 2x^2 - x - 4$ and $g(x) = x^2 + 5x + 2$, find an expression for

   a $f(x) + xg(x)$ b $[f(x)]^2$ c $f^2(x)$ d $gf(x)$

3.2 Division of polynomials

To divide a polynomial by another polynomial you first need to remember how to do long division with numbers.

The steps for calculating $5508 \div 17$ are:

```
          3 2 4
  1 7 ) 5 5 0 8
        5 1
          4 0
          3 4
            6 8
            6 8
              0
```

Divide 55 by 17
$3 \times 17 = 51$
$55 - 51 = 4$, bring down the 0 from the next column
Divide 40 by 17, $2 \times 17 = 34$
$40 - 34 = 6$, bring down the 8 from the next column
Divide 68 by 17, $4 \times 17 = 68$
$68 - 68 = 0$

So $5508 \div 17 = 324$

dividend   divisor   quotient

The same process can be applied to the division of polynomials.

WORKED EXAMPLE 2

Divide $x^3 - 5x^2 + 8x - 4$ by $x - 2$.

Answers

Step 1:

```
                 x²
  x - 2 ) x³ - 5x² + 8x - 4
          x³ - 2x²
             -3x² + 8x
```

divide the first term of the polynomial by x, $x^3 \div x = x^2$
multiply $(x - 2)$ by x^2, $x^2(x - 2) = x^3 - 2x^2$
subtract, $(x^3 - 5x^2) - (x^3 - 2x^2) = -3x^2$, bring down the $8x$ from the next column

Step 2: Repeat the process.

```
                 x² - 3x
  x - 2 ) x³ - 5x² + 8x - 4
          x³ - 2x²
             -3x² + 8x
             -3x² + 6x
                    2x - 4
```

divide $-3x^2$ by x, $-3x^2 \div x = -3x$
multiply $(x - 2)$ by $-3x$, $-3x(x - 2) = -3x^2 + 6x$
subtract, $(-3x^2 + 8x) - (-3x^2 + 6x) = 2x$, bring down the -4 from the next column

CONTINUED

Step 3: Repeat the process.

$$
\require{enclose}
\begin{array}{r}
x^2 - 3x + 2 \\
x - 2 \enclose{longdiv}{x^3 - 5x^2 + 8x - 4} \\
\underline{x^3 - 2x^2} \\
-3x^2 + 8x \\
\underline{-3x^2 + 6x} \\
2x - 4 \\
\underline{2x - 4} \\
0
\end{array}
$$

          divide $2x$ by x, $2x \div x = 2$

          multiply $(x - 2)$ by 2, $2(x - 2) = 2x - 4$

          subtract, $(2x - 4) - (2x - 4) = 0$

So $(x^3 - 5x^2 + 8x - 4) \div (x - 2) = x^2 - 3x + 2$

WORKED EXAMPLE 3

Divide $2x^3 - x + 51$ by $x + 3$.

Answers

There are no x^2 terms in $2x^3 - x + 51$, so we write it as $2x^3 + 0x^2 - x + 51$.

Step 1:

$$
\begin{array}{r}
2x^2 \\
x + 3 \enclose{longdiv}{2x^3 + 0x^2 - x + 51} \\
\underline{2x^3 + 6x^2} \\
-6x^2 - x
\end{array}
$$

divide the first term of the polynomial by x, $2x^3 \div x = 2x^2$

multiply $(x + 3)$ by $2x^2$, $2x^2(x + 3) = 2x^3 + 6x^2$

subtract, $(2x^3 + 0x^2) - (2x^3 + 6x^2) = -6x^2$, bring down the $-x$ from the next column

Step 2: Repeat the process.

$$
\begin{array}{r}
2x^2 - 6x \\
x + 3 \enclose{longdiv}{2x^3 + 0x^2 - x + 51} \\
\underline{2x^3 + 6x^2} \\
-6x^2 - x \\
\underline{-6x^2 - 18x} \\
17x + 51
\end{array}
$$

divide $-6x^2$ by x, $-6x^2 \div x = -6x$

multiply $(x + 3)$ by $-6x$, $-6x(x + 3) = -6x^2 - 18x$

subtract, $(-6x^2 - x) - (-6x^2 - 18x) = 17x$, bring down the 51 from the next column

Step 3: Repeat the process.

$$
\begin{array}{r}
2x^2 - 6x + 17 \\
x + 3 \enclose{longdiv}{2x^3 + 0x^2 - x + 51} \\
\underline{2x^3 + 6x^2} \\
-6x^2 - x \\
\underline{-6x^2 - 18x} \\
17x + 51 \\
\underline{17x + 51} \\
0
\end{array}
$$

divide $17x$ by x, $17x \div x = 17$

multiply $(x + 3)$ by 17, $17(x + 3) = 17x + 51$

subtract, $(17x + 51) - (17x + 51) = 0$

So $(2x^3 - x + 51) \div (x + 3) = 2x^2 - 6x + 17$

Exercise 3.2

1 Simplify each of the following.

    a $(x^3 + 3x^2 - 46x - 48) \div (x + 1)$ **b** $(x^3 - x^2 - 3x + 2) \div (x - 2)$

    c $(x^3 - 20x^2 + 100x - 125) \div (x - 5)$ **d** $(x^3 - 3x - 2) \div (x - 2)$

    e $(x^3 - 3x^2 - 33x + 35) \div (x - 7)$ **f** $(x^3 + 2x^2 - 9x - 18) \div (x + 2)$

2 Simplify each of the following.

    a $(3x^3 + 8x^2 + 3x - 2) \div (x + 2)$ **b** $(6x^3 + 11x^2 - 3x - 2) \div (3x + 1)$

    c $(3x^3 - 11x^2 + 20) \div (x - 2)$ **d** $(3x^3 - 21x^2 + 4x - 28) \div (x - 7)$

3 Simplify.

    a $\dfrac{3x^3 - 3x^2 - 4x + 4}{x - 1}$ **b** $\dfrac{2x^3 + 9x^2 + 25}{x + 5}$

    c $\dfrac{3x^3 - 50x + 8}{3x^2 + 12x - 2}$ **d** $\dfrac{x^3 - 14x - 15}{x^2 - 3x - 5}$

4   **a**   Divide $x^4 - 1$ by $(x + 1)$.   **b**   Divide $x^3 - 8$ by $(x - 2)$.

3.3 The factor theorem

In **Worked example 2** you found that $x - 2$ divided exactly into $(x^3 - 5x^2 + 8x - 4)$.

$(x^3 - 5x^2 + 8x - 4) \div (x - 2) = x^2 - 3x + 2$

This can also be written as:

$(x^3 - 5x^2 + 8x - 4) = (x - 2)(x^2 - 3x + 2)$

If a polynomial $P(x)$ is divided exactly by a linear factor $x - c$ to give the polynomial $Q(x)$, then

$P(x) = (x - c)Q(x)$.

Substituting $x = c$ into this formula gives $P(c) = 0$.

Hence:

> If, for a polynomial $P(x)$, $P(c) = 0$, then $x - c$ is a factor of $P(x)$.

This is known as the **factor theorem**.

For example, when $x = 2$,

$4x^3 - 8x^2 - x + 2 = 4(2)^3 - 8(2)^2 - 2 + 2 = 32 - 32 - 2 + 2 = 0$.

Therefore $x - 2$ is a factor of $4x^3 - 8x^2 - x + 2$.

The factor theorem can be extended to:

> If, for a polynomial P(x), P$\left(\dfrac{b}{a}\right)$ = 0, then $ax - b$ is a factor of P(x).

For example, when $x = \dfrac{1}{2}$,

$$4x^3 - 2x^2 + 8x - 4 = 4\left(\dfrac{1}{2}\right)^3 - 2\left(\dfrac{1}{2}\right)^2 + 8\left(\dfrac{1}{2}\right) - 4 = \dfrac{1}{2} - \dfrac{1}{2} + 4 - 4 = 0$$

Therefore $2x - 1$ is a factor of $4x^3 - 2x^2 + 8x - 4$.

CLASS DISCUSSION

Discuss with your classmates which of the following expressions are exactly divisible by $x - 2$.

$x^3 - x^2 - x - 2$	$2x^3 + 5x^2 - 4x - 3$	$x^3 - 4x^2 + 8x - 8$
$2x^4 - x^3 + 3x^2 - 2x - 5$	$x^3 - 8$	$3x^3 - 8x - 8$
$6x^3 - 10x^2 - 18$	$x^3 + x^2 - 4x - 4$	$x^3 + x + 10$

WORKED EXAMPLE 4

Show that $x - 3$ is a factor of $x^3 - 6x^2 + 11x - 6$ by

a algebraic division

b the factor theorem.

Answers

a Divide $x^3 - 6x^2 + 11x - 6$ by $x - 3$.

$$
\begin{array}{r}
x^2 - 3x + 2 \\
x - 3 \overline{\smash{\big)}\, x^3 - 6x^2 + 11x - 6} \\
\underline{x^3 - 3x^2} \\
-3x^2 + 11x \\
\underline{-3x^2 + 9x} \\
2x - 6 \\
\underline{2x - 6} \\
0
\end{array}
$$

The remainder = 0, so $x - 3$ is a factor of $x^3 - 6x^2 + 11x - 6$.

b Let f(x) = $x^3 - 6x^2 + 11x - 6$ if f(3) = 0, then $x - 3$ is a factor.

$$\begin{aligned}
\text{f}(3) &= (3)^3 - 6(3)^2 + 11(3) - 6 \\
&= 27 - 54 + 33 - 6 \\
&= 0
\end{aligned}$$

So $x - 3$ is a factor of $x^3 - 6x^2 + 11x - 6$

WORKED EXAMPLE 5

$2x^2 + x - 1$ is a factor of $2x^3 - x^2 + ax + b$.

Find the value of a and the value of b.

Answers

Let $f(x) = 2x^3 - x^2 + ax + b$.

If $2x^2 + x - 1 = (2x - 1)(x + 1)$ is a factor of $f(x)$, then $2x - 1$ and $x + 1$ are also factors of $f(x)$.

Using the factor theorem $f\left(\dfrac{1}{2}\right) = 0$ and $f(-1) = 0$.

$f\left(\dfrac{1}{2}\right) = 0$ gives $2\left(\dfrac{1}{2}\right)^3 - \left(\dfrac{1}{2}\right)^2 + a\left(\dfrac{1}{2}\right) + b = 0$

$$\frac{1}{4} - \frac{1}{4} + \frac{a}{2} + b = 0$$

$$a = -2b \qquad (1)$$

$f(-1) = 0$ gives $2(-1)^3 - (-1)^2 + a(-1) + b = 0$

$$-2 - 1 - a + b = 0$$

$$a = b - 3 \qquad (2)$$

$(2) = (1)$ gives $b - 3 = -2b$

$$3b = 3$$

$$b = 1$$

Substituting in (2) gives $a = -2$.

So, $a = -2$, $b = 1$

Exercise 3.3

1   Use the factor theorem to show:

   a $x - 4$ is a factor of $x^3 - 3x^2 - 6x + 8$

   b $x + 1$ is a factor of $x^3 - 3x - 2$

   c $x - 2$ is a factor of $5x^3 - 17x^2 + 28$

   d $3x + 1$ is a factor of $6x^3 + 11x^2 - 3x - 2$

2   Find the value of a in each of the following.

   a $x + 1$ is a factor of $6x^3 + 27x^2 + ax + 8$

   b $x + 7$ is a factor of $x^3 - 5x^2 - 6x + a$

   c $2x + 3$ is a factor of $4x^3 + ax^2 + 29x + 30$

3   $x - 2$ is a factor of $x^3 + ax^2 + bx - 4$.

   Express b in terms of a.

4   Find the value of a and the value of b in each of the following.

    a $x^2 + 3x - 10$ is a factor of $x^3 + ax^2 + bx + 30$.

    b $2x^2 - 11x + 5$ is a factor of $ax^3 - 17x^2 + bx - 15$.

    c $4x^2 - 4x - 15$ is a factor of $4x^3 + ax^2 + bx + 30$.

5   It is given that $x^2 - 5x + 6$ and $x^3 - 6x^2 + 11x + a$ have a common factor.
    Find the possible value of a.

6   $x - 2$ is a common factor of $3x^3 - (a - b)x - 8$ and $x^3 - (a + b)x + 30$.
    Find the value of a and the value of b.

7   $x - 3$ and $2x - 1$ are factors of $2x^3 - px^2 - 2qx + q$.

    a   Find the value of p and the value of q.

    b   Explain why $x + 3$ is also a factor of the expression.

8   $x + a$ is a factor of $x^3 + 8x^2 + 4ax - 3a$.

    a   Show that $a^3 - 4a^2 + 3a = 0$.

    b   Find the possible values of a.

3.4 Cubic expressions and equations

Consider factorising $x^3 - 5x^2 + 8x - 4$ completely.

In **Worked example 2** you found that $(x^3 - 5x^2 + 8x - 4) \div (x - 2) = x^2 - 3x + 2$

This can be rewritten as: $x^3 - 5x^2 + 8x - 4 = (x - 2)(x^2 - 3x + 2)$

Factorising completely gives: $x^3 - 5x^2 + 8x - 4 = (x - 2)(x - 2)(x - 1)$

Hence, if you know one factor of a cubic expression it is possible to then factorise the expression completely. The next example illustrates three different methods for doing this.

> ### WORKED EXAMPLE 6
>
> Factorise $x^3 - 3x^2 - 13x + 15$ completely.
>
> **Answers**
>
> Let $f(x) = x^3 - 3x^2 - 13x + 15$.
>
> The positive and negative factors of 15 are ± 1, ± 3, ± 5 and ± 15.
>
> $f(1) = (1)^3 - 3 \times (1)^2 - 13 \times (1) + 15 = 0$
>
> So $x - 1$ is a factor of $f(x)$.
>
> The other factors can be found by any of the following methods.

CONTINUED

Method 1 (by trial and error)

$f(x) = x^3 - 3x^2 - 13x + 15$

$f(1) = (1)^3 - 3 \times (1)^2 - 13 \times (1) + 15 = 0$

So $x - 1$ is a factor of $f(x)$.

$f(-3) = (-3)^3 - 3 \times (-3)^2 - 13 \times (-3) + 15 = 0$

So $x + 3$ is a factor of $f(x)$.

$f(5) = (5)^3 - 3 \times (5)^2 - 13 \times (5) + 15 = 0$

So $x - 5$ is a factor of $f(x)$.

Hence $f(x) = (x - 1)(x - 5)(x + 3)$

Method 2 (by long division)

$$
\require{enclose}
\begin{array}{r}
x^2 - 2x - 15 \\[-3pt]
x - 1 \enclose{longdiv}{x^3 - 3x^2 - 13x + 15} \\
\underline{x^3 - x^2} \\
-2x^2 - 13x \\
\underline{-2x^2 + 2x} \\
-15x + 15 \\
\underline{-15x + 15} \\
0
\end{array}
$$

$f(x) = (x - 1)(x^2 - 2x - 15)$

$ = (x - 1)(x - 5)(x + 3)$

Method 3 (by equating coefficients)

Since $x - 1$ is a factor, $x^3 - 3x^2 - 13x + 15$ can be written as:

$\qquad x^3 - 3x^2 - 13x + 15 = (x - 1)(ax^2 + bx + c)$

coefficient of x^3 is 1, so $a = 1$ since $1 \times 1 = 1$	constant term is -15, so $c = -15$ since $-1 \times -15 = 15$

$\qquad x^3 - 3x^2 - 13x + 15 = (x - 1)(x^2 + bx - 15)$ *expand and collect like terms*

$\qquad x^3 - 3x^2 - 13x + 15 = x^3 + (b - 1)x^2 + (-b - 15)x + 15$

Equating coefficients of x^2: $b - 1 = -3$

$\qquad\qquad\qquad\qquad\qquad\qquad b = -2$

$\qquad f(x) = (x - 1)(x^2 - 2x - 15)$

$\qquad = (x - 1)(x - 5)(x + 3)$

WORKED EXAMPLE 7

Solve $2x^3 - 3x^2 - 18x - 8 = 0$.

Answers

Let $f(x) = 2x^3 - 3x^2 - 18x - 8$.

The positive and negative factors of 8 are ± 1, ± 2, ± 4 and ± 8.

$f(-2) = 2(-2)^3 - 3 \times (-2)^2 - 18 \times (-2) - 8 = 0$
So $x + 2$ is a factor of $f(x)$.

$$2x^3 - 3x^2 - 18x - 8 = (x + 2)(ax^2 + bx + c)$$

coefficient of x^3 is 2, so $a = 2$ since $1 \times 2 = 2$	constant term is -8, so $c = -4$ since $2 \times -4 = -8$

$2x^3 - 3x^2 - 18x - 8 = (x + 2)(2x^2 + bx - 4)$ expand and collect like terms

$2x^3 - 3x^2 - 18x - 8 = 2x^3 + (b + 4)x^2 + (2b - 4)x - 8$

Equating coefficients of x^2: $b + 4 = -3$

$$b = -7$$

$f(x) = (x + 2)(2x^2 - 7x - 4)$
$\quad\ = (x + 2)(2x + 1)(x - 4)$

Hence $(x + 2)(2x + 1)(x - 4) = 0$.

So $x = -2$ or $x = -\dfrac{1}{2}$ or $x = 4$.

WORKED EXAMPLE 8

Solve $2x^3 + 7x^2 - 2x - 1 = 0$.

Answers

Let $f(x) = 2x^3 + 7x^2 - 2x - 1$.

The positive and negative factors of -1 are ± 1.

$f(-1) = 2(-1)^3 + 7 \times (-1)^2 - 2 \times (-1) - 1 \neq 0$
$f(1) = 2(1)^3 + 7 \times (1)^2 - 2 \times (1) - 1 \neq 0$

So $x - 1$ and $x + 1$ are not factors of $f(x)$.

By inspection, $f\left(\dfrac{1}{2}\right) = 2\left(\dfrac{1}{2}\right)^3 + 7 \times \left(\dfrac{1}{2}\right)^2 - 2 \times \left(\dfrac{1}{2}\right) - 1 = 0$

So $2x - 1$ is a factor of:

$$2x^3 + 7x^2 - 2x - 1 = (2x - 1)(ax^2 + bx + c)$$

coefficient of x^3 is 2, so $a = 1$ since $2 \times 1 = 2$	constant term is -1, so $c = 1$ since $-1 \times 1 = -1$

CONTINUED

$$2x^3 + 7x^2 - 2x - 1 = (2x - 1)(x^2 + bx + 1)$$

$$2x^3 + 7x^2 - 2x - 1 = 2x^3 + (2b - 1)x^2 + (2 - b)x - 1$$

Equating coefficients of x^2: $2b - 1 = 7$

$$b = 4$$

So $2x^3 + 7x^2 - 2x - 1 = (2x - 1)(x^2 + 4x + 1)$.

$$x = \frac{1}{2} \text{ or } x = \frac{-4 \pm \sqrt{4^2 - 4 \times 1 \times 1}}{2 \times 1}$$

$$x = \frac{1}{2} \text{ or } x = \frac{-4 \pm 2\sqrt{3}}{2}$$

$$x = \frac{1}{2} \text{ or } x = -2 + \sqrt{3} \text{ or } x = -2 - \sqrt{3}$$

Not all cubic expressions can be factorised into 3 linear factors.

Consider the cubic expression $x^3 + x^2 - 36$.

Let $f(x) = x^3 + x^2 - 36$.

$$f(3) = (3)^3 + (3)^2 - 36 = 0$$

So $x - 3$ is a factor of $f(x)$.

$$x^3 + x^2 - 36 = (x - 3)(ax^2 + bx + c)$$

coefficient of x^3 is 1, so $a = 1$ since $1 \times 1 = 1$	constant term is -36, so $c = 12$ since $-3 \times 12 = -36$

$$x^3 + x^2 - 36 = (x - 3)(x^2 + bx + 12)$$

$$x^3 + x^2 - 36 = x^3 + (b - 3)x^2 + (12 - 3b)x - 36$$

Equating coefficients of x^2: $b - 3 = 1$

$$b = 4$$

So $x^3 + x^2 - 36 = (x - 3)(x^2 + 4x + 12)$

Note that $x^2 + 4x + 12$ cannot be factorised into two further linear factors, since the discriminant < 0.

Exercise 3.4

1 **a** Show that $x - 1$ is a factor of $2x^3 - x^2 - 2x + 1$.

   b Hence factorise $2x^3 - x^2 - 2x + 1$ completely.

2 Factorise these cubic expressions completely.

a $x^3 + 2x^2 - 3x - 10$ b $x^3 + 4x^2 - 4x - 16$

c $2x^3 - 9x^2 - 18x$ d $x^3 - 8x^2 + 5x + 14$

e $2x^3 - 13x^2 + 17x + 12$ f $3x^3 + 2x^2 - 19x + 6$

g $4x^3 - 8x^2 - x + 2$ h $2x^3 + 3x^2 - 32x + 15$

3 Solve the following equations.

a $x^3 - 3x^2 - 33x + 35 = 0$ b $x^3 - 6x^2 + 11x - 6 = 0$

c $3x^3 + 17x^2 + 18x - 8 = 0$ d $2x^3 + 3x^2 - 17x + 12 = 0$

e $2x^3 - 3x^2 - 11x + 6 = 0$ f $2x^3 + 7x^2 - 5x - 4 = 0$

g $4x^3 + 12x^2 + 5x - 6 = 0$ h $2x^3 - 3x^2 - 29x + 60 = 0$

4 Solve the following equations.

Express roots in the form $a \pm b\sqrt{c}$, where necessary.

a $x^3 + 5x^2 - 4x - 2 = 0$ b $x^3 + 8x^2 + 12x - 9 = 0$

c $x^3 + 2x^2 - 7x - 2 = 0$ d $2x^3 + 3x^2 - 17x + 12 = 0$

5 Solve the equation $2x^3 + 9x^2 - 14x - 9 = 0$.

Express roots in the form $a \pm b\sqrt{c}$, where necessary.

6 Solve the equation $x^3 + 8x^2 + 12x = 9$.

Write your answers correct to 2 decimal places where necessary.

7 a Show that $x - 2$ is a factor of $x^3 - x^2 - x - 2$.

b Hence show that $x^3 - x^2 - x - 2 = 0$ has only one real root and state the value of this root.

8 f(x) is a cubic polynomial where the coefficient of x^3 is 1.

Find f(x) when the roots of f$(x) = 0$ are

a -2, 1 and 5 b -5, -2 and 4 c -3, 0 and 2.

9 f(x) is a cubic polynomial where the coefficient of x^3 is 2.

Find f(x) when the roots of f$(x) = 0$ are

a -0.5, 2 and 4 b 0.5, 1 and 2 c -1.5, 1 and 5.

10 f(x) is a cubic polynomial where the coefficient of x^3 is 1.

The roots of f$(x) = 0$ are -3, $1 + \sqrt{2}$ and $1 - \sqrt{2}$.

Express f(x) as a cubic polynomial in x with integer coefficients.

11 f(x) is a cubic polynomial where the coefficient of x^3 is 2.

The roots of f$(x) = 0$ are $\frac{1}{2}$, $2 + \sqrt{3}$ and $2 - \sqrt{3}$.

Express f(x) as a cubic polynomial in x with integer coefficients.

12 $2x + 3$ is a factor of $2x^4 + (a^2 + 1)x^3 - 3x^2 + (1 - a^3)x + 3$.

a Show that $4a^3 - 9a^2 + 4 = 0$. b Find the possible values of a.

3.5 The remainder theorem

Consider $f(x) = 2x^3 - 4x^2 + 7x - 37$.

Substituting $x = 3$ in the polynomial gives $f(3) = 2(3)^3 - 4(3)^2 + 7(3) - 37 = 2$.

When $2x^3 - 4x^2 + 7x - 37$ is divided by $x - 3$, there is a remainder.

$$
\begin{array}{r}
2x^2 + 2x + 13 \\
x - 3 \overline{\big)\, 2x^3 - 4x^2 + 7x - 37} \\
\underline{2x^3 - 6x^2} \\
2x^2 + 7x \\
\underline{2x^2 - 6x} \\
13x - 37 \\
\underline{13x - 39} \\
2
\end{array}
$$

The remainder is 2. This is the same value as $f(3)$.

$f(x) = 2x^3 - 4x^2 + 7x - 36$, can be written as

$f(x) = (x - 3)(2x^2 + 2x + 13) + 2$.

In general:

If a polynomial $P(x)$ is divided by $x - c$ to give the polynomial $Q(x)$ and a remainder R, then

$P(x) = (x - c)Q(x) + R$.

Substituting $x = c$ into this formula gives $P(c) = R$.

This leads to the **remainder theorem**:

> If a polynomial $P(x)$ is divided by $x - c$, the remainder is $P(c)$.

The Remainder Theorem can be extended to:

> If a polynomial $P(x)$ is divided by $ax - b$, the remainder is $P\left(\frac{b}{a}\right)$.

WORKED EXAMPLE 9

Find the remainder when $7x^3 + 6x^2 - 40x + 17$ is divided by $(x + 3)$ by using

a   algebraic division

b   the factor theorem.

CONTINUED

Answers

a Divide $7x^3 + 6x^2 - 40x + 17$ by $(x + 3)$.

$$
\begin{array}{r}
7x^2 - 15x + 5 \\
x + 3 \overline{)\ 7x^3 + 6x^2 - 40x + 17} \\
\underline{7x^3 + 21x^2} \\
-15x^2 - 40x \\
\underline{-15x^2 - 45x} \\
5x + 17 \\
\underline{5x + 15} \\
2
\end{array}
$$

The remainder is 2.

b Let $f(x) = 7x^3 + 6x^2 - 40x + 17$.

$$
\begin{aligned}
\text{Remainder} &= f(-3) \\
&= 7(-3)^3 + 6(-3)^2 - 40(-3) + 17 \\
&= -189 + 54 + 120 + 17 \\
&= 2
\end{aligned}
$$

WORKED EXAMPLE 10

$f(x) = 2x^3 + ax^2 - 9x + b$

When $f(x)$ is divided by $x - 1$, the remainder is 1.

When $f(x)$ is divided by $x + 2$, the remainder is 19.

Find the value of a and of b.

Answers

$f(x) = 2x^3 + ax^2 - 9x + b$

When $f(x)$ is divided by $x - 1$, the remainder is 1 means that: $f(1) = 1$.

$$
\begin{aligned}
2(1)^3 + a(1)^2 - 9(1) + b &= 1 \\
2 + a - 9 + b &= 1 \\
a + b &= 8 \qquad (1)
\end{aligned}
$$

When $f(x)$ is divided by $x + 2$, the remainder is 19 means that: $f(-2) = 19$.

$$
\begin{aligned}
2(-2)^3 + a(-2)^2 - 9(-2) + b &= 19 \\
-16 + 4a + 18 + b &= 19 \\
4a + b &= 17 \qquad (2)
\end{aligned}
$$

(2) − (1) gives $3a = 9$

$$a = 3$$

Substituting $a = 3$ in equation (2) gives $b = 5$.

$a = 3$ and $b = 5$

Exercise 3.5

1   Find the remainder when

    a $x^3 + 2x^2 - x + 3$ is divided by $x - 1$

    b $x^3 - 6x^2 + 11x - 7$ is divided by $x - 2$

    c $x^3 - 3x^2 - 33x + 30$ is divided by $x + 2$

    d $2x^3 - x^2 - 18x + 11$ is divided by $2x - 1$.

2   **a**   When $x^3 + x^2 + ax - 2$ is divided by $x - 1$, the remainder is 5.
        Find the value of a.

    b   When $2x^3 - 6x^2 + 7x + b$ is divided by $x + 2$, the remainder is 3.
        Find the value of b.

    c   When $2x^3 + x^2 + cx - 10$ is divided by $2x - 1$, the remainder is -4.
        Find the value of c.

3   $f(x) = x^3 + ax^2 + bx - 5$
    $f(x)$ has a factor of $x - 1$ and leaves a remainder of 3 when divided by $x + 2$.
    Find the value of a and of b.

4   $f(x) = x^3 + ax^2 + 11x + b$
    $f(x)$ has a factor of $x - 2$ and leaves a remainder of 24 when divided by $x - 5$.
    Find the value of a and of b.

5   $f(x) = x^3 - 2x^2 + ax + b$
    $f(x)$ has a factor of $x - 3$ and leaves a remainder of 15 when divided by $x + 2$.

    a   Find the value of a and of b.

    b   Solve the equation $f(x) = 0$.

6   $f(x) = 4x^3 + 8x^2 + ax + b$
    $f(x)$ has a factor of $2x - 1$ and leaves a remainder of 48 when divided by $x - 2$.

    a   Find the value of a and of b.

    b   Find the remainder when $f(x)$ is divided by $x - 1$.

7   $f(x) = 2x^3 + (a + 1)x^2 - ax + b$
    When $f(x)$ is divided by $x - 1$, the remainder is 5.
    When $f(x)$ is divided by $x - 2$, the remainder is 14.
    Show that $a = -4$ and find the value of b.

8   $f(x) = ax^3 + bx^2 + 5x - 2$
    When $f(x)$ is divided by $x - 1$, the remainder is 6.
    When $f(x)$ is divided by $2x + 1$, the remainder is -6.
    Find the value of a and of b.

9   $f(x) = x^3 - 5x^2 + ax + b$

$f(x)$ has a factor of $x - 2$.

a   Express b in terms of a.

b   When $f(x)$ is divided by $x + 1$, the remainder is -9.
    Find the value of a and of b.

10 $f(x) = x^3 + ax^2 + bx + c$

The roots of $f(x) = 0$ are 2, 3, and k.

When $f(x)$ is divided by $x - 1$, the remainder is -8.

a   Find the value of k.

b   Find the remainder when $f(x)$ is divided by $x + 1$.

11 $f(x) = 4x^3 + ax^2 + 13x + b$

$f(x)$ has a factor of $2x - 1$ and leaves a remainder of 21 when divided by $x - 2$.

a   Find the value of a and of b.

b   Find the remainder when the expression is divided by $x + 1$.

12 $f(x) = x^3 - 8x^2 + kx - 20$

When $f(x)$ is divided by $x - 1$, the remainder is R.

When $f(x)$ is divided by $x - 2$, the remainder is $4R$.

Find the value of k.

13 $f(x) = x^3 + 2x^2 - 6x + 9$

When $f(x)$ is divided by $x + a$, the remainder is R.

When $f(x)$ is divided by $x - a$, the remainder is $2R$.

a   Show that $3a^3 - 2a^2 - 18a - 9 = 0$.

b   Solve the equation in **part a** completely.

14 $f(x) = x^3 + 6x^2 + kx - 15$

When $f(x)$ is divided by $x - 1$, the remainder is R.

When $f(x)$ is divided by $x + 4$, the remainder is $-R$.

a   Find the value of k.

b   Hence find the remainder when the expression is divided by $x + 2$.

15 $P(x) = 5(x - 1)(x - 2)(x - 3) + a(x - 1)(x - 2) + b(x - 1) + c$

It is given that when $P(x)$ is divided by each of $x - 1$, $x - 2$ and $x - 3$ the remainders are 7, 2 and 1 respectively. Find the values of a, b, and c.

16 **CHALLENGE QUESTION**

$f(x) = x^3 + ax^2 + bx + c$

The roots of $f(x) = 0$ are 1, k, and $k + 1$.

When $f(x)$ is divided by $x - 2$, the remainder is 20.

a   Show that $k^2 - 3k - 18 = 0$.

b   Hence find the possible values of k.

REFLECTION

Without referring back to your textbook, explain to a classmate the difference between the factor theorem and the remainder theorem.

Do you think you did this successfully?

SUMMARY

The factor theorem:
If, for a polynomial $P(x)$, $P(c) = 0$, then $x - c$ is a factor of $P(x)$.
If, for a polynomial $P(x)$, $P\left(\dfrac{b}{a}\right) = 0$, then $ax - b$ is a factor of $P(x)$.

The remainder theorem:
If a polynomial $P(x)$ is divided by $x - c$, the remainder is $P(c)$.
If a polynomial $P(x)$ is divided by $ax - b$, the remainder is $P\left(\dfrac{b}{a}\right)$.

Past paper questions

Worked example

The polynomial $p(x) = ax^3 - 9x^2 + bx - 6$, where a and b are constants, has a factor of $x - 2$.
The polynomial has a remainder of 66 when divided by $x - 3$.

a   Find the value of a and of b. [4]

b   Using your values of a and b, show that $p(x) = (x - 2)q(x)$, where $q(x)$
     is a quadratic factor to be found. [2]

c   Hence show that the equation $p(x) = 0$ has only one real solution. [2]

Cambridge IGCSE Additional Mathematics 0606 Paper 11 Q3 Jun 2021

Answers

a $p(x) = ax^3 - 9x^2 + bx - 6$

   If $x - 2$ is a factor, then $p(2) = 0$.

   $a(2)^3 - 9(2)^2 + b(2) - 6 = 0$

   $\qquad\qquad 8a + 2b = 42$

   $\qquad\qquad 4a + b = 21 \qquad$ (1)

   Remainder = 66 when divided by $x - 3$, means that $p(3) = 66$.

   $a(3)^3 - 9(3)^2 + b(3) - 6 = 66$

   $\qquad 27a - 81 + 3b - 6 = 66$

   $\qquad\qquad 27a + 3b = 153$

   $\qquad\qquad 9a + b = 51 \qquad$ (2)

$(2) - (1)$ gives:

$5a = 30$

$a = 6$

So $a = 6$, $b = -3$

b $p(x) = 6x^3 - 9x^2 - 3x - 6$

$= (x - 2)(6x^2 + 3x + 3)$

c $p(x) = 0$

$6x^3 - 9x^2 - 3x - 6 = 0$

$(x - 2)(6x^2 + 3x + 3) = 0$

$x - 2 = 0$ or $6x^2 + 3x + 3 = 0$

$x = 2$ or $2x^2 + x + 1 = 0$

$x = 2$ or $x = \dfrac{(-1) \pm \sqrt{(1)^2 - 4(2)(1)}}{2(1)}$

$x = 2$ or $x = \dfrac{-1 \pm \sqrt{-7}}{2}$

Hence, $x = 2$ is the only real solution (because discriminant < 0).

1 **DO NOT USE A CALCULATOR IN THIS QUESTION.**

It is given that $x + 4$ is a factor of $p(x) = 2x^3 + 3x^2 + ax - 12$. When $p(x)$ is divided by $x - 1$ the remainder is b.

i Show that $a = -23$ and find the value of the constant b. [2]

ii Factorise $p(x)$ completely and hence state all the solutions of $p(x) = 0$. [4]

Cambridge IGCSE Additional Mathematics 0606 Paper 21 Q4 Jun 2018

2 $p(x) = 2x^3 + 5x^2 + 4x + a$

$q(x) = 4x^2 + 3ax + b$

Given that $p(x)$ has a remainder of 2 when divided by $2x + 1$ and that $q(x)$ is divisible by $x + 2$,

i find the value of each of the constants a and b. [3]

Given that $r(x) = p(x) - q(x)$ and using your values of a and b,

ii find the exact remainder when $r(x)$ is divided by $3x - 2$. [3]

Cambridge IGCSE Additional Mathematics 0606 Paper 11 Q2 Nov 2018

3 The polynomial $p(x) = (2x - 1)(x + k) - 12$, where k is a constant.

i Write down the value of $p(-k)$. [1]

When $p(x)$ is divided by $x + 3$, the remainder is 23.

ii Find the value of k. [2]

iii Using your value of k, show that the equation $p(x) = -25$ has no real solutions. [3]

Cambridge IGCSE Additional Mathematics 0606 Paper 11 Q3 Jun 2019

4 **DO NOT USE A CALCULATOR IN THIS QUESTION.**

$p(x) = 2x^3 - 3x^2 - 23x + 12$

a Find the value of $p\left(\frac{1}{2}\right)$. [1]

b Write $p(x)$ as the product of three linear factors and hence solve $p(x) = 0$. [5]

Cambridge IGCSE Additional Mathematics 0606 Paper 21 Q7 Nov 2020

5 The polynomial $f(x) = ax^3 + 7x^2 - 9x + b$ is divisible by $2x - 1$.
The remainder when $f(x)$ is divided by $x - 2$ is 5 times the remainder when $f(x)$ is divided by $x + 1$.

i Show that $a = 6$ and find the value of b. [4]

ii Using the values from part (i), show that $f(x) = (2x - 1)(cx^2 + dx + e)$, where c, d and e are integers to be found. [2]

iii Hence factorise $f(x)$ completely. [2]

Cambridge IGCSE Additional Mathematics 0606 Paper 12 Q7 Mar 2016

6 The polynomial $p(x)$ is $x^4 - 2x^3 - 3x^2 + 8x - 4$.

i Show that $p(x)$ can be written as $(x - 1)(x^3 - x^2 - 4x + 4)$. [1]

ii Hence write $p(x)$ as a product of its linear factors, showing all your working. [4]

Cambridge IGCSE Additional Mathematics 0606 Paper 22 Q3 Mar 2017

7 The remainder obtained when the polynomial $p(x) = x^3 + ax^2 - 3x + b$ is divided by $x + 3$ is twice the remainder obtained when $p(x)$ is divided by $x - 2$.
Given also that $p(x)$ is divisible by $x + 1$, find the value of a and of b. [5]

Cambridge IGCSE Additional Mathematics 0606 Paper 12 Q1 Mar 2018

8 $p(x) = ax^3 + 3x^2 + bx - 12$ has a factor of $2x + 1$.
When $p(x)$ is divided by $x - 3$ the remainder is 105.

a Find the value of a and of b. [5]

b Using your values of a and b, write $p(x)$ as a product of $2x + 1$ and a quadratic factor. [2]

c Hence solve $p(x) = 0$. [2]

Cambridge IGCSE Additional Mathematics 0606 Paper 12 Q7 Mar 2020

> Chapter 4

Equations, inequalities and graphs

THIS SECTION WILL SHOW YOU HOW TO:

- solve graphically or algebraically equations of the type $|ax + b| = |cx + d|$
- solve graphically or algebraically inequalities of the type $|ax + b| > c$ when $c \geqslant 0$, $|ax + b| \leqslant c$ when $c > 0$, and $|ax + b| \leqslant |cx + d|$
- solve cubic inequalities in the form $k(x - a)(x - b)(x - c) \leqslant d$ graphically
- sketch the graphs of cubic polynomials and their moduli, when given in factorised form
- use substitution to form and solve quadratic equations.

PRE-REQUISITE KNOWLEDGE

Before you start…

Where it comes from	What you should be able to do	Check your skills

Chapter 1	Solve algebraically equations of the type $\lvert ax + b \rvert = c$ and $\lvert ax + b \rvert = cx + d$	1 Solve algebraically:
		   a $\lvert 3x - 1 \rvert = 5$
		   b $\lvert x + 4 \rvert = 2x + 1$

Chapter 1	Solve graphically equations of the type $\lvert ax + b \rvert = c$ and $\lvert ax + b \rvert = cx + d$	2 Solve using a graph:
		   a $\lvert 2x + 1 \rvert = 5$
		   b $\lvert x - 4 \rvert = 2x + 1$

Cambridge IGCSE/O Level Mathematics	Solve quadratic equations.	3 Solve $x^2 + 2x - 35 = 0$.

4.1 Solving equations of the type $\lvert ax + b \rvert = \lvert cx + d \rvert$

CLASS DISCUSSION

Using the fact that $\lvert p \rvert^2 = p^2$ and $\lvert q \rvert^2 = q^2$ you can say that:

$$p^2 - q^2 = \lvert p \rvert^2 - \lvert q \rvert^2$$

Using the difference of two squares then gives:

$$p^2 - q^2 = (\lvert p \rvert - \lvert q \rvert)(\lvert p \rvert + \lvert q \rvert)$$

Using the statement above, explain how these three important results can be obtained:

(The symbol $\Leftrightarrow$ means 'is equivalent to'.)

- $\lvert p \rvert = \lvert q \rvert \Leftrightarrow p^2 = q^2$
- $\lvert p \rvert > \lvert q \rvert \Leftrightarrow p^2 > q^2$
- $\lvert p \rvert < \lvert q \rvert \Leftrightarrow p^2 < q^2$, is $q \neq 0$

The next worked example shows you how to solve equations of the form $\lvert ax + b \rvert = \lvert cx + d \rvert$ using algebra. To solve this type of equation you can use the techniques that you learned in Chapter 1 or you can use the rule:

$$\lvert p \rvert = \lvert q \rvert \Leftrightarrow p^2 = q^2$$

WORKED EXAMPLE 1

Solve the equation $|x - 5| = |x + 1|$ using an algebraic method.

Answers

Method 1

$|x - 5| = |x + 1|$

$x - 5 = x + 1$ or $x - 5 = -(x + 1)$

$\quad 0 = 6 \quad$ or $\quad 2x = 4 \qquad\qquad 0 = 6$ is false

$\qquad\qquad\qquad\quad x = 2$

CHECK: $|2 - 5| = |2 + 1|$ ✓

The solution is $x = 2$

Method 2

$\quad |x - 5| = |x + 1| \qquad$ use $|p| = |q| \Leftrightarrow p^2 = q^2$

$\quad (x - 5)^2 = (x + 1)^2 \qquad$ expand

$x^2 - 10x + 25 = x^2 + 2x + 1 \quad$ simplify

$\qquad\qquad 12x = 24$

$\qquad\qquad\quad x = 2$

The equation $|x - 5| = |x + 1|$ could also have been solved graphically.

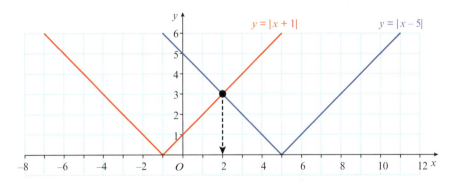

The solution is the x-coordinate where the two graphs intersect.

WORKED EXAMPLE 2

Solve the equation $|2x + 1| = |x - 3|$ using an algebraic method.

Answers

Method 1

$|2x + 1| = |x - 3|$

$2x + 1 = x - 3$ or $2x + 1 = -(x - 3)$

$x = -4$ or $3x = 2$

$$x = \frac{2}{3}$$

CHECK: $|2(-4) + 1| = |-4 - 3|$ ✓, $\left|2 \times \frac{2}{3} + 1\right| = \left|\frac{2}{3} - 3\right|$ ✓

Solution is $x = \frac{2}{3}$ or $x = -4$

Method 2

$|2x + 1| = |x - 3|$ use $|p| = |q| \Leftrightarrow p^2 = q^2$

$(2x + 1)^2 = (x - 3)^2$ expand

$4x^2 + 4x + 1 = x^2 - 6x + 9$ simplify

$3x^2 + 10x - 8 = 0$ factorise

$(3x - 2)(x + 4) = 0$

$x = \frac{2}{3}$ or $x = -4$

The equation $|2x + 1| = |x - 3|$ could also be solved graphically.

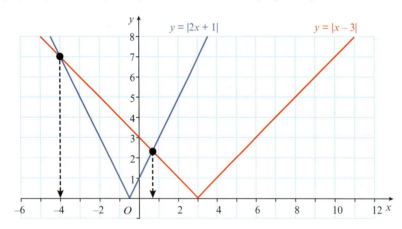

From the graph, one of the answers is clearly $x = -4$. The exact value of the other answer is not so obvious and algebra is needed to find this exact value.

The graph of $y = |x - 3|$ can be written as:

$y = x - 3$ if $x \geqslant 3$

$y = -(x - 3)$ if $x < 3$

The graph of $y = |2x + 1|$ can be written as:

$y = 2x + 1$ if $x \geqslant -\dfrac{1}{2}$

$y = -(2x + 1)$ if $x < -\dfrac{1}{2}$

The second answer is found by finding the x value at the point where $y = -(x - 3)$ and $y = 2x + 1$ intersect.

$2x + 1 = -(x - 3)$

$2x + 1 = -x + 3$

$\quad 3x = 2$

$\quad\; x = \dfrac{2}{3}$

Hence the solution is $x = -4$ or $x = \dfrac{2}{3}$

WORKED EXAMPLE 3

Solve $|x + 4| + |x - 5| = 11$.

Answers

$|x + 4| + |x - 5| = 11$	subtract $|x - 5|$ from both sides
$\qquad |x + 4| = 11 - |x - 5|$	split the equation into two parts
$\qquad\quad x + 4 = 11 - |x - 5| \qquad$ (1)	
$\qquad\quad x + 4 = |x - 5| - 11 \qquad$ (2)	

Using equation (1)

$|x - 5| = 7 - x$	split this equation into two parts
$\quad x - 5 = 7 - x \quad$ or $\quad x - 5 = -(7 - x)$	
$\quad\;\; 2x = 12 \qquad$ or $\qquad 0 = -2$	$0 = -2$ is false
$\quad\;\;\; x = 6$	

Using equation (2)

$|x - 5| = x + 15$	split this equation into two parts
$\quad x - 5 = x + 15 \;$ or $\;\; x - 5 = -(x + 15)$	
$\quad\;\; 0 = 20 \qquad$ or $\qquad 2x = -10$	$0 = 20$ is false
$\qquad\qquad\qquad\qquad\quad x = -5$	

CHECK: $|6 + 4| + |6 - 5| = 11$ ✓, $|-5 + 4| + |-5 - 5| = 11$ ✓

The solution is $x = 6$ or $x = -5$.

Exercise 4.1

1   Solve.

    a $|2x - 1| = |x|$ **b** $|x + 5| = |x - 4|$

    c $|2x - 3| = |4 - x|$ **d** $|5x + 1| = |1 - 3x|$

    e $|1 - 4x| = |2 - x|$ **f** $\left|1 - \dfrac{x}{2}\right| = |3x + 2|$

    g $|3x - 2| = |2x + 5|$ **h** $|2x - 1| = 2|3 - x|$

    i $|2 - x| = 5\left|\dfrac{1}{2}x + 1\right|$

2   Solve the simultaneous equations $y = |x - 5|$ and $y = |8 - x|$.

3   Solve the equation $6|x + 2|^2 + 7|x + 2| - 3 = 0$.

4   **a** Solve the equation $x^2 - 6|x| + 8 = 0$.

    b Use graphing software to draw the graph of $f(x) = x^2 - 6|x| + 8$.

    c Use your graph in **part b** to find the range of the function f.

5   **CHALLENGE QUESTION**

    Solve the equation $|x + 1| + |2x - 3| = 8$.

6   **CHALLENGE QUESTION**

    Solve the simultaneous equations $y = |x - 5|$ and $y = |3 - 2x| + 2$.

7   **CHALLENGE QUESTION**

    Solve the equation $2|3x + 4y - 2| + 3\sqrt{25 - 5x + 2y} = 0$.

4.2 Solving modulus inequalities

Two useful properties that can be used when solving modulus inequalities are:

$$|p| \leqslant q \Leftrightarrow -q \leqslant p \leqslant q \qquad \text{and} \qquad |p| \geqslant q \Leftrightarrow p \leqslant -q \text{ or } p \geqslant q$$

The following examples illustrate the different methods that can be used when solving modulus inequalities.

WORKED EXAMPLE 4

Solve $|2x - 1| < 3$.

Answers

Method 1 (using algebra)

$$|2x - 1| < 3$$

use $|p| < q \Leftrightarrow -q < p < q$

$$-3 < 2x - 1 < 3$$
$$-2 < 2x < 4$$
$$-1 < x < 2$$

Method 2 (using a graph)

The graphs of $y = |2x - 1|$ and $y = 3$ intersect at the points A and B.

$$|2x - 1| = \begin{cases} 2x - 1 & \text{if } x \geqslant \dfrac{1}{2} \\ -(2x - 1) & \text{if } x < \dfrac{1}{2} \end{cases}$$

At A, the line $y = -(2x - 1)$ intersects the line $y = 3$.

$$-(2x - 1) = 3$$
$$-2x + 1 = 3$$
$$2x = -2$$
$$x = -1$$

At B, the line $y = 2x - 1$ intersects the line $y = 3$.

$$2x - 1 = 3$$
$$2x = 4$$
$$x = 2$$

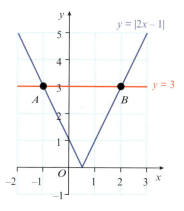

To solve the inequality $|2x - 1| < 3$ you must find where the graph of the function $y = |2x - 1|$ is below the graph of $y = 3$.

Hence, the solution is $-1 < x < 2$.

WORKED EXAMPLE 5

Solve $|2x + 3| > 4$.

Answers

Method 1 (using algebra)

$|2x + 3| > 4$ $\qquad\qquad$ use $|p| > q \Leftrightarrow p < -q$ or $p > q$

$2x + 3 < -4$ or $2x + 3 > 4$

$\quad 2x < -7$ or $\quad\quad 2x > 1$

$\quad\quad x < -\dfrac{7}{2}$ or $\quad\quad x > \dfrac{1}{2}$

Method 2 (using a graph)

The graphs of $y = |2x + 3|$ and $y = 4$ intersect at the points A and B.

$|2x + 3| = \begin{cases} 2x + 3 & \text{if } x \geqslant -1\frac{1}{2} \\ -(2x + 3) & \text{if } x < -1\frac{1}{2} \end{cases}$

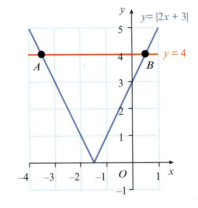

At A, the line $y = -(2x + 3)$ intersects the line $y = 4$.

$-(2x + 3) = 4$

$\quad -2x - 3 = 4$

$\quad\quad 2x = -7$

$\quad\quad\quad x = -\dfrac{7}{2}$

At B, the line $y = 2x + 3$ intersects the line $y = 4$.

$2x + 3 = 4$

$\quad 2x = 1$

$\quad\quad x = \dfrac{1}{2}$

To solve the inequality $|2x + 3| > 4$ you must find where the graph of the function $y = |2x + 3|$ is above the graph of $y = 4$.

Hence, the solution is $x < -\dfrac{7}{2}$ or $x > \dfrac{1}{2}$.

WORKED EXAMPLE 6

Solve the inequality $|2x + 1| \geqslant |3 - x|$.

Answers

Method 1 (using algebra)

$$|2x + 1| \geqslant |3 - x| \qquad \text{use } |p| \geqslant |q| \Leftrightarrow p^2 \geqslant q^2$$

$$(2x + 1)^2 \geqslant (3 - x)^2$$

$$4x^2 + 4x + 1 \geqslant 9 - 6x + x^2$$

$$3x^2 + 10x - 8 \geqslant 0 \qquad \text{factorise}$$

$$(3x - 2)(x + 4) \geqslant 0$$

Critical values are $\dfrac{2}{3}$ and -4.

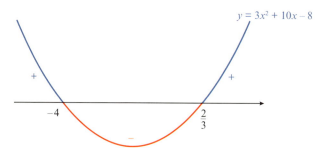

$$y = 3x^2 + 10x - 8$$

Hence, $x \leqslant -4$ or $x \geqslant \dfrac{2}{3}$.

Method 2 (using a graph)

The graphs of $y = |2x + 1|$ and $y = |3 - x|$ intersect at the points A and B.

$$|2x + 1| = \begin{cases} 2x + 1 & \text{if } x \geqslant -\dfrac{1}{2} \\ -(2x + 1) & \text{if } x < -\dfrac{1}{2} \end{cases}$$

$$|3 - x| = |x - 3| = \begin{cases} x - 3 & \text{if } x \geqslant 3 \\ -(x - 3) & \text{if } x < 3 \end{cases}$$

At A, the line $y = -(2x + 1)$ intersects the line $y = -(x - 3)$.

$$2x + 1 = x - 3$$

$$x = -4$$

At B, the line $y = 2x + 1$ intersects the line $y = -(x - 3)$.

$$2x + 1 = -(x - 3)$$

$$3x = 2$$

$$x = \dfrac{2}{3}$$

CONTINUED

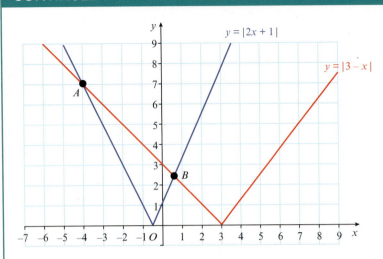

To solve the inequality $|2x + 1| \geqslant |3 - x|$ you must find where the graph of the function $y = |2x + 1|$ is above the graph of $y = |3 - x|$.

Hence, $x \leqslant -4$ or $x \geqslant \dfrac{2}{3}$.

Exercise 4.2

1

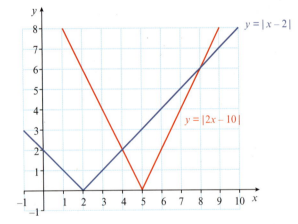

The graphs of $y = |x - 2|$ and $y = |2x - 10|$ are shown on the grid.

Write down the set of values of x that satisfy the inequality $|x - 2| > |2x - 10|$.

2 **a** On the same axes sketch the graphs of $y = |3x - 6|$ and $y = |4 - x|$.

   b Solve the inequality $|3x - 6| \geqslant |4 - x|$.

3 Solve.

    a $|2x - 3| > 5$ b $|4 - 5x| \leqslant 9$ c $|8 - 3x| < 2$

    d $|2x - 7| > 3$ e $|3x + 1| > 8$ f $|5 - 2x| \leqslant 7$

4 Solve.

    a $|2x - 3| \leqslant x - 1$ b $|5 + x| > 7 - 2x$ c $|x - 2| - 3x \leqslant 1$

5 Solve.

    a $|2x - 1| \leqslant |3x|$ b $|x + 1| > |x|$ c $|x| > |3x - 2|$

    d $|4x + 3| > |x|$ e $|x + 3| \geqslant |2x|$ f $|2x| < |x - 3|$

6 Solve.

    a $|x + 1| > |x - 4|$ b $|x - 2| \geqslant |x + 5|$ c $|x + 1| \leqslant |3x + 5|$

    d $|2x + 3| \leqslant |x - 3|$ e $|x + 2| < \left|\frac{1}{2}x - 5\right|$ f $|3x - 2| \geqslant |x + 4|$

7 Solve.

    a $2|x - 3| > |3x + 1|$ b $3|x - 1| < |2x + 1|$ c $|2x - 5| \leqslant 3|2x + 1|$

8 Solve the inequality $|x + 2k| \geqslant |x - 3k|$ where k is a positive constant.

9 Solve the inequality $|x + 3k| < 4|x - k|$ where k is a positive constant.

10 **CHALLENGE QUESTION**

    Solve $|3x + 2| + |3x - 2| \leqslant 8$.

REFLECTION

In this section you have learned how to solve quadratic inequalities by both algebraic and graphical methods.

a   Which method did you prefer?

b   What are the advantages of having a choice of methods available?

4.3 Sketching graphs of cubic polynomials and their moduli

In this section you will learn how to sketch graphs of functions of the form

$y = k(x - a)(x - b)(x - c)$ and their moduli.

When sketching graphs of this form you should show the general shape of the curve and all of the axis intercepts.

To help find the general shape of the curve you need to consider what happens to

- y as x tends to positive infinity (i.e. as $x \to +\infty$)

- y as x tends to negative infinity (i.e. as $x \to -\infty$)

WORKED EXAMPLE 7

a Sketch the graph of the function $y = (2x - 1)(2 - x)(x + 1)$.

b Hence sketch the graph of $y = |(2x - 1)(2 - x)(x + 1)|$

Answers

a When $x = 0$, $y = -1 \times 2 \times 1 = -2$.

∴ The curve intercepts the y-axis at $(0, -2)$.

When $y = 0$, $(2x - 1)(2 - x)(x + 1) = 0$

$2x - 1 = 0 \qquad 2 - x = 0 \qquad x + 1 = 0$

$x = \dfrac{1}{2} \qquad\qquad x = 2 \qquad\qquad x = -1$

∴ The curve intercepts the x-axis at $\left(\dfrac{1}{2}, 0\right)$, $(2, 0)$ and $(-1, 0)$.

As $x \to +\infty$, $y \to -\infty$

As $x \to -\infty$, $y \to +\infty$

The graph of the function $y = (2x - 1)(2 - x)(x + 1)$ is:

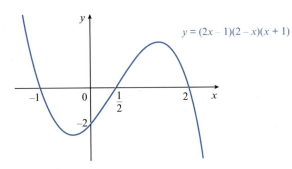

b To sketch the curve $y = |(2x - 1)(2 - x)(x + 1)|$ you reflect in the x-axis the parts of the curve $y = (2x - 1)(2 - x)(x + 1)$ that are below the x-axis.

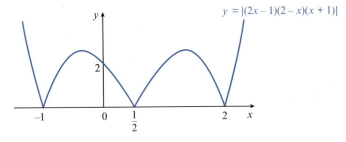

WORKED EXAMPLE 8

a Sketch the graph of the function $y = (x - 1)^2(x + 1)$.

b Hence sketch the graph of $y = |(x - 1)^2(x + 1)|$.

Answers

a When $x = 0$, $y = (-1)^2 \times 1 = 1$.

∴ The curve intercepts the y-axis at $(0, 1)$.

When $y = 0$, $(x - 1)(x - 1)(x + 1) = 0$

$x - 1 = 0$ $x - 1 = 0$ $x + 1 = 0$

$\quad x = 1$ (repeated root) $x = -1$

∴ The curve intercepts the x-axis at $(1, 0)$ and $(-1, 0)$.

As $x \to +\infty$, $y \to +\infty$

As $x \to -\infty$, $y \to -\infty$

The graph of the function $y = (x - 1)^2(x + 1)$ is:

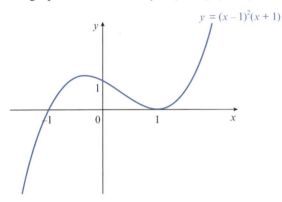

$y = (x - 1)^2(x + 1)$

b To sketch the curve $y = |(x - 1)^2(x + 1)|$ you reflect in the x-axis the part of the curve $y = (x - 1)^2(x + 1)$ that is below the x-axis.

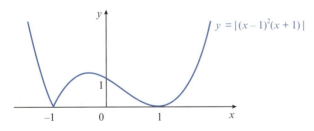

$y = |(x - 1)^2(x + 1)|$

In **Worked examples 7 and 8** you considered the value of y as $x \to \pm\infty$ to determine the shape of the cubic curve. It is often easier to remember that for a curve with equation

$y = k(x - a)(x - b)(x - c)$ the shape of the graph is:

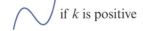 if k is positive   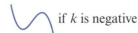 if k is negative

Exercise 4.3

1 Find the coordinates of the points A, B and C where the curve intercepts the x-axis and the point D where the curve intercepts the positive y-axis.

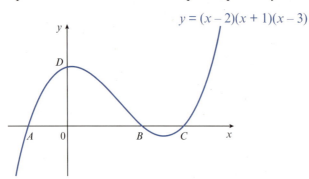

$$y = (x - 2)(x + 1)(x - 3)$$

2 Sketch each of these curves and clearly indicate the axis intercepts.

   a $y = (x - 2)(x - 4)(x + 3)$

   b $y = (x + 2)(x + 1)(3 - x)$

   c $y = (2x + 1)(x + 2)(x - 2)$

   d $y = (3 - 2x)(x - 1)(x + 2)$

3 Find the coordinates of the point A and the point B, where A is the point where the curve intercepts the positive x-axis and B is the point where the curve intercepts the positive y-axis.

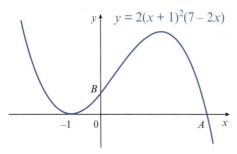

$$y = 2(x + 1)^2(7 - 2x)$$

4 Sketch each of these curves and indicate clearly the axis intercepts.

   a $y = x^2(x + 2)$

   b $y = x^2(5 - 2x)$

   c $y = (x + 1)^2(x - 2)$

   d $y = (x - 2)^2(10 - 3x)$

5 Sketch each of these curves and indicate clearly the axis intercepts.

   a $y = |(x + 1)(x - 2)(x - 3)|$

   b $y = |2(5 - 2x)(x + 1)(x + 2)|$

   c $y = |x(9 - x^2)|$

   d $y = |3(x - 1)^2(x + 1)|$

6 Factorise each of these functions and then sketch the graph of each function indicating clearly the axis intercepts.

   a $y = 9x - x^3$

   b $y = x^3 + 4x^2 + x - 6$

   c $y = 2x^3 + x^2 - 25x + 12$

   d $y = 2x^3 + 3x^2 - 29x - 60$

7   **a**   On the same axes, sketch the graphs of $y = x(x - 5)(x - 7)$ and $y = x(7 - x)$, showing clearly the points at which the curves meet the coordinate axes.

   b   Use algebra to find the coordinates of all the points where the graphs intersect.

8   **a**   On the same axes, sketch the graphs of $y = (2x - 1)(x + 2)(x + 1)$ and $y = (x + 1)(4 - x)$, showing clearly the points at which the curves meet the coordinate axes.

   b   Use algebra to find the coordinates of all the points where the graphs intersect.

9   **CHALLENGE QUESTION**

The diagram shows the graph of $y = k(x - a)^2(x - b)$.

Find the values of a, b and k.

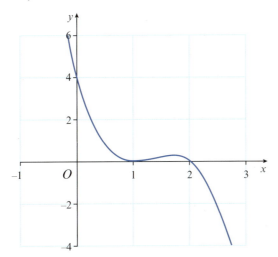

10   **CHALLENGE QUESTION**

The diagram shows the graph of $y = |k(x - a)(x - b)(x - c)|$ where $a < b < c$.

Find the values of a, b, c and k.

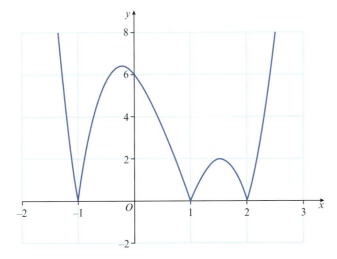

4.4 Solving cubic inequalities graphically

In this section you will learn how to use a graphical method to solve inequalities of the form $k(x - a)(x - b)(x - c) \leqslant d$.

WORKED EXAMPLE 9

The diagram shows part of the graph of $y = \dfrac{1}{6}(x - 3)(x - 2)(x + 2)$.

Use the graph to solve the inequality $(x - 3)(x - 2)(x + 2) \leqslant 6$.

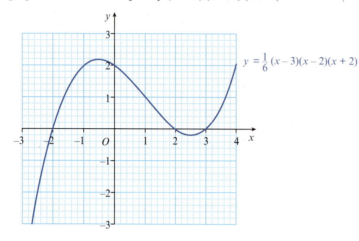

Answers

$(x - 3)(x - 2)(x + 2) \leqslant 6$ divide both sides by 6

$\dfrac{1}{6}(x - 3)(x - 2)(x + 2) \leqslant 1$

You need to find where the curve $y = \dfrac{1}{6}(x - 3)(x - 2)(x + 2)$ is below the line $y = 1$.

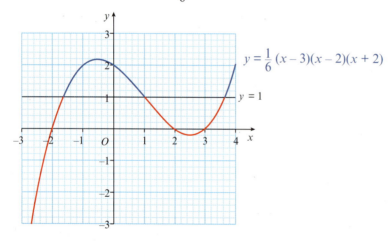

The red sections of the graph represent where the curve

$y = \dfrac{1}{6}(x - 3)(x - 2)(x + 2)$ is below the line $y = 1$.

The solution is $x \leqslant -1.6$ or $1 \leqslant x \leqslant 3.6$

Exercise 4.4

1   The diagram shows part of the graph
of $y = x(x - 2)(x + 1)$.

Use the graph to solve each of the
following inequalities.

a $x(x - 2)(x + 1) \leqslant 0$

b $x(x - 2)(x + 1) \geqslant 1$

c $x(x - 2)(x + 1) \leqslant -2$

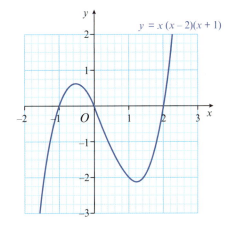

2   The diagram shows part of the graph
of $y = (x + 1)^2(2 - x)$.

Use the graph to solve each of
the following inequalities.

a $(x + 1)^2(2 - x) \geqslant 0$

b $(x + 1)^2(2 - x) \leqslant 4$

c $(x + 1)^2(2 - x) \leqslant 3$

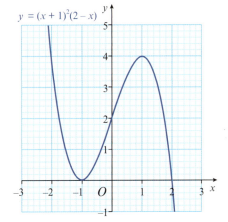

3   The diagram shows part of the graph of
$y = (1 - x)(x - 2)(x + 1)$.

Use the graph to solve each of the
following inequalities.

a $(1 - x)(x - 2)(x + 1) \leqslant -3$

b $(1 - x)(x - 2)(x + 1) \leqslant 0$

c $(1 - x)(x - 2)(x + 1) \geqslant -1$

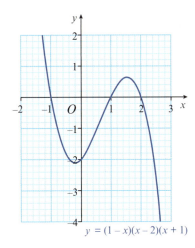

4.5 Solving more complex quadratic equations

You may be asked to solve an equation that is quadratic in some function of x.

WORKED EXAMPLE 10

Solve the equation $4x^4 - 17x^2 + 4 = 0$.

Answers

Method 1 (substitution method)

$4x^4 - 17x^2 + 4 = 0$ use the substitution $y = x^2$

$4y^2 - 17y + 4 = 0$ factorise

$(4y - 1)(y - 4) = 0$ solve

$4y - 1 = 0$ or $y - 4 = 0$

      $y = \dfrac{1}{4}$ or $y = 4$ substitute x^2 for y

      $x^2 = \dfrac{1}{4}$ or $x^2 = 4$

      $x = \pm\dfrac{1}{2}$ or $x = \pm 2$

Method 2 (factorise directly)

   $4x^4 - 17x^2 + 4 = 0$

$(4x^2 - 1)(x^2 - 4) = 0$

$4x^2 - 1 = 0$ or $x^2 - 4 = 0$

      $x^2 = \dfrac{1}{4}$ or $x^2 = 4$

      $x = \pm\dfrac{1}{2}$ or $x = \pm 2$

WORKED EXAMPLE 11

Use the substitution $y = x^{\frac{1}{3}}$ to solve the equation $3x^{\frac{2}{3}} - 5x^{\frac{1}{3}} + 2 = 0$.

Answers

   $3x^{\frac{2}{3}} - 5x^{\frac{1}{3}} + 2 = 0$ let $y = x^{\frac{1}{3}}$

   $3y^2 - 5y + 2 = 0$ factorise

$(3y - 2)(y - 1) = 0$ solve

$3y - 2 = 0$ or $y - 1 = 0$

      $y = \dfrac{2}{3}$ or $y = 1$ substitute $x^{\frac{1}{3}}$ for y

      $x^{\frac{1}{3}} = \dfrac{2}{3}$ or $x^{\frac{1}{3}} = 1$ cube both sides

      $x = \left(\dfrac{2}{3}\right)^3$ or $x = (1)^3$

      $x = \dfrac{8}{27}$ or $x = 1$

WORKED EXAMPLE 12

Solve the equation $x - 4\sqrt{x} - 21 = 0$.

Answers

$x - 4\sqrt{x} - 21 = 0$	use the substitution $y = \sqrt{x}$
$y^2 - 4y - 21 = 0$	
$(y - 7)(y + 3) = 0$	
$y - 7 = 0$ or $y + 3 = 0$	
$y = 7$ or $y = -3$	substitute $\sqrt{x}$ for y
$\sqrt{x} = 7$ or $\sqrt{x} = -3$	$\sqrt{x} = -3$ has no real solutions
$\therefore x = 49$	

WORKED EXAMPLE 13

Solve the equation $8(4^x) - 33(2^x) + 4 = 0$.

Answers

$8(4^x) - 33(2^x) + 4 = 0$	4^x can be written as $(2^2)^x = (2^x)^2$
$8(2^x)^2 - 33(2^x) + 4 = 0$	let $y = 2^x$
$8y^2 - 33y + 4 = 0$	
$(8y - 1)(y - 4) = 0$	
$y = \dfrac{1}{8}$ or $y = 4$	substitute 2^x for y
$2^x = \dfrac{1}{8}$ or $2^x = 4$	$\dfrac{1}{8} = 2^{-3}$ and $4 = 2^2$
$x = -3$ or $x = 2$	

Exercise 4.5

1 Find the real values of x satisfying the following equations.

a $x^4 - 5x^2 + 4 = 0$ b $x^4 + x^2 - 6 = 0$ c $x^4 - 20x^2 + 64 = 0$

d $x^4 + 2x^2 - 8 = 0$ e $x^4 - 4x^2 - 21 = 0$ f $2x^4 - 17x^2 - 9 = 0$

g $4x^4 + 6 = 11x^2$ h $\dfrac{2}{x^4} + \dfrac{1}{x^2} = 6$ i $\dfrac{8}{x^6} + \dfrac{7}{x^3} = 1$

2 Use the quadratic formula to solve these equations.
Write your answers correct to 3 significant figures.

a $x^4 - 8x^2 + 1 = 0$ b $x^4 - 5x^2 - 2 = 0$ c $2x^4 + x^2 - 5 = 0$

d $2x^6 - 3x^3 - 8 = 0$ e $3x^6 - 5x^3 - 2 = 0$ f $2x^8 - 7x^4 - 3 = 0$

3 Solve.

a $x - 7\sqrt{x} + 10 = 0$ b $x - \sqrt{x} - 12 = 0$ c $x + 5\sqrt{x} - 24 = 0$

d $\sqrt{x}(2 + \sqrt{x}) = 35$ e $8x - 18\sqrt{x} + 9 = 0$ f $6x + 11\sqrt{x} - 35 = 0$

g $2x + 4 = 9\sqrt{x}$ h $3\sqrt{x} + \dfrac{5}{\sqrt{x}} = 16$ i $2\sqrt{x} + \dfrac{4}{\sqrt{x}} = 9$

4 Solve the equation $2x^{\frac{2}{3}} - 7x^{\frac{1}{3}} + 6 = 0$.

5 The curve $y = \sqrt{x}$ and the line $5y = x + 4$ intersect at the points P and Q.

a Write down an equation satisfied by the x-coordinates of P and Q.

b Solve your equation in **part a** and hence find the coordinates of P and Q.

6 Solve.

a $2^{2x} - 6(2^x) + 8 = 0$ b $3^{2x} - 10(3^x) + 9 = 0$

c $2(2^{2x}) - 9(2^x) + 4 = 0$ c $3^{2x+1} - 28(3^x) + 9 = 0$

d $2^{2x+2} - 33(2^x) + 8 = 0$ e $3^{2x+2} + 3(3^x) - 2 = 0$

7 $f(x) = 2^x$ and $g(x) = 4x^2 + 7x$. Solve $gf(x) = 2$.

8 $f(x) = x^3 - 2$ and $g(x) = x^2 - 5x$. Solve $gf(x) = 6$.

9 $f(x) = x^2 + 3x$ and $g(x) = x^2 - 4x$. Solve $gf(x) = 0$.

SUMMARY

Solving modulus equations

To solve modulus equations, you can use the property:

$|a| = |b| \Leftrightarrow a^2 = b^2$

Solving modulus inequalities

To solve modulus inequalities, you can use the properties:

$|a| \leqslant b \ \Leftrightarrow -b \leqslant a \leqslant b$

$|a| \geqslant b \ \Leftrightarrow a \leqslant -b$ or $a \geqslant b$

$|a| > |b| \Leftrightarrow a^2 > b^2$

$|a| < |b| \Leftrightarrow a^2 < b^2, \ b \neq 0$

The graph of $y = k(x - a)(x - b)(x - c)$

The x-axis intercepts are $(a, 0)$, $(b, 0)$ and $(c, 0)$.

The shape of the graph is ⌣⌢ if k is positive ⌢⌣ if k is negative

The graph of $y = |k(x - a)(x - b)(x - c)|$

To sketch the curve $y = |k (x - a)(x - b)(x - c)|$ you reflect in the x-axis the parts of the curve $y = k(x - a)(x - b)(x - c)$ that are below the x-axis.

Past paper questions and practice questions

Worked example

a On the same axes, sketch the graphs of $y = x^2(x - 2)$ and $y = x(6 - x)$, showing clearly the points at which the curves meet the coordinate axes. [5]

b Use algebra to find the coordinates of the points where the graphs intersect. [6]

Practice question

Answers

a The graph of $y = x^2(x - 2)$:

When $x = 0$, $y = 0^2 \times (-2) = 0$.

∴ The curve intercepts the y-axis at $(0, 0)$.

When $y = 0$, $x^2(x - 2) = 0$

$x = 0$ $x = 0$ $x - 2 = 0$

$x = 0$ (repeated root) $x = 2$

∴ The curve intercepts the x-axis at $(0, 0)$ and $(2, 0)$.

As $x \rightarrow +\infty$, $y \rightarrow +\infty$

As $x \rightarrow -\infty$, $y \rightarrow -\infty$

The graph of $y = x(6 - x)$:

When $x = 0$, $y = 0 \times 6 = 0$.

∴ The curve intercepts the y-axis at $(0, 0)$.

When $y = 0$, $x(6 - x) = 0$

$x = 0$ $6 - x = 0$

$x = 0$ $x = 6$

∴ The curve intercepts the x-axis at $(0, 0)$ and $(6, 0)$.

As $x \rightarrow +\infty$, $y \rightarrow -\infty$

As $x \rightarrow -\infty$, $y \rightarrow -\infty$

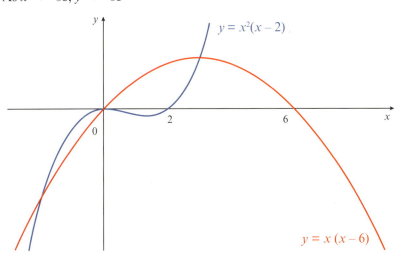

b At the points of intersection:
$$x^2(x - 2) = x(6 - x)$$
$$x^2(x - 2) - x(6 - x) = 0$$
$$x[x(x - 2) - (6 - x)] = 0$$
$$x(x^2 - x - 6) = 0$$
$$x(x - 3)(x + 2) = 0$$
$$x = 0 \quad \text{or} \quad x = 3 \quad \text{or} \quad x = -2$$
$$x = 0 \quad \Rightarrow \quad y = 0$$
$$x = 3 \quad \Rightarrow \quad y = 9$$
$$x = -2 \quad \Rightarrow \quad y = -16$$

The points of intersection are $(-2, -16)$, $(0, 0)$ and $(3, 9)$.

1 Solve the equation $|2x - 3| = |3x - 5|$. [3]

Practice question

2 Solve the inequality $|2x - 1| > 7$. [3]

Practice question

3 Solve the inequality $|7 - 5x| < 3$. [3]

Practice question

4 Solve the inequality $|x| > |3x - 2|$. [4]

Practice question

5 Solve the inequality $|x - 1| \leqslant |x + 2|$. [4]

Practice question

6 Solve the inequality $|x + 2| < \left|\frac{1}{2}x - 1\right|$. [4]

Practice question

7 Solve the inequality $|x + 2k| > |x - k|$ where k is a positive constant. [4]

Practice question

8 **a** Solve the equation $|x - 13| = 14$. [3]
   b Hence solve the equation $|y^3 - 13| = 14$. [1]

Practice question

9 Sketch the graph of $y = x(3 - 2x)(x - 4)$, showing clearly the points at which the curve meets the coordinate axes. [3]

Practice question

10 Sketch the graph of $y = 2(2x - 1)(x - 3)(x + 1)$, showing clearly the points at which the curve meets the coordinate axes. [4]

Practice question

11

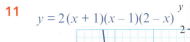

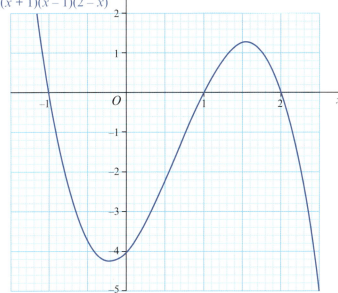

The diagram shows part of the graph of $y = 2(x + 1)(x - 1)(2 - x)$.

Use the graph to solve the inequality $(x + 1)(x - 1)(2 - x) > -1$. [3]

Practice question

12 a Sketch the graph of $y = (x - 4)(x - 1)(x + 2)$, showing clearly the points at which the curve meets the coordinate axes. [4]

b Hence sketch the curve $y = |(x - 4)(x - 1)(x + 2)|$. [1]

Practice question

13 a Factorise completely $x^3 + x^2 - 6x$. [3]

b Hence sketch the curve with equation $y = x^3 + x^2 - 6x$, showing clearly the points at which the curve meets the coordinate axes. [3]

Practice question

14 a On the same axes, sketch the graphs of $y = (x - 3)(x + 1)^2$ and $y = \dfrac{6}{x}$, showing clearly the points at which the curves meet the coordinate axes. [5]

b Hence state the number of real roots of the equation $(x - 3)(x + 1)^2 = \dfrac{6}{x}$. [1]

Practice question

15 a Factorise completely $2x^3 + x^2 - 25x + 12$. [5]

b Hence sketch the curve with equation $y = 2x^3 + x^2 - 25x + 12$, showing clearly the points at which the curve meets the coordinate axes. [4]

Practice question

16 a On the same axes, sketch the graphs of $y = x^2(6 - x)$ and $y = 4x(4 - x)$, showing the points at which the curves meet the coordinate axes. [5]

b Use algebra to find the coordinates of the points where the graphs intersect. [6]

Practice question

17 Solve the inequality $|3x + 2| > 8 + x$. [3]

Cambridge IGCSE Additional Mathematics 0606 Paper 21 Q1 Nov 2020

18 a On the axes below, sketch the graph of $y = -3(x - 2)(x - 4)(x + 1)$, showing the coordinates of the points where the curve intersects the coordinate axes. [3]

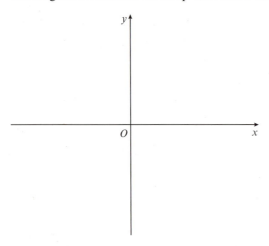

b Hence find the values of x for which $-3(x - 2)(x - 4)(x + 1) > 0$. [2]

Cambridge IGCSE Additional Mathematics 0606 Paper 12 Q1 Mar 2020

19 The diagram shows the graph of a cubic curve $y = f(x)$.

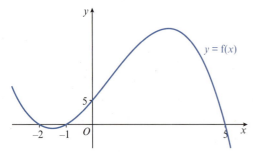

a Find an expression for $f(x)$. [2]

b Solve $f(x) \leq 0$. [2]

Cambridge IGCSE Additional Mathematics 0606 Paper 11 Q1 Jun 2020

20

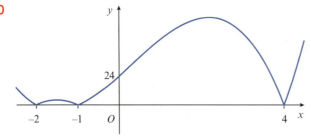

The diagram shows the graph of $y = |p(x)|$, where $p(x)$ is a cubic function. Find the two possible expressions for $p(x)$. [3]

Cambridge IGCSE Additional Mathematics 0606 Paper 11 Q1 Nov 2020

21

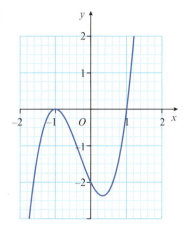

The diagram shows the graph of $y = f(x)$, where $f(x) = a(x + b)^2(x + c)$ and a, b and c are integers.

a   Find the value of each of a, b and c.   [2]

b   Hence solve the inequality $f(x) \leqslant -1$.   [3]

Cambridge IGCSE Additional Mathematics 0606 Paper 22 Q3 Mar 2021

❯ Chapter 4

Curve Fitter

This problem challenges you to find cubic equations which satisfy different conditions. *You may like to use Desmos to help you investigate possible cubics.*

Part 1

Can you find a cubic which passes through (0,0) and the points (1,2) and (2,1)?

Can you find more than one possible cubic?

Part 2 (a)

Can you find a cubic which passes through (0,0) and the points (1,2) and (2,1), and where the point (1,2) is a turning point of the cubic?

Can you find more than one cubic satisfying all the conditions?

Part 2 (b)

Can you find a cubic which passes through (0,0) and the points (1,2) and (2,1), and where the point (2,1) is a turning point of the cubic?

Can you find more than one cubic satisfying all the conditions?

Part 3

Can you find a cubic which passes through (0,0) and where the points (1,2) and (2,1) are both turning points?

> # Chapter 5
Logarithmic and exponential functions

THIS SECTION WILL SHOW YOU HOW TO:

- use simple properties of the logarithmic and exponential functions including $\ln x$ and e^x
- use graphs of the logarithmic and exponential functions including $\ln x$ and e^x and graphs of $ke^{nx} + a$ and $k\ln(ax + b)$ where n, k, a and b are integers
- use the laws of logarithms, including change of base of logarithms
- solve equations of the form $a^x = b$.

PRE-REQUISITE KNOWLEDGE

Before you start…

Where it comes from	What you should be able to do	Check your skills
Cambridge IGCSE/O Level Mathematics	Sketch graphs of exponential functions.	1 Sketch the graph of $y = 2^x$.
Chapter 1	Sketch the graph of the inverse of an exponential function.	2 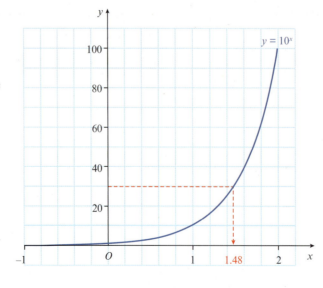 The diagram shows the graph of the function $f(x) = 3^x$. On the same axes, sketch the graph of $y = f^{-1}(x)$.
Cambridge IGCSE/O Level Mathematics	Solve simple exponential equations.	3 Solve: a $2^x = 64$ b $5^{x+1} = 125$ c $8^{2x-3} = 2^{x+6}$

5.1 Logarithms to base 10

Consider the exponential function $f(x) = 10^x$.

To solve $10^x = 30$

You can say $10^1 = 10$ and
$10^2 = 100$

So $1 < x < 2$

The graph of $y = 10^x$ could be used to give a more accurate value for x when $10^x = 30$.

From the graph, $x \approx 1.48$.

There is a function that gives the value of x directly.

If $10^x = 30$ then $x = \log_{10} 30$.

$\log_{10} 30$ is read as 'log 30 to base 10'.

log is short for **logarithm**.

KEY WORDS

logarithm

laws of logarithms

change of base rule

natural logarithms

TIP

$\log_{10} 30$ can also be written as lg 30 or log 30.

On your calculator, for logs to the base 10, you use the $\boxed{\textbf{log}}$ or $\boxed{\textbf{lg}}$ key.

So if $10^x = 30$

then $x = \log_{10} 30$

$x = 1.477$ to 4 sf.

Hence the rule for base 10 is:

> If $y = 10^x$ then $x = \log_{10} y$.

This rule can be described in words as:

> $\log_{10} y$ is the power that 10 must be raised to in order to obtain y.

For example, $\log_{10} 100 = 2$ since $100 = 10^2$.

$y = 10^x$ and $y = \log_{10} x$ are inverse functions.

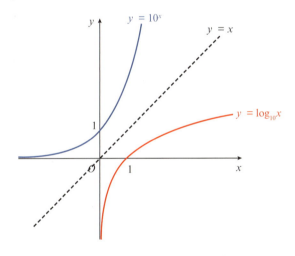

CLASS DISCUSSION

Discuss with your classmates why each of these four statements is true.

$\boxed{\log_{10} 10 = 1}$ $\boxed{\log_{10} 1 = 0}$ $\boxed{\log_{10} 10^x = x \text{ for } x \in \mathbb{R}}$ $\boxed{x = 10^{\log_{10} x} \text{ for } x > 0}$

WORKED EXAMPLE 1

a   Convert $10^x = 45$ to logarithmic form.

b   Solve $10^x = 45$, giving your answer correct to 3 sf.

CONTINUED

Answers

a **Method 1**

$10^x = 45$

Step 1:	Identify the base and index:	The base is 10. The index is x.
Step 2:	Start to write in log form:	In log form the index always goes on its own and the base goes at the base of the logarithm. So $x = \log_{10}?$
Step 3:	Complete the log form:	Fill in the last number. $x = \log_{10} 45$

So $x = \log_{10} 45$

Method 2

$10^x = 45$

$\log_{10} 10^x = \log_{10} 45$ Take logs to base 10 of both sides. $\log_{10} 10^x = x$

$x = \log_{10} 45$

b $10^x = 45$

$x = \log_{10} 45$

$x \approx 1.65$

WORKED EXAMPLE 2

a   Convert $\log_{10} x = 2.9$ to exponential form.

b   Solve $\log_{10} x = 2.9$, giving your answer correct to 3 sf.

Answers

a **Method 1**

$\log_{10} x = 2.9$

Step 1:	Identify the base and index:	The base is 10. The index is 2.9. (In log form the index is always on its own.)
Step 2:	Start to write in exponential form:	Write the base and the index first. So $10^{2.9} = ?$
Step 3:	Complete the exponential form:	$x = 10^{2.9}$

So $x = 10^{2.9}$

Method 2

$\log_{10} x = 2.9$

$10^{\log_{10} x} = 10^{2.9}$ $10^{\log_{10} x} = x$

$x = 10^{2.9}$

b $\log_{10} x = 2.9$

$x = 10^{2.9}$

$x \approx 794$

WORKED EXAMPLE 3

Find the value of

a $\log_{10} 100\,000$ b $\log_{10} 0.001$ c $\log_{10} 100\sqrt{10}$

Answers

a $\log_{10} 100\,000 = \log_{10} 10^5$ write $100\,000$ as a power of 10, $100\,000 = 10^5$
$\phantom{\log_{10} 100\,000} = 5$

b $\log_{10} 0.001 = \log_{10} 10^{-3}$ write 0.001 as a power of 10, $0.001 = 10^{-3}$
$\phantom{\log_{10} 0.001} = -3$

c $\log_{10} 100\sqrt{10} = \log_{10} 10^{2.5}$ write $100\sqrt{10}$ as a power of 10
$\phantom{\log_{10} 100\sqrt{10}} = 2.5$ $100\sqrt{10} = 10^2 \times 10^{0.5} = 10^{2.5}$

Exercise 5.1

1   Convert from exponential form to logarithmic form.

   a $10^3 = 1000$ b $10^2 = 100$

   c $10^6 = 1\,000\,000$ d $10^x = 2$

   e $10^x = 15$ f $10^x = 0.06$

2   Solve each of these equations, giving your answers correct to 3 sf.

   a $10^x = 75$ b $10^x = 300$

   c $10^x = 720$ d $10^x = 15.6$

   e $10^x = 0.02$ f $10^x = 0.005$

3   Convert from logarithmic form to exponential form.

   a $\lg 100\,000 = 5$ b $\lg 10 = 1$

   c $\lg \dfrac{1}{1000} = -3$ d $\lg x = 7.5$

   e $\lg x = 1.7$ f $\lg x = -0.8$

4   Solve each of these equations, giving your answers correct to 3 sf.

   a $\lg x = 5.1$ b $\lg x = 3.16$

   c $\lg x = 2.16$ d $\lg x = -0.3$

   e $\lg x = -1.5$ f $\lg x = -2.84$

5   Without using a calculator, find the value of

   a $\lg 10\,000$ b $\lg 0.01$

   c $\lg \sqrt{10}$ d $\lg \left(\sqrt[3]{10}\right)$

   e $\lg \left(10\sqrt{10}\right)$ f $\lg \left(\dfrac{1000}{\sqrt{10}}\right)$

5.2 Logarithms to base a

In the last section you learned about logarithms to the base of 10.

The same principles can be applied to define logarithms in other bases.

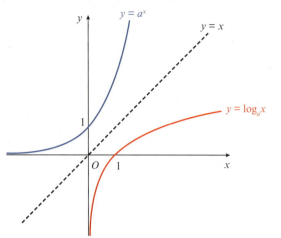

> If $y = a^x$ then $x = \log_a y$
>
> $\log_a a = 1$ $\log_a 1 = 0$
>
> $\log_a a^x = x$ $x = a^{\log_a x}$

The conditions for $\log_a x$ to be defined are:

- $a > 0$ and $a \neq 1$

- $x > 0$

WORKED EXAMPLE 4

Convert $2^4 = 16$ to logarithmic form.

Answers

Method 1

$2^4 = 16$

Step 1: Identify the base and index: The base is 2. The index is 4.

Step 2: Start to write in log form: In log form, the index always goes on its own and the base goes at the base of the logarithm. So $4 = \log_2 ?$

Step 3: Complete the log form: Fill in the last number. $4 = \log_2 16$

So $4 = \log_2 16$.

Method 2

$2^4 = 16$

$\log_2 2^4 = \log_2 16$ Take logs to base 2 of both sides, $\log_2 2^4 = 4$.

$4 = \log_2 16$

WORKED EXAMPLE 5

Convert $\log_7 49 = 2$ to exponential form.

Answers

Method 1

$\log_7 49 = 2$

Step 1: Identify the base and index: The base is 7. The index is 2.

(In log form the index is always on its own.)

Step 2: Start to write in exponential form: Write the base and the index first.
So $7^2 = ?$

Step 3: Complete the exponential form: $7^2 = 49$

So $7^2 = 49$

Method 2

$\log_7 49 = 2$

$\log_7 49 = 7^2$ $7^{\log_7 49} = 49$

$\quad 49 = 7^2$

WORKED EXAMPLE 6

Find the value of

a $\log_3 81$ **b** $\log_2 128$ **c** $\log_4 \dfrac{1}{16}$

Answers

a $\log_3 81 = \log_3 3^4$ write 81 as a power of 3, $81 = 3^4$
$\quad\quad\quad = 4$

b $\log_2 128 = \log_2 2^7$ write 128 as a power of 2, $128 = 2^7$
$\quad\quad\quad = 7$

c $\log_4 \dfrac{1}{16} = \log_4 4^{-2}$ write $\dfrac{1}{16}$ as a power of 4, $\dfrac{1}{16} = 4^{-2}$
$\quad\quad\quad = -2$

Exercise 5.2

1 Convert from exponential form to logarithmic form.

a $4^3 = 64$ **b** $2^5 = 32$ **c** $5^3 = 125$

d $6^2 = 36$ **e** $2^{-5} = \dfrac{1}{32}$ **f** $3^{-4} = \dfrac{1}{81}$

g $a^2 = b$ **h** $x^y = 4$ **i** $a^b = c$

2   Convert from logarithmic form to exponential form.

   a $\log_2 4 = 2$ b $\log_2 64 = 6$ c $\log_5 1 = 0$

   d $\log_3 9 = 2$ e $\log_{36} 6 = \dfrac{1}{2}$ f $\log_8 2 = \dfrac{1}{3}$

   g $\log_x 1 = 0$ h $\log_x 8 = y$ i $\log_a b = c$

3   Solve.

   a $\log_2 x = 4$ b $\log_3 x = 2$ c $\log_5 x = 4$

   d $\log_3 x = \dfrac{1}{2}$ e $\log_x 144 = 2$ f $\log_x 27 = 3$

   g $\log_2 (x - 1) = 4$ h $\log_3 (2x + 1) = 2$ i $\log_5 (2 - 3x) = 3$

4   Find the value of

   a $\log_4 16$ b $\log_3 81$ c $\log_4 64$

   d $\log_2 0.25$ e $\log_3 243$ f $\log_2 (8\sqrt{2})$

   g $\log_5 (25\sqrt{5})$ h $\log_2 \left(\dfrac{1}{\sqrt{8}}\right)$ i $\log_{64} 8$

   j $\log_7 \left(\dfrac{\sqrt{7}}{7}\right)$ k $\log_5 \sqrt[3]{5}$ l $\log_3 \dfrac{1}{\sqrt{3}}$

5   Simplify.

   a $\log_x x^2$ b $\log_x \sqrt[3]{x}$ c $\log_x (x\sqrt{x})$

   d $\log_x \dfrac{1}{x^2}$ e $\log_x \left(\dfrac{1}{x^2}\right)^3$ f $\log_x \left(\sqrt{x^7}\right)$

   g $\log_x \left(\dfrac{x}{\sqrt[3]{x}}\right)$ h $\log_x \left(\dfrac{x\sqrt{x}}{\sqrt[3]{x}}\right)$

6   Solve.

   a $\log_3 (\log_2 x) = 1$ b $\log_2 (\log_5 x) = 2$

5.3 The laws of logarithms

If x and y are both positive and $a > 0$ and $a \neq 1$, then the following **laws of logarithms** can be used:

Multiplication law	**Division law**	**Power law**
$\log_a(xy) = \log_a x + \log_a y$	$\log_a\left(\dfrac{x}{y}\right) = \log_a x - \log_a y$	$\log_a(x)^m = m\log_a x$

Proof:

$\log_a(xy)$

$= \log_a(a^{\log_a x} \times a^{\log_a y})$

$= \log_a(a^{\log_a x + \log_a y})$

$= \log_a x + \log_a y$

$\log_a\left(\dfrac{x}{y}\right)$

$= \log_a\left(\dfrac{a^{\log_a x}}{a^{\log_a y}}\right)$

$= \log_a(a^{\log_a x - \log_a y})$

$= \log_a x - \log_a y$

$\log_a(x)^m$

$= \log_a\left((a^{\log_a x})^m\right)$

$= \log_a(a^{m\log_a x})$

$= m\log_a x$

Using the power law, $\log_a\left(\dfrac{1}{x}\right) = \log_a x^{-1} = -\log_a x$

This gives another useful rule to remember:

$$\log_a\left(\frac{1}{x}\right) = -\log_a x$$

WORKED EXAMPLE 7

Use the laws of logarithms to simplify these expressions.

a $\lg 8 + \lg 2$ **b** $\log_4 15 \div \log_4 5$ **c** $2\log_3 4 + 5\log_3 2$

Answers

a $\lg 8 + \lg 2$

$= \lg(8 \times 2)$

$= \lg 16$

b $\log_4 15 \div \log_4 5$

$= \log_4\left(\dfrac{15}{5}\right)$

$= \log_4 3$

c $2\log_3 4 + 5\log_3 2$

$= \log_3 4^2 + \log_3 2^5$

$= \log_3(16 \times 32)$

$= \log_3 512$

WORKED EXAMPLE 8

Given that $\log_5 p = x$ and $\log_5 q = y$, express in terms of x and/or y

a $\log_5 p + \log_5 q^3$ **b** $\log_5 p^2 - \log_5 \sqrt{q}$ **c** $\log_5\left(\dfrac{q}{5}\right)$

Answers

a $\log_5 p + \log_5 q^3$

$= \log_5 p + 3\log_5 q$

$= x + 3y$

b $\log_5 p^2 - \log_5 \sqrt{q}$

$= 2\log_5 p - \dfrac{1}{2}\log_5 q$

$= 2x - \dfrac{1}{2}y$

c $\log_5\left(\dfrac{q}{5}\right)$

$= \log_5 q - \log_5 5$

$= y - 1$

Exercise 5.3

1 Write as a single logarithm.

a $\log_2 5 + \log_2 3$ **b** $\log_3 12 - \log_3 2$ **c** $3\log_5 2 + \log_5 8$

d $2\log_7 4 - 3\log_7 2$ **e** $\dfrac{1}{2}\log_3 25 + \log_3 4$ **f** $2\log_7\left(\dfrac{1}{4}\right) + \log_7 8$

g $1 + \log_4 3$ **h** $\lg 5 - 2$ **i** $3 - \log_4 10$

2 Write as a single logarithm, then simplify your answer.

a $\log_2 56 - \log_2 7$ **b** $\log_6 12 + \log_6 3$ **c** $\dfrac{1}{2}\log_2 36 - \log_2 3$

d $\log_3 15 - \dfrac{1}{2}\log_3 25$ **e** $\log_4 40 - \dfrac{1}{3}\log_4 125$ **f** $\dfrac{1}{2}\log_3 16 - 2\log_3 6$

3 Simplify.

a $2\log_5 3 - \dfrac{1}{2}\log_5 4 + \log_5 8$ **b** $2 + \dfrac{1}{2}\log_2 49 - \log_2 21$

4   **a**   Express 16 and 0.25 as powers of 2.   **b**   Hence, simplify $\dfrac{\log_3 16}{\log_3 0.25}$

5   Simplify.

a $\dfrac{\log_7 4}{\log_7 2}$ **b** $\dfrac{\log_7 27}{\log_7 3}$ **c** $\dfrac{\log_3 64}{\log_3 0.25}$ **d** $\dfrac{\log_5 100}{\log_5 0.01}$

6   Given that $u = \log_5 x$, find, in simplest form in terms of u

a x **b** $\log_5\left(\dfrac{x}{25}\right)$ **c** $\log_5(5\sqrt{x})$ **d** $\log_5\left(\dfrac{x\sqrt{x}}{125}\right)$

7   Given that $\log_4 p = x$ and $\log_4 q = y$, express in terms of x and/or y

a $\log_4(4p)$ **b** $\log_4\left(\dfrac{16}{p}\right)$ **c** $\log_4 p + \log_4 q^2$ **d** pq

8   Given that $\log_a x = 5$ and $\log_a y = 8$, find

a $\log_a\left(\dfrac{1}{y}\right)$ **b** $\log_a\left(\dfrac{\sqrt{x}}{y}\right)$ **c** $\log_a(xy)$ **d** $\log_a(x^2 y^3)$

9   Given that $\log_a x = 12$ and $\log_a y = 4$, find the value of

a $\log_a\left(\dfrac{x}{y}\right)$ **b** $\log_a\left(\dfrac{x^2}{y}\right)$ **c** $\log_a(x\sqrt{y})$ **d** $\log_a\left(\dfrac{y}{\sqrt[3]{x}}\right)$

5.4 Solving logarithmic equations

You have already learned how to solve simple logarithmic equations.

In this section you will learn how to solve more complicated equations.

It is essential, when solving equations involving logs, that all roots are checked in the original equation.

WORKED EXAMPLE 9

Solve.

a $2\log_8(x + 2) = \log_8(2x + 19)$ **b** $4\log_x 2 - \log_x 4 = 8$

Answers

a $2\log_8(x + 2) = \log_8(2x + 19)$ use the power law
$\log_8(x + 2)^2 = \log_8(2x + 19)$ use equality of logarithms
$(x + 2)^2 = 2x + 19$ expand brackets
$x^2 + 4x + 4 = 2x + 19$
$x^2 + 2x - 15 = 0$
$(x - 3)(x + 5) = 0$
$x = 3$ or $x = -5$

Check when $x = 3$: $2\log_8(x + 2) = 2\log_8 5 = \log_8 25$ is defined
$\log_8(2x + 19) = \log_8 25$ is defined

So $x = 3$ is a solution, since both sides of the equation are defined and equivalent in value.

Check when $x = -5$: $2\log_8(x + 2) = 2\log_8(-3)$ is not defined

So $x = -5$ is not a solution of the original equation.

Hence, the solution is $x = 3$.

CONTINUED

b $4\log_x 2 - \log_x 4 = 2$ use the power law

$\log_x 2^4 - \log_x 2^2 = 2$ use the division law

$\log_x 2^{4-2} = 2$

$\log_x 2^2 = 2$

$\log_x 4 = 2$ convert to exponential form

$x^2 = 4$

$x = \pm 2$

Since logarithms only exist for positive bases, $x = -2$ is not a solution.

Check when $x = 2$: $4\log_2 2 - \log_2 4 = 4 - 2\log_2 2$

$= 4 - 2$

$= 2$

So $x = 2$ satisfies the original equation.

Hence, the solution is $x = 2$.

Exercise 5.4

1   Solve.

   a $\log_2 x + \log_2 4 = \log_2 20$ **b** $\log_4 2x - \log_4 5 = \log_4 3$

   c $\log_4(x - 5) + \log_4 5 = 2\log_4 10$ **d** $\log_3(x + 3) = 2\log_3 4 + \log_3 5$

2   Solve.

   a $\log_6 x + \log_6 3 = 2$ **b** $\lg(5x) - \lg(x - 4) = 1$

   c $\log_2(x + 4) = 1 + \log_2(x - 3)$ **d** $\log_3(2x + 3) = 2 + \log_3(2x - 5)$

   e $\log_5(10x + 3) - \log_5(2x - 1) = 2$ **f** $\lg(4x + 5) + 2\lg 2 = 1 + \lg(2x - 1)$

3   Solve.

   a $\log_5(x + 8) + \log_5(x + 2) = \log_5 20x$ **b** $\log_3 x + \log_3(x - 2) = \log_3 15$

   c $2\log_4 x - \log_4(x + 3) = 1$ **d** $\lg x + \lg(x + 1) = \lg 20$

   e $\log_3 x + \log_3(2x - 5) = 1$ **f** $3 + 2\log_2 x = \log_2(14x - 3)$

   g $\lg(x + 5) + \lg 2x = 2$ **h** $\lg x + \lg(x - 15) = 2$

4   Solve.

   a $\log_x 64 - \log_x 4 = 1$ **b** $\log_x 16 + \log_x 4 = 3$

   c $\log_x 4 - 2\log_x 3 = 2$ **d** $\log_x 15 = 2 + \log_x 5$

5   Solve.

   a $(\log_5 x)^2 - 3\log_5(x) + 2 = 0$ **b** $(\log_5 x)^2 - \log_5(x^2) = 15$

   c $(\log_5 x)^2 - \log_5(x^3) = 18$ **d** $2(\log_2 x)^2 + 5\log_2(x^2) = 72$

6 CHALLENGE QUESTION

Solve the simultaneous equations.

a $xy = 64$

$\log_x y = 2$

b $2^x = 4^y$

$2\lg y = \lg x + \lg 5$

c $\log_4(x + y) = 2\log_4 x$

$\log_4 y = \log_4 3 + \log_4 x$

d $xy = 640$

$2\log_{10} x - \log_{10} y = 2$

e $\log_{10} a = 2\log_{10} b$

$\log_{10}(2a - b) = 1$

f $4^{xy} = 2^{x+5}$

$\log_2 y - \log_2 x = 1$

7 CHALLENGE QUESTION

a Show that $\lg(x^2 y) = 18$ can be written as $2\lg x + \lg y = 18$.

b $\lg(x^2 y) = 18$ and $\lg\left(\dfrac{x}{y^3}\right) = 2$

Find the value of $\lg x$ and $\lg y$.

5.5 Solving exponential equations

In IGCSE/O Level Mathematics you learned how to solve exponential equations whose terms could be converted to the same base. In this section you will learn how to solve exponential equations whose terms cannot be converted to the same base.

WORKED EXAMPLE 10

Solve, giving your answers correct to 3 sf.

a $3^x = 40$

b $5^{2x+1} = 200$

Answers

a $\quad 3^x = 40$ take logs of both sides

$\quad \lg 3^x = \lg 40$ use the power rule

$\quad x \lg 3 = \lg 40$ divide both sides by $\lg 3$

$\quad x = \dfrac{\lg 40}{\lg 3}$

$\quad x \approx 3.36$

b $\quad 5^{2x+1} = 200$ take logs of both sides

$\quad \lg 5^{2x+1} = \lg 200$ use the power rule

$\quad (2x + 1)\lg 5 = \lg 200$ divide both sides by $\lg 5$

$\quad 2x + 1 = \dfrac{\lg 200}{\lg 5}$

$\quad 2x + 1 = 3.292\ldots$

$\quad 2x = 2.292\ldots$

$\quad x \approx 1.15$

WORKED EXAMPLE 11

Solve $3(2^{2x}) - 2^{x+1} - 8 = 0$.

Answers

$3(2^{2x}) - 2^{x+1} - 8 = 0$ replace 2^{x+1} with $2(2^x)$

$3(2^{2x}) - 2(2^x) - 8 = 0$ use the substitution $y = 2^x$

$\quad\quad 3y^2 - 2y - 8 = 0$ factorise

$\quad (y-2)(3y+4) = 0$

When $y = 2$

$\quad\quad 2 = 2^x$

$\quad\quad x = 1$

When $y = -\dfrac{4}{3}$

$\quad -\dfrac{4}{3} = 2^x$ there are no solutions to this equation
since 2^x is always positive

Hence, the solution is $x = 1$.

Exercise 5.5

1 Solve, giving your answers correct to 3 sf.

    a $2^x = 70$ **b** $3^x = 20$ **c** $5^x = 4$ **d** $2^{3x} = 150$

    e $3^{x+1} = 55$ **f** $2^{2x+1} = 20$ **g** $7^{x-5} = 40$ **h** $7^x = 3^{x+4}$

    i $5^{x+1} = 3^{x+2}$ **j** $4^{x-1} = 5^{x+1}$ **k** $3^{2x+3} = 5^{3x+1}$ **l** $3^{4-5x} = 2^{x+4}$

2 **a**   Show that $2^{x+1} - 2^{x-1} = 15$ can be written as $2(2^x) - \dfrac{1}{2}(2^x) = 15$.

    b   Hence find the value of 2^x.

    c   Find the value of x.

3 Solve, giving your answers correct to 3 sf.

    a $2^{x+2} - 2^x = 4$ **b** $2^{x+1} - 2^{x-1} - 8 = 0$

    c $3^{x+1} - 8(3^{x-1}) - 5 = 0$ **d** $2^{x+2} - 2^{x-3} = 12$

    e $5^x - 5^{x+2} + 125 = 0$

4 Use the substitution $y = 3^x$ to solve the equation $3^{2x} + 2 = 5(3^x)$.

5 Solve, giving your answers correct to 3 sf.

    a $3^{2x} - 6 \times 3^x + 5 = 0$ **b** $4^{2x} - 6 \times 4^x - 7 = 0$

    c $2^{2x} - 2^x - 20 = 0$ **d** $5^{2x} - 2(5^x) - 3 = 0$

6 Use the substitution $u = 5^x$ to solve the equation $5^{2x} - 2(5^{x+1}) + 21 = 0$.

7 Solve, giving your answers correct to 3 sf.

    a $2^{2x} + 2^{x+1} - 15 = 0$ **b** $6^{2x} - 6^{x+1} + 7 = 0$

    c $3^{2x} - 2(3^{x+1}) + 8 = 0$ **d** $4^{2x+1} = 17(4^x) - 15$

8 Solve, giving your answers correct to 3 sf.

    a $4^x - 3(2^x) - 10 = 0$ **b** $16^x + 2(4^x) - 35 = 0$

    c $9^x - 2(3^{x+1}) + 8 = 0$ **d** $25^x + 20 = 12(5^x)$

9 CHALLENGE QUESTION

$3^{2x+1} \times 5^{x-1} = 27^x \times 5^{2x}$

Find the value of

a 15^x b x

10 CHALLENGE QUESTION

Solve the equations, giving your answers correct to 3 significant figures.

a $|3^x + 2| = |3^x - 10|$ b $|2^{x+1} + 3| = |2^x + 10|$

c $3^{2|x|} = 5(3^{|x|}) + 24$ d $4^{|x|} = 5(2^{|x|}) + 14$

11 CHALLENGE QUESTION

Solve the inequality $|2^{x+1} - 1| < |2^x - 8|$ giving your answer in exact form.

REFLECTION

Many students find it challenging to convert between logarithmic and exponential form.

Are you confident converting between these two forms?

What advice would you give to someone who is struggling with these conversions?

5.6 Change of base of logarithms

You sometimes need to change the base of a logarithm.

A logarithm in base b can be written with a different base c using the **change of base rule**.

If $a, b, c > 0$ and $b, c \neq 1$, then:

$$\log_b a = \frac{\log_c a}{\log_c b}$$

Proof:

If $x = \log_b a$, then $b^x = a$ take logs of both sides

$\log_c b^x = \log_c a$ use the power rule

$x \log_c b = \log_c a$ divide both sides by $\log_c b$

$x = \dfrac{\log_c a}{\log_c b}$

$\log_b a = \dfrac{\log_c a}{\log_c b}$

If $c = a$ in the change of base rule, then the rule gives:

$$\log_b a = \frac{1}{\log_a b}$$

WORKED EXAMPLE 12

Change $\log_2 7$ to base 10. Hence evaluate $\log_2 7$ correct to 3 sf.

Answer

$\log_2 7 = \dfrac{\lg 7}{\lg 2} \approx 2.81$

Note:

Some calculators have a $\boxed{\log_\square \square}$ key.

This can be used to evaluate $\log_2 7$ directly.

The change of base rule can be used to solve equations involving logarithms with different bases.

WORKED EXAMPLE 13

Solve $\log_3 x = \log_9 (x + 6)$.

Answers

$\log_3 x = \log_9 (x + 6)$ change $\log_9(x + 6)$ to base 3

$\log_3 x = \dfrac{\log_3 (x + 6)}{\log_3 9}$ $\log_3 9 = \log_3 3^2 = 2$

$\log_3 x = \dfrac{\log_3 (x + 6)}{2}$ multiply both sides by 2

$2\log_3 x = \log_3 (x + 6)$ use the power rule

$\log_3 x^2 = \log_3 (x + 6)$ use equality of logs

$x^2 = x + 6$

$x^2 - x - 6 = 0$

$(x - 3)(x + 2) = 0$

$x = 3$ or $x = -2$

Check when $x = 3$: $\log_3 3$ is defined and is equal to 1

                              $\log_9 (3 + 6) = \log_9 9$ is defined and is equal to 1

So $x = 3$ is a solution, since both sides of the equation are defined and equivalent in value.

Check when $x = -2$: $\log_3 (-2)$ is not defined

So $x = -2$ is not a solution of the original equation.

Hence, the solution is $x = 3$.

Exercise 5.6

1. Use the rule $\log_b a = \dfrac{\log_{10} a}{\log_{10} b}$ to evaluate these correct to 3 sf.

   a $\log_2 10$ b $\log_3 33$ c $\log_5 8$ d $\log_7 0.0025$

2. Given that $u = \log_4 x$, find, in simplest form in terms of u

   a $\log_x 4$ b $\log_x 16$ c $\log_x 2$ d $\log_x 8$

3. Given that $\log_9 y = x$, express in terms of x.

   a $\log_y 9$ b $\log_9 (9y)$ c $\log_3 y$ d $\log_3 (81y)$

4 **a** Given that $\log_p x = 20$ and $\log_p y = 5$, find $\log_y x$.

b Given that $\log_p X = 15$ and $\log_p Y = 6$, find the value of $\log_X Y$.

5 Evaluate $\log_p 2 \times \log_8 p$.

6 Solve.

a $\log_9 3 + \log_9(x + 4) = \log_5 25$ **b** $2\log_4 2 + \log_7(2x + 3) = \log_3 27$

7 **a** Express $\log_4 x$ in terms of $\log_2 x$.

b Using your answer of **part a**, and the substitution $u = \log_2 x$, solve the equation $\log_4 x + \log_2 x = 12$.

8 Solve.

a $\log_2 x + 5\log_4 x = 14$ **b** $\log_3 x + 2\log_9 x = 4$

c $5\log_2 x - \log_4 x = 3$ **d** $4\log_3 x = \log_9 x + 2$

9 **a** Express $\log_x 3$ in terms of a logarithm to base 3.

b Using your answer of **part a**, and the substitution $u = \log_3 x$, solve the equation $\log_3 x = 3 - 2\log_x 3$.

10 Solve.

a $\log_3 x = 9\log_x 3$ **b** $\log_5 x + \log_x 5 = 2$

c $\log_4 x - 4\log_x 4 + 3 = 0$ **d** $\log_4 x + 6\log_x 4 - 5 = 0$

e $\log_2 x - 9\log_x 2 = 8$ **f** $\log_5 y = 4 - 4\log_y 5$

11 **a** Express $\log_4 x$ in terms of $\log_2 x$.

b Express $\log_8 y$ in terms of $\log_2 y$.

c Hence solve, the simultaneous equations

$6\log_4 x + 3\log_8 y = 16$

$\log_2 x - 2\log_4 y = 4$.

12 CHALLENGE QUESTION

Solve the simultaneous equations.

$2\log_3 y = \log_5 125 + \log_3 x$

$2^y = 4^x$

5.7 Natural logarithms

There is another type of logarithm to a special base called e.

The number e is an irrational number and e $\approx$ 2.718.

The number e is an important number in mathematics as it has special properties. You will learn about these special properties in Chapters 15 and 16.

Logarithms to the base of e are called **natural logarithms**.

$\ln x$ is used to represent $\log_e x$.

If $y = e^x$ then $x = \ln y$.

$y = \ln x$ is the reflection of $y = e^x$ in the line $y = x$.

$y = \ln x$ and $y = e^x$ are inverse functions.

All the rules of logarithms that you have learned so far also apply for natural logarithms.

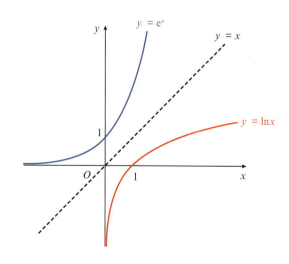

Exercise 5.7

1 Use a calculator to evaluate correct to 3 sf.

   a e^2 **b** $e^{1.5}$ **c** $e^{0.2}$ **d** e^{-3}

2 Use a calculator to evaluate correct to 3 sf.

   a $\ln 4$ **b** $\ln 2.1$ **c** $\ln 0.7$ **d** $\ln 0.39$

3 Without using a calculator find the value of

   a $e^{\ln 5}$ **b** $e^{\frac{1}{2}\ln 64}$ **c** $3e^{\ln 2}$ **d** $-e^{-\ln \frac{1}{2}}$

4 Solve.

   a $e^{\ln x} = 7$ **b** $\ln e^x = 2.5$ **c** $e^{2\ln x} = 36$ **d** $e^{-\ln x} = 20$

5 Solve, giving your answers correct to 3 sf.

   a $e^x = 70$ **b** $e^{2x} = 28$ **c** $e^{x+1} = 16$ **d** $e^{2x-1} = 5$

6 Solve, giving your answers in terms of natural logarithms.

   a $e^x = 7$ **b** $2e^x + 1 = 7$ **c** $e^{2x-5} = 3$ **d** $\dfrac{1}{2}e^{3x-1} = 4$

7 Solve, giving your answers correct to 3 sf.

   a $\ln x = 3$ **b** $\ln x = -2$ **c** $\ln(x + 1) = 7$ **d** $\ln(2x - 5) = 3$

8 Solve, giving your answers correct to 3 sf.

   a $\ln x^3 + \ln x = 5$ **b** $e^{3x+4} = 2e^{x-1}$ **c** $\ln(x + 5) - \ln x = 3$

9 Solve, giving your answers in exact form.

   a $\ln(x - 3) = 2$ **b** $e^{2x-1} = 7$ **c** $e^{2x} - 4e^x = 0$

   d $e^x = 2e^{-x}$ **e** $e^{2x} - 9e^x + 20 = 0$ **f** $e^x + 6e^{-x} = 5$

10 Solve, giving your answers correct to 3 sf.

   a $e^{2x} - 2e^x - 24 = 0$ **b** $e^{2x} - 5e^x + 4 = 0$ **c** $e^x + 2e^{-x} = 80$

11 Solve the simultaneous equations, giving your answers in exact form.

   a $\ln x = 2\ln y$ **b** $e^{5x-y} = 3e^{3x}$

      $\ln y - \ln x = 1$ $e^{2x} = 5e^{x+y}$

12 Solve $5\ln(7 - e^{2x}) = 3$, giving your answer correct to 3 significant figures.

13 Solve $ex - xe^{5x-1} = 0$.

14 CHALLENGE QUESTION

   Solve $5x^2 - x^2e^{2x} + 2e^{2x} = 10$ giving your answers in exact form.

5.8 Practical applications of exponential equations

In this section you will see how exponential equations can be applied to real-life situations.

WORKED EXAMPLE 14

The temperature, $T\,°C$, of a hot drink, t minutes after it is made, is given by

$$T = 75e^{-0.02t} + 20.$$

a Find the temperature of the drink when it was made.

b Find the temperature of the drink when $t = 6$.

c Find the value of t when $T = 65$.

Answers

a When $t = 0$,

$$T = 75e^{-0.02 \times 0} + 20$$
$$= 75e^0 + 20 \qquad \text{use } e^0 = 1$$
$$= 95$$

Temperature of the drink when first made is $95\,°C$.

b When $t = 6$,

$$T = 75e^{-0.02 \times 6} + 20$$
$$= 86.5$$

Temperature of the drink when $t = 6$ is $86.5\,°C$.

c When $T = 65$,

$$65 = 75e^{-0.02t} + 20 \qquad \text{subtract 20 from both sides}$$
$$45 = 75e^{-0.02t} \qquad \text{divide both sides by 75}$$
$$0.6 = e^{-0.02t} \qquad \text{take ln of both sides}$$
$$\ln 0.6 = -0.02t \qquad \text{divide both sides by } -0.02$$
$$t = \frac{\ln 0.6}{-0.02}$$
$$t = 25.5 \text{ to 3 sf}$$

Exercise 5.8

1 At the start of an experiment the number of bacteria was 100.

   This number increases so that after t minutes the number of bacteria, N, is given by the formula

   $$N = 100 \times 2^t.$$

   a Estimate the number of bacteria after 12 minutes.

   b Estimate the time, in minutes, it takes for the number of bacteria to exceed $10\,000\,000$.

2   At the beginning of 2015, the population of a species of animals was estimated at 50 000.

This number decreased so that, after a period of n years, the population was
$$50\,000\,e^{-0.03n}.$$

a   Estimate the population at the beginning of 2020.

b   Estimate the year in which the population would be expected to have first decreased to 5000.

3   The volume of water in a container, $V\,\text{cm}^3$, at time t minutes, is given by the formula
$$V = 2000\,e^{-kt}.$$

When $V = 1000$, $t = 15$.

a   Find the value of k.

b   Find the value of V when $t = 22$.

4   A species of fish is introduced to a lake.

The population, N, of this species of fish after t weeks is given by the formula
$$N = 500\,e^{-0.3t}.$$

a   Find the initial population of these fish.

b   Estimate the number of these fish after 6 weeks.

c   Estimate the number of weeks it takes for the number of these fish to have fallen to $\dfrac{1}{2}$ of the number introduced.

5   The value, $\$V$, of a house n years after it was built is given by the formula
$$V = 250\,000\,e^{an}.$$

When $n = 3$, $V = 350\,000$.

a   Find the initial value of this house.

b   Find the value of a.

c   Estimate the number of years for this house to double in value.

6   The area, $A\,\text{cm}^2$, of a patch of mould is measured daily.

The area, n days after the measurements started, is given by the formula
$$A = A_0 b^n.$$

When $n = 2$, $A = 1.8$ and when $n = 3$, $A = 2.4$.

a   Find the value of b.

b   Find the value of A_0 and explain what A_0 represents.

c   Estimate the number of days for the area of this patch of mould to exceed $7\,\text{cm}^2$.

5.9 The graphs of simple logarithmic and exponential functions

You should already know the properties of the graphs $y = e^x$ and $y = \ln x$.

The graph of $y = e^x$

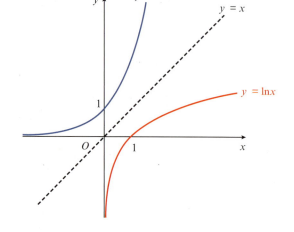

- $y = e^x$ intercepts the y-axis at $(0, 1)$.

- $e^x > 0$ for all values of x.

- When x gets closer to $-\infty$ then y gets closer to 0.

- This can be written: As $x \to -\infty$, then $y \to 0$.

- The graph is said to be asymptotic to the negative x-axis.

- Also, as $x \to +\infty$, then $y \to +\infty$.

The graph of $y = \ln x$

- $y = \ln x$ intercepts the x-axis at $(1, 0)$.

- $\ln x$ only exists for positive values of x.

- As $x \to 0$, then $y \to -\infty$.

- The graph is asymptotic to the negative y-axis.

- As $x \to +\infty$, then $y \to +\infty$.

$y = e^x$ and $y = \ln x$ are inverse functions, so they are mirror images of each other in the line $y = x$.

Exercise 5.9

1   Use a graphing software package to plot each of the following family of curves for
    $k = 3, 2, 1, -1, -2$ and -3.

    a $y = e^{kx}$ **b** $y = k e^x$ **c** $y = e^x + k$

    Describe the properties of each family of curves.

2   Use a graphing software package to plot each of the following family of curves for
    $k = 3, 2, 1, -1, -2$ and -3.

    a $y = \ln kx$ **b** $y = k \ln x$ **c** $y = \ln(x + k)$

    Describe the properties of each family of curves.

5.10 The graphs of $y = k\,e^{nx} + a$ and $y = k \ln(ax + b)$ where n, k, a and b are integers

CLASS DISCUSSION

Consider the function $y = 2e^{3x} + 1$.

Discuss the following with a partner and decide on the missing answers:

1 When $x = 0$, $y = \ldots$

2 The y-intercept is $(\ldots, \ldots)$

3 When $y = 0$, x is $\ldots$

4 As $x \to +\infty$, $y \to \ldots$

5 As $x \to -\infty$, $y \to \ldots$

6 The line $y = \ldots$ is an $\ldots$

Now sketch the graph of $y = 2e^{3x} + 1$ and compare your answer with your classmates.

(Remember to show any axis crossing points and asymptotes on your sketch graph.)

WORKED EXAMPLE 15

Sketch the graph of $y = 3e^{-2x} - 5$.

Answers

When $x = 0$,
$$\begin{aligned} y &= 3e^{0} - 5 \\ &= 3 - 5 \\ &= -2 \end{aligned}$$
hence the y-intercept is $(0, -2)$

When $y = 0$, $0 = 3e^{-2x} - 5$

$$\frac{5}{3} = e^{-2x}$$

$$\ln\left(\frac{5}{3}\right) = -2x$$

$$x = -\frac{1}{2}\ln\left(\frac{5}{3}\right)$$

$$x \approx -0.255$$
hence the x-intercept is $(-0.255, 0)$

As $x \to +\infty$, $e^{-2x} \to 0$ so $y \to -5$ hence the asymptote is $y = -5$

As $x \to -\infty$, $e^{-2x} \to \infty$ so $y \to \infty$

The sketch graph of $y = 3e^{-2x} - 5$ is:

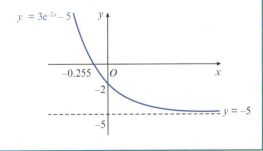

WORKED EXAMPLE 16

Sketch the graph of $y = 4\ln(2x + 5)$.

Answers

When $x = 0$, $y = 4\ln 5$

              ≈ 6.44 hence the y-intercept is $(0, 4\ln 5)$

When $y = 0$, $0 = 4\ln(2x + 5)$

              $0 = \ln(2x + 5)$

              $e^0 = 2x + 5$

              $1 = 2x + 5$

              $x = -2$ hence the x-intercept is $(-2, 0)$

$\ln x$ only exists for positive values of x.

So $4\ln(2x + 5)$ only exists for $2x + 5 > 0$

                         $x > -2.5$

As $x \to +\infty$, $y \to \infty$

As $x \to -2.5$, $y \to -\infty$ hence the asymptote is $x = -2.5$

The sketch graph of $y = 4\ln(2x + 5)$ is:

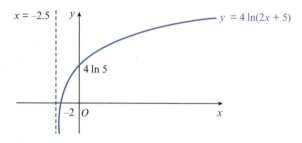

Exercise 5.10

1   Sketch the graphs of each of the following exponential functions.

    [Remember to show the axis crossing points and the asymptotes.]

    a $y = 2e^x - 4$ **b** $y = 3e^x + 6$ **c** $y = 5e^x + 2$

    d $y = 2e^{-x} + 6$ **e** $y = 3e^{-x} - 1$ **f** $y = -2e^{-x} + 4$

    g $y = 4e^{2x} + 1$ **h** $y = 2e^{-5x} + 8$ **i** $y = -e^{4x} + 2$

2   Sketch the graphs of each of the following logarithmic functions.

    [Remember to show the axis crossing points and the asymptotes.]

    a $y = \ln(2x + 4)$ **b** $y = \ln(3x - 6)$ **c** $y = \ln(8 - 2x)$

    d $y = 2\ln(2x + 2)$ **e** $y = 4\ln(2x - 4)$ **f** $y = -3\ln(6x - 9)$

5.11 The inverse of logarithmic and exponential functions

In Chapter 1 you learned how to find the inverse of a one-one function.

This section shows you how to find the inverse of exponential and logarithmic functions.

WORKED EXAMPLE 17

Find the inverse of each function and state its domain.

a $f(x) = 2e^{-4x} + 3$ for $x \in \mathbb{R}$

b $f(x) = 3\ln(2x - 4)$ for $x > 2$

Answers

a $f(x) = 2e^{-4x} + 3$ for $x \in \mathbb{R}$

Step 1: Write the function as $y =$ → $y = 2e^{-4x} + 3$

Step 2: Interchange the x and y variables. → $x = 2e^{-4y} + 3$

Step 3: Rearrange to make y the subject. → $x - 3 = 2e^{-4y}$

$$\frac{x - 3}{2} = e^{-4y}$$

$$\ln\left(\frac{x - 3}{2}\right) = -4y$$

$$y = -\frac{1}{4}\ln\left(\frac{x - 3}{2}\right)$$

$f^{-1}(x) = -\frac{1}{4}\ln\left(\frac{x - 3}{2}\right)$ for $x > 3$

b $f(x) = 3\ln(2x - 4)$, $x > 2$

Step 1: Write the function as $y =$ → $y = 3\ln(2x - 4)$

Step 2: Interchange the x and y variables. → $x = 3\ln(2y - 4)$

Step 3: Rearrange to make y the subject. → $\frac{x}{3} = \ln(2y - 4)$

$$e^{\frac{x}{3}} = 2y - 4$$

$$2y = e^{\frac{x}{3}} + 4$$

$$y = \frac{1}{2}e^{\frac{x}{3}} + 2$$

$f^{-1}(x) = \frac{1}{2}e^{\frac{x}{3}} + 2$ for $x \in \mathbb{R}$

Exercise 5.11

1   The following functions are each defined for $x \in \mathbb{R}$.
Find $f^{-1}(x)$ for each function and state its domain.

a $f(x) = e^x + 4$

b $f(x) = e^x - 2$

c $f(x) = 5e^x - 1$

d $f(x) = 3e^{2x} + 1$

e $f(x) = 5e^{2x} + 3$

f $f(x) = 4e^{-3x} + 5$

g $f(x) = 2 - e^x$

h $f(x) = 5 - 2e^{-2x}$

2 Find $f^{-1}(x)$ for each function.

   a $f(x) = \ln(x + 1),\ x > -1$ **b** $f(x) = \ln(x - 3),\ x > 3$

   c $f(x) = 2\ln(x + 2),\ x > -2$ **d** $f(x) = 2\ln(2x + 1),\ x > -\dfrac{1}{2}$

   e $f(x) = 3\ln(2x - 5),\ x > \dfrac{5}{2}$ **f** $f(x) = -5\ln(3x - 1),\ x > \dfrac{1}{3}$

3 $f(x) = e^{2x} + 1$ for $x \in \mathbb{R}$

   a   State the range of $f(x)$.   **b**   Find $f^{-1}(x)$.

   c   State the domain of $f^{-1}(x)$.   **d**   Find $f^{-1}f(x)$.

4 $f(x) = e^x$ for $x \in \mathbb{R}$ $g(x) = \ln 5x$ for $x > 0$

   a   Find

      i $fg(x)$

      ii $gf(x)$

   b   Solve $g(x) = 3f^{-1}(x)$.

5 $f(x) = e^{3x}$ for $x \in \mathbb{R}$ $g(x) = \ln x$ for $x > 0$

   a   Find

      i $fg(x)$

      ii $gf(x)$

   b   Solve $f(x) = 2g^{-1}(x)$.

6 $f(x) = e^{2x}$ for $x \in \mathbb{R}$ $g(x) = \ln(2x + 1)$ for $x > -\dfrac{1}{2}$

   a   Find $fg(x)$.   **b**   Solve $f(x) = 8g^{-1}(x)$.

SUMMARY

The rules of logarithms

If $y = a^x$ then $x = \log_a y$.

$\log_a a = 1$ $\log_a 1 = 0$

$\log_a a^x = x$ $x = a^{\log_a x}$

Product rule: $\log_a(xy) = \log_a x + \log_a y$

Division rule: $\log_a\left(\dfrac{x}{y}\right) = \log_a x - \log_a y$

Power rule: $\log_a(x)^m = m\log_a x$

$\left[\text{special case: } \log_a\left(\dfrac{1}{x}\right) = -\log_a x\right]$

Change of base: $\log_b a = \dfrac{\log_c a}{\log_c b}$

$\left[\text{special case: } \log_b a = \dfrac{1}{\log_a b}\right]$

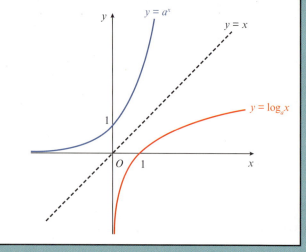

CONTINUED

Natural logarithms

Logarithms to the base of e are called natural logarithms.

$\ln x$ is used to represent $\log_e x$.

If $y = e^x$ then $x = \ln y$.

All the rules of logarithms apply for natural logarithms.

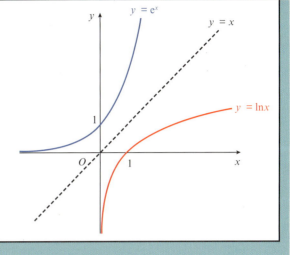

Past paper questions

Worked example

By changing the base of $\log_{2a} 4$, express $(\log_{2a} 4)(1 + \log_a 2)$ as a single logarithm to base a.   [4]

Cambridge IGCSE Additional Mathematics 0606 Paper 21 Q11b Jun 2014

Answers

$\log_{2a} 4 = \dfrac{\log_a 4}{\log_a 2a}$ 　　　　　　　　　　use the product rule on the denominator

$= \dfrac{\log_a 4}{\log_a 2 + \log_a a}$ 　　　　　　　　remember that $\log_a a = 1$

$= \dfrac{\log_a 4}{1 + \log_a 2}$

$(\log_{2a} 4)(1 + \log_a 2) = \left(\dfrac{\log_a 4}{1 + \log_a 2} \right)(1 + \log_a 2)$

$= \dfrac{(\log_a 4)(1 + \log_a 2)}{1 + \log_a 2}$ 　　　divide numerator and denominator by $(1 + \log_a 2)$

$= \log_a 4$

1 a Using the substitution $y = 5^x$, show that the equation $5^{2x+1} - 5^{x+1} + 2 = 2(5^x)$ can be written in the form
   $ay^2 + by + 2 = 0$, where a and b are constants to be found.   [2]

  b Hence solve the equation $5^{2x+1} - 5^{x+1} + 2 = 2(5^x)$.   [4]

Cambridge IGCSE Additional Mathematics 0606 Paper 11 Q4 Nov 2014

2 Solve the following simultaneous equations.

   $\log_2(x + 3) = 2 + \log_2 y$

   $\log_2(x + y) = 3$ [5]

Cambridge IGCSE Additional Mathematics 0606 Paper 21 Q3 Nov 2014

3   a   Write $\log_{27} x$ as a logarithm to base 3 [2]

   b   Given that $\log_a y = 3(\log_a 15 - \log_a 3) + 1$, express y in terms of a. [3]

Cambridge IGCSE Additional Mathematics 0606 Paper 21 Q1 Jun 2015

4   **DO NOT USE A CALCULATOR IN THIS QUESTION.**

   i   Find the value of $-\log_p p^2$. [1]

   ii   Find $\lg\left(\dfrac{1}{10^n}\right)$. [1]

   iii   Show that $\dfrac{\lg 20 - \lg 4}{\log_5 10} = (\lg y)^2$, where y is a constant to be found. [2]

   iv   Solve $\log_r 2x + \log_r 3x = \log_r 600$. [2]

Cambridge IGCSE Additional Mathematics 0606 Paper 21 Q3 Jun 2016

5   a   i   Sketch the graph of $y = e^x - 5$ on the axes below, showing the exact coordinates of any points where the graph meets the coordinate axes. [3]

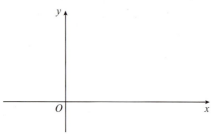

     ii   Find the range of values of k for which the equation $e^x - 5 = k$ has no solutions. [1]

   b   Simplify $\log_a \sqrt{2} + \log_a 8 + \log_a\left(\dfrac{1}{2}\right)$, giving your answer in the form $p \log_a 2$, where p is a constant. [2]

   c   Solve the equation $\log_3 x - \log_9 4x = 1$. [4]

Cambridge IGCSE Additional Mathematics 0606 Paper 22 Q10 Mar 2015

6   a   Solve the following equations to find p and q.

$$8^{q-1} \times 2^{2p+1} = 4^7$$
$$9^{p-4} \times 3^q = 81$$

   b   Solve the equation $\lg(3x - 2) + \lg(x + 1) = 2 - \lg 2$. [5]

Cambridge IGCSE Additional Mathematics 0606 Paper 21 Q5 Nov 2015

7   a   Solve the following equations.

     i $5e^{3x+4} = 14$ [2]

     ii $\lg(2y - 7) + \lg y = 2\lg 3$ [4]

   b   Write $\dfrac{\log_2 p - \log_2 q}{(\log_2 r)(\log_r 2)}$ as a single logarithm to base 2. [2]

Cambridge IGCSE Additional Mathematics 0606 Paper 22 Q8 Mar 2018

8   Solve the equation $\log_5(10x + 5) = 2 + \log_5(x - 7)$ [4]

Cambridge IGCSE Additional Mathematics 0606 Paper 21 Q3 Nov 2017

9   a   Write $(\log_2 p)(\log_3 2) + \log_3 q$ as a single logarithm to base 3. [3]

   b   Given that $(\log_a 5)^2 - 4\log_a 5 + 3 = 0$, find the possible values of a. [3]

Cambridge IGCSE Additional Mathematics 0606 Paper 11 Q6 Jun 2018

10 a Solve $3^{\frac{x}{2}} - 1 = 10$. [3]

b Solve $2\mathrm{e}^{1-2y} = 3\mathrm{e}^{3y+2}$. [4]

Cambridge IGCSE Additional Mathematics 0606 Paper 21 Q2 Nov 2018

11 Find the exact solutions of the equation $3(\ln 5x)^2 + 2\ln 5x - 1 = 0$. [4]

Cambridge IGCSE Additional Mathematics 0606 Paper 12 Q1 Mar 2021

12 In this question, a, b, c and d are positive constants.

    a **i** It is given that $y = \log_a(x + 3) + \log_a(2x - 1)$.

            Explain why x must be greater than $\dfrac{1}{2}$. [1]

      ii Find the exact solution of the equation $\dfrac{\log_a 6}{\log_a(y + 3)} = 2$. [3]

    b Write the expression $\log_a 9 + (\log_a b)(\log_{\sqrt{b}} 9a)$ in the form $c + d\log_a 9$, where c and d are integers. [4]

Cambridge IGCSE Additional Mathematics 0606 Paper 21 Q8 Jun 2021

> Chapter 5

How Does Your Function Grow?

Four enthusiastic mathematicians are asked to think of a function involving the number 100. The challenge is to think of the function which is biggest for big values of n

- Archimedes chooses a logarithm function

  $A(n) = \log(100n)$

- Bernoulli decides to take 100th powers

  $B(n) = n100$

- Copernicus takes powers of 100

  $C(n) = 100n$

- and, finally, de Moivre, who likes to be different, chooses the factorial function which he claims will be quite big enough without any reference to 100 at all

$D(n) = n \times (n - 1) \times (n - 2) \times \ldots \times 2 \times 1$

Which function is biggest for large values of n? Can you determine a value beyond which you know this function will be biggest?

Can you find the switch over value? To find the exact switch-over value will require the clever use of a spreadsheet or computer.

EXTENSION

What could you say if the 100s were replaced by a million? Or billions? Create a convincing argument to prove your results to the mathematicians.

> Chapter 6

Straight-line graphs

THIS SECTION WILL SHOW YOU HOW TO:

- solve questions involving the midpoint and length of a line segment
- use the condition for two lines to be parallel or perpendicular
- interpret the equation of a straight-line graph in the form $y = mx + c$
- transform given relationships, including $y = ax^n$ and $y = ab^x$, to straight-line form and hence determine unknown constants by calculating the gradient or intercept of the transformed graph.

The rules for the prerequisite section can be written formally as follows:

Length of a line segment, gradient and midpoint

P is the point (x_1, y_1) and Q is the point (x_2, y_2).

M is the midpoint of the line segment PQ.

The length of the line segment $PQ = \sqrt{(x_2 - x_1)^2 + (y_2 - y_1)^2}$

The gradient of the line segment $PQ = \dfrac{y_2 - y_1}{x_2 - x_1}$

The coordinates of M are $\left(\dfrac{x_1 + x_2}{2}, \dfrac{y_1 + y_2}{2}\right)$

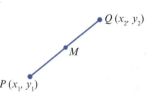

Gradients of parallel lines

If two lines are parallel then their gradients are equal.

Gradients of perpendicular lines

If a line has a gradient of m, a line perpendicular to it has a gradient of $-\dfrac{1}{m}$.

This rule can also be written as:

If the gradients of the two perpendicular lines are m_1 and m_2, then $m_1 \times m_2 = -1$.

The equation of a straight line

The equation of a straight line is $y = mx + c$ where m = the gradient and c = the y-intercept.

6.1 Problems involving the length of a line segment and the midpoint

You need to know how to apply the formulae for the midpoint and the length of a line segment to solve problems.

WORKED EXAMPLE 1

A is the point $(-3, 7)$ and B is the point $(6, -2)$.

a Find the length of AB.   **b** Find the midpoint of AB.

Answers

a $(-3, 7)$ $(6, -2)$ decide which values to use for x_1, y_1, x_2, y_2

  (x_1, y_1) (x_2, y_2)

$$AB = \sqrt{(x_2 - x_1)^2 + (y_2 - y_1)^2}$$
$$= \sqrt{(6 - -3)^2 + (-2 - 7)^2}$$
$$= 9\sqrt{2}$$

b Midpoint $= \left(\dfrac{x_1 + x_2}{2}, \dfrac{y_1 + y_2}{2}\right)$
$$= \left(\dfrac{-3 + 6}{2}, \dfrac{7 + -2}{2}\right)$$
$$= (1.5, 2.5)$$

WORKED EXAMPLE 2

The distance between two points $P(7, a)$ and $Q(a + 1, 9)$ is 15.
Find the two possible values of a.

Answers

$(7, a)$ $(a + 1, 9)$ decide which values to use for x_1, y_1, x_2, y_2

(x_1, y_1) (x_2, y_2)

Using $PQ = \sqrt{(x_2 - x_1)^2 + (y_2 - y_1)^2}$ and $PQ = 15$
$$\sqrt{(a + 1 - 7)^2 + (9 - a)^2} = 15$$
$$\sqrt{(a - 6)^2 + (9 - a)^2} = 15 \quad \text{square both sides}$$
$$(a - 6)^2 + (9 - a)^2 = 225$$
$$a^2 - 12a + 36 + 81 - 18a + a^2 = 225 \quad \text{collect terms on one side}$$
$$2a^2 - 30a - 108 = 0 \quad \text{divide both sides by 2}$$
$$a^2 - 15a - 54 = 0 \quad \text{factorise}$$
$$(a - 18)(a + 3) = 0 \quad \text{solve}$$
$$a - 18 = 0 \text{ or } a + 3 = 0$$

Hence $a = 18$ or $a = -3$.

WORKED EXAMPLE 3

The coordinates of the midpoint of the line segment joining $A(-5, 11)$ and $B(p, q)$ are $(2.5, -6)$.

Find the value of p and the value of q.

Answers

$(-5, 11)$ (p, q) decide which values to use for x_1, y_1, x_2, y_2

   ↑   ↑

(x_1, y_1) (x_2, y_2)

Using $\left(\dfrac{x_1 + x_2}{2}, \dfrac{y_1 + y_2}{2}\right)$ and midpoint $= (2.5, -6)$.

$\left(\dfrac{x_1 + x_2}{2}, \dfrac{y_1 + y_2}{2}\right) = (2.5, -6)$

$\left(\dfrac{-5 + p}{2}, \dfrac{11 + q}{2}\right) = (2.5, -6)$

Equating the x-coordinates gives: $\dfrac{-5 + p}{2} = 2.5$

$$-5 + p = 5$$
$$p = 10$$

Equating the y-coordinates gives: $\dfrac{11 + q}{2} = -6$

$$11 + q = -12$$
$$q = -23$$

Hence $p = 10$ and $q = -23$.

WORKED EXAMPLE 4

Three of the vertices of a parallelogram $ABCD$ are $A(-10, 1)$, $B(6, -2)$ and $C(14, 4)$.

a   Find the midpoint of AC.   **b**   Find the coordinates of D.

Answers

a   Midpoint of $AC = \left(\dfrac{-10 + 14}{2}, \dfrac{1 + 4}{2}\right) = (2, 2.5)$

b   Let the coordinates of D be (m, n).

Since $ABCD$ is a parallelogram, the midpoint of BD is the same as the midpoint of AC.

Midpoint of $BD = \left(\dfrac{6 + m}{2}, \dfrac{-2 + n}{2}\right) = (2, 2.5)$

Equating the x-coordinates gives: $\dfrac{6 + m}{2} = 2$

$$6 + m = 4$$
$$m = -2$$

CONTINUED

Equating the y-coordinates gives: $\dfrac{-2 + n}{2} = 2.5$

$$-2 + n = 5$$

$$n = 7$$

The coordinates of D are $(-2, 7)$.

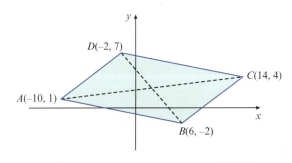

CLASS DISCUSSION

This triangle has sides of length $5\sqrt{3}$ cm, $2\sqrt{6}$ cm and $7\sqrt{2}$ cm.

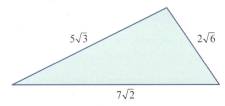

Priya says that the triangle is right-angled.

Discuss whether she is correct.

Explain your reasoning.

Exercise 6.1

1   Find the length of the line segment joining each pair of points.

    a $(2, 0)$ and $(5, 0)$ **b** $(-7, 4)$ and $(-7, 8)$ **c** $(2, 1)$ and $(8, 9)$

    d $(-3, 1)$ and $(2, 13)$ **e** $(5, -2)$ and $(2, -6)$ **f** $(4, 4)$ and $(-20, -3)$

    g $(6, -5)$ and $(1, 2)$ **h** $(-3, -2)$ and $(-1, -5)$ **i** $(-7, 7)$ and $(5, -5)$

2   Calculate the lengths of the sides of the triangle PQR.

    Use your answers to determine whether or not the triangle is right-angled.

    a $P(3, 11)$, $Q(5, 7)$, $R(11, 10)$

    b $P(-7, 8)$, $Q(-1, 4)$, $R(5, 12)$

    c $P(-8, -3)$, $Q(-4, 5)$, $R(-2, -6)$

3   $A(-1, 0)$, $B(1, 6)$ and $C(7, 4)$.

    Show that triangle ABC is a right-angled isosceles triangle.

4   The distance between two points $P(10, 2b)$ and $Q(b, -5)$ is $5\sqrt{10}$.
Find the two possible values of b.

5   The distance between two points $P(6, -2)$ and $Q(2a, a)$ is 5.
Find the two possible values of a.

6   Find the coordinates of the midpoint of the line segment joining each pair of points.

   a $(5, 2)$ and $(7, 6)$ **b** $(4, 3)$ and $(9, 11)$ **c** $(8, 6)$ and $(-2, 10)$

   d $(-1, 7)$ and $(2, -4)$ **e** $(-7, -8)$ and $(-2, 3)$ **f** $(2a, -3b)$ and $(4a, 5b)$.

7   The coordinates of the midpoint of the line segment joining $P(-8, 2)$ and $Q(a, b)$ are $(5, -3)$.
Find the value of a and the value of b.

8   Three of the vertices of a parallelogram $ABCD$ are $A(-7, 6)$, $B(-1, 8)$ and $C(7, 3)$.

   a   Find the midpoint of AC.

   b   Find the coordinates of D.

9   The point $P(2k, k)$ is equidistant from $A(-2, 4)$ and $B(7, -5)$.
Find the value of k.

10   In triangle ABC, the midpoints of the sides AB, BC and AC are $P(2, 3)$, $Q(3, 5)$ and $R(-4, 4)$ respectively. Find the coordinates of A, B and C.

6.2 Parallel and perpendicular lines

You need to know how to apply the rules for gradients to solve problems involving parallel and perpendicular lines.

WORKED EXAMPLE 5

The coordinates of three points are $A(8 - k, 2)$, $B(-2, k)$ and $C(-8, 2k)$.

Find the possible values of k if A, B and C are collinear.

Answers

If A, B and C are collinear then they lie on the same line.

gradient of AB = gradient of BC

$$\frac{k - 2}{-2 - (8 - k)} = \frac{2k - k}{-8 - (-2)}$$

$$\frac{k - 2}{k - 10} = \frac{k}{-6} \qquad \text{cross multiply}$$

$$-6(k - 2) = k(k - 10) \qquad \text{expand brackets}$$

$$-6k + 12 = k^2 - 10k \qquad \text{collect terms on one side}$$

$$k^2 - 4k - 12 = 0 \qquad \text{factorise}$$

$$(k - 6)(k + 2) = 0 \qquad \text{solve}$$

$$k - 6 = 0 \text{ or } k + 2 = 0$$

Hence $k = 6$ or $k = -2$

WORKED EXAMPLE 6

The vertices of triangle ABC are $A(-4, 2)$, $B(5, -5)$ and $C(k, k + 2)$.

Find the possible values of k if angle ACB is 90°.

Answers

Since angle ACB is 90°, the gradient of $AC \times$ gradient of $BC = -1$

$$\frac{(k + 2) - 2}{k - (-4)} \times \frac{(k + 2) - (-5)}{k - 5} = -1$$

$$\frac{k}{k + 4} \times \frac{k + 7}{k - 5} = -1$$

$$k(k + 7) = -(k + 4)(k - 5)$$

$$k^2 + 7k = -(k^2 - k - 20)$$

$$k^2 + 7k = -k^2 + k + 20$$

$$2k^2 + 6k - 20 = 0$$

$$k^2 + 3k - 10 = 0$$

$$(k + 5)(k - 2) = 0$$

$$k + 5 = 0 \text{ or } k - 2 = 0$$

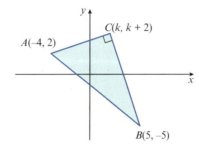

Hence $k = -5$ or $k = 2$

The two possible situations are:

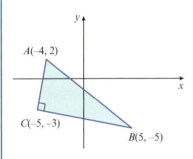
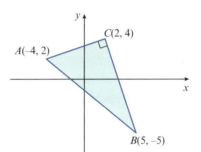

Exercise 6.2

1   Find the gradient of the line segment AB for each of the following pairs of points.

   a $A(1, 2)$ and $B(3, -2)$ b $A(4, 3)$ and $B(5, 0)$

   c $A(-4, 4)$ and $B(7, 4)$ d $A(1, -9)$ and $B(4, 1)$

   e $A(-4, -3)$ and $B(5, 0)$ f $A(6, -7)$ and $B(2, -4)$

2   Write down the gradient of lines perpendicular to a line with gradient

   a 3 b $-\frac{1}{2}$ c $\frac{2}{5}$ d $1\frac{1}{4}$ e $-2\frac{1}{2}$

3   Two vertices of a rectangle $ABCD$ are $A(3, -5)$ and $B(6, -3)$.

   a   Find the gradient of CD.

   b   Find the gradient of BC.

4    $A(-1, -5)$, $B(5, -2)$ and $C(1, 1)$
     $ABCD$ is a trapezium.
     AB is parallel to DC and angle BAD is 90°.
     Find the coordinates of D.

5   The midpoint of the line segment joining $P(-2, 3)$ and $Q(4, -1)$ is M.
     The point C has coordinates $(-1, -2)$.
     Show that CM is perpendicular to PQ.

6    $A(-2, 2)$, $B(3, -1)$ and $C(9, -4)$.
     a   Find the gradient of AB and the gradient of BC.
     b   Use your answer to **part a** to decide whether or not the points A, B and C
          are collinear.

7   The coordinates of three points are $A(-4, 4)$, $B(k, -2)$ and $C(2k + 1, -6)$.
     Find the value of k if A, B and C are collinear.

8   The vertices of triangle ABC are $A(-k, -2)$, $B(k, -4)$ and $C(4, k - 2)$.
     Find the possible values of k if angle ABC is 90°.

9   **CHALLENGE QUESTION**
     A is the point $(-2, 0)$ and B is the point $(2, 6)$.
     Find the point C on the x-axis such that angle ABC is 90°.

6.3 Equations of straight lines

You should already know that the equation of a straight line is

$$y = mx + c$$

where m = the gradient and c = the y-intercept.

There is an alternative formula that can be used when you know the gradient of the
straight line and a point on the line.

Consider a line, with gradient m, which passes through the
known point $A(x_1, y_1)$ and whose general point is $P(x, y)$.

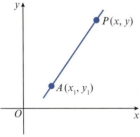

Gradient of $AP = m$, hence $\dfrac{y - y_1}{x - x_1} = m$ multiply both sides by $(x - x_1)$

$$y - y_1 = m(x - x_1)$$

The equation of a straight line, with gradient m,
which passes through the point (x_1, y_1) is:

$$y - y_1 = m(x - x_1)$$

WORKED EXAMPLE 7

Find the equation of the straight line

a with gradient 2 and passing through the point (4, 7)

b passing through the points (−5, 8) and (1, −4).

Answers

a Using $y - y_1 = m(x - x_1)$ with $m = 2$, $x_1 = 4$ and $y_1 = 7$

$$y - 7 = 2(x - 4)$$
$$y - 7 = 2x - 8$$
$$y = 2x - 1$$

b (−5, 8)   (1, −4)   decide which values to use for x_1, y_1, x_2, y_2

(x_1, y_1) (x_2, y_2)

Gradient $= m = \dfrac{y_2 - y_1}{x_2 - x_1} = \dfrac{(-4) - 8}{1 - (-5)} = -2$

Using $y - y_1 = m(x - x_1)$ with $m = -2$, $x_1 = -5$ and $y_1 = 8$.

$$y - 8 = -2(x + 5)$$
$$y - 8 = -2x - 10$$
$$y = -2x - 2$$

WORKED EXAMPLE 8

Find the equation of the perpendicular bisector of the line joining $A\,(3, 2)$ and $B\,(7, 10)$.

Answers

Gradient of $AB = \dfrac{10 - 2}{7 - 3} = \dfrac{8}{4} = 2$ using $\dfrac{y_2 - y_1}{x_2 - x_1}$

Gradient of the perpendicular is $-\dfrac{1}{2}$ using $m_1 \times m_2 = -1$

Midpoint of $AB = \left(\dfrac{3 + 7}{2}, \dfrac{2 + 10}{2}\right) = (5, 6)$ using $\left(\dfrac{x_1 + x_2}{2}, \dfrac{y_1 + y_2}{2}\right)$

So, the perpendicular bisector is the line passing through the point (5, 6) with gradient $-\dfrac{1}{2}$.

Using $y - y_1 = m(x - x_1)$ with $x_1 = 5$, $y_1 = 6$ and $m = -\dfrac{1}{2}$

$$y - 6 = -\dfrac{1}{2}(x - 5)$$
$$y - 6 = -\dfrac{1}{2}x + \dfrac{5}{2}$$
$$y = -\dfrac{1}{2}x + 8.5$$ multiply both sides by 2 and rearrange
$$x + 2y = 17$$

Exercise 6.3

1   Find the equation of the line with
  a   gradient 3 and passing through the point $(6, 5)$
  b   gradient -4 and passing through the point $(2, -1)$
  c   gradient $-\dfrac{1}{2}$ and passing through the point $(8, -3)$.

2   Find the equation of the line passing through
  a $(3, 2)$ and $(5, 7)$
  b $(-1, 6)$ and $(5, -3)$
  c $(5, -2)$ and $(-7, 4)$.

3   Find the equation of the line
  a   parallel to the line $y = 2x + 4$, passing through the point $(6, 2)$
  b   parallel to the line $x + 2y = 5$, passing through the point $(2, -5)$
  c   perpendicular to the line $2x + 3y = 12$, passing through the point $(7, 3)$
  d   perpendicular to the line $4x - y = 6$, passing through the point $(4, -1)$.

4 P is the point $(2, 5)$ and Q is the point $(6, 0)$.
  A line l is drawn through P, perpendicular to PQ, to meet the y-axis at the point R.
  a   Find the equation of the line l.
  b   Find the coordinates of the point R.
  c   Find the area of triangle OPR where O is the origin.

5   Find the equation of the perpendicular bisector of the line segment joining the points
  a $(1, 3)$ and $(-3, 1)$
  b $(-1, -5)$ and $(5, 3)$
  c $(0, -9)$ and $(5, -2)$.

6   The perpendicular bisector of the line joining $A(-1, 4)$ and $B(2, 2)$ intersects the x-axis at P and the y-axis at Q.
  a   Find the coordinates of P and of Q.
  b   Find the length of PQ.
  c   Find the area of triangle OPQ where O is the origin.

7   The line l_1 has equation $3x + 2y = 12$.
  The line l_2 has equation $y = 2x - 1$.
  The lines l_1 and l_2 intersect at the point A.
  a   Find the coordinates of A.
  b   Find the equation of the line through A which is perpendicular to the line l_1.

8   The coordinates of three points are $A(1, 5)$, $B(9, 7)$ and $C(k, -6)$.
  M is the midpoint of AB and MC is perpendicular to AB.
  a   Find the coordinates of M.
  b   Find the value of k.

9   The coordinates of triangle ABC are $A(2, -1)$, $B(3, 7)$ and $C(14, 5)$.

    P is the foot of the perpendicular from B to AC.

    a   Find the equation of BP.

    b   Find the coordinates of P.

    c   Find the lengths of AC and BP.

    d   Use your answers to **part c** to find the area of triangle ABC.

10   CHALLENGE QUESTION

    The coordinates of triangle PQR are $P(-3, -2)$, $Q(5, 10)$ and $R(11, -2)$.

    a   Find the equation of the perpendicular bisectors of

        i PQ **ii** QR.

    b   Find the coordinates of the point which is equidistant from P, Q and R.

> **TIP**
>
> The point is where the perpendicular bisectors of the sides intersect.

6.4 Areas of rectilinear figures

> ## CLASS DISCUSSION
>
> Discuss with your classmates, how you can find the area of triangle ABC.
>
> Try to find as many different methods as possible.
>
> Compare the ease of use of each of these methods.
>
>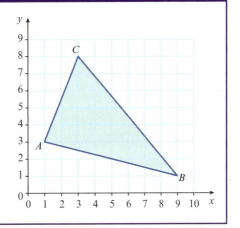

There is a method that you might not have seen before. It is often referred to as the 'shoestring' or 'shoelace' method. The study of this method is not part of the syllabus, however you may benefit from knowing this for broader understanding.

If the vertices of triangle ABC are $A(x_1, y_1)$, $B(x_2, y_2)$ and $C(x_3, y_3)$, then:

> Area of triangle $ABC = \dfrac{1}{2}\left| x_1 y_2 + x_2 y_3 + x_3 y_1 - x_2 y_1 - x_3 y_2 - x_1 y_3 \right|$

This complicated formula can be written as:
$$\frac{1}{2}\left| \begin{array}{cccc} x_1 & x_2 & x_3 & x_1 \\ y_1 & y_2 & y_3 & y_1 \end{array} \right|$$

The products in the direction ↘ are given positive signs and the products in the direction ↗ are given negative signs.

For the triangle in the class discussion
$$\frac{1}{2}\left| \begin{array}{cccc} 1 & 9 & 3 & 1 \\ 3 & 1 & 8 & 3 \end{array} \right|$$
$A(1, 3)$, $B(9, 1)$ and $C(3, 8)$:

area of triangle $ABC = \frac{1}{2}|1 \times 1 + 9 \times 8 + 3 \times 3 - 3 \times 9 - 1 \times 3 - 8 \times 1|$

$$= \frac{1}{2}|1 + 72 + 9 - 27 - 3 - 8|$$

$$= \frac{1}{2}|44|$$

$$= 22 \text{ units}^2$$

This method can be extended for use with polygons with more than 3 sides.

Note:
If you take the vertices in an anticlockwise direction around a shape, then the inside of the modulus sign will be positive. If you take the vertices in a clockwise direction, then the inside of the modulus sign will be negative.

WORKED EXAMPLE 9

The vertices of a pentagon $ABCDE$ are $A(0, -1)$, $B(5, 1)$, $C(3, 4)$, $D(-1, 6)$ and $E(-3, 2)$.

a Find the area of the pentagon using the 'shoestring' method.

b Find the area of the pentagon using the 'boxing in' method.

Answers

a $\dfrac{1}{2}\begin{vmatrix} 0 & 5 & 3 & -1 & -3 & 0 \\ -1 & 1 & 4 & 6 & 2 & -1 \end{vmatrix}$

Area of pentagon $= \dfrac{1}{2}|0 + 20 + 18 + (-2) + 3 - (-5) - 3 - (-4) - (-18) - 0|$

$$= \frac{1}{2}|63|$$

$$= 31.5 \text{ units}^2$$

b For the 'boxing in' method, you draw a rectangle around the outside of the pentagon.

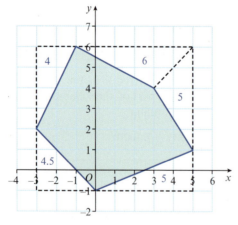

Area of pentagon = area of rectangle − sum of the outside areas

$$= (8 \times 7) - (4 + 6 + 5 + 5 + 4.5)$$

$$= 56 - 24.5$$

$$= 31.5 \text{ units}^2$$

Exercise 6.4

1 Find the area of these triangles.

    a $A(-2, 3)$, $B(0, -4)$, $C(5, 6)$ **b** $P(-3, 1)$, $Q(5, -3)$, $R(2, 4)$

2 Find the area of these quadrilaterals.

    a $A(1, 8)$, $B(-4, 5)$, $C(-2, -3)$, $D(4, -2)$

    b $P(2, 7)$, $Q(-5, 6)$, $R(-3, -4)$, $S(7, 2)$

3 Triangle PQR where $P(1, 4)$, $Q(-3, -4)$ and $R(7, k)$ is right-angled at Q.

    a Find the value of k.   **b** Find the area of triangle PQR.

4 A is the point $(-4, 0)$ and B is the point $(2, 3)$.

    M is the midpoint of the line AB.

    Point C is such that $\overrightarrow{MC} = \begin{pmatrix} 3 \\ -6 \end{pmatrix}$.

    a Find the coordinates of M and C.

    b Show that CM is perpendicular to AB.

    c Find the area of triangle ABC.

5 Angle ABC is 90° and M is the midpoint of the line AB.

    The point C lies on the y-axis.

    a Find the coordinates of B and C.

    b Find the area of triangle ABC.

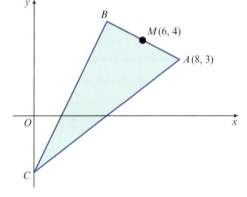

6 A is the point $(-4, 5)$ and B is the point $(5, 8)$.

    The perpendicular to the line AB at the point A crosses the y-axis at the point C.

    a Find the coordinates of C.   **b** Find the area of triangle ABC.

7 AB is parallel to DC and BC is perpendicular to AB.

    a Find the coordinates of C.

    b Find the area of trapezium $ABCD$.

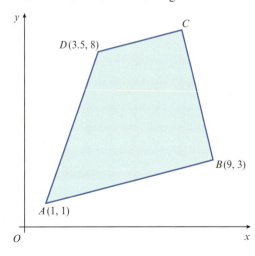

8   *ABCD* is a square.

 A is the point $(-2, 0)$ and *C* is the point $(6, 4)$.

 AC and *BD* are diagonals of the square, which intersect at *M*.

 a   Find the coordinates of *M*, *B* and *D*.

 b   Find the area of *ABCD*.

9   The coordinates of three of the vertices of a parallelogram *ABCD* are $A(-4, 3)$, $B(5, -5)$ and $C(15, -1)$.

 a   Find the coordinates of the points of intersection of the diagonals.

 b   Find the coordinates of the point *D*.

 c   Find the area of parallelogram *ABCD*.

6.5 Converting from a non-linear equation to linear form

Some situations in the real world can be modelled using an equation.

Consider an experiment where a simple pendulum, of length L cm, travels from A to B and back to A in a time of T seconds. The table shows the time taken, T, for different lengths L.

L	5	10	15	20	25	30
T	0.45	0.63	0.78	0.90	1.00	1.10

When the graph of T against L is drawn, the points lie on a curve.

They do not lie on a straight line.

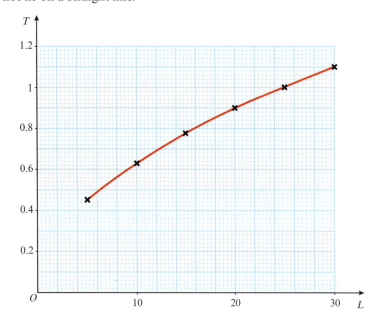

So how can you find the equation of the curve?

If you are told that the rule connecting the variables T and L is believed to be $T = a\sqrt{L}$, you should draw the graph of T against $\sqrt{L}$. To do this, you must first make a table of values for T and $\sqrt{L}$.

$\sqrt{L}$	2.24	3.16	3.87	4.47	5.00	5.48
T	0.45	0.63	0.78	0.90	1.00	1.10

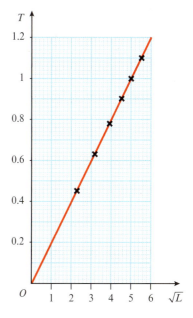

The fact that the points now lie on a straight line confirms the belief that the rule connecting the variables T and L is $T = a\sqrt{L}$.

The gradient of the straight line tells you the value of a.

Using the two end points, gradient $\dfrac{1.1 - 0.45}{5.48 - 2.24} \approx 0.2$

Hence the approximate rule connecting T and L is

$$T = 0.2\sqrt{L}$$

In the example above you have converted from a non-linear graph to a linear graph.

Before you can model more complicated relationships, you must first learn how to choose suitable variables for Y and X to convert from a non-linear equation into the linear form $Y = mX + c$, where m is the gradient of the straight line and c is the Y-intercept.

WORKED EXAMPLE 10

Convert $y = ax + \dfrac{b}{x}$, where a and b are constants, into the form $Y = mX + c$.

Answers

Method 1 $y = ax + \dfrac{b}{x}$

Multiplying both sides of the equation by x gives:

$$xy = ax^2 + b$$

Now compare $xy = ax^2 + b$ with $Y = mX + c$:

$$\underbrace{xy}_{Y} = \underbrace{a}_{m}\underbrace{x^2}_{X} + \underbrace{b}_{c}$$

The non-linear equation $y = ax + \dfrac{b}{x}$ becomes the linear equation:

$Y = mX + c$, where $Y = xy$, $X = x^2$, $m = a$ and $c = b$

Method 2 $y = ax + \dfrac{b}{x}$

Dividing both sides of the equation by x gives:

$$\frac{y}{x} = a + \frac{b}{x^2}$$

> **Note:**
> Use X and Y as the variables in the linear equation to avoid confusion with the original variables x and y in the non-linear equation.

CONTINUED

Now compare $\dfrac{y}{x} = a + \dfrac{b}{x^2}$ with $Y = mX + c$:

$$\boxed{\dfrac{y}{x}} = b\left(\dfrac{1}{x^2}\right) + a$$
$$\uparrow \qquad \uparrow \quad \uparrow \qquad \uparrow$$
$$Y \ = \ m \ \ X \ + \ c$$

The non-linear equation $y = ax + \dfrac{b}{x}$ becomes the linear equation:

$Y = mX + c$, where $Y = \dfrac{y}{x}$, $X = \dfrac{1}{x^2}$, $m = b$ and $c = a$

It is important to note that the variables X and Y in $Y = mX + c$ must contain only the original variables x and y. They must not contain the unknown constants a and b.

Similarly, the constants m and c must contain only the original unknown constants a and b. They must not contain the variables x and y.

WORKED EXAMPLE 11

Convert $y = ae^{-bx}$, where a and b are constants, into the form $Y = mX + c$.

Answers

$$y = ae^{-bx}$$

Taking natural logarithms of both sides gives

$$\ln y = \ln\left(ae^{-bx}\right)$$
$$\ln y = \ln a + \ln e^{-bx}$$
$$\ln y = \ln a - bx$$
$$\ln y = -bx + \ln a$$

Now compare $\ln y = -bx + \ln a$ with $Y = mX + c$:

$$\boxed{\ln y} = -b\boxed{x} + \ln a$$
$$\uparrow \qquad \uparrow \ \ \uparrow \qquad \uparrow$$
$$Y \ = \ m \ X \ + \ c$$

The non-linear equation $y = ae^{-bx}$ becomes the linear equation:

$Y = mX + c$, where $Y = \ln y$, $X = x$, $m = -b$ and $c = \ln a$

Exercise 6.5

1   Convert each of these non-linear equations into the form $Y = mX + c$, where a and b are constants. State clearly what the variables X and Y and the constants m and c represent.

a $\ y = ax^2 + b$ 　　**b** $\ y = ax + \dfrac{b}{x}$ 　　**c** $\ y = ax^2 - bx$ 　　**d** $\ y(a - x) = bx$

e $\ y = a\sqrt{x} + \dfrac{b}{\sqrt{x}}$ 　　**f** $\ y = \dfrac{a}{x^2} + b$ 　　**g** $\ x = axy + by$ 　　**h** $\ \dfrac{1}{y} = a\sqrt{x} - \dfrac{b}{\sqrt{x}}$

> **Note:**
> there may be more than one way to do this.

2   Convert each of these non-linear equations into the form $Y = mX + c$, where a and b are constants. State clearly what the variables X and Y and the constants m and c represent.

a $y = 10^{ax+b}$ b $y = e^{ax-b}$ c $y = ax^b$ d $y = ab^x$

e $x^a y^b = e^2$ f $xa^y = b$ g $a = e^{x^2+by}$ h $y = ae^{bx}$

Note:
there may be more
than one way to
do this.

6.6 Converting from linear form to a non-linear equation

WORKED EXAMPLE 12

Find y in terms of x.

a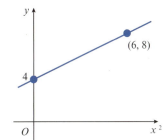

b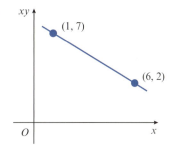

Answers

a   The linear equation is $Y = mX + c$, where $Y = y$ and $X = x^2$.

Gradient $= m = \dfrac{8 - 4}{6 - 0} = \dfrac{2}{3}$

Y-intercept $= c = 4$

Hence $Y = \dfrac{2}{3}X + 4$

The non-linear equation is $y = \dfrac{2}{3}x^2 + 4$

b   The linear equation is $Y = mX + c$, where $Y = xy$ and $X = x$.

Gradient $= m = \dfrac{2 - 7}{6 - 1} = -1$

Using $Y = mX + c$, $m = -1$, $X = 6$ and $Y = 2$

$2 = -1 \times 6 + c$

$c = 8$

Hence $Y = -X + 8$

The non-linear equation is $xy = -x + 8$

$y = -1 + \dfrac{8}{x}$

WORKED EXAMPLE 13

Variables x and y are such that $y = a \times b^x$, where a and b are constants.

The diagram shows the graph of $\lg y$ against x, passing through the points $(2, 5)$ and $(6, 13)$.

Find the value of a and the value of b.

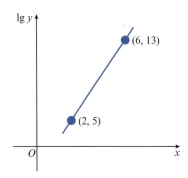

Answers

$$y = a \times b^x$$

Taking logarithms of both sides gives:

$$\lg y = \lg(a \times b^x)$$

$$\lg y = \lg a + \lg b^x$$

$$\lg y = x \lg b + \lg a$$

Now compare $\lg y = x \lg b + \lg a$ with $Y = mX + c$:

$$\boxed{\lg y} = \lg b \,\boxed{x}\, + \lg a$$
$$Y \;=\; m \;\; X \;+\; c$$

Gradient $= m = \dfrac{13 - 5}{6 - 2} = 2$

$$\lg b = 2$$

$$b = 10^2$$

Using $Y = mX + c$, $m = 2$, $X = 2$ and $Y = 5$

$$5 = 2 \times 2 + c$$

$$c = 1$$

$$\lg a = 1$$

$$a = 10^1$$

Hence $a = 10$ and $b = 100$.

Exercise 6.6

1 The graphs show part of a straight line obtained by plotting y against some function of x.

For each graph, express y in terms of x.

a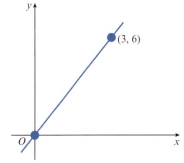

b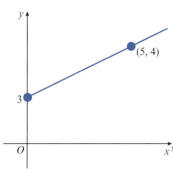

c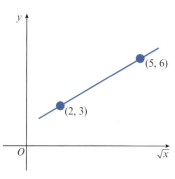

d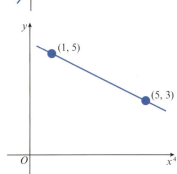

e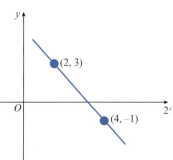

f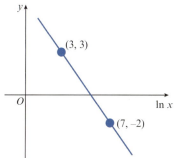

2 For each of the following relations

   i express y in terms of x **ii** find the value of y when $x = 2$.

a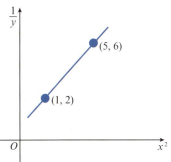

b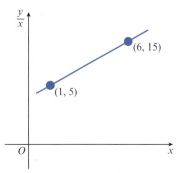

c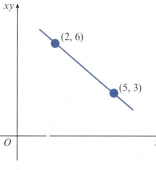

d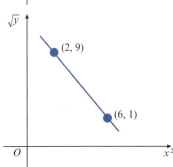

e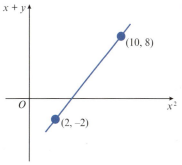

f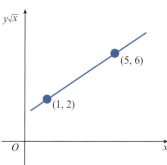

3   Variables x and y are related so that, when $\dfrac{y}{x^2}$ is plotted on the vertical axis and x^3 is plotted on the horizontal axis, a straight-line graph passing through (2, 12) and (6, 4) is obtained.

Express y in terms of x.

4   Variables x and y are related so that, when y^2 is plotted on the vertical axis and 2^x is plotted on the horizontal axis, a straight-line graph which passes through the point (8, 49) with gradient 3 is obtained.

    a   Express y^2 in terms of 2^x.

    b   Find the value of x when $y = 11$.

5   Variables x and y are related so that, when $\dfrac{y}{x}$ is plotted on the vertical axis and x is plotted on the horizontal axis, a straight-line graph passing through the points (2, 4) and (5, −2) is obtained.

    a   Express y in terms of x.

    b   Find the value of x and the value of y such that $\dfrac{y}{x} = 3$.

6   Variables x and y are related so that, when e^y is plotted on the vertical axis and x^2 is plotted on the horizontal axis, a straight-line graph passing through the points (3, 4) and (8, 9) is obtained.

    a   Express e^y in terms of x.

    b   Express y in terms of x.

7   Variables x and y are related so that, when $\lg y$ is plotted on the vertical axis and x is plotted on the horizontal axis, a straight-line graph passing through the points (6, 2) and (10, 8) is obtained.

    a   Express $\lg y$ in terms of x.

    b   Express y in terms of x, giving your answer in the form $y = a \times 10^{bx}$.

8   Variables x and y are related so that, when $\lg y$ is plotted on the vertical axis and $\lg x$ is plotted on the horizontal axis, a straight-line graph passing through the points (4, 8) and (8, 14) is obtained.

    a   Express y in terms of x, giving your answer in the form $y = a \times x^b$.

    b   Find the value of x when $y = 51.2$.

9   Variables x and y are related so that, when $\ln y$ is plotted on the vertical axis and $\ln x$ is plotted on the horizontal axis, a straight-line graph passing through the points (1, 2) and (4, 11) is obtained.

    a   Express $\ln y$ in terms of x.

    b   Express y in terms of x.

10 Variables x and y are such that, when $\ln y$ is plotted on the vertical axis and $\ln x$ is plotted on the horizontal axis, a straight-line graph passing through the points (2.5, 7.7) and (3.7, 5.3) is obtained.

    a   Find the value of $\ln y$ when $\ln x = 0$.

    b   Given that $y = a \times x^b$, find the value of a and the value of b.

6.7 Finding relationships from data

When experimental data is collected for two variables, it is useful if you can then establish the mathematical relationship connecting the two variables.

If the data forms a straight-line when a graph is plotted, it is easy to establish the connection using the equation $y = mx + c$.

It is more usual, however, for the data to lie on a curve and to be connected by a non-linear equation.

In this section, you will learn how to apply what you have just learnt in **sections 6.5** and **6.6** to find the non-linear equation connecting two variables.

WORKED EXAMPLE 14

x	5	10	20	40	80
y	2593	1596	983	605	372

The table shows experimental values of the variables x and y.

a   By plotting a suitable straight-line graph, show that x and y are related by the equation $y = k \times x^n$, where k and n are constants.

b   Use your graph to estimate the value of k and the value of n.

Answers

a $y = k \times x^n$ take logs of both sides

$\lg y = \lg (k \times x^n)$ use the multiplication law

$\lg y = \lg k + \lg x^n$ use the power law

$\lg y = n \lg x + \lg k$

Now compare $\lg y = n \lg x + \lg k$ with $Y = mX + c$:

$$\lg y = n \lg x + \lg k$$
$$Y = m \ X \ + \ c$$

Hence the graph of $\lg y$ against $\lg x$ needs to be drawn where

- gradient $= n$

- intercept on vertical axis $= \lg k$.

The table of values is

$\lg x$	0.699	1.000	1.301	1.602	1.903
$\lg y$	3.414	3.203	2.993	2.782	2.571

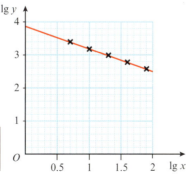

CONTINUED

b The points form an approximate straight line, so x and y are related by the equation

$$y = k \times x^n.$$

n = gradient

$$\approx \frac{2.571 - 3.414}{1.903 - 0.699}$$

$$\approx -0.7$$

$\lg k$ = intercept on vertical axis

$$\lg k \approx 3.9$$

$$k \approx 10^{3.9}$$

$$k \approx 7943$$

WORKED EXAMPLE 15

x	2	4	6	8	10
y	4.75	2.19	1.42	1.05	0.83

The table shows experimental values of the variables x and y.

The variables are known to be related by the equation $y = \dfrac{a + bx}{x^2}$, where a and b are constants.

a Draw the graph of $x^2 y$ against x.

b Use your graph to estimate the value of a and the value of b.

An alternate method for obtaining a straight-line graph for the equation $y = \dfrac{a + bx}{x^2}$ is to plot xy on the vertical axis and $\dfrac{1}{x}$ on the horizontal axis.

c Without drawing a second graph, estimate the gradient and the intercept on the vertical axis of this graph.

Answers

a First make a table of values for of $x^2 y$ and x:

x	2	4	6	8	10
$x^2 y$	19	35.0	51.1	67.2	83

The graph of $x^2 y$ against x is:

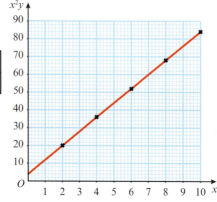

CONTINUED

b Using $y = \dfrac{a + bx}{x^2}$

$$x^2 y = bx + a$$

Now compare $x^2 y = bx + a$ with $Y = mX + c$:

$$\underset{Y}{\boxed{x^2y}} = \underset{m}{b} \; \underset{X}{\boxed{x}} + \underset{c}{a}$$

Hence, b = gradient = $\dfrac{83 - 19}{10 - 2} = 8$ and a = y-intercept = 3

$a = 3$ and $b = 8$

c Using $y = \dfrac{a + bx}{x^2}$

$$xy = \dfrac{a + bx}{x}$$

$$xy = \dfrac{a}{x} + b$$

Now compare $xy = \dfrac{a}{x} + b$ with $Y = mX + c$:

$$\underset{Y}{\boxed{xy}} = \underset{m}{a} \times \underset{X}{\boxed{\dfrac{1}{x}}} + \underset{c}{b}$$

Gradient = $a = 3$ and intercept on vertical axis = $b = 8$.

Exercise 6.7

1

x	0.5	1.0	1.5	2.0	2.5
y	1.00	3.00	3.67	4.00	4.20

The table shows experimental values of the variables x and y.

a Copy and complete the following table.

x					
xy					

b Draw the graph of xy against x.

c Express y in terms of x.

d Find the value of x and the value of y for which $xy = 2$.

2

x	0.1	0.2	0.3	0.4	0.5
y	0.111	0.154	0.176	0.189	0.200

The table shows experimental values of the variables x and y.

a Copy and complete the following table.

$\frac{1}{x}$				
$\frac{1}{y}$				

b Draw the graph of $\frac{1}{y}$ against $\frac{1}{x}$.

c Express y in terms of x.

d Find the value of x when $y = 0.16$.

3

x	1	2	3	4	5
y	12.8	7.6	6.4	6.2	6.4

The table shows experimental values of the variables x and y.

a Draw the graph of xy against x^2.

b Use your graph to express y in terms of x.

c Find the value of x and the value of y for which $xy = 12.288$.

4 The mass, m grams, of a radioactive substance is given by the formula $m = m_0 e^{-kt}$, where t is the time in days after the mass was first recorded and m_0 and k are constants.

The table below shows experimental values of t and m.

t	10	20	30	40	50
m	40.9	33.5	27.4	22.5	18.4

a Draw the graph of $\ln m$ against t.

b Use your graph to estimate the value of m_0 and k.

c Estimate the value of m when $t = 27$.

5

x	10	100	1000	10000
y	15	75	378	1893

The table shows experimental values of the variables x and y.

The variables are known to be related by the equation $y = kx^n$ where k and n are constants.

a Draw the graph of $\lg y$ against $\lg x$.

b Use your graph to estimate the value of k and n.

6

x	2	4	6	8	10
y	12.8	32.8	83.9	214.7	549.8

The table shows experimental values of the variables x and y.

The variables are known to be related by the equation $y = a \times b^x$ where a and b are constants.

a Draw the graph of $\lg y$ against x.

b Use your graph to estimate the value of a and the value of b.

7

x	2	4	6	8	10
y	4.9	13.3	36.2	98.3	267.1

The table shows experimental values of the variables x and y.

The variables are known to be related by the equation $y = a \times e^{nx}$ where a and n are constants.

a   Draw the graph of $\ln y$ against x.

b   Use your graph to estimate the value of a and the value of n.

8

x	2	4	6	8	10
y	30.0	44.7	66.7	99.5	148.4

The table shows experimental values of the variables x and y.

The variables are known to be related by the equation $y = e^{ax+b}$ where a and b are constants.

a   Draw the graph of $\ln y$ against x.

b   Use your graph to estimate the value of a and the value of b.

c   Estimate the value of x when $y = 50$.

9

x	2	4	6	8	10
y	0.10	0.33	1.08	3.48	11.29

The table shows experimental values of the variables x and y.

The variables are known to be related by the equation $y = 10^a \times b^x$, where a and b are constants.

a   Draw the graph of $\lg y$ against x.

b   Use your graph to estimate the value of a and the value of b.

c   Estimate the value of x when $y = 5$.

10

x	0.2	0.4	0.5	0.7	0.9
y	36	12	9	6	4.5

The table shows experimental values of the variables x and y.

The variables are known to be related by the equation $y = \dfrac{a}{x + b}$, where a and b are constants.

a   Draw the graph of y against xy.

b   Use your graph to estimate the value of a and the value of b.

An alternate method for obtaining a straight-line graph for the equation

$y = \dfrac{a}{x + b}$ is to plot x on the vertical axis and $\dfrac{1}{y}$ on the horizontal axis.

c   Without drawing a second graph, estimate the gradient and the intercept on the vertical axis of this graph.

11

x	2	5	15	25	60
y	11.5	5.54	2.30	1.53	0.76

The table shows experimental values of the variables x and y.

a Draw the graph of $\ln y$ against $\ln x$.

b Express y in terms of x.

An alternate method for obtaining the relationship between x and y is to plot $\lg y$ on the vertical axis and $\lg x$ on the horizontal axis.

c Without drawing a second graph, find the gradient and the intercept on the vertical axis of this graph.

REFLECTION

Imagine you are helping a younger brother who is studying for IGCSE Additional Mathematics. He finds it difficult to understand why and how we reduce graphs to linear form. Think about how you might help him understand.

Make notes on how you would explain this topic to him.

SUMMARY

Length of a line segment, gradient and midpoint

Length of $PQ = \sqrt{(x_2 - x_1)^2 + (y_2 - y_1)^2}$

Gradient of $PQ = \dfrac{y_2 - y_1}{x_2 - x_1}$

Midpoint of $PQ = \left(\dfrac{x_1 + x_2}{2}, \dfrac{y_1 + y_2}{2} \right)$

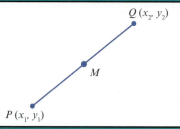

Parallel and perpendicular lines

If two lines are parallel then their gradients are equal.

If a line has a gradient of m, a line perpendicular to it has a gradient of $-\dfrac{1}{m}$.

If the gradients of the two perpendicular lines are m_1 and m_2, then $m_1 \times m_2 = -1$.

The equation of a straight line

$y = mx + c$ where m = the gradient and c = the y-intercept.

$y - y_1 = m(x - x_1)$ where m = the gradient and (x_1, y_1) is a known point on the line.

Non-linear equations

To convert a non-linear equation involving x and y into a linear equation, express the equation in the form $Y = mX + c$, where X and Y are expressions in x and/or y.

Past paper questions

Worked example

When $\lg y$ is plotted against x, a straight line graph passing through the points (2.2, 3.6) and (3.4, 6) is obtained.

i   Given that $y = Ab^x$, find the value of the constants A and b.   [5]

ii   Find x when $y = 900$.   [2]

Cambridge IGCSE Additional Mathematics 0606 Paper 11 Q7 Nov 2019

Answers

i $y = Ab^x$

Taking logarithms of both sides:

$\lg y = \lg(Ab^x)$

$\lg y = \lg A + \lg b^x$

$\lg y = \lg A + x \lg b$

$\lg y = (\lg b)x + \lg A$ (1)

Using (2.2, 3.6) gives:

$3.6 = 2.2(\lg b) + \lg A$ (2)

Using (3.4, 6) gives:

$6 = 3.4(\lg b) + \lg A$ (3)

(3) − (2) gives:

$2.4 = 1.2(\lg b)$

$\lg b = 2$

$b = 10^2$

$b = 100$

Substituting for b in (3) gives:

$6 = 3.4(2) + \lg A$

$-0.8 = \lg A$

$A = 10^{-0.8}$

$A = 0.158$ to 3 sf

ii   Using the values for A and b and substituting $y = 900$ in equation (1) gives:

$\lg 900 = 2x - 0.8$

$x = \dfrac{0.8 + \lg 900}{2}$

$x = 1.88$ to 3 sf

1 The point P lies on the line joining $A(-2, 3)$ and $B(10, 19)$ such that $AP:PB = 1:3$.

   a Show that the x-coordinate of P is 1 and find the y-coordinate of P.   [2]

   b Find the equation of the line through P which is perpendicular to AB.   [3]

The line through P which is perpendicular to AB meets the y-axis at the point Q.

   c Find the area of the triangle AQB.   [3]

Cambridge IGCSE Additional Mathematics 0606 Paper 11 Q8 Nov 2014

2 The table shows values of variables V and p.

V	10	50	100	200
p	95.0	8.5	3.0	1.1

   a By plotting a suitable straight line graph, show that V and p are related by the equation $p = kV^n$, where k and n are constants.   [4]

Use your graph to find

   b the value of n [2]

   c the value of p when $V = 35$.   [2]

Cambridge IGCSE Additional Mathematics 0606 Paper 11 Q8i,ii,iii Jun 2014

3 **Solutions to this question by accurate drawing will not be accepted.**

The points A and B have coordinates $(2, -1)$ and $(6, 5)$ respectively.

   i Find the equation of the perpendicular bisector of AB, giving your answer in the form $ax + by = c$, where a, b and c are integers.   [4]

The point C has coordinates $(10, -2)$.

   ii Find the equation of the line through C which is parallel to AB.   [2]

   iii Calculate the length of BC.   [2]

   iv Show that triangle ABC is isosceles.   [1]

Cambridge IGCSE Additional Mathematics 0606 Paper 22 Q8 Mar 2015

4 The curve $y = xy + x^2 - 4$ intersects the line $y = 3x - 1$ at the points A and B.

Find the equation of the perpendicular bisector of the line AB.   [8]

Cambridge IGCSE Additional Mathematics 0606 Paper 11 Q5 Jun 2015

5 **Solutions to this question by accurate drawing will not be accepted.**

Two points A and B have coordinates $(-3, 2)$ and $(9, 8)$ respectively.

   i Find the coordinates of C, the point where the line AB cuts the y-axis.   [3]

   ii Find the coordinates of D, the mid-point of AB.   [1]

   iii Find the equation of the perpendicular bisector of AB.   [2]

The perpendicular bisector of AB cuts the y-axis at the point E.

   iv Find the coordinates of E.   [1]

   v Show that the area of triangle ABE is four times the area of triangle ECD.   [3]

Cambridge IGCSE Additional Mathematics 0606 Paper 21 Q8 Nov 2015

6 The line AB is such that the points A and B have coordinates $(-4, 6)$ and $(2, 14)$ respectively.

   a The point C, with coordinates $(7, a)$ lies on the perpendicular bisector of AB. Find the value of a.   [4]

   b Given that the point D also lies on the perpendicular bisector of AB, find the coordinates of D such that the line AB bisects the line CD.   [2]

Cambridge IGCSE Additional Mathematics 0606 Paper 12 Q3 Mar 2021

7 When $\lg y$ is plotted against x^2 a straight line graph is obtained which passes through the points (2, 4) and (6, 16).

    i Show that $y = 10^{A+Bx^2}$, where A and B are constants.   [4]

    ii Find y when $x = \dfrac{1}{\sqrt{3}}$.   [2]

    iii Find the positive value of x when $y = 2$.   [3]

Cambridge IGCSE Additional Mathematics 0606 Paper 11 Q10 Jun 2019

8 Variables x and y are such that, when $\sqrt[4]{y}$ is plotted against $\dfrac{1}{x}$, a straight line graph passing through the points (0.5, 9) and (3, 34) is obtained. Find y as a function of x.   [4]

Cambridge IGCSE Additional Mathematics 0606 Paper 21 Q1 Jun 2020

9 The relationship between experimental values of two variables, x and y, is given by $y = Ab^x$, where A and b are constants.

    i Transform the relationship $y = Ab^x$ into straight line form.   [2]

The diagram shows $\ln y$ plotted against x for ten different pairs of values of x and y.

The line of best fit has been drawn.

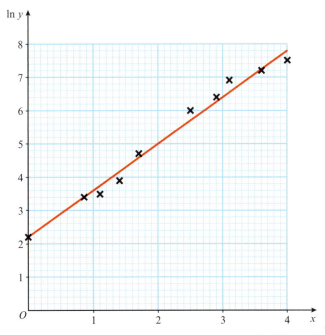

    ii Find the equation of the line of best fit and the value, correct to 1 significant figure, of A and of b.   [4]

    iii Find the value, correct to 1 significant figure, of y when $x = 2.7$.   [2]

Cambridge IGCSE Additional Mathematics 0606 Paper 22 Q6 Mar 2019

> # Chapter 7
Coordinate geometry of the circle

THIS SECTION WILL SHOW YOU HOW TO:

- understand that the equation $(x - a)^2 + (y - b)^2 = r^2$ represents the circle with centre (a, b) and radius r
- use algebraic methods to solve problems involving lines and circles.

PRE-REQUISITE KNOWLEDGE

Before you start…

Where it comes from	You should be able to…	Check your skills
Chapter 6	Find the gradient of a line and state the gradient of a line that is perpendicular to the line.	**1** **a** Find the gradient of the line joining $A(-2, 7)$ and $B(5, 4)$. **b** State the gradient of the line that is perpendicular to the line AB.
Chapter 6	Find the midpoint of a line segment.	**2** Find the midpoint of the line joining $A(-2, 8)$ and $B(6, 4)$.
Chapter 6	Interpret and use equations of lines of the form $y = mx + c$.	**3** The equation of a line is $y = \frac{4}{5}x - 8$. Write down **a** the gradient of the line **b** the y-intercept **c** the x-intercept.
Chapter 2	Complete the square and solve quadratic equations.	**4** **a** Complete the square for $x^2 - 6x + 3$. **b** Solve $x^2 - 6x + 3 = 0$.

7.1 The equation of a circle

In this section you will learn about the equation of a circle. A circle is defined as the locus of all the points in a plane that are a fixed distance (the radius) from a given point (the centre).

KEY WORDS

completed square form

general form

CLASS DISCUSSION

1 Use graphing software to draw each of the following circles. From your graphs, find the coordinates of the centre and the radius of each circle and complete your own copy of the table below.

	Equation of circle	Centre	Radius
a	$x^2 + y^2 = 4$		
b	$(x - 1)^2 + (y - 3)^2 = 1$		
c	$(x - 5)^2 + (y + 2)^2 = 25$		
d	$(x + 8)^2 + (y - 6)^2 = 49$		
e	$(x + 2)^2 + (y + 3)^2 = 9$		
f	$x^2 + (y - 3)^2 = 16$		
g	$(x + 6)^2 + y^2 = 32$		

2 Discuss your results with your classmates and explain how you can find the coordinates of the centre of a circle and the radius of a circle just by looking at the equation of the circle.

To find the equation of a circle we let $P(x, y)$ be any point on the circumference of a circle with centre $C(a, b)$ and radius r.

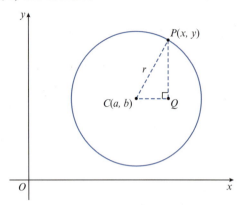

Using Pythagoras' theorem on triangle CQP gives $CQ^2 + PQ^2 = r^2$.

Substituting $CQ = x - a$ and $PQ = y - b$ into $CQ^2 + PQ^2 = r^2$ gives $(x - a)^2 + (y - b)^2 = r^2$.

> The equation of a circle with centre (a, b) and radius r can be written in **completed square form** as:
>
> $$(x - a)^2 + (y - b)^2 = r^2$$

WORKED EXAMPLE 1

Write down the coordinates of the centre and the radius of each of these circles.

a $x^2 + y^2 = 25$

b $(x + 4)^2 + (y - 7)^2 = 81$

c $(x + 2)^2 + (y + 5)^2 = 8$

Answers

a   Centre = $(0, 0)$ Radius = $\sqrt{25} = 5$

b   Centre = $(-4, 7)$ Radius = $\sqrt{81} = 9$

c   Centre = $(-2, -5)$ Radius = $\sqrt{8} = 2\sqrt{2}$

WORKED EXAMPLE 2

Find the equation of the circle with centre $(5, -4)$ and radius 3.

Answers

Equation of circle is $(x - a)^2 + (y - b)^2 = r^2$ where $a = 5$, $b = -4$ and $r = 3$.

$$(x - 5)^2 + (y - (-4))^2 = 3^2$$

$$(x - 5)^2 + (y + 4)^2 = 9$$

WORKED EXAMPLE 3

A is the point $(4, 0)$ and B is the point $(10, -6)$.

Find the equation of the circle that has AB as a diameter.

Answers

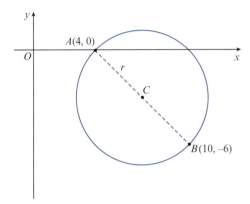

The centre of the circle, C, is the midpoint of AB.

$$C = \left(\frac{4 + 10}{2}, \frac{0 + (-6)}{2} \right) = (7, -3)$$

Radius of circle, r, is equal to AC.

$$r = \sqrt{(7 - 4)^2 + (-3 - 0)^2} = \sqrt{18}$$

Equation of circle is $(x - a)^2 + (y - b)^2 = r^2$ where $a = 7, b = -3$ and $r = \sqrt{18}$.

$$(x - 7)^2 + (y - (-3))^2 = (\sqrt{18})^2$$

$$(x - 7)^2 + (y + 3)^2 = 18$$

Expanding the equation $(x - a)^2 + (y - b)^2 = r^2$ gives:

$$x^2 - 2ax + a^2 + y^2 - 2by + b^2 = r^2$$

Rearranging gives:

$$x^2 + y^2 - 2ax - 2by + (a^2 + b^2 - r^2) = 0$$

When the equation of a circle is written in this form, some of the important characteristics of the equation of a circle are highlighted. For example

* the coefficients of x^2 and y^2 are equal
* there is no xy term.

We often write the expanded form of a circle as follows:

> The expanded form of a circle is $x^2 + y^2 + 2gx + 2fy + c = 0$ where $(-g, -f)$ is the centre and $\sqrt{g^2 + f^2 - c}$ is the radius.
>
> This is the equation of a circle in expanded **general form**.

> **WORKED EXAMPLE 4**
>
> Find the centre and the radius of the circle $x^2 + y^2 - 8x + 6y - 11 = 0$.
>
> **Answers**
>
> $x^2 - 8x + y^2 + 6y - 11 = 0$ complete the squares
>
> $(x - 4)^2 - 4^2 + (y + 3)^2 - 3^2 - 11 = 0$ collect constant terms together
>
> $(x - 4)^2 + (y + 3)^2 = 36$ compare with $(x - a)^2 + (y - b)^2 = r^2$
>
> $a = 4$ $b = -3$ $r^2 = 36$
>
> Centre $= (4, -3)$ and radius $= 6$

It is useful to remember the three following right-angle facts for circles.

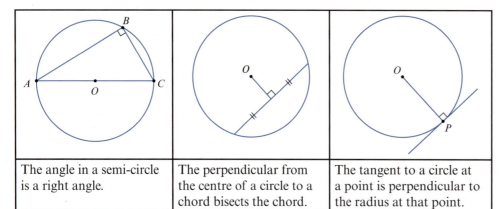

The angle in a semi-circle is a right angle.	The perpendicular from the centre of a circle to a chord bisects the chord.	The tangent to a circle at a point is perpendicular to the radius at that point.

From these statements you can conclude that:

* If triangle ABC is right-angled at B, then the points A, B and C lie on the circumference of a circle with AC as diameter.

* The perpendicular bisector of a chord passes through the centre of the circle.

* If a radius and a line at a point P on the circumference are at right angles, then the line must be a tangent to the curve.

> **CLASS DISCUSSION**
>
> **1**   Sara says:
>
> 'Given two distinct points in the plane, there is a unique circle that passes through these two points.'
>
> Discuss with your classmates whether this statement is
>
> * ALWAYS true
> * SOMETIMES true
> * NEVER true
>
> **2**   If you are given two distinct points on the circumference of a circle, what can you say about the position of the centre of the circle?

WORKED EXAMPLE 5

A circle passes through the points $P(-1, -1)$ and $Q(7, -5)$ and has radius 5.

Find the two possible equations for this circle.

Answers

Midpoint of $PQ = \left(\dfrac{-1 + 7}{2}, \dfrac{-1 + (-5)}{2} \right) = (3, -3)$

Gradient of $PQ = \dfrac{-5 - (-1)}{7 - (-1)} = -\dfrac{1}{2}$

Gradient of perpendicular bisector of $PQ = 2$.

Equation of perpendicular bisector of PQ:

$$y = 2x + c$$

$$-3 = 2(3) + c$$

$$c = -9$$

$$y = 2x - 9$$

The center (a, b) lies on the perpendicular bisector of the line PQ.

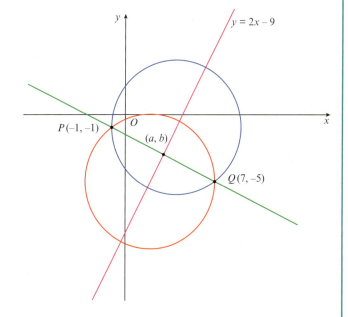

Hence $b = 2a - 9$ (1)

Using $P(-1, -1)$, P lies on the circle:

$$(x - a)^2 + (y - b)^2 = 5^2$$

$$(-1 - a)^2 + (-1 - b)^2 = 5^2$$

$$a^2 + 2a + 1 + b^2 + 2b + 1 = 25$$

$$a^2 + b^2 + 2a + 2b = 23 \qquad (2)$$

Substituting for b from (1) in (2):

$$a^2 + (2a - 9)^2 + 2a + 2(2a - 9) = 23$$

$$a^2 + 4a^2 - 36a + 81 + 2a + 4a - 18 = 23$$

$$5a^2 - 30a + 40 = 0$$

$$a^2 - 6a + 8 = 0$$

$$(a - 2)(a - 4) = 0$$

$$a = 2 \text{ or } a = 4$$

When $a = 2$, $b = -5$ and when $a = 4$, $b = -1$

Hence the equations of the circles are:

$(x - 2)^2 + (y + 5)^2 = 25$ and $(x - 4)^2 + (y + 1)^2 = 25$

CLASS DISCUSSION

1 Philip says:

'Given three distinct non-linear points in the plane, there is a unique circle that passes through these three points.'

Discuss with your classmates whether this statement is

- ALWAYS true
- SOMETIMES true
- NEVER true

2 Given three points on the circumference of a circle, what can you say about the position of the centre of the circle?

WORKED EXAMPLE 6

A circle passes through the points $P(-1, 4)$, $Q(1, 6)$ and $R(5, 4)$.

Find the equation of the circle.

Answers

The centre of the circle lies on the perpendicular bisector of PQ and on the perpendicular bisector of QR.

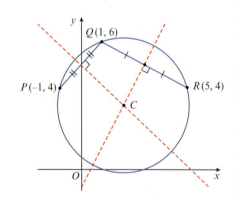

Midpoint of $PQ = \left(\dfrac{-1 + 1}{2}, \dfrac{4 + 6}{2}\right) = (0, 5)$

Gradient of $PQ = \dfrac{6 - 4}{1 - (-1)} = 1$

Gradient of perpendicular bisector of $PQ = -1$

Equation of perpendicular bisector of PQ is:

$y = -x + c$

$5 = -1(0) + c$

$c = 5$

$y = -x + 5$ (1)

Midpoint of $QR = \left(\dfrac{1 + 5}{2}, \dfrac{6 + 4}{2}\right) = (3, 5)$

Gradient of $QR = \dfrac{4 - 6}{5 - 1} = -\dfrac{1}{2}$

Gradient of perpendicular bisector of $QR = 2$

Equation of perpendicular bisector of QR is:

$y = 2x + d$

$5 = 2(3) + d$

$d = -1$

$y = 2x - 1$ (2)

CONTINUED

Solving equations (1) and (2) gives

$x = 2, y = 3$

Centre of circle = (2, 3)

Radius = $CR = \sqrt{(5-2)^2 + (4-3)^2} = \sqrt{10}$

Hence, the equation of the circle is $(x-2)^2 + (y-3)^2 = 10$

CLASS DISCUSSION

The method used in **Worked example 6** involves a geometric approach. This involves finding the intersection of two perpendicular bisectors.

Vivi says that she can solve the problem using an algebraic approach, which involves writing down and solving three simultaneous equations.

1   Discuss with your classmates what Vivi's three equations might be.

2   Working in small groups, solve your three equations and compare your answers with the answers obtained in **Worked example 6**.

3   Discuss with your classmates which method you prefer.

Exercise 7.1

1   Find the centre and the radius of each of the following circles.

   a $x^2 + y^2 = 25$
   b $x^2 + y^2 = 81$
   c $x^2 + y^2 = 48$
   d $2x^2 + 2y^2 = 64$
   e $x^2 + (y-3)^2 = 1$
   f $(x-5)^2 + y^2 = 9$
   g $(x+7)^2 + y^2 = 18$
   h $(x+1)^2 + (y-2)^2 = 100$
   i $(x-2)^2 + (y+4)^2 = 9$
   j $(x+3)^2 + (y+4)^2 = 27$

2   Find the equation of each of the following circles.

   a   centre (0, 0), radius 4
   b   centre (3, −1), radius 6
   c   centre (2, −5), radius $2\sqrt{5}$
   d   centre $\left(-\frac{1}{2}, -\frac{3}{2}\right)$, radius 4

3   Find the centre and the radius of each of the following circles.

   a $x^2 + y^2 - 4x + 10y + 25 = 0$
   b $x^2 + y^2 + 14x - 8y + 40 = 0$
   c $x^2 + y^2 + 4x + 12y + 4 = 0$
   d $x^2 + y^2 - 2x - 16y + 33 = 0$
   e $4x^2 + 4y^2 - 4x - 16y + 1 = 0$
   f $4x^2 + 4y^2 + 24x - 12y + 13 = 0$

4   Find the equation of the circle with centre (3, 2) passing through the point (7, 5).

5   A diameter of a circle has its end points at $A(5, -3)$ and $B(15, 5)$. Find the equation of the circle.

6   Sketch the circle $(x-4)^2 + (y-5)^2 = 16$.

7   Find the equation of the circle that touches the y-axis and whose centre is (−3, 2).

8   Show that $x^2 + y^2 + 8x - 14y = 10$ can be written in the form $(x - a)^2 + (y - b)^2 = r^2$ where a, b and r are constants to be found. Hence, write down the coordinates of the centre of the circle and also the radius of the circle.

9   The points $A(-9, -1)$ and $B(3, -5)$ lie on the circumference of a circle. Show that the centre of the circle lies on the line $y = 3x + 6$.

10 Find the equation of the circle that passes through the point $(5, -2)$ and has the same centre as the circle $x^2 + y^2 - 4x + 8y - 40 = 0$.

11 A circle passes through the points $P(-2, 0)$, $Q(3, 5)$ and $R(6, 2)$. Show that PR is a diameter of the circle and find the equation of this circle.

12 A circle passes through the points $(-1, -4)$ and $(3, 2)$ and has radius $\sqrt{26}$. Find the two possible equations for this circle.

13 A circle has radius 5 units and passes through the point $(2, -6)$. The y-axis is a tangent to the circle. Find the possible equations of the circle.

14 The equation of a circle is $(x - 4)^2 + (y + 3)^2 = 41$. Show that the point $P(8, 2)$ lies on the circle and find the equation of the tangent to the circle at the point P.

15 Find the equation of the circle which passes through the points $(7, 4)$ and $(0, 5)$ and has its centre lying on the line $x + 2y = 5$.

16 The points $A(-1, -4)$, $B(10, 7)$ and $C(13, 4)$ are joined to form a triangle.

   a   Show that angle ABC is a right angle.

   b   Find the equation of the circle that passes through the points A, B and C.

17 For each set of three points, find the equation of the circle that passes through them:

   a $(4, 3), (4, 7), (6, 9)$

   b $(0, 5), (5, 0), (0, -7)$

   c $(0, 0), (4, 2), (4, 8)$

   d $(3, 4), (1, -2), (6, 3)$

   e $(-6, 5), (-7, -2), (-3, 6)$

REFLECTION

Did you draw a sketch for each part of Question 17? Might you have found some parts easier if you had drawn a sketch first?

18

   Use graphing software to draw the Olympic rings.

19 The design is made from four red circles and one blue circle.

   The radius of each red circle is 1 unit.

   a   Find the radius of the blue circle.

   b   Use graphing software to draw the design.

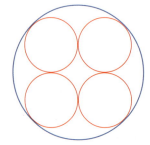

20 CHALLENGE QUESTION

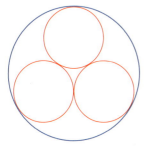

The design is made from 3 red circles and 1 blue circle.

The radius of each red circle is 1 unit.

Use graphing software to draw the design.

21 CHALLENGE QUESTION

a   Use each of the following methods to show that the four points $A(-4, -2)$, $B(-2, 2)$, $C(5, 1)$ and $D(4, -6)$ lie on a circle.

> Methods

- Use the property that the perpendicular bisector of two chords on a circle intersect at the centre of the circle to find the equation of the circle passing through three of the points and then show that the fourth point lies on this circle.

- Use three of the points to write down three simultaneous equations. Solve the equations and then show that the fourth point lies on this circle.

- Show that opposite angles in the quadrilateral $ABCD$ add up to 180° which means that $ABCD$ is a cyclic quadrilateral. Hence the four points lie on a circle.

- Show that angle $BAD = 90°$. Hence BD must be a diameter of the circle (angles in a semi-circle) and the centre of the circle is the midpoint of BD. Then show that the fourth point , C, lies on this circle.

b   Find out about Ptolemy's theorem and use it to show that the points A, B, C and D lie on a circle.

REFLECTION

Look back at your five methods in Question 21 for showing that A, B, C and D lie on a circle.

Which method do you consider to be easiest for this particular question?

Which method do you consider to be hardest for this particular question?

Which method do you consider to be most efficient for this particular question?

What advice would you give someone for tackling a question of this nature?

7.2 Problems involving the intersection of lines and circles

In **Chapter 2** you learned that the points of intersection of a line and a curve can be found by solving their equations simultaneously. You also learned that if the resulting equation is of the form $ax^2 + bx + c = 0$, then $b^2 - 4ac$ gives information about the line and the curve.

$b^2 - 4ac$	Nature of roots	Line and parabola
> 0	2 distinct real roots	two distinct points of intersection
$= 0$	2 equal real roots	one point of intersection (line is a tangent)
< 0	no real roots	no points of intersection

In this section you will solve problems involving the intersection of lines and circles.

When given the equations of two circles, their radii can be used to determine whether they intersect, touch or do not meet at all.

We will now consider each of the possible situations.

Non-intersecting circles

Case 1

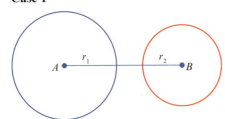

The distance between the centres is greater than the sum of the radii.

$AB > r_1 + r_2$

Case 2

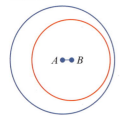

$AB + r_2 < r_1$

Hence $AB < r_1 - r_2$

This means that the distance between the centres is less than the difference between the lengths of the two radii.

Touching circles

Case 1

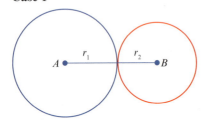

The distance between the centres is the same as the sum of the radii.

$AB = r_1 + r_2$

Case 2

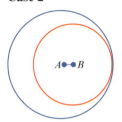

The distance between the centres is the same as the difference between the radii.

$AB = r_1 - r_2$

Intersecting circles

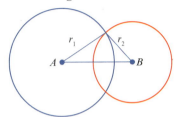

The distance between the centres is less than the sum of the radii.

$AB < r_1 + r_2$

Common chords

When two circles intersect at two points, the line joining the two points of intersection is called the common chord and the length of this chord can be found using Pythagoras' theorem.

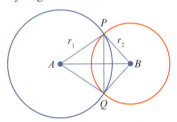

WORKED EXAMPLE 7

The line $x + 2y = 10$ intersects the circle $(x - 2)^2 + (y + 1)^2 = 25$ at the points A and B.

a   Find the coordinates of the points A and B.

b   The perpendicular bisector of AB intersects the circle at the points P and Q.
Find the exact coordinates of P and Q.

Answers

a $\qquad (x - 2)^2 + (y + 1)^2 = 25$ substitute $10 - 2y$ for x

$\quad (10 - 2y - 2)^2 + (y + 1)^2 = 25$

$\qquad (8 - 2y)^2 + (y + 1)^2 = 25$ expand and simplify

$\qquad\qquad 5y^2 - 30y + 40 = 0$

$\qquad\qquad\quad y^2 - 6y + 8 = 0$ factorise

$\qquad\qquad (y - 2)(y - 4) = 0$

$\qquad\qquad\qquad y = 2$ or $y = 4$

When $y = 2$, $x = 6$ and when $y = 4$, $x = 2$

A and B are the points $(2, 4)$ and $(6, 2)$

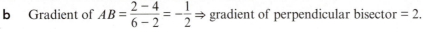

CONTINUED

b   Gradient of $AB = \dfrac{2-4}{6-2} = -\dfrac{1}{2} \Rightarrow$ gradient of perpendicular bisector = 2.

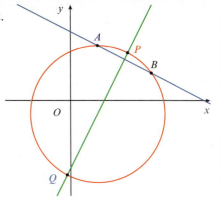

Midpoint of $AB = \left(\dfrac{2+6}{2}, \dfrac{4+2}{2}\right) = (4, 3)$

$y = mx + c$ use $m = 2$, $x = 4$ and $y = 3$

$3 = 2(4) + c$

$c = -5$

Perpendicular bisector of AB is $y = 2x - 5$.
Substitute $2x - 5$ for y in equation of circle:

$$(x - 2)^2 + (y + 1)^2 = 25$$

$$(x - 2)^2 + (2x - 4)^2 = 25$$

$$5x^2 - 20x - 5 = 0$$

$$x^2 - 4x - 1 = 0$$

$$x = \frac{-(-4) \pm \sqrt{(-4)^2 - 4(1)(-1)}}{2(1)}$$

$$x = \frac{4 \pm \sqrt{20}}{2}$$

$$x = 2 \pm \sqrt{5}$$

When $x = 2 + \sqrt{5}$, $y = 2\sqrt{5} - 1$

When $x = 2 - \sqrt{5}$, $y = -1 - 2\sqrt{5}$

P and Q are the points $(2 + \sqrt{5}, 2\sqrt{5} - 1)$ and $(2 - \sqrt{5}, -1 - 2\sqrt{5})$

WORKED EXAMPLE 8

Show that the line $x + 2y = 30$ is a tangent to the circle $x^2 + y^2 - 8x - 6y = 55$.

Answers

$$x^2 + y^2 - 8x - 6y = 55 \qquad \text{substitute } 30 - 2y \text{ for } x$$

$$(30 - 2y)^2 + y^2 - 8(30 - 2y) - 6y = 55 \qquad \text{expand and simplify}$$

$$5y^2 - 110y + 605 = 0$$

$$y^2 - 22y + 121 = 0 \qquad \text{factorise}$$

$$(y - 11)(y - 11) = 0$$

$$y = 11 \text{ or } y = 11$$

The equation has one repeated root, hence $x + 2y = 30$ is a tangent.

WORKED EXAMPLE 9

Show that the circle $x^2 + y^2 = 25$ lies inside the circle $x^2 + y^2 - 8x = 84$.

Answer

By inspection the circle $x^2 + y^2 = 25$ has centre $(0, 0)$ and radius 5.

$$x^2 + y^2 - 8x = 84 \qquad \text{complete the square}$$

$$(x - 4)^2 + y^2 - (-4)^2 = 84$$

$$(x - 4)^2 + y^2 = 100$$

Hence the circle $x^2 + y^2 - 8x = 84$ has centre $(4, 0)$ and radius 10.

The centre of each circle lies on the x-axis and the distance between these centres is 4.

The difference between the lengths of the radii $= 10 - 5 = 5$.

The distance between the centres is less than the difference between the lengths of the radii.

Hence the circle $x^2 + y^2 = 25$ lies inside the circle $x^2 + y^2 - 8x = 84$.

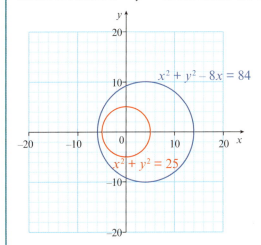

Exercise 7.2

1   Find the points of intersection of the line and the circle:

   a $y = 4$ and $x^2 + y^2 - 4x - 6y + 8 = 0$

   b $y = x - 1$ and $x^2 + y^2 + 5x - y - 6 = 0$

   c $y = 2x + 3$ and $x^2 + y^2 - 4x + 6y - 12 = 0$

   d $y = 2x + 1$ and $x^2 + y^2 + 5x - 3y - 12 = 0$

2   By considering the radii of each circle and the distance between their centres, determine whether each pair of circles intersect, touch or do not intersect.

   a $x^2 + y^2 + 25$ and $x^2 + y^2 - 15x + 8y - 75 = 0$

   b $x^2 + y^2 + 16$ and $x^2 + y^2 + 3x - 5y - 6 = 0$

   c $x^2 + y^2 + 25$ and $x^2 + y^2 - 24x - 18y + 125 = 0$

   d $(x - 8)^2 + y^2 = 4$ and $(x - 2)^2 + (y + 5)^2 = 1$

   e $x^2 + (y - 3)^2 = 9$ and $(x - 8)^2 + (y - 5)^2 = 144$

3   The line $y = x - 3$ intersects the circle $(x - 3)^2 + (y + 2)^2 = 20$ at two points P and Q. Find the length of PQ.

4   Explain why the line $4y = x + 26$ is a tangent to the circle $x^2 + y^2 + 10x - 2y + 9 = 0$ and find the coordinates of the point where the line touches the curve.

5   Explain why the circles $(x + 7)^2 + (y + 2)^2 = 4$ and $(x + 7)^2 + (y + 5)^2 = 16$ touch at one point. Determine whether they touch internally or externally and find the coordinates of the point at which they touch.

6   The circles $(x - 10)^2 + (y - 5)^2 = 25$ and $(x - 20)^2 + (y - 10)^2 = 100$ intersect at the points P and Q.

   a   Find the length of the line PQ.

   b   Find the equation of the common chord PQ.

7   Show that the line $y = 2x - 1$ and the circle $x^2 + y^2 + 15x - 11y + 9 = 0$ do not intersect.

8   The equation of a circle is $x^2 + y^2 - 6x - 8y = 0$.

   a   Show that the point $P(7, 7)$ lies on this circle.

   b   Find the equation of the tangent to the circle at the point P.

9   The line $2x + y = 6$ intersects the circle $x^2 + y^2 - 12x - 8y + 27 = 0$ at the points A and B.

   a   Find the coordinates of the points A and B.

   b   Find the equation of the perpendicular bisector of AB.

   c   The perpendicular bisector of AB intersects the circle at the points P and Q. Find the exact coordinates of P and Q.

   d   Find the exact area of quadrilateral $APBQ$.

10 Show that these pairs of circles touch each other and find the coordinates of the point where they touch. In each case, sketch the circles to show how they touch.

   a $x^2 + y^2 - 8x - 4y + 19 = 0$ and $(x - 5)^2 + (y - 2)^2 = 4$

   b $x^2 + y^2 + 6x - 4y + 3 = 0$ and $x^2 + y^2 - 12x + 2y - 3 = 0$

11 **CHALLENGE QUESTION**

   Find the set of values of m for which the line $y = mx + 3$ intersects the circle $x^2 + y^2 - 10x - 8y + 28 = 0$ at two distinct points.

12 CHALLENGE QUESTION

Two circles have the following properties:

- the x-axis is a common tangent to the circles
- the point $(14, 2)$ lies on both circles
- the centre of each circle lies on the line $x + 2y = 28$

Find the equation of each circle.

SUMMARY

The equation of a circle is
$(x - a)^2 + (y - b)^2 = r^2$ where (a, b) is the centre and r is the radius.
$x^2 + y^2 + 2gx + 2fy + c = 0$ where $(-g, -f)$ is the centre and $\sqrt{g^2 + f^2 - c}$ is the radius.

Practice questions

Worked example

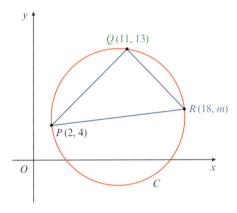

The points $P(2, 4)$, $Q(11, 13)$ and $R(18, m)$ lie on the circle C.

PR is a diameter of the circle.

a   Show that $m = 6$.   [3]

b   Find the equation of circle C.   [5]

Practice question

Answers

a   Triangle PQR is right-angled at Q.

$$PR^2 = PQ^2 + PR^2$$

$$(18 - 2)^2 + (m - 4)^2 = [(11 - 2)^2 + (13 - 4)^2] + [(18 - 11)^2 + (m - 13)^2]$$

$$256 + m^2 - 8m + 16 = 81 + 81 + 49 + m^2 - 26m + 169$$

$$272 - 8m = 380 - 26m$$

$$18m = 108$$

$$m = 6$$

b $P(2, 4)$ and $R(18, 6)$

Midpoint of $PR = \left(\dfrac{2 + 18}{2}, \dfrac{4 + 6}{2}\right) = (10, 5)$

$$PR^2 = (18 - 2)^2 + (6 - 4)^2$$
$$PR^2 = 260$$
$$PR = 2\sqrt{65}$$

Radius of circle $\dfrac{1}{2}PR = \sqrt{65}$

Equation of circle is:

$(x - a)^2 + (y - b)^2 = r^2$ where $a = 10$, $b = 5$ and $r^2 = 65$

$(x - 10)^2 + (y - 5)^2 = 65$

1 The line $2y = mx + 15$ intersects the circle $x^2 + y^2 - 4x - 2y - 69 = 0$ at the point $P(9, 6)$.

    a Find the coordinates of the point Q where the line meets the curve again. [4]

    b Find the equation of the perpendicular bisector of the line PQ. [3]

    c Find the x-coordinates of the points where this perpendicular bisector intersects the circle.
Give your answers in exact form. [4]

Practice question

2 The points $A(2, 2)$ and $B(7, 0)$ lie on a circle with centre $C(p, p - 5)$.

    a Find the equation of the perpendicular bisector of the line segment AB. [4]

    b Use your answer to **part a** to find the value of p. [1]

    c Find the equation of the circle. [4]

Practice question

3 A is the point $(-3, 4)$ and B is the point $(1, 12)$.

    a Find the equation of the line through A and B. [3]

    b Show that the perpendicular bisector of the line AB is $x + 2y = 15$. [3]

    c A circle passes through A and B and has its centre on the line $x = 9$.
Find the equation of this circle. [4]

Practice question

4 A diameter of a circle has its endpoints at $P(0, -9)$ and $Q(10, 1)$.
Find the equation of the circle. [4]

Practice question

5 A circle C has centre $A(3, 4)$ and passes through the point $B(7, 10)$.

    a Find the length AB giving your answer in exact form. [2]

    b Write down the equation of circle C. [2]

The line L is a tangent to circle C at the point B.

    c Find the equation for L, giving your answer in the form $ax + by + c = 0$ where a, b and c are integers. [4]

Practice question

6   The equation of a circle is $x^2 + y^2 - 4x - 10y - 21 = 0$.

    a   Find the radius of the circle and the coordinates of its centre.   [4]

    b   Find the y-coordinates of the points where the circle crosses the y-axis, giving your answers
        in exact form.   [3]

Practice question

7   Prove that the circles $x^2 + y^2 = 25$ and $x^2 + y^2 - 24x - 18y + 125 = 0$ touch each other and state the
coordinates of the point where the two circles touch.   [6]

Practice question

› Chapter 7

Baby Circle

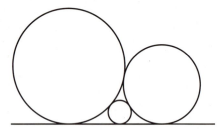

A circle with radius 1 and a circle with radius 2 touch at a point. A third circle fits between these two circles so that all three touch each other and all three have a common tangent.

What is the radius of the smallest circle?

> Chapter 8
Circular measure

THIS SECTION WILL SHOW YOU HOW TO:

- use radian measure
- solve problems involving the arc length and sector area of a circle.

Before you start…

Where it comes from	What you should be able to do	Check your skills
Cambridge IGCSE/O Level Mathematics	Calculate the length of an arc of a circle and the area of a sector of a circle.	**1** **a** Find the length of arc AB. **b** Find the area of sector OAB. 

8.1 Circular measure

Have you ever wondered why there are 360 degrees in one complete revolution?

The original reason for choosing the degree as a unit of angular measure is not known, but there are a number of different theories:

- ancient astronomers claimed that the Sun advanced in its path by one degree each day and that a solar year consisted of 360 days

- the ancient Babylonians divided the circle into 6 equilateral triangles and then subdivided each angle at O into 60 further parts, resulting in 360 divisions in one complete revolution

- 360 has many factors which makes division of the circle so much easier.

Degrees are not the only way in which you can measure angles. In this section you will learn how to use **radian** measure. This is sometimes referred to as the natural unit of angular measure and it is used extensively in mathematics because it can simplify many formulae and calculations.

In the diagram, the magnitude of angle AOB is 1 radian (1 radian is written as 1 rad or 1^c).

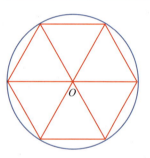

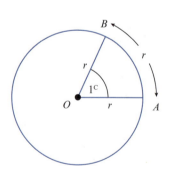

> An arc equal in length to the radius of a circle subtends an angle of 1 radian at the centre.

It follows that the circumference (an arc of length $2\pi r$) subtends an angle of 2π radians at the centre.

$$2\pi \text{ radians} = 360°$$

> $$\pi \text{ radians} = 180°$$

When an angle is written in terms of π, the radian symbol is usually omitted.

Hence, $\pi = 180°$.

Converting from degrees to radians

Since $180° = \pi$, then $90° = \dfrac{\pi}{2}$, $45° = \dfrac{\pi}{4}$, etc.

Angles that are not simple fractions of $180°$ can be converted using the following rule:

> To convert from degrees to radians, multiply by $\dfrac{\pi}{180}$.

Converting from radians to degrees

Since $\pi = 180°$, $\dfrac{\pi}{6} = 30°$, $\dfrac{\pi}{10} = 18°$ etc.

Angles that are not simple fractions of π can be converted using the following rule:

> To change from radians to degrees, multiply by $\dfrac{180}{\pi}$.

(It is useful to remember that 1 radian $= 1 \times \dfrac{180}{\pi} \approx 57°$)

WORKED EXAMPLE 1

a   Change $60°$ to radians, giving your answer in terms of π.

b   Change $\dfrac{3\pi}{5}$ radians to degrees.

Answers

a   **Method 1:**

$$180° = \pi \text{ radians}$$

$$\left(\dfrac{180}{3}\right)° = \dfrac{\pi}{3} \text{ radians}$$

$$60° = \dfrac{\pi}{3} \text{ radians}$$

Method 2:

$$60° = \left(60 \times \dfrac{\pi}{180}\right) \text{ radians}$$

$$60° = \dfrac{\pi}{3} \text{ radians}$$

b   **Method 1:**

$$\pi \text{ radians} = 180°$$

$$\dfrac{\pi}{5} \text{ radians} = 36°$$

$$\dfrac{3\pi}{5} \text{ radians} = 108°$$

Method 2:

$$\dfrac{3\pi}{5} \text{ radians} = \left(\dfrac{3\pi}{5} \times \dfrac{180}{\pi}\right)°$$

$$\dfrac{3\pi}{5} \text{ radians} = 108°$$

Exercise 8.1

1   Convert these angles to radians, in terms of π.

a $10°$	**b** $20°$	**c** $40°$	**d** $50°$	**e** $15°$
f $120°$	**g** $135°$	**h** $225°$	**i** $360°$	**j** $720°$
k $80°$	**l** $300°$	**m** $9°$	**n** $75°$	**o** $210°$

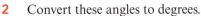
2 Convert these angles to degrees.

a $\dfrac{\pi}{2}$ b $\dfrac{\pi}{6}$ c $\dfrac{\pi}{12}$ d $\dfrac{\pi}{9}$ e $\dfrac{2\pi}{3}$

f $\dfrac{4\pi}{5}$ g $\dfrac{7\pi}{10}$ h $\dfrac{5\pi}{12}$ i $\dfrac{3\pi}{20}$ j $\dfrac{9\pi}{10}$

k $\dfrac{6\pi}{5}$ l 3π m $\dfrac{7\pi}{4}$ n $\dfrac{8\pi}{3}$ o $\dfrac{9\pi}{2}$

3 Write each of these angles in radians, correct to 3 sf.

a $32°$ b $55°$ c $84°$ d $123°$ e $247°$

4 Write each of these angles in degrees correct to 1 decimal place.

a 1.3 rad b 2.5 rad c 1.02 rad d 1.83 rad e 0.58 rad

5 Copy and complete the tables, giving your answers in terms of π.

a

Degrees	0	45	90	135	180	225	270	315	360
Radians	0				π				2π

b

Degrees	0	30	60	90	120	150	180	210	240	270	300	330	360
Radians	0						π						2π

6 Use your calculator to find

a $\sin 1.3\,\text{rad}$ b $\tan 0.8\,\text{rad}$ c $\sin 1.2\,\text{rad}$

d $\sin\dfrac{\pi}{2}$ e $\cos\dfrac{\pi}{3}$ f $\tan\dfrac{\pi}{4}$

TIP

In question 6, you don't need to convert each angle to degrees. You should set the angle mode on your calculator to radians.

---

7 CHALLENGE QUESTION

Anna is told the size of angle BAC in degrees and she is then asked to calculate the length of the line segment BC. She uses her calculator but forgets that her calculator is in radian mode. Luckily, she still manages to obtain the correct answer.

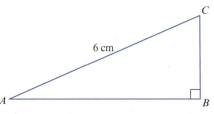

Given that angle BAC is between $10°$ and $15°$, use graphing software to help you find the size of angle BAC in degrees correct to 2 decimal places.

CLASS DISCUSSION

You should already be familiar with the following mathematical words that are used in circle questions.

(chord)   (arc)   (sector)   (segment)

Discuss and explain, with the aid of diagrams, the meaning of each of these words.

CONTINUED

Explain what is meant by:

- minor arc and major arc
- minor sector and major sector
- minor segment and major segment.

If you know the radius, r cm, and the angle θ (in degrees) at the centre of the circle, describe how you would find:

- arc length
- perimeter of sector
- perimeter of segment
- area of sector
- length of chord
- area of segment.

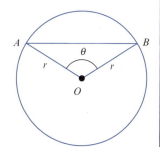

8.2 Length of an arc

From the definition of a radian, the arc that subtends an angle of 1 radian at the centre of a circle is of length r. Hence, if an arc subtends an angle of θ radians at the centre, the length of the arc is $r\theta$.

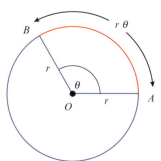

Arc length = $r\theta$

WORKED EXAMPLE 2

An arc subtends an angle of $\dfrac{\pi}{6}$ radians at the centre of a circle with radius 8 cm. Find the length of the arc in terms of π.

Answers

$$\begin{aligned} \text{Arc length} &= r\theta \\ &= 8 \times \frac{\pi}{6} \\ &= \frac{4\pi}{3}\,\text{cm} \end{aligned}$$

WORKED EXAMPLE 3

A sector has an angle of 2 radians and an arc length of 9.6 cm.

Find the radius of the sector.

Answers

Arc length $= r\theta$

$\qquad 9.6 = r \times 2$

$\qquad\quad r = 4.8$ cm

WORKED EXAMPLE 4

The circle has radius 5 cm and centre O.

PQ is a tangent to the circle at the point P.

QRO is a straight line. Find

a angle POQ, in radians

b QR

c the perimeter of the shaded region.

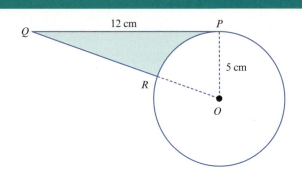

Answers

a $\tan POQ = \dfrac{12}{5}$ triangle QPO is right-angled since PQ is a tangent

$\quad$ Angle $POQ = \tan^{-1}\left(\dfrac{12}{5}\right)$ remember to have your calculator in radian mode

$\qquad\qquad\quad = 1.176\ldots$

$\qquad\qquad\quad = 1.18$ radians

b $OQ^2 = 12^2 + 5^2$

$\quad OQ^2 = 169$ using Pythagoras' theorem

$\quad\ OQ = 13$

$\quad$ Hence $QR = 8$ cm.

c Perimeter $= PQ + QR + $ arc PR use arc $PR = r\theta$

$\qquad\qquad\quad = 12 + 8 + (5 \times 1.176\ldots)$

$\qquad\qquad\quad = 25.9$ cm

Exercise 8.2

1 Find, in terms of π, the arc length of a circular sector of

$\quad$ **a** radius 6 cm and angle $\dfrac{\pi}{4}$ $\qquad\qquad$ **b** radius 5 cm and angle $\dfrac{2\pi}{5}$

$\quad$ **c** radius 10 cm and angle $\dfrac{3\pi}{8}$ $\qquad\qquad$ **d** radius 18 cm and angle $\dfrac{5\pi}{6}$

2   Find the arc length of a circular sector of

    a   radius 8 cm and angle 1.2 radians   **b**   radius 2.5 cm and angle 0.8 radians.

3   Find, in radians, the angle of a circular sector of

    a   radius 4 cm and arc length 5 cm   **b** · radius 9 cm and arc length 13.5 cm.

4   Find the perimeter of each of these circular sectors.

a   **b** 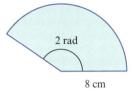   **c**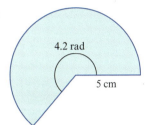

5   *ABCD* is a rectangle with
$AB = 6$ cm and $BC = 16$ cm.

O is the midpoint of *BC*.

OAED is a sector of a circle, centre *O*.
Find

    a   *AO*

    b   angle *AOD*, in radians

    c   the perimeter of the shaded region.

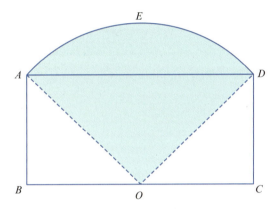

6   Find

    a   the length of arc *AB*

    b   the length of chord *AB*

    c   the perimeter of the shaded segment.

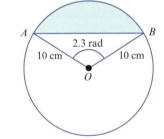

7   Triangle *EFG* is isosceles with
$EG = FG = 16$ cm.

GH is an arc of a circle, centre *F*,
with angle *HFG* = 0.85 radians.
Find

    a   the length of arc *GH*

    b   the length of *EF*

    c   the perimeter of the shaded region.

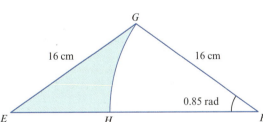

8.3 Area of a sector

To find the formula for the area of a sector you use the ratio:

$$\frac{\text{area of sector}}{\text{area of circle}} = \frac{\text{angle in the sector}}{\text{complete angle at the centre}}$$

When θ is measured in radians, the ratio becomes

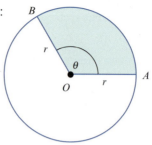

$$\frac{\text{area of sector}}{\pi r^2} = \frac{\theta}{2\pi}$$

$$\text{area of sector} = \frac{\theta}{2\pi} \times \pi r^2$$

$$\text{Area of sector} = \frac{1}{2}r^2\theta$$

WORKED EXAMPLE 5

Find the area of a sector of a circle with radius 6 cm and angle $\frac{2\pi}{3}$ radians.

Give your answer in terms of π.

Answers

$$\begin{aligned}
\text{Area of sector} &= \frac{1}{2}r^2\theta \\
&= \frac{1}{2} \times 6^2 \times \frac{2\pi}{3} \\
&= 12\pi \text{ cm}^2
\end{aligned}$$

WORKED EXAMPLE 6

Calculate the area of the shaded segment correct to 3 sf.

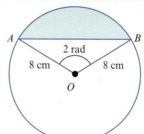

Answers

$$\begin{aligned}
\text{Area of triangle } AOB &= \frac{1}{2} \times 8 \times 8 \times \sin(2^c) \\
&= 29.0975\ldots \quad\quad \text{using area of triangle} = \frac{1}{2}ab\sin C
\end{aligned}$$

$$\begin{aligned}
\text{Area of sector } AOB &= \frac{1}{2} \times 8 \times 8 \times 2 \\
&= 64 \quad\quad \text{using area of sector} = \frac{1}{2}r^2\theta
\end{aligned}$$

$$\begin{aligned}
\text{Area of segment} &= \text{area of sector } AOB - \text{area of triangle } AOB \\
&= 64 - 29.0975\ldots \\
&= 34.9 \text{ cm}^2
\end{aligned}$$

Exercise 8.3

1 Find, in terms of π, the area of a circular sector of

   a radius 6 cm and angle $\dfrac{\pi}{3}$ **b** radius 15 cm and angle $\dfrac{3\pi}{5}$

   c radius 10 cm and angle $\dfrac{7\pi}{10}$ **d** radius 9 cm and angle $\dfrac{5\pi}{6}$.

2 Find the area of a circular sector of

   a radius 4 cm and angle 1.3 radians   **b** radius 3.8 cm and angle 0.6 radians.

3 Find, in radians, the angle of a circular sector of

   a radius 3 cm and area 5 cm^2 **b** radius 7 cm and area 30 cm^2.

4 POQ is the sector of a circle, centre O, radius 10 cm.

   The length of arc PQ is 8 cm.
   Find

   a angle POQ, in radians   **b** the area of the sector POQ.

5 A sector of a circle, radius r cm, has a perimeter of 150 cm.

   Find an expression, in terms of r, for the area of the sector.

6 $ABCD$ is a rectangle with
$AB = 9$ cm and $BC = 18$ cm.

   O is the midpoint of BC.

   $OAED$ is a sector of a circle, centre O.
   Find

   a AO

   b angle AOD, in radians

   c the area of the shaded region.

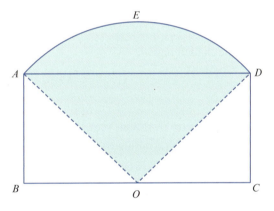

7 The circle has radius 12 cm and centre O.

   PQ is a tangent to the circle at the point P.

   QRO is a straight line.
   Find

   a angle POQ, in radians

   b the area of sector POR

   c the area of the shaded region.

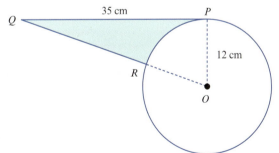

8 *AOB* is the sector of a circle, centre *O*, radius 8 cm.

AC is a tangent to the circle at the point *A*.

CBO is a straight line and the area of sector *AOB* is 32 cm².

Find

   a   angle *AOB*, in radians

   b   the area of triangle *AOC*

   c   the area of the shaded region.

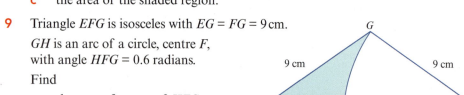

9 Triangle *EFG* is isosceles with *EG* = *FG* = 9 cm.

GH is an arc of a circle, centre *F*, with angle *HFG* = 0.6 radians.

Find

   a   the area of sector of *HFG*

   b   the area of triangle *EFG*

   c   the area of the shaded region.

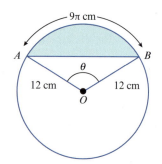

10 The diagram shows a circle, centre *O*, radius 12 cm.

Angle *AOB* = θ radians.

Arc *AB* = 9π cm.

   a   Show that $\theta = \dfrac{3\pi}{4}$

   b   Find the area of the shaded region.

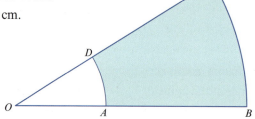

11 *AOD* is a sector of a circle, centre *O*, radius 4 cm.

BOC is a sector of a circle, centre *O*, radius 10 cm.

The shaded region has a perimeter of 18 cm.

Find

   a   angle *AOD*, in radians

   b   the area of the shaded region.

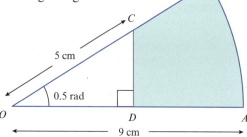

12 *AOB* is a sector of a circle, centre *O*, with radius 9 cm.

Angle *COD* = 0.5 radians and angle *ODC* is a right angle.

OC = 5 cm.

Find

   a   *OD*

   b   *CD*

   c   the perimeter of the shaded region

   d   the area of the shaded region.

13 *FOG* is a sector of a circle, centre *O*, with angle *FOG* = 1.2 radians.

EOH is a sector of a circle, centre *O*, with radius 5 cm.

The shaded region has an area of 71.4 cm^2.

Find the perimeter of the shaded region.

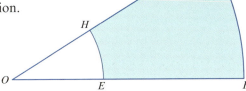

14 CHALLENGE QUESTION

The diagram shows a semicircle, centre *O*, radius 10 cm.

FH is the arc of a circle, centre *E*.

Find the area of

a   triangle *EOF*

b   sector *FOG*

c   sector *FEH*

d   the shaded region.

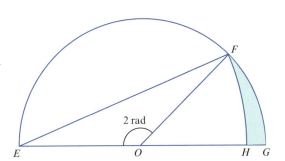

15 CHALLENGE QUESTION

The diagram shows a circle inscribed inside a square of side length 10 cm. A quarter circle of radius 10 cm is drawn with the vertex of the square as its centre.

Find the shaded area.

REFLECTION

In this chapter you have learned about using radians as a unit of angular measure.

a   Without looking back in the textbook can you explain to a friend what a radian is?

b   What are the advantages of using radians as a unit of angular measure?

SUMMARY

One radian (1^c) is the size of the angle subtended at the centre of a circle, radius r, by an arc of length r.

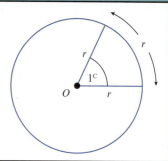

When θ is measured in radians:

- the length of arc $AB = r\theta$
- the area of sector $AOB = \dfrac{1}{2}r^2\theta$

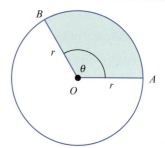

Past paper questions

Worked example

In this question all lengths are in centimetres.

The diagram shows a shaded shape. The arc AB is the major arc of a circle, centre O, radius 10.
The line AB is of length 15, the line OC is of length 25 and the lengths of AC and BC are equal.

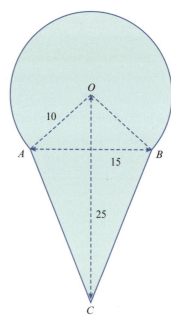

a Show that the angle AOB is 1.70 radians correct to 2 decimal places. [2]

b Find the perimeter of the shaded shape. [4]

c Find the area of the shaded shape. [5]

Cambridge IGCSE Additional Mathematics 0606 Paper 11 Q10 Jun 2021

Answers

a $\sin AOC = \dfrac{7.5}{10}$

  $\sin AOC = 0.75$

  $AOC = \sin^{-1} 0.75$

  $AOB = 2 \times AOC = 2 \times \sin^{-1} 0.75$

  $= 1.6961\ldots$

  $= 1.70$ to 2 dp

b   Perimeter of shaded shape = arc $AB + AC + BC$

arc AB: arc $AB = r\theta$ where θ is the angle in the major sector $= 2\pi - A\widehat{O}B$

  $= 10 \times (2\pi - 1.696\ldots)$

  $= 45.8706\ldots$

AC: $AC^2 = 10^2 + 25^2 - 2 \times 10 \times 25 \times \cos A\widehat{O}C$

  $AC = 19.8565\ldots$

BC: $BC = AC = 19.8565\ldots$

Perimeter of shaded shape = arc $AB + AC + BC$

  $= 45.8706\ldots + 19.8565\ldots + 19.8565\ldots$

  $= 85.58\ldots$ cm

  $= 85.6$ cm to 3 sf

c   Area of shaded shape = $2 \times$ area of triangle OAC + area of major sector OAB

  $= 2 \times \dfrac{1}{2} \times 10 \times 25 \times \sin A\widehat{O}C + \dfrac{1}{2}r^2\theta$

  $= 2 \times \dfrac{1}{2} \times 10 \times 25 \times 0.75 + \dfrac{1}{2} \times 10^2 \times (2\pi - A\widehat{O}B)$

  $= 2 \times \dfrac{1}{2} \times 10 \times 25 \times 0.75 + \dfrac{1}{2} \times 10^2 \times (2\pi - 1.6961\ldots)$

  $= 187.5 + 229.35$

  $= 417$ cm^2 to 3 sf

1   The diagram shows a sector OPQ of a circle with centre O and radius x cm.

Angle POQ is 0.8 radians. The point S lies on OQ such that $OS = 5$ cm.

The point R lies on OP such that angle ORS is a right angle.

Given that the area of triangle ORS is one-fifth of the area of sector OPQ, find

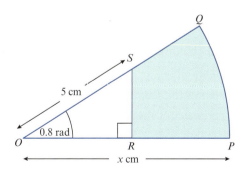

a the area of sector OPQ in terms of x and hence show that the value of x is 8.837, correct to 4 significant figures [5]

b the perimeter of $PQSR$ [3]

c the area of $PQSR$. [2]

Cambridge IGCSE Additional Mathematics 0606 Paper 21 Q11 Nov 2014

2

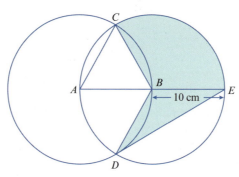

The diagram shows two circles, centres A and B, each of radius 10 cm. The point B lies on the circumference of the circle with centre A. The two circles intersect at the points C and D. The point E lies on the circumference of the circle centre B such that ABE is a diameter.

i Explain why triangle ABC is equilateral. [1]

ii Write down, in terms of π, angle CBE. [1]

iii Find the perimeter of the shaded region. [5]

iv Find the area of the shaded region. [3]

Cambridge IGCSE Additional Mathematics 0606 Paper 11 Q10 Nov 2015

3

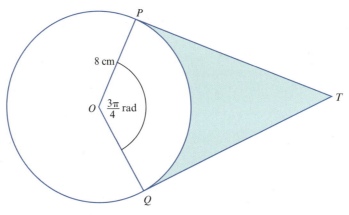

The diagram shows a circle, centre O, radius 8 cm. The points P and Q lie on the circle.

The lines PT and QT are tangents to the circle and angle $POQ = \dfrac{3\pi}{4}$ radians.

i Find the length of PT. [2]

ii Find the area of the shaded region. [3]

iii Find the perimeter of the shaded region. [2]

Cambridge IGCSE Additional Mathematics 0606 Paper 21 Q4 Jun 2015

4

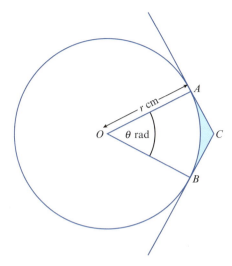

The diagram shows a circle, centre O, radius r cm. Points A, B and C are such that A and B lie on the circle and the tangents at A and B meet at C. Angle $AOB = \theta$ radians.

i   Given that the area of the major sector AOB is 7 times the area of the minor sector AOB, find the value of θ.   [2]

ii   Given also that the perimeter of the minor sector AOB is 20 cm, show that the value of r, correct to 2 decimal places, is 7.18.   [2]

iii Using the values of θ and r from **parts i** and **ii**, find the perimeter of the shaded region ABC.   [3]

iv   Find the area of the shaded region ABC.   [3]

Cambridge IGCSE Additional Mathematics 0606 Paper 12 Q9 Mar 2016

5

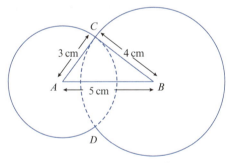

The diagram shows a shape consisting of two circles of radius 3 cm and 4 cm with centres A and B which are 5 cm apart. The circles intersect at C and D as shown. The lines AC and BC are tangents to the circles, centres B and A respectively. Find

a   the angle CAB in radians   [2]

b   the perimeter of the whole shape   [4]

c   the area of the whole shape.   [4]

Cambridge IGCSE Additional Mathematics 0606 Paper 21 Q12 Nov 2020

6   *AOB* is a sector of a circle with centre *O* and radius 16 cm.

Angle *AOB* is $\frac{2\pi}{7}$ radians.

The point *C* lies on *OB* such that *OC* is of length 7.5 cm and *AC* is a straight line.

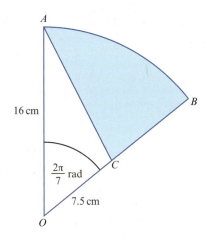

   a   Find the perimeter of the shaded region. [3]

   b   Find the area of the shaded region. [3]

Cambridge IGCSE Additional Mathematics 0606 Paper 22 Q6 Mar 2021

7   a   A circle has a radius 6 cm. A sector of this circle has a perimeter of $2(6 + 5\pi)$ cm.

      Find the area of this sector. [4]

   b

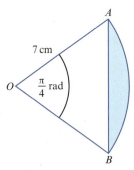

The diagram shows the sector *AOB* of a circle with centre *O* and radius 7 cm.

Angle $AOB = \frac{\pi}{4}$ radians. Find the perimeter of the shaded region. [3]

Cambridge IGCSE Additional Mathematics 0606 Paper 22 Q6 Mar 2020

8   The diagram shows a right-angled triangle *ABC* with *AB* = 8 cm and angle $ABC = \frac{\pi}{2}$ radians.

The points *D* and *E* lie on *AC* and *BC* respectively. *BAD* and *ECD* are sectors of the circles with centres *A* and *C* respectively. Angle $BAD = \frac{2\pi}{9}$ radians.

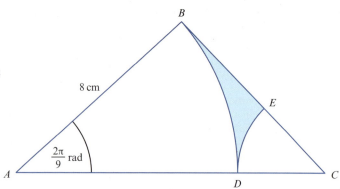

   i   Find the area of the shaded region. [6]

   ii Find the perimeter of the shaded region. [3]

Cambridge IGCSE Additional Mathematics 0606 Paper 21 Q8 Jun 2019

> # Chapter 9
> # Trigonometry

THIS SECTION WILL SHOW YOU HOW TO:

- find the trigonometric ratios of angles of any magnitude

- determine the amplitude and period of trigonometric functions

- describe the relationship between trigonometric graphs

- sketch graphs of $y = |f(x)|$, where $f(x)$ is a trigonometric function

- draw and use the graphs of $y = a\sin bx + c$, $y = a\cos bx + c$, $y = a\tan bx + c$ where a is a positive integer, b is a simple fraction or integer and c is an integer

- use trigonometric relationships

- solve simple trigonometric equations

- prove simple trigonometric identities.

PRE-REQUISITE KNOWLEDGE

Before you start…

Where it comes from	What you should be able to do	Check your skills
Cambridge IGCSE/O Level Mathematics	Use Pythagoras' theorem and trigonometry on right-angled triangles.	**1** Find each of the following in terms of r. **a** AB **b** $\sin x$ **c** $\cos x$ **d** $\tan x$
Chapter 8	Convert between degrees and radians.	**2 a** Convert these angles to radians. **i** 30° **ii** 135° **iii** 540° **b** Convert these angles to degrees. **i** $\dfrac{\pi}{3}$ **ii** $\dfrac{7\pi}{6}$ **iii** $\dfrac{4\pi}{9}$
Cambridge IGCSE/O Level Mathematics	Solve quadratic equations.	**3 a** Solve $x^2 + 6x = 0$. **b** Solve $2x^2 + 9x - 5 = 0$

9.1 Angles between 0° and 90°

You should already know the following trigonometric ratios:

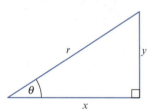

$$\sin \theta = \frac{\text{opposite}}{\text{hypotenuse}} \qquad \cos \theta = \frac{\text{adjacent}}{\text{hypotenuse}} \qquad \tan \theta = \frac{\text{opposite}}{\text{adjacent}}$$

$$\sin \theta = \frac{y}{r} \qquad\qquad\quad \cos \theta = \frac{x}{r} \qquad\qquad\quad \tan \theta = \frac{y}{x}$$

KEY WORDS

periodic function

period

amplitude

trigonometric identity

cosecant

secant

cotangent

WORKED EXAMPLE 1

Given that $\sin \theta = \dfrac{2}{\sqrt{5}}$ and that θ is acute, find the exact values of

a $\sin^2 \theta$ **b** $\cos \theta$ **c** $\tan \theta$ **d** $\dfrac{\sin \theta}{\tan \theta - \cos \theta}$

Answers

a $\sin^2 \theta = \sin \theta \times \sin \theta$ $\sin^2 \theta$ means $(\sin \theta)^2$

$= \dfrac{2}{\sqrt{5}} \times \dfrac{2}{\sqrt{5}}$

$= \dfrac{4}{5}$

b $\sin \theta = \dfrac{2}{\sqrt{5}}$

The right-angled triangle to represent θ is:

Using Pythagoras' theorem, $x = \sqrt{(\sqrt{5})^2 - 2^2}$

$x = 1$

Hence, $\cos \theta = \dfrac{1}{\sqrt{5}} = \dfrac{\sqrt{5}}{5}$

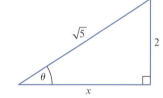

c $\tan \theta = \dfrac{2}{1} = 2$

d $\dfrac{\sin \theta}{\tan \theta - \cos \theta} = \dfrac{\dfrac{2}{\sqrt{5}}}{2 - \dfrac{1}{\sqrt{5}}}$ multiply numerator and denominator by $\sqrt{5}$

$= \dfrac{2}{2\sqrt{5} - 1}$ multiply numerator and denominator by $(2\sqrt{5} + 1)$

$= \dfrac{2(2\sqrt{5} + 1)}{(2\sqrt{5} - 1)(2\sqrt{5} + 1)}$

$= \dfrac{2(2\sqrt{5} + 1)}{19}$

$= \dfrac{2 + 4\sqrt{5}}{19}$

The sine, cosine and tangent of 30°, 45° and 60° (or $\frac{\pi}{6}, \frac{\pi}{4}$ and $\frac{\pi}{3}$) can be obtained exactly from the following two triangles:

Consider a right-angled isosceles triangle whose two equal sides are of length 1 unit.

The third side is found using Pythagoras' theorem: $\sqrt{1^2 + 1^2} = \sqrt{2}$

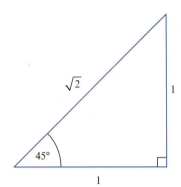

$$\sin 45° = \frac{1}{\sqrt{2}} \left(= \frac{\sqrt{2}}{2}\right) \qquad \sin \frac{\pi}{4} = \frac{1}{\sqrt{2}} \left(= \frac{\sqrt{2}}{2}\right)$$

$$\cos 45° = \frac{1}{\sqrt{2}} \left(= \frac{\sqrt{2}}{2}\right) \qquad \cos \frac{\pi}{4} = \frac{1}{\sqrt{2}} \left(= \frac{\sqrt{2}}{2}\right)$$

$$\tan 45° = 1 \qquad\qquad\qquad \tan \frac{\pi}{4} = 1$$

Consider an equilateral triangle whose sides are of length 2 units.

The perpendicular bisector to the base splits the equilateral triangle into two congruent right-angled triangles.

The height of the triangle can be found using Pythagoras' theorem: $\sqrt{2^2 - 1^2} = \sqrt{3}$

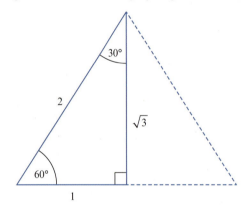

$$\sin 60° = \frac{\sqrt{3}}{2} \qquad\qquad \sin \frac{\pi}{3} = \frac{\sqrt{3}}{2}$$

$$\cos 60° = \frac{1}{2} \qquad\qquad \cos \frac{\pi}{3} = \frac{1}{2}$$

$$\tan 60° = \sqrt{3} \qquad\qquad \tan \frac{\pi}{3} = \sqrt{3}$$

$$\sin 30° = \frac{1}{2} \qquad\qquad \sin \frac{\pi}{6} = \frac{1}{2}$$

$$\cos 30° = \frac{\sqrt{3}}{2} \qquad\qquad \cos \frac{\pi}{6} = \frac{\sqrt{3}}{2}$$

$$\tan 30° = \frac{1}{\sqrt{3}} \left(= \frac{\sqrt{3}}{3}\right) \qquad \tan \frac{\pi}{6} = \frac{1}{\sqrt{3}} \left(= \frac{\sqrt{3}}{3}\right)$$

Note:
You do not need to learn these triangles and ratios for your course, but you may find it useful to know them.

WORKED EXAMPLE 2

Find the exact value of

a $\sin 45° \sin 60°$ **b** $\cos^2 45°$ **c** $\dfrac{\tan\dfrac{\pi}{3} + \sin\dfrac{\pi}{4}}{1 - \cos\dfrac{\pi}{3}}$

Answers

a $\sin 45° \sin 60° = \dfrac{1}{\sqrt{2}} \times \dfrac{\sqrt{3}}{2}$

                 $= \dfrac{\sqrt{3}}{2\sqrt{2}}$ rationalise the denominator

                 $= \dfrac{\sqrt{3} \times \sqrt{2}}{2\sqrt{2} \times \sqrt{2}}$

                 $= \dfrac{\sqrt{6}}{4}$

b $\cos^2 45° = \cos 45° \times \cos 45°$ $\cos^2 45°$ means $(\cos 45°)^2$

          $= \dfrac{1}{\sqrt{2}} \times \dfrac{1}{\sqrt{2}}$

          $= \dfrac{1}{2}$

c $\dfrac{\tan\dfrac{\pi}{3} + \sin\dfrac{\pi}{4}}{1 - \cos\dfrac{\pi}{3}} = \dfrac{\sqrt{3} + \dfrac{1}{\sqrt{2}}}{1 - \dfrac{1}{2}}$ the denominator simplifies to $\dfrac{1}{2}$

                 $= \left(\sqrt{3} + \dfrac{1}{\sqrt{2}}\right) \times 2$ expand brackets

                 $= 2\sqrt{3} + \dfrac{2}{\sqrt{2}}$ rationalise the denominator

                 $= 2\sqrt{3} + \sqrt{2}$

Exercise 9.1

1 Given that $\tan\theta = \dfrac{2}{3}$ and that θ is acute, find the exact values of

    a $\sin\theta$ **b** $\cos\theta$ **c** $\sin^2\theta$

    d $\sin^2\theta + \cos^2\theta$ **e** $\dfrac{2 + \sin\theta}{3 - \cos\theta}$

2 Given that $\sin\theta = \dfrac{\sqrt{2}}{5}$ and that θ is acute, find the exact values of

    a $\cos\theta$ **b** $\tan\theta$ **c** $1 - \sin^2\theta$

    d $\sin\theta + \cos\theta$ **e** $\dfrac{\cos\theta - \sin\theta}{\tan\theta}$

3 Given that $\cos\theta = \dfrac{1}{7}$ and that θ is acute, find the exact values of

    a $\sin\theta$ **b** $\tan\theta$ **c** $\tan\theta \cos\theta$

    d $\sin^2\theta + \cos^2\theta$ **e** $\dfrac{\cos\theta - \tan\theta}{1 - \cos^2\theta}$

4 Find the exact value of each of the following.

    a $\tan 45° \cos 60°$ **b** $\tan^2 60°$ **c** $\dfrac{\tan 30°}{\cos 30°}$

    d $\sin 45° + \cos 30°$ **e** $\dfrac{\cos^2 30°}{\cos 45° + \cos 60°}$ **f** $\dfrac{\tan 45° - \sin 30°}{1 + \sin^2 60°}$

5 Find the exact value of each of the following.

    a $\sin \dfrac{\pi}{4} \cos \dfrac{\pi}{3}$ **b** $\sin^2 \dfrac{\pi}{4}$ **c** $\dfrac{\tan \dfrac{\pi}{6}}{\cos \dfrac{\pi}{4}}$

    d $\dfrac{5 - \tan \dfrac{\pi}{3}}{\sin \dfrac{\pi}{3}}$ **e** $\dfrac{1}{\sin \dfrac{\pi}{6}} - \dfrac{1}{\cos \dfrac{\pi}{4}}$ **f** $\dfrac{\tan \dfrac{\pi}{4} - \sin \dfrac{\pi}{4}}{\tan \dfrac{\pi}{6} \sin \dfrac{\pi}{6}}$

9.2 The general definition of an angle

You need to be able to use the three basic trigonometric functions for any angle.

To do this you need a general definition for an angle:

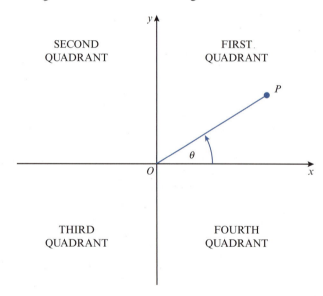

An angle is a measure of the rotation of a line OP about a fixed point O.

The angle is measured from the positive x-direction.

An anticlockwise rotation is taken as positive and a clockwise rotation is taken as negative.

The Cartesian plane is divided into four quadrants and the angle θ is said to be in the quadrant where OP lies. In the diagram above, θ is in the first quadrant.

WORKED EXAMPLE 3

Draw a diagram showing the quadrant in which the rotating line OP lies for each of the following angles. In each case find the acute angle that the line OP makes with the x-axis.

a 240° **b** −70° **c** 490° **d** $\dfrac{2\pi}{3}$

Answers

a 240° is an anticlockwise rotation

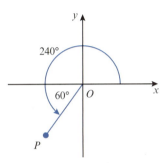

acute angle = 60°

b −70° is a clockwise rotation

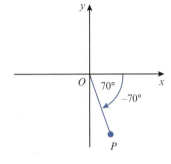

acute angle = 70°

c 490° is an anticlockwise rotation
$490° = 360° + 130°$

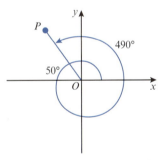

acute angle = 50°

d $\dfrac{2\pi}{3}$ is an anticlockwise rotation

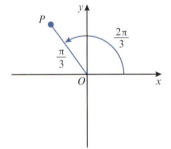

acute angle = $\dfrac{\pi}{3}$

Exercise 9.2

1 Draw a diagram showing the quadrant in which the rotating line OP lies for each of the following angles. In each question, indicate clearly the direction of rotation and find the acute angle that the line OP makes with the x-axis.

a 110° **b** −60° **c** 220° **d** −135° **e** −300°

f $\dfrac{3\pi}{4}$ **g** $\dfrac{7\pi}{6}$ **h** $-\dfrac{5\pi}{3}$ **i** $\dfrac{13\pi}{9}$ **j** $-\dfrac{5\pi}{8}$

2 State the quadrant that OP lies in when the angle that OP makes with the positive x-axis is

a 110° **b** 300° **c** −160° **d** 245° **e** −500°

f $\dfrac{\pi}{4}$ **g** $\dfrac{11\pi}{6}$ **h** $-\dfrac{5\pi}{6}$ **i** $\dfrac{13\pi}{6}$ **j** $\dfrac{9\pi}{4}$

9.3 Trigonometric ratios of general angles

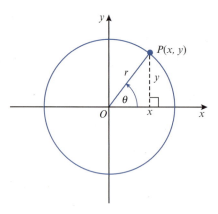

In general, trigonometric ratios of any angle θ in any quadrant are defined as:

$$\sin\theta = \frac{y}{r}, \qquad \cos\theta = \frac{x}{r}, \qquad \tan\theta = \frac{y}{x}, \qquad x \neq 0$$

where x and y are the coordinates of the point P and r is the length of OP and $r = \sqrt{x^2 + y^2}$.

You need to know the signs of the three trigonometric ratios in the four quadrants.

In the first quadrant, all three ratios are positive. (Since x, y and r are all positive.)

By considering the sign of x and y you can find the sign of each of the three trigonometric ratios in the other three quadrants.

CLASS DISCUSSION

$\sin\theta = \dfrac{y}{r}$	By considering the sign of y in the second, third and fourth quadrants, determine the signs of the sine ratio in each of these quadrants.
$\cos\theta = \dfrac{x}{r}$	By considering the sign of x in the second, third and fourth quadrants, determine the signs of the cosine ratio in each of these quadrants.
$\tan\theta = \dfrac{y}{x}$	By considering the sign of x and y in the second, third and fourth quadrants, determine the signs of the tangent ratio in each of these quadrants. What happens to the tangent ratio when $x = 0$?

On a copy of the diagram, record which ratios are positive in each quadrant.

The first quadrant has been completed for you.

(All three ratios are positive in the first quadrant.)

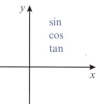

The results of the class discussion can be summarised as:

You can memorise this diagram using a mnemonic such as

'**A**ll **S**tudents **T**rust **C**ambridge'.

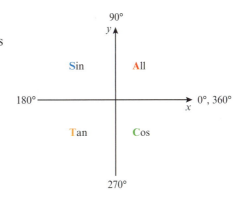

REFLECTION

The mnemonic '**A**ll **S**tudents **T**rust **C**ambridge' was suggested for remembering the sign of trigonometric functions in each quadrant of the plane.

a   Explain to a friend how this mnemonic works.

b   Make up your own mnemonic for remembering these facts.

WORKED EXAMPLE 4

Express in terms of trigonometric ratios of acute angles.

a $\cos(-110°)$

b $\sin 125°$

Answers

a   The acute angle made with the positive
    x-axis is 70°.

    In the third quadrant only tan is positive,
    so cos is negative.

    $\cos(-110°) = -\cos 70°$

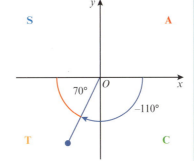

b   The acute angle made with the positive
    x-axis is 55°.

    In the second quadrant sin is positive.

    $\sin 125° = \sin 55°$

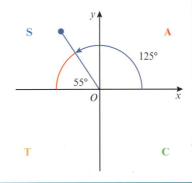

WORKED EXAMPLE 5

Given that $\sin\theta = -\dfrac{3}{5}$ and that $180° \leqslant \theta \leqslant 270°$, find the value of $\tan\theta$ and the value of $\cos\theta$.

Answers

θ is in the third quadrant.

tan is positive and cos is negative in this quadrant.

$x^2 + (-3)^2 = 5^2$

$x^2 = 25 - 9 = 16$

Since $x < 0$, $x = -4$

$\tan\theta = \dfrac{-3}{-4} = \dfrac{3}{4}$ and $\cos\theta = \dfrac{-4}{5} = -\dfrac{4}{5}$

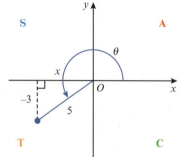

Exercise 9.3

1 Express the following as trigonometric ratios of acute angles.

 a $\sin 220°$ b $\cos 325°$ c $\tan 140°$ d $\cos(-25°)$

 e $\tan 600°$ f $\sin\dfrac{4\pi}{5}$ g $\tan\dfrac{7\pi}{4}$ h $\cos\left(-\dfrac{11\pi}{6}\right)$

 i $\tan\dfrac{2\pi}{3}$ j $\sin\dfrac{9\pi}{4}$

2 Given that $\cos\theta = \dfrac{2}{5}$ and that $270° \leqslant \theta \leqslant 360°$, find the value of

 a $\tan\theta$ b $\sin\theta$

3 Given that $\tan\theta = -\sqrt{3}$ and that $90° \leqslant \theta \leqslant 180°$, find the value of

 a $\sin\theta$ b $\cos\theta$

4 Given that $\sin\theta = \dfrac{5}{13}$ and that θ is obtuse, find the value of

 a $\cos\theta$ b $\tan\theta$

5 Given that $\tan\theta = \dfrac{2}{3}$ and that θ is reflex, find the value of

 a $\sin\theta$ b $\cos\theta$

6 Given that $\tan A = \dfrac{4}{3}$ and $\cos B = -\dfrac{1}{\sqrt{3}}$, where A and B are in the same quadrant, find the value of

 a $\sin A$ b $\cos A$ c $\sin B$ d $\tan B$

7 Given that $\sin A = -\dfrac{12}{13}$ and $\cos B = \dfrac{3}{5}$, where A and B are in the same quadrant, find the value of

 a $\cos A$ b $\tan A$ c $\sin B$ d $\tan B$

9.4 Graphs of trigonometric functions

CLASS DISCUSSION

Consider taking a ride on a Ferris wheel, with radius 50 metres, which rotates at a constant speed.

You enter the ride from a platform that is level with the centre of the wheel and the wheel turns in an anticlockwise direction.

Sketch the following two graphs and discuss their properties:

• a graph of your **vertical displacement from the centre of the wheel** plotted against **angle turned through**

• a graph of your **horizontal displacement from the centre of the wheel** plotted against **angle turned through**.

The graphs of $y = \sin x$ and $y = \cos x$

Suppose that OP makes an angle of x with the positive horizontal axis and that P moves around the unit circle, through one complete revolution. The coordinates of P will be $(\cos x, \sin x)$.

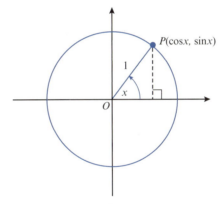

The height of P above the horizontal axis changes from $0 \rightarrow 1 \rightarrow 0 \rightarrow -1 \rightarrow 0$.

The graph of $\sin x$ against x for $0° \leqslant x \leqslant 360°$ is:

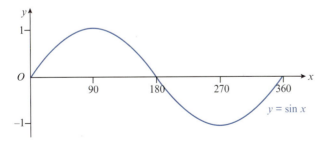

The distance of P from the vertical axis changes from $1 \to 0 \to -1 \to 0 \to 1$.

The graph of $\cos x$ against x for $0° \leqslant x \leqslant 360°$ is:

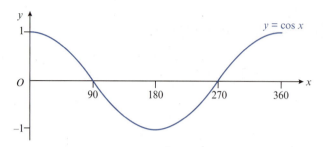

The graphs of $y = \sin x$ and $y = \cos x$ can be continued beyond $0° \leqslant x \leqslant 360°$:

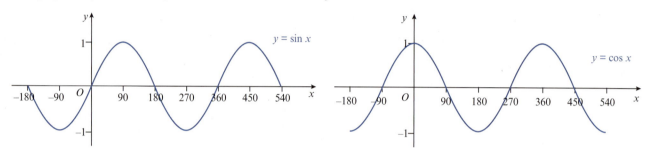

The sine and cosine functions are called **periodic functions** because they repeat themselves over and over again.

The **period** of a periodic function is defined as the length of one repetition or cycle.

The basic sine and cosine functions repeat every $360°$.

We say they have a period of $360°$. (Period $= 2\pi$, if working in radians.)

The **amplitude** of a periodic function is defined as the distance between a maximum (or minimum) point and the principal axis.

The basic sine and cosine functions have amplitude 1.

The graph of $y = \tan x$

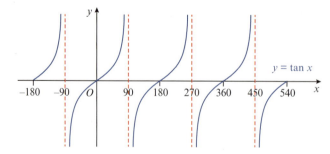

The tangent function behaves very differently to the sine and cosine functions.

The tangent function repeats its cycle every $180°$ so its period is $180°$.

The red dashed lines at $x = \pm 90°$, $x = 270°$ and $x = 450°$ are asymptotes. The branches of the graph get closer and closer to the asymptotes without ever reaching them.

The tangent function does not have an amplitude.

The graphs of $y = a \sin bx + c$, $y = a \cos bx + c$ and $y = a \tan bx + c$

You have already learned about the graphs of $y = \sin x$, $y = \cos x$ and $y = \tan x$.

In this section you will learn how to sketch the graphs of $y = a \sin bx + c$, $y = a \cos bx + c$ and $y = a \tan bx + c$, where a and b are positive integers and c is an integer.

You can use graphing software to observe how the values of a, b and c affect the trigonometric functions.

The graph of $y = a \sin x$

Using graphing software, the graphs of $y = \sin x$ and $y = 2 \sin x$ are:

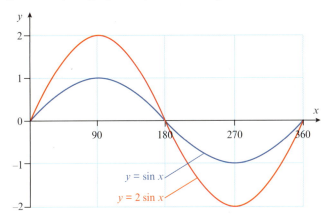

The graph of $y = 2 \sin x$ is a stretch of the graph of $y = \sin x$.

It has been stretched from the x-axis with a stretch factor of 2.

The amplitude of $y = 2 \sin x$ is 2 and the period is 360°.

Similarly, it can be shown that the graph of $y = 3 \sin x$ is a stretch of $y = \sin x$ from the x-axis with stretch factor 3. The amplitude of $y = 3 \sin x$ is 3 and the period is 360°.

ACTIVITY

Use graphing software, to confirm that:

- $y = 2 \cos x$ is a stretch of $y = \cos x$ from the x-axis with stretch factor 2

- $y = 3 \cos x$ is a stretch of $y = \cos x$ from the x-axis with stretch factor 3

and

- $y = 2 \tan x$ is a stretch of $y = \tan x$ from the x-axis with stretch factor 2

- $y = 3 \tan x$ is a stretch of $y = \tan x$ from the x-axis with stretch factor 3.

The graph of $y = \sin bx$

Using graphing software, the graphs of $y = \sin x$ and $y = \sin 2x$ are:

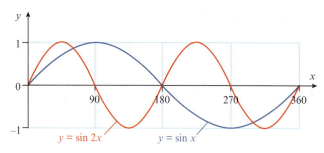

The graph of $y = \sin 2x$ is a stretch of the graph of $y = \sin x$.

It has been stretched from the y-axis with a stretch factor of $\dfrac{1}{2}$

The amplitude of $y = \sin 2x$ is 1 and the period is 180°.

Similarly, the graph of $y = \sin 3x$ is a stretch, from the y-axis, of $y = \sin x$ with stretch factor $\dfrac{1}{3}$.

The amplitude of $y = \sin 3x$ is 1 and the period is 120°.

ACTIVITY

Use graphing software, confirm that:

- $y = \cos 2x$ is a stretch of $y = \cos x$ from the y-axis with stretch factor $\dfrac{1}{2}$

- $y = \cos 3x$ is a stretch of $y = \cos x$ from the y-axis with stretch factor $\dfrac{1}{3}$

and

- $y = \tan 2x$ is a stretch of $y = \tan x$ from the y-axis with stretch factor $\dfrac{1}{2}$

- $y = \tan 3x$ is a stretch of $y = \tan x$ from the y-axis with stretch factor $\dfrac{1}{3}$

The graph of $y = \sin x + c$

Using graphing software, the graphs of $y = \sin x$ and $y = \sin x + 1$ are:

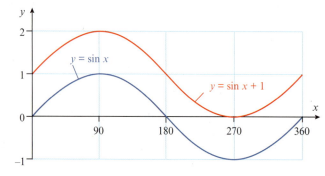

The graph of $y = \sin x + 1$ is a translation of the graph of $y = \sin x$.

It has been translated by the vector $\begin{pmatrix} 0 \\ 1 \end{pmatrix}$.

The amplitude of $y = \sin x + 1$ is 1 and the period is 360°.

Similarly, the graph of $y = \sin x + 2$ is a translation of $y = \sin x$ by the vector $\begin{pmatrix} 0 \\ 2 \end{pmatrix}$

The amplitude of $y = \sin x + 2$ is 1 and the period is 360°.

ACTIVITY

Use graphing software, confirm that:

- $y = \cos x + 1$ is a translation of $y = \cos x$ by the vector $\begin{pmatrix} 0 \\ 1 \end{pmatrix}$

- $y = \cos x + 2$ is a translation of $y = \cos x$ by the vector $\begin{pmatrix} 0 \\ 2 \end{pmatrix}$

- $y = \cos x - 3$ is a translation of $y = \cos x$ by the vector $\begin{pmatrix} 0 \\ -3 \end{pmatrix}$

and

- $y = \tan x + 1$ is a translation of $y = \tan x$ by the vector $\begin{pmatrix} 0 \\ 1 \end{pmatrix}$

- $y = \tan x - 2$ is a translation of $y = \tan x$ by the vector $\begin{pmatrix} 0 \\ -2 \end{pmatrix}$

In conclusion,

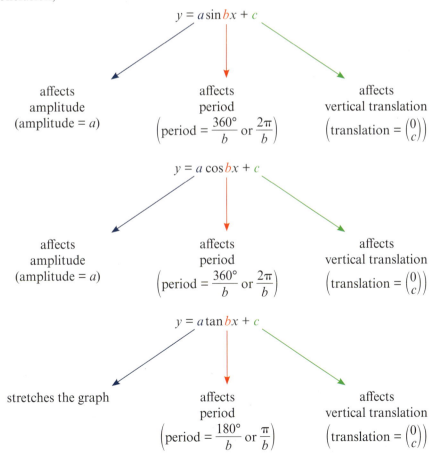

$y = a \sin bx + c$

affects amplitude (amplitude = a)

affects period $\left(\text{period} = \dfrac{360°}{b} \text{ or } \dfrac{2\pi}{b} \right)$

affects vertical translation $\left(\text{translation} = \begin{pmatrix} 0 \\ c \end{pmatrix} \right)$

$y = a \cos bx + c$

affects amplitude (amplitude = a)

affects period $\left(\text{period} = \dfrac{360°}{b} \text{ or } \dfrac{2\pi}{b} \right)$

affects vertical translation $\left(\text{translation} = \begin{pmatrix} 0 \\ c \end{pmatrix} \right)$

$y = a \tan bx + c$

stretches the graph

affects period $\left(\text{period} = \dfrac{180°}{b} \text{ or } \dfrac{\pi}{b} \right)$

affects vertical translation $\left(\text{translation} = \begin{pmatrix} 0 \\ c \end{pmatrix} \right)$

Note:
there is no amplitude for a tangent function.

Sketching trigonometric functions

The sketch graph of a trigonometric function, such as $y = 2\cos 3x - 1$ for $0° \leqslant x \leqslant 360°$, can be built up in steps.

Step 1: Start with a sketch of $y = \cos x$:

Period = 360°

Amplitude = 1

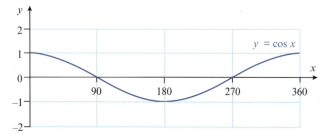

Step 2: Sketch the graph of $y = \cos 3x$:

Stretch $y = \cos x$ from the y-axis with stretch factor $\dfrac{1}{3}$

Period = $\dfrac{360°}{3} = 120°$

Amplitude = 1

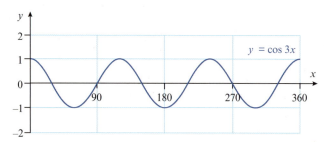

Step 3: Sketch the graph of $y = 2\cos 3x$:

Stretch $y = \cos 3x$ from the x-axis with stretch factor 2.

Period = $\dfrac{360°}{3} = 120°$

Amplitude = 2

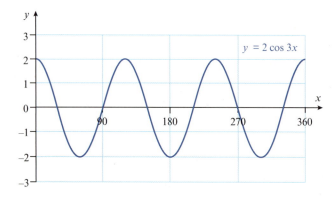

Step 4: Sketch the graph of $y = 2\cos 3x - 1$:

Translate $y = 2\cos 3x$ by $\begin{pmatrix} 0 \\ -1 \end{pmatrix}$

Period $= \dfrac{360°}{3} = 120°$

Amplitude $= 2$

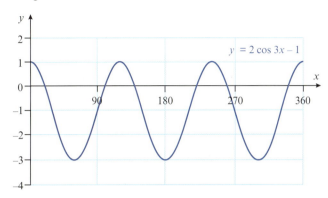

WORKED EXAMPLE 6

$f(x) = 3\sin 2x$ for $0° \le x \le 360°$

a Write down the period of f.

b Write down the amplitude of f.

c Write down the coordinates of the maximum and minimum points on the curve $y = f(x)$.

d Sketch the graph of $y = f(x)$.

e Use your answer to **part d** to sketch the graph of $y = 3\sin 2x + 1$.

Answers

a Period $= \dfrac{360°}{2} = 180°$

b Amplitude $= 3$

c $y = \sin x$ has its maximum and minimum points at:
  $(90°, 1)$, $(270°, -1)$, $(450°, 1)$ and $(630°, -1)$
  Hence, $f(x) = 3\sin 2x$ has its maximum and minimum points at:
  $(45°, 3)$, $(135°, -3)$, $(225°, 3)$ and $(315°, -3)$

d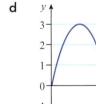

CONTINUED

e $y = 3\sin 2x + 1$ is a translation of the graph $y = 3\sin 2x$ by the vector $\begin{pmatrix} 0 \\ 1 \end{pmatrix}$

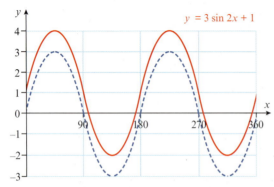

WORKED EXAMPLE 7

a On the same grid, sketch the graphs of $y = \sin 2x$ and $y = 1 + \cos 2x$ for $0° \le x \le 360°$.

b State the number of roots of the equation $\sin 2x = 1 + \cos 2x$ for $0° \le x \le 360°$.

Answers

a

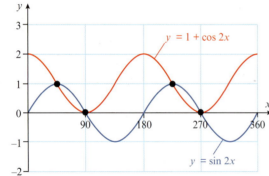

b The graphs of $y = \sin 2x$ and $y = 1 + \cos 2x$ intersect each other at 4 points in the interval.

Hence, the number of roots of $\sin 2x = 1 + \cos 2x$ is 4.

Exercise 9.4

1 **a** The following functions are defined for $0° \le x \le 360°$.

For each function, write down the amplitude, the period and the coordinates of the maximum and minimum points.

i $f(x) = 7\cos x$ **ii** $f(x) = 2\sin 2x$ **iii** $f(x) = 2\cos 3x$

iv $f(x) = 3\sin \dfrac{1}{2}x$ **v** $f(x) = 4\cos x + 1$ **vi** $f(x) = 5\sin 2x - 2$

b Sketch the graph of each function in **part a** and use graphing software to check your answers.

2 **a** The following functions are defined for $0 \leqslant x \leqslant 2\pi$.

For each function, write down the amplitude, the period and the coordinates of the maximum and minimum points.

    i $f(x) = 4\sin x$ **ii** $f(x) = \cos 3x$ **iii** $f(x) = 2\sin 3x$

    iv $f(x) = 3\cos \frac{1}{2}x$ **v** $f(x) = \sin 2x + 3$ **vi** $f(x) = 4\cos 2x - 1$

  b Sketch the graph of each function in **part a** and use graphing software to check your answers.

3

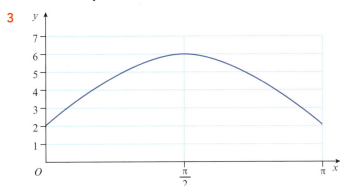

This is the graph of $y = a + b\sin cx$, for $0 \leqslant x \leqslant \pi$.

Write down the value of a, the value of b and the value of c.

4

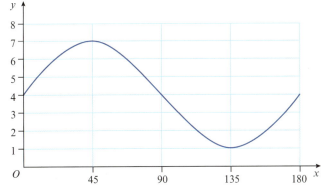

Here is part of the graph of $y = a\sin bx + c$.

Write down the value of a, the value of b and the value of c.

5

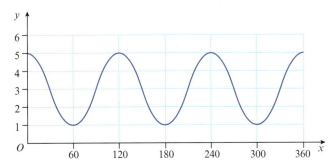

This is the graph of $y = a + b\cos cx$, for $0° \leqslant x \leqslant 360°$.

Write down the value of a, the value of b and the value of c.

6 **a** The following functions are defined for $0° \leqslant x \leqslant 360°$.

For each function, write down the period and the equations of the asymptotes.

    i $f(x) = \tan 2x$ **ii** $f(x) = 3 \tan \dfrac{1}{2} x$ **iii** $f(x) = 2 \tan 3x + 1$

    b Sketch the graph of each function and use graphing software to check your answers.

7 **a** The following functions are defined for $0 \leqslant x \leqslant 2\pi$.

For each function, write down the period and the equations of the asymptotes.

    i $f(x) = \tan 4x$ **ii** $f(x) = 2 \tan 3x$ **iii** $f(x) = 5 \tan 2x - 3$

    b Sketch the graph of each function and use graphing software to check your answers.

8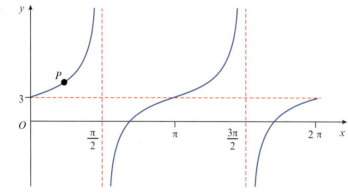

Here is part of the graph of $y = A \tan Bx + C$.

The graph passes through the point $P\left(\dfrac{\pi}{4}, 4\right)$.

Find the value of A, the value of B and the value of C.

9 $f(x) = a + b \sin cx$

The maximum value of f is 13, the minimum value of f is 5 and the period is 60°.

Find the value of a, the value of b and the value of c.

10 $f(x) = A + 3 \cos Bx$ for $0° \leqslant x \leqslant 360°$

The maximum value of f is 5 and the period is 72°.

    a Write down the value of A and the value of B.

    b Write down the amplitude of f.

    c Sketch the graph of f.

11 $f(x) = A + B \sin Cx$ for $0° \leqslant x \leqslant 360°$

The amplitude of f is 3, the period is 90° and the minimum value of f is −2.

    a Write down the value of A, the value of B and the value of C.

    b Sketch the graph of f.

12 **a** On the same grid, sketch the graphs of $y = \sin x$ and $y = 1 + \sin 2x$ for $0° \leqslant x \leqslant 360°$.

    b State the number of roots of the equation $\sin 2x - \sin x + 1 = 0$ for $0° \leqslant x \leqslant 360°$.

13 a On the same grid, sketch the graphs of $y = \sin x$ and $y = 1 + \cos 2x$ for $0° \le x \le 360°$.

b State the number of roots of the equation $\sin x = 1 + \cos 2x$ for $0° \le x \le 360°$.

14 a On the same grid, sketch the graphs of $y = 3\cos 2x$ and $y = 2 + \sin x$ for $0° \le x \le 360°$.

b State the number of roots of the equation $3\cos 2x = 2 + \sin x$ for $0° \le x \le 360°$.

9.5 Graphs of $y = |\mathrm{f}(x)|$, where $\mathrm{f}(x)$ is a trigonometric function

You have already learned how to sketch graphs of $y = |\mathrm{f}(x)|$ where $\mathrm{f}(x)$ is either linear or quadratic.

In this section you will learn how to sketch graphs of $y = |\mathrm{f}(x)|$ where $\mathrm{f}(x)$ is a trigonometric function.

WORKED EXAMPLE 8

a Sketch the graph of $\mathrm{f}(x) = |\sin x|$ for $0 \le x \le 2\pi$.

b State the range of the function f.

Answers

a **Step 1:** Sketch the graph of $y = \sin x$

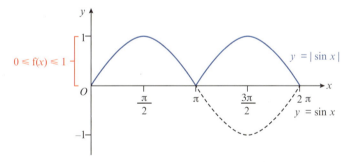

Step 2: Reflect in the x-axis the part of the curve $y = \sin x$ that is below the x-axis.

b The range of the function f is $0 \le \mathrm{f}(x) \le 1$.

WORKED EXAMPLE 9

a On the same grid, sketch the graphs of $y = |\sin 2x|$ and $y = \cos x$ for $0° \leqslant x \leqslant 360°$.

b State the number of roots of the equation $|\sin 2x| = \cos x$ for $0° \leqslant x \leqslant 360°$.

Answers

a For $y = |\sin 2x|$, sketch the graph of $y = \sin 2x$ and then reflect in the x-axis the part of the curve $y = \sin 2x$ that is below the x-axis.

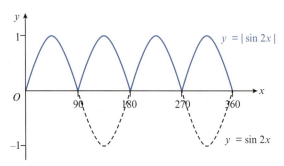

Hence, the graphs of $y = |\sin 2x|$ and $y = \cos x$ are:

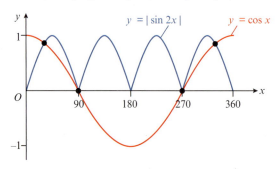

b The graphs of $y = |\sin 2x|$ and $y = \cos x$ intersect each other at 4 points in the interval.

Hence, the number of roots of $|\sin 2x| = \cos x$ is 4.

Exercise 9.5

Use graphing software to check your graphs in this exercise.

1 Sketch the graphs of each of the following functions, for $0° \leqslant x \leqslant 360°$, and state the range of each function.

    a $f(x) = |\tan x|$ **b** $f(x) = |\cos 2x|$ **c** $f(x) = |3 \sin x|$

    d $f(x) = \left|\sin \dfrac{1}{2}x\right|$ **e** $f(x) = \left|2 \cos \dfrac{1}{2}x\right|$ **f** $f(x) = |2 \sin 2x|$

    g $f(x) = |\sin x - 2|$ **h** $f(x) = |5 \sin x + 1|$ **i** $f(x) = |4 \cos x - 3|$

2   a   Sketch the graph of $y = 2\sin x - 1$ for $0° \leqslant x \leqslant 180°$.

    b   Sketch the graph of $y = |2\sin x - 1|$ for $0° \leqslant x \leqslant 180°$.

    c   Write down the number of solutions of the equation $|2\sin x - 1| = 0.5$ for
        $0° \leqslant x \leqslant 180°$.

3   a   Sketch the graph of $y = 2 + 5\cos x$ for $0° \leqslant x \leqslant 180°$.

    b   Sketch the graph of $y = |2 + 5\cos x|$ for $0° \leqslant x \leqslant 180°$.

    c   Write down the number of solutions of the equation $|2 + 5\cos x| = 1$ for
        $0° \leqslant x \leqslant 180°$.

4   a   Sketch the graph of $y = 2 + 3\cos x$ for $0° \leqslant x \leqslant 180°$.

    b   Sketch the graph of $y = |2 + 3\cos x|$ for $0° \leqslant x \leqslant 180°$.

    c   Write down the number of solutions of the equation $|2 + 3\sin 2x| = 1$ for
        $0° \leqslant x \leqslant 180°$.

5   a   On the same grid, sketch the graphs of $y = |\tan x|$ and $y = \cos x$ for
        $0° \leqslant x \leqslant 360°$.

    b   State the number of roots of the equation $|\tan x| = \cos x$ for $0° \leqslant x \leqslant 360°$.

6   a   On the same grid, sketch the graphs of $y = |\sin 2x|$ and $y = \tan x$ for $0 \leqslant x \leqslant 2\pi$.

    b   State the number of roots of the equation $|\sin 2x| = \tan x$ for $0 \leqslant x \leqslant 2\pi$.

7   a   On the same grid, sketch the graphs of $y = |0.5 + \sin x|$ and $y = \cos x$ for
        $0° \leqslant x \leqslant 360°$.

    b   State the number of roots of the equation $|0.5 + \sin x| = \cos x$ for $0° \leqslant x \leqslant 360°$.

8   a   On the same grid, sketch the graphs of $y = |1 + 4\cos x|$ and $y = 2 + \cos x$ for
        $0° \leqslant x \leqslant 360°$.

    b   State the number of roots of the equation $|1 + 4\cos x| = 2 + \cos x$ for
        $0° \leqslant x \leqslant 360°$.

9   The equation $|3\cos x - 2| = k$, has 2 roots for the interval $0 \leqslant x \leqslant 2\pi$.
    Find the possible values of k.

10 **CHALLENGE QUESTION**

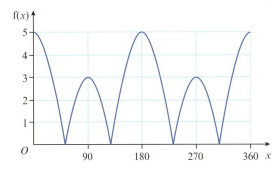

The diagram shows the graph of $f(x) = |a + b\cos cx|$, where a, b and c are positive integers.
Find the value of a, the value of b and the value of c.

9.6 Trigonometric equations

Consider the right-angled triangle:

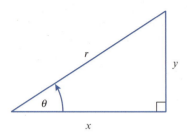

$$\sin \theta = \frac{y}{r} \qquad \cos \theta = \frac{x}{r} \qquad \tan \theta = \frac{y}{x}$$

The following rules can be found from this triangle:

$\tan \theta = \frac{y}{x}$ divide numerator and denominator by r

$= \dfrac{\left(\frac{y}{r}\right)}{\left(\frac{x}{r}\right)}$ use $\dfrac{y}{r} = \sin \theta$ and $\dfrac{x}{r} = \cos \theta$

$$\tan \theta = \frac{\sin \theta}{\cos \theta}$$

$x^2 + y^2 = r^2$ divide both sides by r^2

$\left(\dfrac{x}{r}\right)^2 + \left(\dfrac{y}{r}\right)^2 = 1$ use $\dfrac{x}{r} = \cos \theta$ and $\dfrac{y}{r} = \sin \theta$

$$\cos^2 \theta + \sin^2 \theta = 1$$

These two important rules will be needed to solve some trigonometric equations later in this section.

Consider solving the equation: $\sin x = 0.5$

$$x = \sin^{-1}(0.5)$$

A calculator will give the answer: $x = 30°$

There are, however, many more values of x for which $\sin x = 0.5$

Consider the graph of $y = \sin x$ for $-360° \leqslant x \leqslant 360°$:

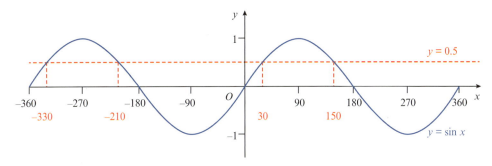

The sketch graph shows there are four values of x, between $-360°$ and $360°$, for which $\sin x = 0.5$.

You can use the calculator value of $x = 30°$, together with the symmetry of the curve to find the remaining values.

Hence the solution of $\sin x = 0.5$ for $-360° \leqslant x \leqslant 360°$ is:

$$x = -330°, -210°, 30° \text{ or } 150°$$

WORKED EXAMPLE 10

Solve $\cos x = -0.4$ for $0° \leqslant x \leqslant 360°$.

Answers

$\cos x = -0.4$ use a calculator to find $\cos^{-1}(-0.4)$ to 1 decimal place

$\quad x = 113.6°$

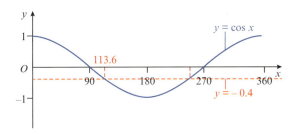

The sketch graph shows there are two values of x, between $0°$ and $360°$, for which $\cos x = -0.4$.

Using the symmetry of the curve, the second value is $(360° - 113.6°) = 246.4°$

Hence the solution of $\cos x = -0.4$ for $0° \leqslant x \leqslant 360°$ is

$x = 113.6°$ or $246.4°$

WORKED EXAMPLE 11

Solve $\tan 2A = -1.8$ for $0° \leqslant A \leqslant 180°$.

Answers

$\tan 2A = -1.8$ let $x = 2A$

$\tan x = -1.8$ use a calculator to find $\tan^{-1}(-1.8)$

$x = -60.95°$

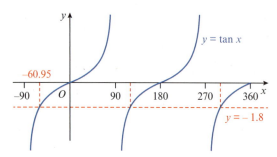

Using the symmetry of the curve:

$x = -60.95$ $x = (-60.95 + 180)$ $x = (119.05 + 180)$

                  $= 119.05$ $= 299.05$

Using $x = 2A$,

    $2A = -60.95$ $2A = 119.05$ $2A = 299.05$

      $A = -30.5$ $A = 59.5$ $A = 149.5$

Hence the solution of $\tan 2A = -1.8$ for $0° \leqslant x \leqslant 180°$ is

$A = 59.5°$ or $A = 149.5°$

WORKED EXAMPLE 12

Solve $\sin\left(2A - \dfrac{\pi}{3}\right) = 0.6$ for $0 \le A \le \pi$.

Answers

$\sin\left(2A - \dfrac{\pi}{3}\right) = 0.6$ let $x = 2A - \dfrac{\pi}{3}$

$\quad\quad \sin x = 0.6$ use a calculator to find $\sin^{-1}0.6$

$\quad\quad\quad x = 0.6435$ radians

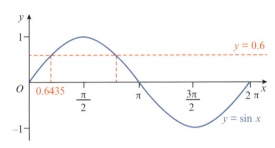

Using the symmetry of the curve:

$x = 0.6435 \quad\quad x = \pi - 0.6435$

$\quad\quad\quad\quad\quad\quad\quad = 2.498$

Using $x = 2A - \dfrac{\pi}{3}$

$2A - \dfrac{\pi}{3} = 0.6435 \quad\quad\quad 2A - \dfrac{\pi}{3} = 2.498$

$\quad A = \dfrac{1}{2}\left(0.6435 + \dfrac{\pi}{3}\right) \quad\quad A = \dfrac{1}{2}\left(2.498 + \dfrac{\pi}{3}\right)$

$\quad A = 0.845 \quad\quad\quad\quad\quad\quad A = 1.77$

Hence the solution of $\sin\left(2A - \dfrac{\pi}{3}\right) = 0.6$ for $0 \le A \le \pi$ is

$A = 0.845$ or 1.77 radians

WORKED EXAMPLE 13

Solve $\sin^2 x - 2\sin x\cos x = 0$ for $0° \le x \le 360°$.

Answers

$\sin^2 x - 2\sin x\cos x = 0$ factorise

$\sin x(\sin x - 2\cos x) = 0$

$\sin x = 0 \quad\quad\quad$ or $\sin x - 2\cos x = 0$

$\quad x = 0°, 180°, 360° \quad\quad\quad \sin x = 2\cos x$

$\quad\quad\quad\quad\quad\quad\quad\quad\quad\quad \tan x = 2$

$\quad\quad\quad\quad\quad\quad\quad\quad\quad\quad x = 63.4$ or $180 + 63.4$

$\quad\quad\quad\quad\quad\quad\quad\quad\quad\quad x = 63.4°$ or $243.4°$

CONTINUED

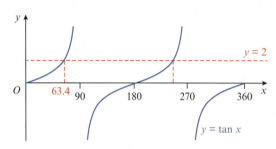

The solution of $\sin^2 x - 2\sin x \cos x = 0$ for $0° \leqslant x \leqslant 360°$ is

$x = 0°, 63.4°, 180°, 243.4°$ or $360°$

WORKED EXAMPLE 14

Solve $1 + \cos x = 3\sin^2 x$ for $0° \leqslant x \leqslant 360°$.

Answers

$$1 + \cos x = 3\sin^2 x \qquad \text{use } \sin^2 x = 1 - \cos^2 x$$

$$1 + \cos x = 3(1 - \cos^2 x) \qquad \text{expand brackets and collect terms}$$

$$3\cos^2 x + \cos x - 2 = 0 \qquad \text{factorise}$$

$$(3\cos x - 2)(\cos x + 1) = 0$$

$$3\cos x - 2 = 0 \qquad \text{or} \qquad \cos x + 1 = 0$$

$$\cos x = \frac{2}{3} \qquad\qquad \cos x = -1$$

$$x = 48.2° \text{ or } 360 - 48.2 \qquad x = 180°$$

$$x = 48.2° \text{ or } 311.8°$$

The solution of $1 + \cos x = 3\sin^2 x$ for $0° \leqslant x \leqslant 360°$ is

$x = 48.2°, 180°$ or $311.8°$

Exercise 9.6

1 Solve each of these equations for $0° \leqslant x \leqslant 360°$.

    a $\sin x = 0.3$ **b** $\cos x = 0.2$ **c** $\tan x = 2$

    d $\sin x = -0.6$ **e** $\tan x = -1.4$ **f** $\sin x = -0.8$

    g $4\sin x - 3 = 0$ **h** $2\cos x + 1 = 0$

2 Solve each of these equations for $0 \leqslant x \leqslant 2\pi$.

    a $\cos x = 0.5$ **b** $\tan x = 0.2$ **c** $\sin x = 2$

    d $\tan x = -3$ **e** $\sin x = -0.75$ **f** $\cos x = -0.55$

    g $4\sin x = 1$ **h** $5\sin x + 2 = 0$

3 Solve each of these equations for $0° \leqslant x \leqslant 180°$.

    a $\sin 2x = 0.8$ **b** $\cos 2x = -0.6$ **c** $\tan 2x = 2$

    d $\sin 2x = -0.6$ **e** $5\cos 2x = 4$ **f** $7\sin 2x = -2$

    g $1 + 3\tan 2x = 0$ **h** $2 - 3\sin 2x = 0$

4 Solve each of these equations for the given domains.

    a $\cos(x - 30°) = -0.5$ for $0° \leqslant x \leqslant 360°$

    b $6\sin(2x + 45°) = -5$ for $0° \leqslant x \leqslant 180°$

    c $2\cos\left(\dfrac{2x}{3}\right) + \sqrt{3} = 0$ for $0° \leqslant x \leqslant 540°$

    d $\cos\left(x + \dfrac{\pi}{6}\right) = -0.5$ for $0 < x < 2\pi$ radians

    e $\sin(2x - 3) = 0.6$ for $0 < x < \pi$ radians

    f $\sqrt{2}\sin\left(\dfrac{x}{2} + \dfrac{\pi}{6}\right) = 1$ for $0 < x < 4\pi$ radians

5 Solve each of these equations for $0° \leqslant x \leqslant 360°$.

    a $4\sin x = \cos x$ **b** $3\sin x + 4\cos x = 0$

    c $5\sin x - 3\cos x = 0$ **d** $5\cos 2x - 4\sin 2x = 0$

6 Solve $4\sin(2x - 0.4) - 5\cos(2x - 0.4) = 0$ for $0 \leqslant x \leqslant \pi$.

7 Solve each of these equations for $0° \leqslant x \leqslant 360°$.

    a $\sin x \tan(x - 30°) = 0$ **b** $5\tan^2 x - 4\tan x = 0$

    c $3\cos^2 x = \cos x$ **d** $\sin^2 x + \sin x \cos x = 0$

    e $5\sin x \cos x = \cos x$ **f** $\sin x \tan x = \sin x$

8 Solve each of these equations for $0° \leqslant x \leqslant 360°$.

    a $4\sin^2 x = 1$ **b** $25\tan^2 x = 9$

9 Solve each of these equations for $0° \leqslant x \leqslant 360°$.

    a $\tan^2 x + 2\tan x - 3 = 0$ **b** $2\sin^2 x + \sin x - 1 = 0$

    c $3\cos^2 x - 2\cos x - 1 = 0$ **d** $2\sin^2 x - \cos x - 1 = 0$

    e $3\cos^2 x - 3 = \sin x$ **f** $\sin x + 5 = 6\cos^2 x$

    g $2\cos^2 x - \sin^2 x - 2\sin x - 1 = 0$ **h** $1 + \tan x \cos x = 2\cos^2 x$

    i $3\cos x = 8\tan x$

10 $f(x) = \sin x$ for $0 \leqslant x \leqslant \dfrac{\pi}{2}$ $g(x) = 2x - 1$ for $x \in \mathbb{R}$

    Solve $gf(x) = 0.5$

9.7 Trigonometric identities

$\sin^2 x + \cos^2 x = 1$ is called a **trigonometric identity** because it is true for all values of x.

In this section you will learn how to prove more complicated identities involving $\sin x$, $\cos x$ and $\tan x$.

When proving an identity, it is usual to start with the more complicated side of the identity and prove that it simplifies to the less complicated side. This is illustrated in the next example.

> **Note:**
> LHS means left-hand side and RHS means right-hand side.

WORKED EXAMPLE 15

Prove the identity $(1 + \sin x)^2 + (1 - \sin x)^2 + 2\cos^2 x = 4$

Answers

$$\begin{aligned}
\text{LHS} &= (1 + \sin x)^2 + (1 - \sin x)^2 + 2\cos^2 x \\
&= (1 + \sin x)(1 + \sin x) + (1 - \sin x)(1 - \sin x) + 2\cos^2 x \quad \text{expand brackets} \\
&= 1 + 2\sin x + \sin^2 x + 1 - 2\sin x + \sin^2 x + 2\cos^2 x \quad \text{collect like terms} \\
&= 2 + 2\sin^2 x + 2\cos^2 x \\
&= 2 + 2(\sin^2 x + \cos^2 x) \quad \text{use } \sin^2 x + \cos^2 x = 1 \\
&= 2 + 2 \times 1 \\
&= 4 \\
&= \text{RHS}
\end{aligned}$$

WORKED EXAMPLE 16

Prove the identity $\dfrac{1}{\cos x} - \cos x = \sin x \tan x$

Answers

$$\begin{aligned}
\text{LHS} &= \frac{1}{\cos x} - \cos x \\
&= \frac{1}{\cos x} - \frac{\cos^2 x}{\cos x} \\
&= \frac{1 - \cos^2 x}{\cos x} \quad \text{use } 1 - \cos^2 x = \sin^2 x \\
&= \frac{\sin^2 x}{\cos x} \\
&= \sin x \times \frac{\sin x}{\cos x} \quad \text{use } \frac{\sin x}{\cos x} = \tan x \\
&= \sin x \tan x \\
&= \text{RHS}
\end{aligned}$$

CLASS DISCUSSION

Equivalent trigonometric expressions

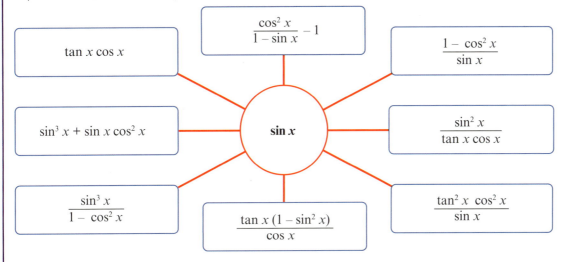

Discuss why each of the trigonometric expressions in the blue boxes simplify to $\sin x$.

Create as many trigonometric expressions of your own which simplify to $\tan x$.

(Your expressions must contain at least two different trigonometric ratios.)

Compare your answers with your classmates.

Exercise 9.7

1 Prove each of these identities.

a $\dfrac{\sin x}{\tan x} = \cos x$

b $\dfrac{\cos x \sin x}{\tan x} = 1 - \sin^2 x$

c $\dfrac{1 - \sin^2 x}{\cos x} = \cos x$

d $\dfrac{\cos^2 x - \sin^2 x}{\cos x + \sin x} + \sin x = \cos x$

e $(\sin x + \cos x)^2 = 1 + 2 \sin x \cos x$

f $\tan^2 x - \sin^2 x = \tan^2 x \sin^2 x$

2 Prove each of these identities.

a $\cos^2 x - \sin^2 x = 2 \cos^2 x - 1$

b $\cos^2 x - \sin^2 x = 1 - 2 \sin^2 x$

c $\cos^4 x + \sin^2 x \cos^2 x = \cos^2 x$

d $2(1 + \cos x) - (1 + \cos x)^2 = \sin^2 x$

e $2 - (\sin x + \cos x)^2 = (\sin x - \cos x)^2$

f $\cos^4 x + \sin^2 x = \sin^4 x + \cos^2 x$

3 Prove each of these identities.

a $\dfrac{\cos^2 x - \sin^2 x}{\cos x - \sin x} = \cos x + \sin x$

b $\dfrac{\sin x}{1 + \cos x} + \dfrac{1 + \cos x}{\sin x} = \dfrac{2}{\sin x}$

c $\dfrac{\cos^4 x - \sin^4 x}{\cos^2 x} = 1 - \tan^2 x$

d $\dfrac{\sin^2 x (1 - \cos^2 x)}{\cos^2 x (1 - \sin^2 x)} = \tan^4 x$

9.8 Further trigonometric equations

The cosecant, secant and cotangent ratios

There are a total of six trigonometric ratios. You have already met the ratios sine, cosine and tangent. In this section you will learn about the other three ratios, which are **cosecant** (cosec), **secant** (sec) and **cotangent** (cot). These three ratios are defined as:

$$\operatorname{cosec} \theta = \frac{1}{\sin \theta} \qquad \sec \theta = \frac{1}{\cos \theta} \qquad \cot \theta = \frac{1}{\tan \theta} \left(= \frac{\cos \theta}{\sin \theta} \right)$$

Consider the right-angled triangle:

$$\sin \theta = \frac{y}{r} \qquad \cos \theta = \frac{x}{r} \qquad \tan \theta = \frac{y}{x}$$

$$\operatorname{cosec} \theta = \frac{r}{y} \qquad \sec \theta = \frac{r}{x} \qquad \cot \theta = \frac{x}{y}$$

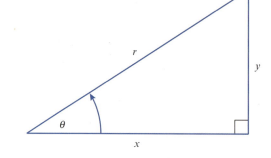

The following rules can be found from this triangle:

$x^2 + y^2 = r^2$ divide both sides by x^2

$1 + \left(\dfrac{y}{x}\right)^2 = \left(\dfrac{r}{x}\right)^2$ use $\dfrac{y}{x} = \tan \theta$ and $\dfrac{r}{x} = \sec \theta$

$$1 + \tan^2 \theta = \sec^2 \theta$$

$x^2 + y^2 = r^2$ divide both sides by y^2

$\left(\dfrac{x}{y}\right)^2 + 1 = \left(\dfrac{r}{y}\right)^2$ use $\dfrac{x}{y} = \cot \theta$ and $\dfrac{r}{y} = \operatorname{cosec} \theta$

$$\cot^2 \theta + 1 = \operatorname{cosec}^2 \theta$$

These important identities will be needed to solve trigonometric equations in this section.

WORKED EXAMPLE 17

Solve $2\cosec^2 x + \cot x - 8 = 0$ for $0° \leqslant x \leqslant 360°$

Answers

$2\cosec^2 x + \cot x - 8 = 0$ use $1 + \cot^2 x = \cosec^2 x$

$2(1 + \cot^2 x) + \cot x - 8 = 0$ expand brackets and collect terms

$2\cot^2 x + \cot x - 6 = 0$ factorise

$(2\cot x - 3)(\cot x + 2) = 0$

$2\cot x - 3 = 0$ or $\cot x + 2 = 0$

$\cot x = \dfrac{3}{2}$ $\qquad\qquad$ $\cot x = -2$

$\tan x = \dfrac{2}{3}$ $\qquad\qquad$ $\tan x = -\dfrac{1}{2}$

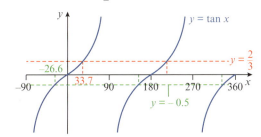

$x = 33.7$ or $(33.7 + 180)$ $\qquad$ $x = -26.6$ or $(-26.6 + 180)$ or $(-26.6 + 360)$

$x = 33.7°$ or $213.7°$ $\qquad$ $x = 153.4°$ or $333.4°$ (since $-26.6°$ is out of range)

The solution of $2\cosec^2 x + \cot x - 8 = 0$ for $0° \leqslant x \leqslant 360°$ is

$$x = 33.7°, 153.4°, 213.7° \text{ or } 333.4°$$

Exercise 9.8

1 Solve each of these equations for $0° \leqslant x \leqslant 360°$.

   a $\cot x = 0.3$ $\quad$ **b** $\sec x = 4$ $\quad$ **c** $\cosec x = -2$ $\quad$ **d** $3\sec x - 5 = 0$

2 Solve each of these equations for $0 \leqslant x \leqslant 2\pi$.

   a $\cosec x = 5$ $\quad$ **b** $\cot x = 0.8$ $\quad$ **c** $\sec x = -4$ $\quad$ **d** $2\cot x + 3 = 0$

3 Solve each of these equations for $0° \leqslant x \leqslant 180°$.

   a $\sec 2x = 1.6$ $\quad$ **b** $\cosec 2x = 5$ $\quad$ **c** $\cot 2x = -1$ $\quad$ **d** $5\cosec 2x = -7$

4 Solve each of these equations for the given domains.

   a $\sec(x - 30°) = 3$ for $0° \leqslant x \leqslant 360°$

   b $\cosec(2x + 45°) = -5$ for $0° \leqslant x \leqslant 180°$

   c $\cot\left(x + \dfrac{\pi}{3}\right) = 2$ for $0 < x < 2\pi$

   d $3\sec(2x + 3) = 4$ for $0 < x < \pi$

5 Solve each of these equations for $0° \leqslant x \leqslant 360°$.

   a $\sec^2 x = 4$ $\qquad\qquad$ **b** $9\cot^2 x = 4$ $\qquad\qquad$ **c** $16\cot^2 \dfrac{1}{2}x = 9$

6 Solve each of these equations for $0° \leqslant x \leqslant 360°$.

a $3\tan^2 x - \sec x - 1 = 0$ **b** $4\tan^2 x + 8\sec x = 1$

c $2\sec^2 x = 5\tan x + 5$ **d** $2\cot^2 x - 5\csc x - 1 = 0$

e $6\cos x + 6\sec x = 13$ **f** $\cot x + 6\sin x - 2\cos x = 3$

g $3\cot x = 2\sin x$ **h** $12\sec x - 10\cos x - 9\tan x = 0$

9.9 Further trigonometric identities

In this section you will learn how to prove trigonometric identities that involve any of the six trigonometric ratios.

CLASS DISCUSSION

Odd one out

Find the trigonometric expression that does not match the other six expressions.

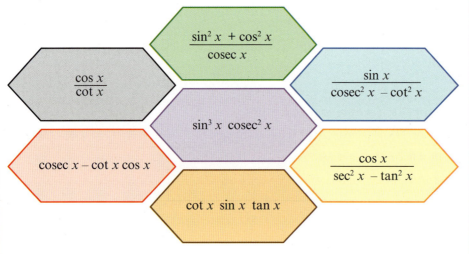

Create as many expressions of your own to match the 'odd one out'.

(Your expressions must contain at least two different trigonometric ratios.)

Compare your answers with your classmates.

WORKED EXAMPLE 18

Prove the identity $\dfrac{1 + \sec x}{\tan x + \sin x} = \operatorname{cosec} x$

Answers

$\text{LHS} = \dfrac{1 + \sec x}{\tan x + \sin x}$ use $\sec x = \dfrac{1}{\cos x}$ and $\tan x = \dfrac{\sin x}{\cos x}$

$= \dfrac{1 + \dfrac{1}{\cos x}}{\dfrac{\sin x}{\cos x} + \sin x}$ multiply numerator and denominator by $\cos x$

$= \dfrac{\cos x + 1}{\sin x + \sin x \cos x}$ factorise the denominator

$= \dfrac{\cos x + 1}{\sin x \,(\cos x + 1)}$ divide numerator and denominator by $(\cos x + 1)$

$= \dfrac{1}{\sin x}$ use $\operatorname{cosec} x = \dfrac{1}{\sin x}$

$= \operatorname{cosec} x$

$= \text{RHS}$

Exercise 9.9

1   Prove each of these identities.

   a $\tan x + \cot x = \sec x \operatorname{cosec} x$ **b** $\sin x + \cos x \cot x = \operatorname{cosec} x$

   c $\operatorname{cosec} x - \sin x = \cos x \cot x$ **d** $\sec x \operatorname{cosec} x - \cot x = \tan x$

2   Prove each of these identities.

   a $(1 + \sec x)(\operatorname{cosec} x - \cot x) = \tan x$

   b $(1 + \sec x)(1 - \cos x) = \sin x \tan x$

   c $\tan^2 x - \sec^2 x + 2 = \operatorname{cosec}^2 x - \cot^2 x$

   d $(\cot x + \tan x)(\cot x - \tan x) = \operatorname{cosec}^2 x - \sec^2 x$

3   Prove each of these identities.

   a $\dfrac{1}{\tan x + \cot x} = \sin x \cos x$ **b** $\dfrac{\sin^2 x + \cos^2 x}{\cos^2 x} = \sec^2 x$

   c $\dfrac{\sin^2 x \cos x + \cos^3 x}{\sin x} = \cot x$ **d** $\dfrac{1 - \cos^2 x}{\sec^2 x - 1} = 1 - \sin^2 x$

   e $\dfrac{1 + \tan^2 x}{\tan x} = \sec x \operatorname{cosec} x$ **f** $\dfrac{\sin x}{1 - \cos^2 x} = \operatorname{cosec} x$

   g $\dfrac{\sin x \tan^2 x}{1 + \tan^2 x} = \sin^3 x$ **h** $\dfrac{1 - \sec^2 x}{1 - \operatorname{cosec}^2 x} = \tan^4 x$

   i $\dfrac{1 + \sin x}{1 - \sin x} = (\tan x + \sec x)^2$ **j** $\dfrac{\cos x \cot x}{\cos x + \cot x} = \sec x - \tan x$

4 Prove each of these identities.

a $\dfrac{\sin x}{\cos x} + \dfrac{\cos x}{\sin x} = \sec x \operatorname{cosec} x$

b $\dfrac{1}{1 - \sin x} - \dfrac{1}{1 + \sin x} = 2 \tan x \sec x$

c $\dfrac{1}{1 + \cos x} + \dfrac{1}{1 - \cos x} = 2 \operatorname{cosec}^2 x$

d $\dfrac{\cos x}{1 - \tan x} + \dfrac{\sin x}{1 - \cot x} = \sin x + \cos x$

e $\dfrac{\cos x}{1 + \sin x} + \dfrac{\cos x}{1 - \sin x} = 2 \sec x$

f $\dfrac{\cos x}{\operatorname{cosec} x + 1} + \dfrac{\cos x}{\operatorname{cosec} x - 1} = 2 \tan x$

5 Show that $(3 + 2\sin x)^2 + (3 - 2\sin x)^2 + 8\cos^2 x$ has a constant value for all x and state this value.

6 **a** Express $5\sin^2 x - 2\cos^2 x$ in the form $a + b\sin^2 x$.

b State the range of the function $f(x) = 5\sin^2 x - 2\cos^2 x$ for $0 \leqslant x \leqslant 2\pi$.

7 **CHALLENGE QUESTION**

a Express $\sin^2 \theta + 4\cos\theta + 2$ in the form $a - (\cos\theta - b)^2$.

b Hence state the maximum and minimum values of $\sin^2 \theta + 4\cos\theta + 2$.

SUMMARY

Positive and negative angles

Angles measured anticlockwise from the positive x-direction are positive.

Angles measured clockwise from the positive x-direction are negative.

Diagram showing where sin, cos and tan are positive

Useful mnemonic: '**A**ll **S**tudents **T**rust **C**ambridge'

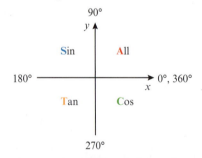

Cosecant, secant and cotangent

$\operatorname{cosec}\theta = \dfrac{1}{\sin\theta} \qquad \sec\theta = \dfrac{1}{\cos\theta} \qquad \cot\theta = \dfrac{1}{\tan\theta}$

Trigonometric identities

$\tan x = \dfrac{\sin x}{\cos x} \qquad \sin^2 x + \cos^2 x = 1 \qquad 1 + \tan^2 x = \sec^2 x \qquad 1 + \cot^2 x = \operatorname{cosec}^2 x$

Past paper questions

Worked example

a   Given that $y = 7\cos 10x - 3$, where the angle x is measured in degrees, state

    i   the period of y [1]

    ii   the amplitude of y.   [1]

b

Find the equation of the curve shown, in the form $y = ag(bx) + c$, where $g(x)$ is a trigonometric function and a, b and c are integers to be found.   [4]

Cambridge IGCSE Additional Mathematics 0606 Paper 21 Q4 Jun 2017

Answers

a   i   Period $= \dfrac{360°}{10} = 36°$ ii   Amplitude $= 7$

b   The graph is a translation of a sine curve by the vector $\begin{pmatrix} 0 \\ 7 \end{pmatrix}$

    Amplitude $= \dfrac{12 - 2}{2} = 5$

    Period $= 90° = \dfrac{360°}{b}$

      $b = 4$

Hence the graph is $y = 5\sin 4x + 7$

1   a   Sketch the curve $y = 3\cos 2x - 1$ for $0° \leqslant x \leqslant 180°$.   [3]

    b   i   State the amplitude of $1 - 4\sin 2x$.   [1]

        ii   State the period of $5\tan 3x + 1$.   [1]

Adapted from Cambridge IGCSE Additional Mathematics 0606 Paper 11 Q2 Nov 2014

2   a   Solve $2\cos 3x = \cot 3x$ for $0° \leqslant x \leqslant 90°$.   [5]

    b   Solve $\sec\left(y + \dfrac{\pi}{2}\right) = -2$ for $0 \leqslant y \leqslant \pi$ radians.   [4]

Cambridge IGCSE Additional Mathematics 0606 Paper 11 Q11 Nov 2014

3   a   Prove that $\sec x \operatorname{cosec} x - \cot x = \tan x$.   [4]

    b   Use the result from **part a** to solve the equation $\sec x \operatorname{cosec} x = 3\cot x$ for $0° < x < 360°$.   [4]

Cambridge IGCSE Additional Mathematics 0606 Paper 21 Q10 Nov 2014

4 a   Solve $4\sin x = \operatorname{cosec} x$ for $0° \leqslant x \leqslant 360°$.   [3]

   b   Solve $\tan^2 3y - 2\sec 3y - 2 = 0$ for $0° \leqslant y \leqslant 180°$.   [6]

   c   Solve $\tan\left(z - \dfrac{\pi}{3}\right) = \sqrt{3}$ for $0 \leqslant z \leqslant 2\pi$ radians.   [3]

Cambridge IGCSE Additional Mathematics 0606 Paper 11 Q10 Jun 2015

5 Show that $\sqrt{\sec^2 \theta - 1} + \sqrt{\operatorname{cosec}^2 \theta - 1} = \sec \theta \operatorname{cosec} \theta$ [5]

Cambridge IGCSE Additional Mathematics 0606 Paper 11 Q3 Nov 2015

6 Solve the following equations.

   i $4\sin 2x + 5\cos 2x = 0$ for $0° \leqslant x \leqslant 180°$ [3]

   ii $\cot^2 y + 3\operatorname{cosec} y = 3$ for $0° \leqslant y \leqslant 360°$ [5]

   iii $\cos\left(z + \dfrac{\pi}{4}\right) = -\dfrac{1}{2}$ for $0 \leqslant z \leqslant 2\pi$ radians, giving each answer as a multiple of π [4]

Cambridge IGCSE Additional Mathematics 0606 Paper 21 Q9 Nov 2015

7 a $f(x) = a\cos bx + c$ has a period of $60°$, an amplitude of 10 and is such $f(0) = 14$.
       State the values of a, b and c.   [2]

   b   Sketch the graph of $y = 3\sin 4x - 2$ for $0° \leqslant x \leqslant 180°$ on the axes below.   [3]

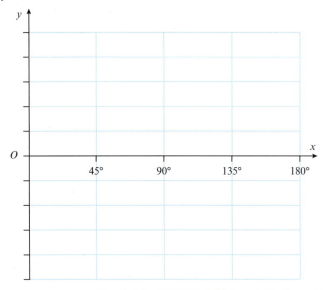

Cambridge IGCSE Additional Mathematics 0606 Paper 22 Q4 Mar 2016

8 i   Show that $2\cos x \cot x + 1 = \cot x + 2\cos x$ can be written in the form
        $(a\cos x - b)(\cos x - \sin x) = 0$ where a and b are constants to be found.   [4]

   ii   Hence, or otherwise, solve $2\cos x \cot x + 1 = \cot x + 2\cos x$ for $0 < x < \pi$.   [3]

Cambridge IGCSE Additional Mathematics 0606 Paper 11 Q9 Jun 2016

9 i   Prove that $\dfrac{\cos x}{1 + \tan x} - \dfrac{\sin x}{1 + \cot x} = \cos x - \sin x$.   [4]

   ii   Hence solve the equation $\dfrac{\cos x}{1 + \tan x} - \dfrac{\sin x}{1 + \cot x} = 3\sin x - 4\cos x$ for $-180° < x < 180°$.   [4]

Cambridge IGCSE Additional Mathematics 0606 Paper 21 Q6 Nov 2016

10 i On the axes below, sketch, for $0° \leqslant x \leqslant 360°$, the graph of $y = 1 + 3\cos 2x$. [3]

ii Write down the coordinates of the point where this graph first has a minimum value. [1]

Cambridge IGCSE Additional Mathematics 0606 Paper 12 Q2 Mar 2017

11 a Solve $3\operatorname{cosec} 2x - 4\sin 2x = 0$ for $0° \leqslant x \leqslant 180°$. [4]

b Solve $3\tan\left(y - \dfrac{\pi}{4}\right) = \sqrt{3}$ for $0 \leqslant y \leqslant 2\pi$ radians, giving your answers in terms of π. [4]

Cambridge IGCSE Additional Mathematics 0606 Paper 11 Q10 Nov 2017

12 a Solve $\sin x \cos x = \dfrac{1}{2}\tan x$ for $0° \leqslant x \leqslant 180°$. [3]

b i Show that $\sec\theta - \dfrac{\sin\theta}{\cot\theta} = \cos\theta$. [3]

ii Hence solve $\sec 3\theta - \dfrac{\sin 3\theta}{\cot 3\theta} = \dfrac{1}{2}$ for $-\dfrac{2\pi}{3} \leqslant \theta \leqslant \dfrac{2\pi}{3}$, where θ is in radians. [4]

Cambridge IGCSE Additional Mathematics 0606 Paper 12 Q11 Mar 2019

13 a i Show that $\dfrac{\operatorname{cosec}\theta - \cot\theta}{\sin\theta} = \dfrac{1}{1 + \cos\theta}$. [4]

ii Hence solve $\dfrac{\operatorname{cosec}\theta - \cot\theta}{\sin\theta} = \dfrac{5}{2}$ for $180° < \theta < 360°$. [2]

b Solve $\tan(3\varphi - 4) = -\dfrac{1}{2}$ for $0 \leqslant \varphi \leqslant \dfrac{\pi}{2}$ radians. [3]

Cambridge IGCSE Additional Mathematics 0606 Paper 21 Q11 Jun 2019

Chapter 10

Permutations and combinations

THIS SECTION WILL SHOW YOU HOW TO:

- find the number of arrangements of n distinct items
- find the number of permutations of r items from n distinct items
- find the number of combinations of r items from n distinct items
- solve problems using permutations and combinations.

10.1 Factorial notation

$5 \times 4 \times 3 \times 2 \times 1$ is called '5 factorial' and is written as 5!

It is useful to remember that $n! = n \times (n-1)!$

For example, $5! = 5 \times 4!$

WORKED EXAMPLE 1

Find the value of

a $\dfrac{8!}{5!}$

b $\dfrac{11!}{8!3!}$

Answers

a $\dfrac{8!}{5!} = \dfrac{8 \times 7 \times 6 \times \cancel{5} \times \cancel{4} \times \cancel{3} \times \cancel{2} \times \cancel{1}}{\cancel{5} \times \cancel{4} \times \cancel{3} \times \cancel{2} \times \cancel{1}}$

$= 8 \times 7 \times 6$

$= 336$

b $\dfrac{11!}{8!3!} = \dfrac{11 \times 10 \times 9 \times \cancel{8} \times \cancel{7} \times \cancel{6} \times \cancel{5} \times \cancel{4} \times \cancel{3} \times \cancel{2} \times \cancel{1}}{\cancel{8} \times \cancel{7} \times \cancel{6} \times \cancel{5} \times \cancel{4} \times \cancel{3} \times \cancel{2} \times \cancel{1} \times 3 \times 2 \times 1}$

$= \dfrac{990}{6}$

$= 165$

Exercise 10.1

1   Without using a calculator, find the value of each of the following.

   Use the $x!$ key on your calculator to check your answers.

   a $7!$
   b $\dfrac{4!}{2!}$
   c $\dfrac{7!}{3!}$
   d $\dfrac{8!}{5!}$

   d $\dfrac{4!}{2!2!}$
   e $\dfrac{6!}{3!2!}$
   f $\dfrac{6!}{(3!)^2}$
   g $\dfrac{5!}{3!} \times \dfrac{7!}{4!}$

2   Rewrite each of the following using factorial notation.

   a 2×1
   b $6 \times 5 \times 4 \times 3 \times 2 \times 1$
   c $5 \times 4 \times 3$

   d $17 \times 16 \times 15 \times 14$
   e $\dfrac{10 \times 9 \times 8}{3 \times 2 \times 1}$
   f $\dfrac{12 \times 11 \times 10 \times 9 \times 8}{4 \times 3 \times 2 \times 1}$

3   **CHALLENGE QUESTION**

   Rewrite each of the following using factorial notation.

   a $n(n-1)(n-2)(n-3)$
   b $n(n-1)(n-2)(n-3)(n-4)(n-5)$

   c $\dfrac{n(n-1)(n-2)}{5 \times 4 \times 3 \times 2 \times 1}$
   d $\dfrac{n(n-1)(n-2)(n-3)(n-4)}{3 \times 2 \times 1}$

10.2 Arrangements

To find the number of ways of arranging the letters A, B and C in a line you can use two methods.

Method 1

List all the possible arrangements.

These are: ABC ACB BAC BCA CAB and CBA.

There are 6 different arrangements.

Method 2

Consider filling 3 spaces.

The first space can be filled in **3** ways with either A or B or C.

For each of these **3** ways of filling the first space, there are **2** ways of filling the second space.

There are **3 × 2** ways of filling the first and second spaces.

For each of the ways of filling the first and second spaces there is just **1** way of filling the third space.

There are **3 × 2 × 1** ways of filling the three spaces.

The number of arrangements = 3 × 2 × 1 = 6

3 × 2 × 1 is called '3 factorial' and can be written as 3!

In the class discussion, you should have found that there were 24 different ways of arranging the 4 books.

$4! = 4 \times 3 \times 2 \times 1 = 24$

> The number of ways of arranging n distinct items in a line = $n!$

WORKED EXAMPLE 2

G R A D I E N T S

a Find the number of different arrangements of these nine cards, if there are no restrictions.

b Find the number of arrangements that begin with **GRAD**.

c Find the number of arrangements that begin with **G** and end with **S**.

Answers

a There are 9 different cards.

number of arrangements = 9! = 362 880

b The first four letters are **GRAD**, so there are now only 5 letters left to be arranged.

number of arrangements = 5! = 120

c The first and last letters are fixed, so there are now 7 letters to arrange between the **G** and the **S**.

number of arrangements = 7! = 5040

WORKED EXAMPLE 3

a Find the number of different arrangements of these seven objects, if there are no restrictions.

b Find the number of arrangements where the squares and circles alternate.

c Find the number of arrangements where all the squares are together.

d Find the number of arrangements where the squares are together and the circles are together.

Answers

a There are 7 different objects.

number of arrangements = 7! = 5040

b If the squares and circles alternate, a possible arrangement is:

There are 4! different ways of arranging the four squares.

There are 3! different ways of arranging the three circles.

So, the total number of possible arrangements = 4! × 3! = 24 × 6 = 144

CONTINUED

c   If the squares are all together, a possible arrangement is:

The number of ways of arranging the 1 block of four squares and the 3 circles = 4!

There are 4! ways of arranging the four squares within the block of squares.

So, the total number of possible arrangements = 4! × 4! = 24 × 24 = 576

d   If the squares are together and the circles are together, a possible arrangement is:

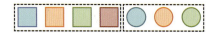

There are 4! × 3! ways of having the squares at the start and the circles at the end.

Another possible arrangement is:

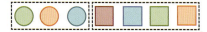

There are 3! × 4! ways of having the circles at the start and the squares at the end.

total number of arrangements = 4! × 3! + 3! × 4! = 144 + 144 = 288

Exercise 10.2

1   Find the number of different arrangements of

   a   4 people sitting in a row on a bench

   b   7 different books on a shelf.

2   Find the number of different arrangements of letters in each of the following words.

   a   TIGER   b   OLYMPICS   c   PAINTBRUSH

3   a   Find the number of different four-digit numbers that can be formed using the digits 3, 5, 7 and 8 without repetition.

   b   How many of these four-digit numbers are

      i   even   ii   greater than 8000?

4   A shelf holds 7 different books.

   Four of the books are cookery books and three of the books are history books.

   a   Find the number of ways the books can be arranged if there are no restrictions.

   b   Find the number of ways the books can be arranged if the 4 cookery books are kept together.

5   Five-digit numbers are to be formed using the digits 2, 3, 4, 5 and 6.

   Each digit may be used only once in any number.

   a   Find how many different five-digit numbers can be formed.

How many of these five-digit numbers are

b   even

c   greater than 40 000

d   even and greater than 40 000?

6   Three girls and two boys are to be seated in a row.

Find the number of different ways that this can be done if

a   the girls and boys sit alternately

b   a girl sits at each end of the row

c   the girls sit together and the boys sit together.

7   **a**   Find the number of different arrangements of the letters in the word ORANGE.

Find the number of these arrangements that

b   begin with the letter O

c   have the letter O at one end and the letter E at the other end.

8   **a**   Find the number of different six-digit numbers which can be made using the digits 0, 1, 2, 3, 4 and 5 without repetition. Assume that a number cannot begin with 0.

b   How many of the six-digit numbers in **part a** are even?

9   Six girls and two boys are to be seated in a row.

Find the number of ways that this can be done if the two boys must have exactly four girls seated between them.

10.3 Permutations

In the last section, you learned that if you had three letters A, B and C and 3 spaces to fill, then the number of ways of filling the spaces was $3 \times 2 \times 1 = 3!$

Now consider having 8 letters A, B, C, D, E, F, G, H and 3 spaces to fill.

The first space can be filled in **8** ways.

For each of these **8** ways of filling the first space, there are **7** ways of filling the second space.

There are **8 × 7** ways of filling the first and second spaces.

For each of the ways of filling the first and second spaces there are **6** ways of filling the third space.

There are **8 × 7 × 6** ways of filling the three spaces.

The number of different ways of arranging three letters chosen from eight letters

$= 8 \times 7 \times 6 = 336$.

The different arrangements of the letters are called **permutations**.

The notation $^8P_3$ is used to represent the number of permutations of 3 items chosen from 8 items.

Note that $8 \times 7 \times 6$ can also be written as $\dfrac{8 \times 7 \times 6 \times 5 \times 4 \times 3 \times 2 \times 1}{5 \times 4 \times 3 \times 2 \times 1} = \dfrac{8!}{5!} = \dfrac{8!}{(8-3)!}$

So $^8P_3 = \dfrac{8!}{(8-3)!} = \dfrac{8!}{5!} = 336$

> The general rule for finding the number of permutations of r items from n distinct items is $^nP_r = \dfrac{n!}{(n-r)!}$

Note:

- In permutations, order matters.
- By definition, $0! = 1$

To explain why $0! = 1$, consider finding the number of permutations of 5 letters taken from 5 letters.

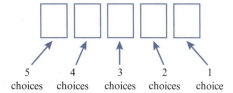

5 choices	4 choices	3 choices	2 choices	1 choice

The number of ways of filling the 5 spaces with the 5 letters $= 5 \times 4 \times 3 \times 2 \times 1 = 120$.

But, $^5P_5 = \dfrac{5!}{(5-5)!} = \dfrac{5!}{0!} = \dfrac{120}{0!}$

So $\dfrac{120}{0!} = 120$. Hence $0!$ must be equal to 1.

WORKED EXAMPLE 4

A security code consists of 3 letters selected from A, B, C, D, E, F followed by 2 digits selected from 5, 6, 7, 8, 9.

Find the number of possible security codes if no letter or number can be repeated.

Answers

Method 1

There are 6 letters and 5 digits to select from.

Number of arrangements of 3 letters from 6 letters $= {}^6P_3$

Number of arrangements of 2 digits from 5 digits $= {}^5P_2$

So, the number of possible security codes $= {}^6P_3 \times {}^5P_2 = 120 \times 20 = 2400$.

Method 2

There are 6 letters and 5 digits to select from.

6 choices	5 choices	4 choices	5 choices	4 choices

CONTINUED

The first three spaces must be filled with three of the six letters.

There is a choice of 6 for the first space, 5 for the second space and 4 for the third space.

The last two spaces must be filled with two of the 5 digits.

There is a choice of 5 for the first space and 4 for the second space.

So, the number of possible security codes = $6 \times 5 \times 4 \times 5 \times 4 = 2400$.

WORKED EXAMPLE 5

Find how many even numbers between 3000 and 4000 can be formed using the digits 1, 3, 5, 6, 7 and 9 if no number can be repeated.

Answers

Method 1

The first number must be a 3 and the last number must be a 6.

| 3 | * | * | 6 |

There are now two spaces to fill using two of the remaining four digits 1, 5, 7 and 9.

Number of ways of filling the remaining two spaces = $^4P_2 = \dfrac{4!}{(4-2)!} = \dfrac{4!}{2!} = 12$

There are 12 different numbers that satisfy the conditions.

Method 2

Consider the number of choices for filling each of the four spaces.

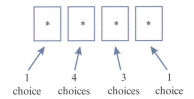

Number of ways of filling the four spaces = $1 \times 4 \times 3 \times 1$

There are 12 different numbers that satisfy the conditions.

Method 3

In this example it is not impractical to list all the possible permutations.

These are: 3156   3516   3176   3716   3196   3916

       3576   3756   3196   3916   3796   3976

There are 12 different numbers that satisfy the conditions.

WORKED EXAMPLE 6

a   Solve $^nP_2 = 30$.

b   Solve $^nP_5 = 72n(n-1)(n-2)$.

Answers

a
$$^nP_2 = 30$$
$$\frac{n(n-1)(n-2)!}{(n-2)!} = 30$$
$$n(n-1) = 30$$
$$n^2 - n - 30 = 0$$
$$(n-6)(n+5) = 0$$
$$n = 6 \text{ or } n = -5 \qquad n \text{ is a positive integer}$$
$$n = 6$$

b
$$^nP_5 = 72n(n-1)(n-2)$$
$$\frac{n(n-1)(n-2)(n-3)(n-4)(n-5)!}{(n-5)!} = 72n(n-1)(n-2)$$
$$n(n-1)(n-2)(n-3)(n-4) = 72n(n-1)(n-2) \quad \text{divide both sides by } n(n-1)(n-2)$$
$$(n-3)(n-4) = 72$$
$$n2 - 7n - 60 = 0$$
$$(n-12)(n+5) = 0$$
$$n = 12 \text{ or } n = -5 \qquad n \text{ is a positive integer}$$
$$n = 12$$

Exercise 10.3

1   Without using a calculator, find the value of each of the following.

Use the ^nP_r key on your calculator to check your answers.

   a $^8P_5$ **b** $^6P_4$ **c** $^{11}P_8$ **d** $^7P_7$

2   Find the number of different ways that 4 books chosen from 6 books be arranged on a shelf.

3   How many different five-digit numbers can be formed from the digits 1, 2, 3, 4, 5, 6, 7, 8, 9 if no digit can be repeated?

4   There are 8 competitors in a long jump competition.

In how many different ways can the first, second and third prizes be awarded?

5   Find how many different four-digit numbers greater than 4000 that can be formed using the digits 1, 2, 3, 4, 5, 6 and 7 if no digit can be used more than once.

6   Find how many even numbers between 5000 and 6000 can be formed from the digits 2, 4, 5, 7, 8, if no digit can be used more than once.

7   A four-digit number is formed using four of the eight digits 1, 2, 3, 4, 5, 6, 7 and 8.

     No digit can be used more than once.

     Find how many different four-digit numbers can be formed if

     a   there are no restrictions

     b   the number is odd

     c   the number is greater than 6000

     d   the number is odd and greater than 6000.

8   Numbers are formed using the digits 3, 5, 6, 8 and 9.

     No digit can be used more than once.

     Find how many different

     a   three-digit numbers can be formed

     b   numbers using three or more digits can be formed.

9   Find how many different even four-digit numbers greater than 2000 can be formed using the digits 1, 2, 3, 4, 5, 6, 7, 8 if no digit may be used more than once.

10   **a**   Solve $^nP_5 = 720$

     b   Solve $^nP_3 = 7n(n-1)$

     c   Solve $\dfrac{^nP_5}{^nP_4} = 3$

CLASS DISCUSSION

You have already investigated the number of ways of arranging 4 different books in a line.

You are now going to consider the number of ways you can select 3 books from the 4 books where the order of selection does not matter.

If the order does not matter, then the selection **BRO** is the same as **OBR**.

Find the number of different ways of selecting 3 books from the 4 books.

A **combination** is a selection of some items where the order of selection does not matter.

So, for combinations order does *not* matter.

Consider the set of 5 crayons.

To find the number of different ways of choosing 3 crayons from the set of 5 crayons you can use two methods.

Method 1

List all the possible selections.

There are 10 different ways of choosing 3 crayons from 5.

Method 2

The number of combinations of 3 from $5 = \dfrac{5!}{3!2!} = \dfrac{5 \times 4}{2!} = 10$

10.4 Combinations

> The general rule for finding the number of combinations of r items from n distinct items is:
>
> $$^nC_r = \frac{n!}{r!(n-r)!}$$

Note:

- In combinations, order does not matter.

- ^nC_r is read as 'n choose r'.

- So $^5C_3$ is read as '5 choose 3'.

- ^nC_r can also be written as $\binom{n}{r}$.

- $^{10}C_3$ is the same as $^{10}C_7$.

The number of ways of choosing 3 from 10 is the same as the number of ways of choosing 7 from 10. This is because when you choose a group of 3 from 10, you are automatically left with a group of 7.

WORKED EXAMPLE 7

A team of 6 swimmers is to be selected from a group of 20 swimmers.

Find the number of different ways in which the team can be selected.

Answers

Number of ways of selecting the team $= {}^{20}C_6 = \dfrac{20!}{6!14!} = 38\,760$

WORKED EXAMPLE 8

The diagram shows 2 different tents A and B.

A    B

Tent A holds 3 people and tent B holds 4 people.

Find the number of ways in which 7 people can be assigned to the two tents.

Answers

Number of ways of choosing 3 people from 7 for tent A = $^7C_3$ = 35.

So, the number of ways of assigning the 7 people to the two tents = 35.

WORKED EXAMPLE 9

3 coats and 2 dresses are to be selected from 9 coats and 7 dresses.

Find the number of different selections that can be made.

Answers

Number of ways of choosing 3 coats from 9 coats = $^9C_3$.

Number of ways of choosing 2 dresses from 7 dresses = $^7C_2$.

So, the number of possible selections = $^9C_3 \times {}^7C_2$ = 84 × 21 = 1764

WORKED EXAMPLE 10

A quiz team of 5 students is to be selected from 6 boys and 4 girls.

Find the number of possible teams that can be selected in which there are more boys than girls.

Answers

If there are more boys than girls there could be:

5 boys and 0 girls   number of ways = $^6C_5 \times {}^4C_0$ = 6 × 1 = 6

4 boys and 1 girl   number of ways = $^6C_4 \times {}^4C_1$ = 15 × 4 = 60

3 boys and 2 girls   number of ways = $^6C_3 \times {}^4C_2$ = 20 × 6 = 120

The total number of possible teams = 6 + 60 + 120 = 186

WORKED EXAMPLE 11

Sofia has to play 5 pieces of music for her music examination.

She has 13 pieces of music to choose from.

There are 7 pieces written by Chopin, 4 written by Mozart and 2 written by Bach.

Find the number of ways the 5 pieces can be chosen if

a there are no restrictions

b there must be 2 pieces by Chopin, 2 pieces by Mozart and 1 piece by Bach

c there must be at least one piece by each composer.

Answers

a Number of ways of choosing 5 from $13 = {}^{13}C_5 = 1287$

b Number of ways of choosing 2 from 7 pieces by Chopin $= {}^7C_2$

Number of ways of choosing 2 from 4 pieces by Mozart $= {}^4C_2$

Number of ways of choosing 1 from 2 pieces by Bach $= {}^2C_1$

So, number of possible selections $= {}^7C_2 \times {}^4C_2 \times {}^2C_1 = 21 \times 6 \times 2 = 252$

c If there is at least one piece by each composer there could be:

3 Chopin	1 Mozart	1 Bach	number of ways $= {}^7C_3 \times {}^4C_1 \times {}^2C_1 = 35 \times 4 \times 2 = 280$
1 Chopin	3 Mozart	1 Bach	number of ways $= {}^7C_1 \times {}^4C_3 \times {}^2C_1 = 7 \times 4 \times 2 = 56$
2 Chopin	2 Mozart	1 Bach	number of ways $= {}^7C_2 \times {}^4C_2 \times {}^2C_1 = 21 \times 6 \times 2 = 252$
2 Chopin	1 Mozart	2 Bach	number of ways $= {}^7C_2 \times {}^4C_1 \times {}^2C_2 = 21 \times 4 \times 1 = 84$
1 Chopin	2 Mozart	2 Bach	number of ways $= {}^7C_1 \times {}^4C_2 \times {}^2C_2 = 7 \times 6 \times 1 = 42$

Total number of ways $= 280 + 56 + 252 + 84 + 42 = 714$

WORKED EXAMPLE 12

Given that $7 \times {}^nC_5 = (n - 18) \times {}^{n+1}C_6$, find the value of n.

Answer

$$7 \times {}^nC_5 = 7 \times \frac{n!}{5!(n - 5)!}$$

$$(n - 18) \times {}^{n+1}C_6 = (n - 18) \times \frac{(n + 1)!}{6!(n + 1 - 6)!}$$

$$= (n - 18) \times \frac{(n + 1)!}{6!(n - 5)!} \qquad \text{replace } (n + 1)! \text{ with } (n + 1) \times n!$$

$$= (n - 18) \times \frac{(n + 1) \times n!}{6 \times 5!(n - 5)!} \qquad \text{replace } 6! \text{ with } 6 \times 5!$$

CONTINUED

Hence $7 \times n!/5!(n-5)! = (n-18) \times \dfrac{(n+1) \times n!}{6 \times 5!(n-5)!}$ divide both sides by $\dfrac{n!}{5!(n-5)!}$

$$7 = \dfrac{(n-18)(n+1)}{6}$$

$$42 = (n-18)(n+1)$$

$$42 = n^2 - 17n - 18$$

$$n^2 - 17n - 60 = 0$$

$$(n-20)(n+3) = 0$$

$$n = 20 \text{ or } n = -3 \qquad n \text{ is a positive integer}$$

$$n = 20$$

Exercise 10.4

1 Without using a calculator, find the value of each of the following, and then use the ^nC_r key on your calculator to check your answers.

a $^5C_1$ b $^6C_3$ c $^4C_4$ d $\binom{8}{4}$ e $\binom{5}{5}$ f $\binom{7}{4}$

2 Show that $^8C_3 = {}^8C_5$.

3 How many different ways are there of selecting

a 3 photographs from 10 photographs

b 5 books from 7 books

c a team of 11 footballers from 14 footballers?

4 How many different combinations of 3 letters can be chosen from the letters P, Q, R, S, T?

5 The diagram shows 2 different boxes, A and B.

8 different toys are to be placed in the boxes.

Find the number of ways in which the 8 toys can be placed in the boxes so that 5 toys are in box A and 3 toys are in box B.

6 4 pencils and 3 pens are to be selected from a collection of 8 pencils and 5 pens.
Find the number of different selections that can be made.

7 Four of the letters of the word PAINTBRUSH are selected at random.
Find the number of different combinations if

a there is no restriction on the letters selected

b the letter T must be selected.

8 A test consists of 30 questions.
Each answer is either correct or incorrect.
Find the number of different ways in which it is possible to answer

a exactly 10 questions correctly

b exactly 25 questions correctly.

9   An athletics club has 10 long distance runners, 8 sprinters and 5 jumpers.

   A team of 3 long distance runners, 5 sprinters and 2 jumpers is to be selected.

   Find the number of ways in which the team can be selected.

10 A team of 5 members is to be chosen from 5 men and 3 women.

   Find the number of different teams that can be chosen

   a   if there are no restrictions

   b   that consist of 3 men and 2 women

   c   that consist of no more than 1 woman.

11 A committee of 5 people is to be chosen from 6 women and 7 men.

   Find the number of different committees that can be chosen

   a   if there are no restrictions

   b   if there are more men than women.

12 A committee of 6 people is to be chosen from 6 men and 7 women.

   The committee must contain at least 1 man.

   Find the number of different committees that can be formed.

13 A school committee of 5 people is to be selected from a group of 4 teachers and 7 students.

   Find the number of different ways that the committee can be selected if

   a   there are no restrictions

   b   there must be at least 1 teacher and there must be more students than teachers.

14 A test consists of 10 different questions.

   4 of the questions are on trigonometry and 6 questions are on algebra.

   Students are asked to answer 8 questions.

   a   Find the number of ways in which students can select 8 questions if there are no restrictions.

   b   Find the number of these selections which contain at least 4 algebra questions.

15 Rafiu has downloaded 10 movies to his smartphone.

   4 of the movies are comedies, 3 are thrillers and 3 are science fiction.

   He is going to select 5 of these movies.

   Find the number of ways he can make his selection if

   a   there are no restrictions

   b   his selection must contain his favourite thriller movie.

   c   his selection must contain at least 3 comedy movies.

16 In a group of 15 entertainers, there are 6 singers, 5 guitarists and 4 comedians.

   A show is to be given by 6 of these entertainers.

   In the show, there must be at least 1 guitarist and 1 comedian.

   There must also be more singers than guitarists.

   Find the number of ways that the 6 entertainers can be selected.

17 Given that $45 \times {}^{n}C_4 = (n + 1) \times {}^{n+1}C_5$, find the value of n.

> **REFLECTION**
>
> In this chapter you have learned about permutations and combinations.
>
> Explain to a friend the difference between a permutation and a combination.

SUMMARY

Arrangements in a line

The number of ways of arranging n distinct items in a line is

$$n \times (n - 1) \times (n - 2) \times \ldots \times 3 \times 2 \times 1 = n!$$

Permutations

The number of permutations of r items from n distinct items is

$$^{n}P_{r} = \frac{n!}{(n - r)!}$$

In permutations, order matters.

Combinations

The number of combinations of r items from n distinct items is

$$^{n}C_{r} = \binom{n}{r} = \frac{n!}{r!(n - r)!}$$

In combinations, order does not matter.

Past paper questions

Worked example

1   a   i   Find how many different 4-digit numbers can be formed using the digits 1, 2, 3, 4, 5 and 6 if
no digit is repeated. [1]

   ii   How many of the 4-digit numbers found in **part i** are greater than 6000? [1]

   iii   How many of the 4-digit numbers found in **part i** are greater than 6000 and are odd? [1]

   b   A quiz team of 10 players is to be chosen from a class of 8 boys and 12 girls.

   i   Find the number of different teams that can be chosen if the team has to have equal numbers of
girls and boys. [3]

   ii   Find the number of different teams that can be chosen if the team has to include the youngest
and oldest boy and the youngest and oldest girl. [2]

Cambridge IGCSE Additional Mathematics 0606 Paper 11 Q10 Nov 2014

Answers

1   a   i   Number of 4-digit numbers = $^{6}P_{4}$ = 360

   ii   **Method 1**

The first number must be a 6.

6	*	*	*

There are now three spaces to fill using three of the remaining five digits 1, 2, 3, 4 and 5.

Number of ways of filling the remaining three spaces $= {}^5P_3 = \dfrac{5!}{(5-3)!} = \dfrac{5!}{2!} = 60$

There are 60 different numbers that satisfy the conditions.

Method 2

Consider the number of choices for filling each of the four spaces.

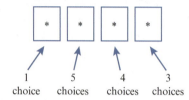

Number of ways of filling the four spaces $= 1 \times 5 \times 4 \times 3$.

There are 60 different numbers that satisfy the conditions.

iii Method 1

The first number must be a 6.

The last number must be a 1, 3 or 5.

The middle two spaces must then be filled using two of the remaining four numbers.

Number of ways of filling the four spaces $= 1 \times {}^4P_2 \times 3 = 1 \times \dfrac{4!}{(4-2)!} \times 3 = 3 \times \dfrac{4!}{2!} = 36$

There are 36 different numbers that satisfy the conditions.

Method 2

Consider the number of choices for filling each of the four spaces.

The first number must be a 6.

The last number must be a 1, 3 or 5.

When the first and last spaces have been filled there will be four numbers left to choose from.

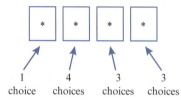

Number of ways of filling the four spaces $= 1 \times 4 \times 3 \times 3$.

There are 36 different numbers that satisfy the conditions.

b i There must be 5 boys and 5 girls.

Number of ways of choosing 5 boys from $8 = {}^8C_5$

Number of ways of choosing 5 girls from $12 = {}^{12}C_5$

Number of possible teams $= {}^8C_5 \times {}^{12}C_5 = 56 \times 792 = 44\,352$

ii The team includes the youngest and oldest boy and the youngest and oldest girl.

There are now 6 places left to fill and 16 people left to choose from.

Number of ways of choosing 6 from $16 = {}^{16}C_6 = 8008$

Number of possible teams $= 8008$

1   a   How many even numbers less than 500 can be formed using the digits 1, 2, 3, 4 and 5?

   Each digit may be used only once in any number. [4]

   b   A committee of 8 people is to be chosen from 7 men and 5 women.

   Find the number of different committees that could be selected if

   i   the committee contains at least 3 men and at least 3 women, [4]

   ii   the oldest man or the oldest woman, but not both, must be included in the committee. [2]

   Cambridge IGCSE Additional Mathematics 0606 Paper 11 Q10 Jun 2014

2   a   Jean has nine different flags.

   i   Find the number of different ways in which Jean can choose three flags from her nine flags. [1]

   ii   Jean has five flagpoles in a row. She puts one of her nine flags on each flagpole.

   Calculate the number of different five-flag arrangements she can make. [1]

   b   The six digits of the number 738 925 are rearranged so that the resulting six-digit number is even.
   Find the number of different ways in which this can be done. [2]

   Cambridge IGCSE Additional Mathematics 0606 Paper 22 Q2 Mar 2015

3   a   A lock can be opened using only the number 4351. State whether this is a permutation or a combination of digits, giving a reason for your answer. [1]

   b   There are twenty numbered balls in a bag. Two of the balls are numbered 0, six are numbered 1, five are numbered 2 and seven are numbered 3, as shown in the table below.

Number on ball	0	1	2	3
Frequency	2	6	5	7

   Four of these balls are chosen at random, without replacement. Calculate the number of ways this can be done so that

   i   the four balls all have the same number, [2]

   ii   the four balls all have different numbers, [2]

   iii   the four balls have numbers that total 3. [3]

   Cambridge IGCSE Additional Mathematics 0606 Paper 21 Q5 Jun 2015

4   a   6 books are to be chosen at random from 8 different books.

   i   Find the number of different selections of 6 books that could be made. [1]

   A clock is to be displayed on a shelf with 3 of the 8 different books on each side of it.
   Find the number of ways this can be done if

   ii   there are no restrictions on the choice of books, [1]

   iii   3 of the 8 books are music books which have to be kept together. [2]

   b   A team of 6 tennis players is to be chosen from 10 tennis players consisting of 7 men and 3 women.
   Find the number of different teams that could be chosen if the team must include at least 1 woman. [3]

   Cambridge IGCSE Additional Mathematics 0606 Paper 11 Q4 Nov 2015

5   a   A 6-character password is to be chosen from the following 9 characters.

   letters   A   B   E   F

   numbers   5   8   9

   symbols   * $

   Each character may be used only once in any password.

   Find the number of different 6-character passwords that may be chosen if

    i   there are no restrictions, [1]

    ii   the password must consist of 2 letters, 2 numbers and 2 symbols, in that order, [2]

    iii the password must start and finish with a symbol. [2]

  b An examination consists of a section A, containing 10 short questions, and a section B, containing 5 long questions. Candidates are required to answer 6 questions from section A and 3 questions from section B. Find the number of different selections of questions that can be made if

    i   there are no further restrictions [2]

    ii   candidates must answer the first 2 questions in section A and the first question in section B. [2]

Cambridge IGCSE Additional Mathematics 0606 Paper 12 Q5 Mar 2016

6 **a** The letters of the word THURSDAY are arranged in a straight line. Find the number of different arrangements of these letters if

    i   there are no restrictions, [1]

    ii   the arrangement must start with the letter T and end with the letter Y, [1]

    iii the second letter in the arrangement must be Y. [1]

  b 7 children have to be divided into two groups, one of 4 children and the other of 3 children. Given that there are 3 girls and 4 boys, find the number of different ways this can be done if

    i   there are no restrictions, [1]

    ii   all the boys are in one group, [1]

    iii one boy and one girl are twins and must be in the same group. [3]

Cambridge IGCSE Additional Mathematics 0606 Paper 12 Q6 Mar 2017

7 **a** A football club has 30 players. In how many different ways can a captain and a vice-captain be selected at random from these players? [1]

  b A team of 11 teachers is to be chosen from 2 mathematics teachers, 5 computing teachers and 9 science teachers. Find the number of different teams that can be chosen if

    i   the team must have exactly 1 mathematics teacher, [2]

    ii   the team must have exactly 1 mathematics teacher and at least 4 computing teachers. [4]

Cambridge IGCSE Additional Mathematics 0606 Paper 21 Q8 Jun 2017

8 A group of five people consists of two women, Alice and Betty, and three men, Carl, David and Ed.

  i Three of these five people are chosen at random to be a chairperson, a treasurer and a secretary. Find the number of ways in which this can be done if the chairperson and treasurer are both men. [2]

These five people sit in a row of five chairs. Find the number of different possible seating arrangements if

  ii David must sit in the middle, [1]

  iii Alice and Carl must sit together. [2]

Cambridge IGCSE Additional Mathematics 0606 Paper 22 Q3 Mar 2018

9 A 7-character password is to be selected from the 12 characters shown in the table. Each character may be used only once.

	Characters			
Upper-case letters	A	B	C	D
Lower-case letters	e	f	g	h
Digits	1	2	3	4

Find the number of different passwords

i if there are no restrictions, [1]

ii that start with a digit, [1]

iii that contain 4 upper-case letters and 3 lower-case letters such that all the upper-case letters are together and all the lower-case letters are together. [3]

Cambridge IGCSE Additional Mathematics 0606 Paper 21 Q3 Jun 2018

10 A 5-digit code is to be formed from the digits 1, 2, 3, 4, 5, 6, 7, 8, 9. Each digit can be used once only in any code. Find how many codes can be formed if

i the first digit of the code is 6 and the other four digits are odd, [2]

ii each of the first three digits is even, [2]

iii the first and last digits are prime. [2]

Cambridge IGCSE Additional Mathematics 0606 Paper 21 Q6 Nov 2018

11 a i Find how many 4-digit numbers can be formed using the digits 1, 3, 4, 6, 7 and 9.

Each digit may be used once only in any 4-digit number. [1]

ii How many of these 4-digit numbers are even and greater than 6000? [3]

b A committee of 5 people is to be formed from 6 doctors, 4 dentists and 3 nurses.

Find the number of different committees that could be formed if

i there are no restrictions, [1]

ii the committee contains at least one doctor, [2]

iii the committee contains all the nurses. [1]

Cambridge IGCSE Additional Mathematics 0606 Paper 11 Q5 Nov 2020

> # Chapter 11
Series

THIS SECTION WILL SHOW YOU HOW TO:

- use the binomial theorem for expansion of $(a + b)^n$ for positive integral n
- use the general term $\binom{n}{r} a^{n-r} b^r$ for a binomial expansion
- recognise arithmetic and geometric progressions
- use the formula for the nth term and for the sum of the first n terms to solve problems involving arithmetic and geometric progressions
- use the condition for the convergence of a geometric progression, and the formula for the sum to infinity of a convergent geometric progression.

PRE-REQUISITE KNOWLEDGE

Before you start…

Where it comes from	What you should be able to do	Check your skills
Cambridge IGCSE/O Level Mathematics	Expand brackets.	**1** Expand: **a** $(2x - 5)^2$ **b** $(1 + 3x)(1 - 2x + 5x^2)$
Cambridge IGCSE/O Level Mathematics	Simplify indices.	**2** Simplify: **a** $(2x^3)^4$ **b** $(-3x^4)^5$
Cambridge IGCSE/O Level Mathematics	Find the nth term of a linear sequence.	**3** Find the nth term of these linear sequences. **a** $-5, -2, 1, 4, 7, \ldots$ **b** $7, 5, 3, 1, -1, \ldots$

11.1 Pascal's triangle

The word 'binomial' means 'two terms'.

The word is used in algebra for expressions such as $x + 5$ and $2x - 3y$.

You should already know that $(a + b)^2 = (a + b)(a + b) = a^2 + 2ab + b^2$.

The expansion of $(a + b)^2$ can be used to expand $(a + b)^3$.

$(a + b)^3 = (a + b)(a + b)^2$

$\qquad = (a + b)(a^2 + 2ab + b^2)$

$\qquad = a^3 + 2a^2b + ab^2 + a^2b + 2ab^2 + b^3$

$\qquad = a^3 + 3a^2b + 3ab^2 + b^3$

Similarly, it can be shown that $(a + b)^4 = a^4 + 4a^3b + 6a^2b^2 + 4ab^3 + b^4$.

Writing out the expansions of $(a + b)^n$ in order:

$(a + b)^1 = \qquad\qquad 1a\ +\ 1b$

$(a + b)^2 = \qquad\qquad 1a^2\ +\ 2ab\ +\ 1b^2$

$(a + b)^3 = \qquad 1a^3\ +\ 3a^2b\ +\ 3ab^2\ +\ 1b^3$

$(a + b)^4 = 1a^4\ +\ 4a^3b\ +\ 6a^2b^2\ +\ 4ab^3\ +\ 1b^4$

If you look at the expansion of $(a + b)^4$, you should notice that the powers of a and b form a pattern.

- The first term is a^4 and then the power of a decreases by 1 in each successive term, while the power of b increases by 1.

- All of the terms have a total index of 4 (a^4, a^3b, a^2b^2, ab^3 and b^4).

KEY WORDS

Pascal's triangle

binomial theorem

progression

arithmetic progression

common difference

series

geometric progression

common ratio

convergent

There is a similar pattern in the other expansions.

The coefficients also form a pattern that is known as **Pascal's triangle**.

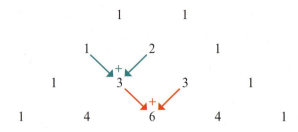

Note:

- Each row always starts and finishes with a 1.
- Each number is the sum of the two numbers in the row above it.

The next row would be:

1   5   10   10   5   1

This row can then be used to write down the expansion of $(a + b)^5$.

$(a + b)^5 = 1a^5 + 5a^4b + 10a^3b^2 + 10a^2b^3 + 5ab^4 + 1b^5$

CLASS DISCUSSION

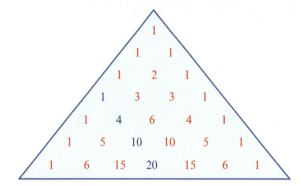

There are many number patterns in Pascal's triangle.

For example, the numbers 1, 4, 10 and 20 have been highlighted.

1   4   10   20

These numbers are called tetrahedral numbers.

Which other number patterns can you find in Pascal's triangle?

What do you notice if you find the total of each row in Pascal's triangle?

WORKED EXAMPLE 1

Use Pascal's triangle to find the expansion of

a $(2 + 5x)^3$ **b** $(2x - 3)^4$

Answers

a $(2 + 5x)^3$

The index = 3 so use the third row in Pascal's triangle.

The third row of Pascal's triangle is 1, 3, 3 and 1.

$(2 + 5x)^3 = 1(2)^3 + 3(2)^2(5x) + 3(2)(5x)^2 + 1(5x)^3$ use the expansion of $(a + b)^3$

$= 8 + 60x + 150x^2 + 125x^3$

b $(2x - 3)^4$

The index = 4 so use the fourth row in Pascal's triangle.

The fourth row of Pascal's triangle is 1, 4, 6, 4 and 1.

$(2x - 3)^4 = 1(2x)^4 + 4(2x)^3(-3) + 6(2x)^2(-3)^2$ use the expansion of $(a + b)^4$

$\qquad\qquad + 4(2x)(-3)^3 + 1(-3)^4$

$= 16x^4 - 96x^3 + 216x^2 - 216x^3 + 81$

WORKED EXAMPLE 2

a   Expand $(2 - x)^5$.

b   Find the coefficient of x^3 in the expansion of $(1 + 3x)(2 - x)^5$.

Answers

a $(2 - x)^5$

The index = 5 so use the fifth row in Pascal's triangle.

The fifth row of Pascal's triangle is 1, 5, 10, 10, 5 and 1.

$(2 - x)^5 = 1(2)^5 + 5(2)^4(-x) + 10(2)^3(-x)^2 + 10(2)^2(-x)^3 + 5(2)(-x)^4 + 1(-x)^5$

$= 32 - 80x + 80x^2 - 40x^3 + 10x^4 - x^5$

b $(1 + 3x)(2 - x)^5 = (1 + 3x)(32 - 80x + 80x^2 - 40x^3 + 10x^4 - x^5)$

The term in x^3 comes from the products:

$(1 + 3x)(32 - 80x + 80x^2 - 40x^3 + 10x^4 - x^5)$

$1 \times (-40x^3) = -40x^3$ and $3x \times 80x^2 = 240x^3$

So the coefficient of x^3 is $-40 + 240 = 200$

Exercise 11.1

1 Write down the sixth and seventh rows of Pascal's triangle.

2 Use Pascal's triangle to find the expansions of

   a $(1 + x)^3$ b $(1 - x)^4$ c $(p + q)^4$ d $(2 + x)^3$

   e $(x + y)^5$ f $(y + 4)^3$ g $(a - b)^3$ h $(2x + y)^4$

   i $(x - 2y)^3$ j $(3x - 4)^4$ k $\left(x + \dfrac{2}{x}\right)^2$ l $\left(\dfrac{x^2 - 1}{2x^3}\right)^3$

3 Find the coefficient of x^3 in the expansions of

   a $(x + 4)^4$ b $(1 + x)^5$ c $(3 - x)^4$ d $(3 + 2x)^3$

   e $(x - 2)^5$ f $(2x + 5)^4$ g $(4x - 3)^5$ h $\left(3 - \dfrac{1}{2}x\right)^4$

4 $(4 + x)^5 + (4 - x)^5 = A + Bx^2 + Cx^4$

   Find the value of A, the value of B and the value of C.

5 Expand $(1 + 2x)(1 + 3x)^4$.

6 The coefficient of x in the expansion of $(2 + ax)^3$ is 96.

   Find the value of the constant a.

7 a Expand $(3 + x)^4$.

   b Use your answer to **part a** to express $(3 + \sqrt{5})^4$ in the form $a + b\sqrt{5}$.

8 a Expand $(1 + x)^5$.

   b Use your answer to **part a** to express

      i $(1 + \sqrt{3})^5$ in the form $a + b\sqrt{3}$

      ii $(1 - \sqrt{3})^5$ in the form $c + d\sqrt{3}$.

   c Use your answers to **part b** to simplify $(1 + \sqrt{3})^5 + (1 - \sqrt{3})^5$.

9 a Expand $(2 - x^2)^4$.

   b Find the coefficient of x^6 in the expansion of $(1 + 3x^2)(2 - x^2)^4$.

10 Find the coefficient of x in the expansion $\left(x - \dfrac{3}{x}\right)^5$.

11 Find the term independent of x in the expansion of $\left(x^2 + \dfrac{1}{2x}\right)^3$.

12 CHALLENGE QUESTION

   a Find the first three terms, in ascending powers of y, in the expansion of $(2 + y)^5$.

   b By replacing y with $3x - 4x^2$, find the coefficient of x^2 in the expansion of $(2 + 3x - 4x^2)^5$.

13 CHALLENGE QUESTION

   The coefficient of x^3 in the expansion of $(3 + ax)^5$ is 12 times the coefficient of x^2 in the expansion of $\left(1 + \dfrac{ax}{2}\right)^4$. Find the value of a.

14 CHALLENGE QUESTION

   a Given that $\left(x^2 + \dfrac{4}{x}\right)^3 - \left(x^2 - \dfrac{4}{x}\right)^3 = ax^3 + \dfrac{b}{x^3}$, find the value of a and the value of b.

b Hence, without using a calculator, find the exact value of

$$\left(2 + \frac{4}{\sqrt{2}}\right)^3 - \left(2 - \frac{4}{\sqrt{2}}\right)^3.$$

15 CHALLENGE QUESTION

Given that $y = x + \frac{1}{x}$, express

a $x^3 + \frac{1}{x^3}$ in terms of y

b $x^5 + \frac{1}{x^5}$ in terms of y.

CLASS DISCUSSION

The stepping stone game

The rules are that you can move East ➡ or South ↓ from any stone.

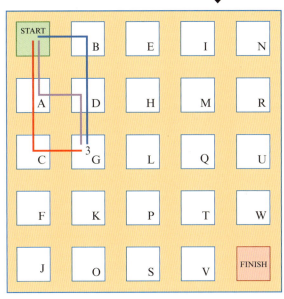

The diagram shows there are 3 routes from the START stone to stone G.

1 Find the number of routes from the START stone to each of the following stones.

    a   **i**   A   **ii**   B

    b   **i**   C   **ii**   D   **iii**   E

    c   **i**   F   **ii**   G   **iii**   H   **iv**   I

    What do you notice about your answers to **parts a**, **b** and **c**?

2 There are 6 routes from the START to stone L.

    How could you have calculated that there are 6 routes without drawing or visualising them?

3 What do you have to do to find the number of routes to any stone?

4 How many routes are there from the START stone to the FINISH stone?

In the class discussion you should have found that the number of routes from the START stone to stone Q is 10.

To move from START to Q you must move east (E) 3 and south (S) 2, in any order.

Hence the number of routes is the same as the number of different combinations of 3 E's and 2 S's.

The combinations are:

    EEESS   EESES   EESSE   ESESE   ESEES

    ESSEE   SSEEE   SESEE   SEESE   SEEES

So, the number of routes is 10.

This is the same as $^5C_3$ (or $^5C_2$).

11.2 The binomial theorem

Pascal's triangle can be used to expand $(a + b)^n$ for any positive integer n, but if n is large it can take a long time. Combinations can be used to help expand binomial expressions more quickly.

Using a calculator:

$$^5C_0 = 1 \qquad ^5C_1 = 5 \qquad ^5C_2 = 10 \qquad ^5C_3 = 10 \qquad ^5C_4 = 5 \qquad ^5C_5 = 1$$

These numbers are the same as the numbers in the fifth row of Pascal's triangle.

So the expansion of $(a + b)^5$ is:

$$(a + b)^5 = {}^5C_0 a^5 + {}^5C_1 a^4 b + {}^5C_2 a^3 b^2 + {}^5C_3 a^2 b^3 + {}^5C_4 ab^4 + {}^5C_5 b^5$$

This can be written more generally as:

$$(a + b)^n = {}^nC_0 a^n + {}^nC_1 a^{n-1} b + {}^nC_2 a^{n-2} b^2 + {}^nC_3 a^{n-3} b^3 + \ldots + {}^nC_r a^{n-r} b^r + \ldots + {}^nC_n b^n$$

But $^nC_0 = 1$ and $^nC_n = 1$, so the formula simplifies to:

$$(a + b)^n = a^n + {}^nC_1 a^{n-1} b + {}^nC_2 a^{n-2} b^2 + {}^nC_3 a^{n-3} b^3 + \ldots + {}^nC_r a^{n-r} b^r + \ldots + b^n$$

or

$$(a + b)^n = a^n + \binom{n}{1} a^{n-1} b + \binom{n}{2} a^{n-2} b^2 + \binom{n}{3} a^{n-3} b^3 + \ldots + \binom{n}{r} a^{n-r} b^r + \ldots + b^n$$

These formulae are known as the **binomial theorem**.

WORKED EXAMPLE 3

Use the binomial theorem to expand $(3 + 4x)^5$.

Answers

$$(3 + 4x)^5 = 3^5 + {}^5C_1 3^4 (4x) + {}^5C_2 3^3 (4x)^2 + {}^5C_3 3^2 (4x)^3 + {}^5C_4 3 (4x)^4 + (4x)^5$$

$$= 243 + 1620x + 4320x^2 + 5760x^3 + 3840x^4 + 1024x^5$$

WORKED EXAMPLE 4

Find the coefficient of x^{20} in the expansion of $(2 - x)^{25}$.

Answers

$(2 - x)^{25} = 2^{25} + {}^{25}C_1 2^{24}(-x) + {}^{25}C_2 2^{23}(-x)^2 + \ldots + {}^{25}C_{20} 2^5 (-x)^{20} + \ldots + (-x)^{25}$

The term containing x^{20} is ${}^{25}C_{20} \times 2^5 \times (-x)^{20}$

$= 53\,130 \times 32 \times x^{20}$

$= 1\,700\,160\,x^{20}$

So the coefficient of x^{20} is $1\,700\,160$.

Using the binomial theorem,

$(1 + x)^7 = 1^7 + {}^7C_1 1^6 x + {}^7C_2 1^5 x^2 + {}^7C_3 1^4 x^3 + {}^7C_4 1^3 x^4 + \ldots$

$= 1 + {}^7C_1 x + {}^7C_2 x^2 + {}^7C_3 x^3 + {}^7C_4 x^4 + \ldots$

But ${}^7C_1, {}^7C_2, {}^7C_3$ and ${}^7C_4$ can also be written as:

$${}^7C_1 = \frac{7!}{1!6!} = 7 \qquad {}^7C_2 = \frac{7!}{2!5!} = \frac{7 \times 6}{2!} \qquad {}^7C_3 = \frac{7!}{3!4!} = \frac{7 \times 6 \times 5}{3!}$$

$${}^7C_4 = \frac{7!}{4!3!} = \frac{7 \times 6 \times 5 \times 4}{4!}$$

So, $(1 + x)^7 = 1 + 7x + \frac{7 \times 6}{2!} x^2 + \frac{7 \times 6 \times 5}{3!} x^3 + \frac{7 \times 6 \times 5 \times 4}{4!} x^4 + \ldots$

This leads to an alternative formula for binomial expansions:

$$(1 + x)^n = 1 + nx + \frac{n(n-1)}{2!} x^2 + \frac{n(n-1)(n-2)}{3!} x^3 + \frac{n(n-1)(n-2)(n-3)}{4!} x^4 + \ldots$$

The following example illustrates how this alternative formula can be applied.

WORKED EXAMPLE 5

Find the first four terms of the binomial expansion to

a $(1 + 3y)^7$ 　　　　　　　　b $(2 - y)^6$

Answers

a $(1 + 3y)^7 = 1 + 7(3y) + \frac{7 \times 6}{2!}(3y)^2 + \frac{7 \times 6 \times 5}{3!}(3y)^3 + \ldots$　Replace x by $3y$ and n by 7 in the formula.

$= 1 + 21y + 189y^2 + 945y^3 + \ldots$

b $(2 - y)^6 = \left[2\left(1 - \frac{y}{2}\right)\right]^6$　　　　　　The formula is for $(1 + x)^n$ so take out a factor of 2.

$= 2^6 \left(1 - \frac{y}{2}\right)^6$

$= 2^6 \left[1 + 6\left(-\frac{y}{2}\right) + \frac{6 \times 5}{2!}\left(-\frac{y}{2}\right)^2 + \frac{6 \times 5 \times 4}{3!}\left(-\frac{y}{2}\right)^3 + \ldots\right]$　Replace x by $-\frac{y}{2}$ and n by 6 in the formula.

$= 2^6 \left[1 - 3y + \frac{15}{4}y^2 - \frac{5}{2}y^3 + \ldots\right]$　Multiply terms in brackets by 2^6.

$= 64 - 192y + 240y^2 - 160y^3 + \ldots$

Exercise 11.2

1 Write the following rows of Pascal's triangle using combination notation.

   a   row 3   **b**   row 4   **c**   row 5

2 Use the binomial theorem to find the expansions of

   a $(1 + x)^4$ **b** $(1 - x)^5$ **c** $(1 + 2x)^4$ **d** $(3 + x)^3$

   e $(x + y)^4$ **f** $(2 - x)^5$ **g** $(a - 2b)^4$ **h** $(2x + 3y)^4$

   i $\left(\dfrac{1}{2}x - 3\right)^4$ **j** $\left(1 - \dfrac{x}{10}\right)^5$ **k** $\left(x - \dfrac{3}{x}\right)^5$ **l** $\left(x^2 + \dfrac{1}{2x^2}\right)^6$.

3 Find the term in x^3 for each of the following expansions.

   a $(2 + x)^5$ **b** $(5 + x)^8$ **c** $(1 + 2x)^6$ **d** $(3 + 2x)^5$

   e $(1 - x)^6$ **f** $(2 - x)^9$ **g** $(10 - 3x)^7$ **h** $(4 - 5x)^{15}$.

4 Use the binomial theorem to find the first three terms in each of these expansions.

   a $(1 + x)^{10}$ **b** $(1 + 2x)^8$ **c** $(1 - 3x)^7$ **d** $(3 + 2x)^6$

   e $(3 - x)^9$ **f** $\left(2 + \dfrac{1}{2}x\right)^8$ **g** $(5 - x^2)^9$ **h** $(4x - 5y)^{10}$.

5 **a**   Write down, in ascending powers of x, the first 4 terms in the expansion of $(1 + 2x)^6$.

     b   Find the coefficient of x^3 in the expansions of $\left(1 - \dfrac{x}{3}\right)(1 + 2x)^6$.

6 **a**   Write down, in ascending powers of x, the first 4 terms in the expansion of $\left(1 + \dfrac{x}{2}\right)^{13}$.

     b   Find the coefficient of x^3 in the expansions of $(1 + 3x)\left(1 + \dfrac{x}{2}\right)^{13}$.

7 **a**   Write down, in ascending powers of x, the first 4 terms in the expansion of $(1 - 3x)^{10}$.

     b   Find the coefficient of x^3 in the expansions of $(1 - 4x)(1 - 3x)^{10}$.

8 **a**   Find, in ascending powers of x, the first 3 terms of the expansion of $(1 + 2x)^7$.

     b   Hence find the coefficient of x^2 in the expansion of $(1 + 2x)^7(1 - 3x + 5x^2)$.

9 **a**   Find, in ascending powers of x, the first 4 terms of the expansion of $(1 + x)^7$.

     b   Hence find the coefficient of y^3 in the expansion of $(1 + y - y^2)^7$.

10 Find the coefficient of x in the binomial expansion of $\left(x - \dfrac{3}{x}\right)^7$.

11 Find the term independent of x in the binomial expansion of $\left(x + \dfrac{1}{2x^2}\right)^9$.

12 CHALLENGE QUESTION

    When $(1 + ax)^n$ is expanded, the coefficients of x^2 and x^3 are equal.

    Find a in terms of n.

> **REFLECTION**
>
> In Exercise 11.2 question 11 you were asked to find the term independent of x in the expansion. Did you fully expand the brackets to answer the question? Explain to a partner how you can find the answer without fully expanding the brackets.

11.3 Arithmetic progressions

At IGCSE level you learned that a number sequence is an ordered set of numbers that satisfy a rule and that the numbers in the sequence are called the terms of the sequence. A number sequence is also called a **progression**.

The sequence 5, 8, 11, 14, 17, … is called an **arithmetic progression**. Each term differs from the term before by a constant. This constant is called the **common difference**.

The notation used for arithmetic progressions is:

a = first term d = common difference l = last term

The first five terms of an arithmetic progression whose first term is a and whose common difference is d are:

a $a + d$ $a + 2d$ $a + 3d$ $a + 4d$

term 1   term 2   term 3   term 4   term 5

This leads to the formula:

$$n\text{th term} = a + (n - 1)\,d$$

WORKED EXAMPLE 6

Find the number of terms in the arithmetic progression $-17, -14, -11, -8, …, 58$.

Answers

$n\text{th term} = a + (n-1)\,d$ use $a = -17$, $d = 3$ and nth term $= 58$

$\quad 58 = -17 + 3(n-1)$ solve

$\quad n - 1 = 25$

$\qquad n = 26$

WORKED EXAMPLE 7

The fifth term of an arithmetic progression is 4.4 and the ninth term is 7.6. Find the first term and the common difference.

Answers

fifth term $= 4.4 \Rightarrow a + 4d = 4.4$ (1)

ninth term $= 7.6 \Rightarrow a + 8d = 7.6$ (2)

(2) − (1), gives $4d = 3.2$

$\qquad\qquad\qquad d = 0.8$

Substituting in (1) gives $a + 3.2 = 4.4$

$\qquad\qquad\qquad\qquad a = 1.2$

First term = 1.2, common difference = 0.8

WORKED EXAMPLE 8

The nth term of an arithmetic progression is $11 - 3n$. Find the first term and the common difference.

Answers

first term $= 11 - 3(1) = 8$ \qquad substitute $n = 1$ into nth term $= 11 - 3n$

second term $= 11 - 3(2) = 5$ \qquad substitute $n = 2$ into nth term $= 11 - 3n$

Common difference = second term − first term $= -3$

The sum of an arithmetic progression

When the terms in a sequence are added together the resulting sum is called a **series**.

CLASS DISCUSSION

$1 + 2 + 3 + 4 + \dots + 97 + 98 + 99 + 100 = ?$

It is said that at the age of eight, the famous mathematician Carl Gauss was asked to find the sum of the numbers from 1 to 100. His teacher expected this task to keep him occupied for some time but Gauss surprised his teacher by writing down the correct answer after just a couple of seconds. His method involved adding the numbers in pairs: $1 + 100 = 101, 2 + 99 = 101,$ $3 + 98 = 101, \dots$

1 Can you complete his method to find the answer?

2 Use Gauss's method to find the sum of

   a $2 + 4 + 6 + 8 + \dots + 394 + 396 + 398 + 400$

   b $3 + 6 + 9 + 12 + \dots + 441 + 444 + 447 + 450$

   c $17 + 24 + 31 + 38 + \dots + 339 + 346 + 353 + 360.$

3 Use Gauss's method to find an expression, in terms of n, for the sum
$1 + 2 + 3 + 4 + \dots + (n - 3) + (n - 2) + (n - 1) + n.$

It can be shown that the sum of an arithmetic progression, S_n, can be written as:

$$S_n = \frac{n}{2}(a + l) \quad \text{or} \quad S_n = \frac{n}{2}[2a + (n - 1)d]$$

Proof:

$$S_n = a + (a + d) + (a + 2d) + \dots + (l - 2d) + (l - d) + l$$

Reversing: $\quad S_n = l + (l - d) + (l - 2d) + \dots + (a + 2d) + (a + d) + a$

Adding: $\quad \overline{2S_n = (a + l) + (a + l) + (a + l) + \dots + (a + l) + (a + l) + (a + l)}$

$$2S_n = n(a + l)$$

$$S_n = \frac{n}{2}(a + l)$$

Using $l = a + (n - 1)d$, gives $S_n = \frac{n}{2}[2a + (n - 1)d]$

It is useful to remember the following rule that applies for all progressions:

$$n\text{th term} = S_n - S_{n-1}$$

WORKED EXAMPLE 9

In an arithmetic progression, the first term is 25, the 19th term is -38 and the last term is -87. Find the sum of all the terms in the progression.

Answers

$n\text{th term} = a + (n-1)d$ use $n\text{th term} = -38$ when $n = 19$ and $a = 25$

$\quad -38 = 25 + 18d$ solve

$\qquad d = -3.5$

$n\text{th term} = a + (n-1)d$ use $n\text{th term} = -87$ when $a = 25$ and $d = -3.5$

$\quad -87 = 25 - 3.5(n-1)$ solve

$\quad n - 1 = 32$

$\qquad n = 33$

$S_n = \dfrac{n}{2}(a + l)$ use $a = 25$, $l = -87$ and $n = 33$

$S_{33} = \dfrac{33}{2}(25 - 87)$

$\quad = -1023$

WORKED EXAMPLE 10

The 12th term in an arithmetic progression is 8 and the sum of the first 13 terms is 78.

Find the first term of the progression and the common difference.

Answers

$n\text{th term} = a + (n-1)d$ use $n\text{th term} = 8$ when $n = 12$

$\quad 8 = a + 11d$ (1)

$S_n = \dfrac{n}{2}[2a + (n-1)d]$ use $n = 13$ and $S_{13} = 78$

$\quad 78 = \dfrac{13}{2}(2a + 12d)$ simplify

$\quad 6 = a + 6d$ (2)

(1) − (2) gives $5d = 2$

$\qquad\qquad\qquad d = 0.4$

Substituting $d = 0.4$ in equation (1) gives $a = 3.6$

First term = 3.6, common difference = 0.4

WORKED EXAMPLE 11

The sum of the first n terms, S_n, of a particular arithmetic progression is given by $S_n = 5n^2 - 3n$.

a Find the first term and the common difference.

b Find an expression for the nth term.

Answers

a $S_1 = 5(1)^2 - 3(1) = 2 \Rightarrow$ first term $= 2$

$S_2 = 5(2)^2 - 3(2) = 14 \Rightarrow$ first term + second term $= 14$

second term $= 14 - 2 = 12$

First term $= 2$, common difference $= 10$.

b **Method 1:**

nth term $= a + (n - 1)d$ use $a = 2$, $d = 10$

$= 2 + 10(n - 1)$

$= 10n - 8$

Method 2:

nth term $= S_n - S_{n-1} = 5n^2 - 3n - [5(n - 1)^2 - 3(n - 1)]$

$= 5n^2 - 3n - (5n^2 - 10n + 5 - 3n + 3)$

$= 10n - 8$

Exercise 11.3

1 The first term in an arithmetic progression is a and the common difference is d.
Write down expressions, in terms of a and d, for the fifth term and the 14th term.

2 Find the sum of each of these arithmetic series.

    a $2 + 9 + 16 + \ldots$ (15 terms)   **b** $20 + 11 + 2 + \ldots$ (20 terms)

    c $8.5 + 10 + 11.5 + \ldots$ (30 terms)   **d** $-2x - 5x - 8x - \ldots$ (40 terms)

3 Find the number of terms and the sum of each of these arithmetic series.

    a $23 + 27 + 31 \ldots + 159$ **b** $28 + 11 - 6 - \ldots -210$

4 The first term of an arithmetic progression is 2 and the sum of the first 12 terms is 618.
Find the common difference.

5 In an arithmetic progression, the first term is -13, the 20th term is 82 and the last term is 112.

    a Find the common difference and the number of terms.

    b Find the sum of the terms in this progression.

6 The first two terms in an arithmetic progression are 57 and 46. The last term is -207. Find the sum of all the terms in this progression.

7   The first two terms in an arithmetic progression are −2 and 5. The last term in the progression is the only number in the progression that is greater than 200. Find the sum of all the terms in the progression.

8   The first term of an arithmetic progression is 8 and the last term is 34. The sum of the first six terms is 58. Find the number of terms in this progression.

9   Find the sum of all the integers between 100 and 400 that are multiples of 6.

10   The first term of an arithmetic progression is 7 and the eleventh term is 32. The sum of all the terms in the progression is 2790. Find the number of terms in the progression.

11   Rafiu buys a boat for \$15 500. He pays for this boat by making monthly payments that are in arithmetic progression. The first payment that he makes is \$140 and the debt is fully repaid after 31 payments. Find the fifth payment.

12   The eighth term of an arithmetic progression is −10 and the sum of the first twenty terms is −350.

   a   Find the first term and the common difference.

   b   Given that the nth term of this progression is −97, find the value of n.

13   The sum of the first n terms, S_n, of a particular arithmetic progression is given by $S_n = 4n^2 + 2n$. Find the first term and the common difference.

14   The sum of the first n terms, S_n, of a particular arithmetic progression is given by $S_n = -3n^2 - 2n$. Find the first term and the common difference.

15   The sum of the first n terms, S_n, of a particular arithmetic progression is given by $S_n = \dfrac{n}{12}(4n + 5)$. Find an expression for the nth term.

16   A circle is divided into twelve sectors. The sizes of the angles of the sectors are in arithmetic progression. The angle of the largest sector is 6.5 times the angle of the smallest sector. Find the angle of the smallest sector.

17   An arithmetic sequence has first term a and common difference d. The sum of the first 25 terms is 15 times the sum of the first 4 terms.

   a   Find a in terms of d.

   b   Find the 55th term in terms of a.

18   The eighth term in an arithmetic progression is three times the third term. Show that the sum of the first eight terms is four times the sum of the first four terms.

19   CHALLENGE QUESTION

The first term of an arithmetic progression is $\cos^2 x$ and the second term is 1.

   a   Write down an expression, in terms of $\cos x$, for the seventh term of this progression.

   b   Show that the sum of the first twenty terms of this progression is $20 + 170\sin^2 x$.

20   CHALLENGE QUESTION

The sum of the digits in the number 56 is 11. (5 + 6 = 11)

   a   Show that the sum of the digits of the integers from 15 to 18 is 30.

   b   Find the sum of the digits of the integers from 1 to 100.

11.4 Geometric progressions

The sequence 7, 14, 28, 56, 112, ... is called a **geometric progression**. Each term is double the preceding term. The constant multiple is called the **common ratio**.

Other examples of geometric progressions are:

Progression	Common ratio
1, −2, 4, −8, 16, −32, ...	−2
81, 54, 36, 24, 16, $10\frac{2}{3}$, ...,	$\frac{2}{3}$
−8, 4, −2, 1, $-\frac{1}{2}$, $\frac{1}{4}$, ...	$-\frac{1}{2}$

The notation used for a geometric progression is:

a = first term $\qquad\qquad$ r = common ratio

The first five terms of a geometric progression whose first term is a and whose common ratio is r are:

$$a \qquad ar \qquad ar^2 \qquad ar^3 \qquad ar^4$$

term 1 $\quad$ term 2 $\quad$ term 3 $\quad$ term 4 $\quad$ term 5

This leads to the formula:

$$n\text{th term} = ar^{n-1}$$

WORKED EXAMPLE 12

The third term of a geometric progression is 144 and the common ratio is $\frac{3}{2}$.

Find the seventh term and an expression for the nth term.

Answers

$n\text{th term} = ar^{n-1}$ $\qquad\qquad$ use nth term = 144 when $n = 3$ and $r = \frac{3}{2}$

$$144 = a\left(\frac{3}{2}\right)^2$$

$$a = 64$$

$$\text{seventh term} = 64\left(\frac{3}{2}\right)^6 = 729$$

$$n\text{th term} = ar^{n-1} = 64\left(\frac{3}{2}\right)^{n-1}$$

WORKED EXAMPLE 13

The second and fourth terms in a geometric progression are 108 and 48 respectively. Given that all the terms are positive, find the first term and the common ratio. Hence, write down an expression for the nth term.

Answers

$108 = ar$ (1)

$48 = ar^3$ (2)

(2) $\div$ (1) gives $\dfrac{ar^3}{ar} = \dfrac{48}{108}$

$\qquad\qquad\qquad r^2 = \dfrac{4}{9}$

$\qquad\qquad\qquad r = \pm\dfrac{2}{3}$ all terms are positive $\Rightarrow r > 0$

$\qquad\qquad\qquad r = \dfrac{2}{3}$

Substituting $r = \dfrac{2}{3}$ into equation (1) gives $a = 162$.

First term $= 162$, common ratio $= \dfrac{2}{3}$, nth term $= 162\left(\dfrac{2}{3}\right)^{n-1}$

WORKED EXAMPLE 14

The nth term of a geometric progression is $30\left(-\dfrac{1}{2}\right)^n$. Find the first term and the common ratio.

Answers

first term $= 30\left(-\dfrac{1}{2}\right)^1 = -15$

second term $= 30\left(-\dfrac{1}{2}\right)^2 = 7.5$

Common ratio $= \dfrac{\text{2nd term}}{\text{1st term}} = \dfrac{7.5}{-15} = -\dfrac{1}{2}$

First term $= -15$, common ratio $= -\dfrac{1}{2}$

WORKED EXAMPLE 15

In the geometric sequence 2, 6, 18, 54, …, which is the first term to exceed $1\,000\,000$?

Answers

$$n\text{th term} = ar^{n-1} \qquad \text{use } a = 2 \text{ and } r = 3$$

$$2 \times 3^{n-1} > 1000\,000 \qquad \text{divide by 2 and take logs}$$

$$\log_{10} 3^{n-1} > \log_{10} 500\,000 \qquad \text{use the power rule for logs}$$

$$(n-1)\log_{10} 3 > \log_{10} 500\,000 \qquad \text{divide both sides by } \log_{10} 3$$

$$n - 1 > \frac{\log_{10} 500\,000}{\log_{10} 3}$$

$$n - 1 > 11.94\ldots$$

$$n > 12.94\ldots$$

The 13th term is the first to exceed $1\,000\,000$.

CLASS DISCUSSION

In this class discussion you are not allowed to use a calculator.

1   Consider the sum of the first 10 terms, S_{10}, of a geometric progression with $a = 1$ and $r = 5$.

$$S_{10} = 1 + 5 + 5^2 + 5^3 + \ldots + 5^7 + 5^8 + 5^9$$

  a   Multiply both sides of the equation above by the common ratio, 5, and complete the following statement.
$$5S_{10} = 5 + 5^2 + 5^{\cdots} + 5^{\cdots} + \ldots + 5^{\cdots} + 5^{\cdots} + 5^{\cdots}$$

  b   What happens when you subtract the equation for S_{10} from the equation for $5S_{10}$?

  c   Can you find an alternative way of expressing the sum S_{10}?

2   Use the method from **question 1** to find an alternative way of expressing each of the following.

  a $3 + 3 \times 2 + 3 \times 2^2 + 3 \times 2^3 + \ldots$ (12 terms)

  b $32 + 32 \times \dfrac{1}{2} + 32 \times \left(\dfrac{1}{2}\right)^2 + 32 \times \left(\dfrac{1}{2}\right)^3 + \ldots$ (15 terms)

  c $27 - 18 + 12 - 8 + \ldots$ (20 terms)

It can be shown that the sum of a geometric progression, S_n, can be written as:

$$S_n = \frac{a(1 - r^n)}{1 - r} \quad \text{or} \quad S_n = \frac{a(r^n - 1)}{r - 1}$$

Note:
For these formulae,
$r \neq 1$

Either formula can be used but it is usually easier to

* use the first formula when $-1 < r < 1$

* use the second formula when $r > 1$ or when $r < -1$.

Proof:

$$S_n = a + ar + ar^2 + \ldots + ar^{n-3} + ar^{n-2} + ar^{n-1} \quad (1)$$

$r \times (1):$
$$rS_n = ar + ar^2 + \ldots + ar^{n-3} + ar^{n-2} + ar^{n-1} + ar^n \quad (2)$$

$(2) - (1):$
$$rS_n - S_n = ar^n - a$$
$$(r - 1)S_n = a(r^n - 1)$$
$$S_n = \frac{a(r^n - 1)}{r - 1}$$

Multiplying numerator and denominator by -1 gives the alternative formula

$$S_n = \frac{a(1 - r^n)}{1 - r}.$$

WORKED EXAMPLE 16

Find the sum of the first ten terms of the geometric series $2 + 6 + 18 + 54 + \ldots$

Answers

$$S_n = \frac{a(r^n - 1)}{r - 1} \qquad \text{use } a = 2, r = 3 \text{ and } n = 10$$

$$S_{12} = \frac{2(3^{10} - 1)}{3 - 1} \qquad \text{simplify}$$

$$= 59\,048$$

WORKED EXAMPLE 17

The second term of a geometric progression is 9 less than the first term. The sum of the second and third terms is 30. Given that all the terms in the progression are positive, find the first term.

Answers

second term = first term -9

$$ar = a - 9 \qquad \text{rearrange to make } a \text{ the subject}$$

$$a = \frac{9}{1 - r} \qquad (1)$$

second term + third term = 30

$$ar + ar^2 = 30 \qquad \text{factorise}$$

$$ar(1 + r) = 30 \qquad (2)$$

$(2) \div (1)$ gives $\dfrac{ar(1 + r)}{a} = \dfrac{30(1 - r)}{9} \qquad$ simplify

$$3r^2 + 13r - 10 = 0 \qquad \text{factorise and solve}$$

$$(3r - 2)(r + 5) = 0$$

$$r = \frac{2}{3} \text{ or } r = -5 \qquad \text{all terms are positive} \Rightarrow r > 0$$

$$r = \frac{2}{3}$$

Substituting $r = \dfrac{2}{3}$ into (1) gives $a = 27$.

First term is 27.

Exercise 11.4

1 Identify whether or not the following sequences are geometric. If they are geometric, write down the common ratio and the eighth term.

   a   1, 2, 4, 6, …

   b $-1, 4, -16, 64, …$

   c   81, 27, 9, 3, …

   d $\dfrac{2}{11}, \dfrac{3}{11}, \dfrac{5}{11}, \dfrac{8}{11}, …$

   e   2, 0.4, 0.08, 0.16, …

   f $-5, 5, -5, 5, …$

2 The first term in a geometric progression is a and the common ratio is r.
   Write down expressions, in terms of a and r, for the ninth term and the 20th term.

3 The third term of a geometric progression is 108 and the sixth term is -32.
   Find the common ratio and the first term.

4 The first term of a geometric progression is 75 and the third term is 27.
   Find the two possible values for the fourth term.

5 The second term of a geometric progression is 12 and the fourth term is 27.
   Given that all the terms are positive, find the common ratio and the first term.

6 The sixth and 13th terms of a geometric progression are $\dfrac{5}{2}$ and 320 respectively.

Find the common ratio, the first term and the 10th term of this progression.

7 The sum of the second and third terms in a geometric progression is 30. The second term is 9 less than the first term. Given that all the terms in the progression are positive, find the first term.

8 Three consecutive terms of a geometric progression are x, $x + 6$ and $x + 9$. Find the value of x.

9 In the geometric sequence $\dfrac{1}{4}, \dfrac{1}{2}, 1, 2, 4, \ldots$ which is the first term to exceed $500\,000$?

10 In the geometric sequence $256, 128, 64, 32, \ldots$ which is the first term that is less than 0.001?

11 Find the sum of the first eight terms of each of these geometric series.

    a $\quad 4 + 8 + 16 + 32 + \ldots$ **b** $\quad 729 + 243 + 81 + 27 + \ldots$

    c $\quad 2 - 6 + 18 - 54 + \ldots$ **d** $\quad -5000 + 1000 - 200 + 40 - \ldots \ldots$

12 The first four terms of a geometric progression are 1, 3, 9 and 27. Find the smallest number of terms that will give a sum greater than $2\,000\,000$.

13 A ball is thrown vertically upwards from the ground. The ball rises to a height of $10\,\text{m}$ and then falls and bounces. After each bounce it rises to $\dfrac{4}{5}$ of the height of the previous bounce.

    a Write down an expression, in terms of n, for the height that the ball rises after the nth impact with the ground.

    b Find the total distance that the ball travels from the first throw to the fifth impact with the ground.

14 The third term of a geometric progression is nine times the first term. The sum of the first four terms is k times the first term. Find the possible values of k.

15 John competes in a $10\,\text{km}$ race. He completes the first kilometre in 4 minutes. He reduces his speed in such a way that each kilometre takes him 1.05 times the time taken for the preceding kilometre. Find the total time, in minutes and seconds, John takes to complete the $10\,\text{km}$ race. Give your answer correct to the nearest second.

16 A geometric progression has first term a, common ratio r and sum to n terms, S_n.

    Show that $\dfrac{S_{3n} - S_{2n}}{S_n} = r^{2n}$.

17 CHALLENGE QUESTION

    $1, 1, 3, \dfrac{1}{3}, 9, \dfrac{1}{9}, 27, \dfrac{1}{27}, 81, \dfrac{1}{81}, \ldots$

    Show that the sum of the first $2n$ terms of this sequence is $\dfrac{1}{2}(3^n - 3^{1-n} + 2)$.

18 CHALLENGE QUESTION

    $S_n = 6 + 66 + 666 + 6666 + 66\,666 + \ldots$

    Find the sum of the first n terms of this sequence.

11.5 Infinite geometric series

An infinite series is a series whose terms continue forever.

The geometric series where $a = 2$ and $r = \dfrac{1}{2}$ is: $2 + 1 + \dfrac{1}{2} + \dfrac{1}{4} + \dfrac{1}{8} + \ldots$

For this series, it can be shown that

$S_1 = 2,\ S_2 = 3,\ S_3 = 3\dfrac{1}{2},\ S_4 = 3\dfrac{3}{4},\ S_5 = 3\dfrac{7}{8},\ \ldots$

This suggests that the sum to infinity approaches the number 4.

The diagram of the 2 by 2 square is a visual representation of this series. If the pattern of rectangles inside the square is continued, the total areas of the inside rectangles approaches the value 4.

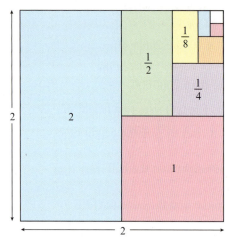

This confirms that the sum to infinity of the series $2 + 1 + \dfrac{1}{2} + \dfrac{1}{4} + \dfrac{1}{8} + \ldots$ is 4.

This is an example of a **convergent** series because the sum to infinity converges on a finite number.

CLASS DISCUSSION

1   Use a spreadsheet to investigate whether the sum of each of these infinite geometric series converge or diverge. If they converge, state their sum to infinity.

   | $a = \dfrac{2}{5}, r = 2$ | $a = -3, r = -\dfrac{1}{2}$ | $a = 5, r = \dfrac{2}{3}$ | $a = \dfrac{1}{2}, r = -5$ |

2   Find other convergent geometric series of your own. In each case find the sum to infinity.

3   Can you find a condition for r for which a geometric series is convergent?

Consider the geometric series $a + ar + ar^2 + ar^3 + \ldots + ar^{n-1}$.

The sum, S_n, is given by the formula $S_n = \dfrac{a(1 - r^n)}{1 - r}$.

If $-1 < r < 1$, then as n gets larger and larger, r^n gets closer and closer to 0.

We say that as $n \to \infty$, $r^n \to 0$.

Hence, as $n \to \infty$, $\dfrac{a(1 - r^n)}{1 - r} \to \dfrac{a(1 - 0)}{1 - r} = \dfrac{a}{1 - r}$.

This gives the result

> **Note:**
> This is not true when $r > 1$ or when $r = -1$.

$$S_\infty = \frac{a}{1 - r} \text{ provided that } -1 < r < 1$$

WORKED EXAMPLE 18

The first three terms of a geometric progression are 25, 15 and 9.

a   Write down the common ratio.

b   Find the sum to infinity.

Answers

a   Common ratio $= \dfrac{\text{second term}}{\text{first term}} = \dfrac{15}{25} = \dfrac{3}{5}$

b $S_\infty = \dfrac{a}{1 - r}$ use $a = 25$ and $r = \dfrac{3}{5}$

$\qquad = \dfrac{25}{1 - \dfrac{3}{5}}$

$\qquad = 62.5$

WORKED EXAMPLE 19

A geometric progression has a common ratio of $-\dfrac{4}{5}$ and the sum of the first four terms is 164.

a   Find the first term of the progression.

b   Find the sum to infinity.

Answers

a $S_4 = \dfrac{a(1 - r^4)}{1 - r}$ use $S_4 = 164$ and $r = -\dfrac{4}{5}$

$\qquad 164 = \dfrac{a\left(1 - \left(-\dfrac{4}{5}\right)^4\right)}{1 - \left(-\dfrac{4}{5}\right)}$ simplify

$\qquad 164 = \dfrac{41}{125}a$ solve

$\qquad\quad a = 500$

CONTINUED

b $S_\infty = \dfrac{a}{1-r}$ use $a = 500$ and $r = -\dfrac{4}{5}$

$= \dfrac{500}{1 - \left(-\dfrac{4}{5}\right)}$

$= 277\dfrac{7}{9}$

Exercise 11.5

1 Find the sum to infinity of each of the following geometric series.

    a $3 + 1 + \dfrac{1}{3} + \dfrac{1}{9} + \ldots$ **b** $1 - \dfrac{1}{2} + \dfrac{1}{4} - \dfrac{1}{8} + \dfrac{1}{16} - \ldots$

    c $8 + \dfrac{8}{5} + \dfrac{8}{25} + \dfrac{8}{125} + \ldots$ **d** $-162 + 108 - 72 + 48 - \ldots$

2 The first term of a geometric progression is 10 and the second term is 8. Find the sum to infinity.

3 The first term of a geometric progression is 300 and the fourth term is $-2\dfrac{2}{5}$. Find the common ratio and the sum to infinity.

4 The first four terms of a geometric progression are $1, 0.8^2, 0.8^4$ and 0.8^6. Find the sum to infinity.

5 **a** Write the recurring decimal $0.4\dot{2}$ as the sum of a geometric progression.

    b Use your answer to **part a** to show that $0.4\dot{2}$ can be written as $\dfrac{14}{33}$.

6 The first term of a geometric progression is -120 and the sum to infinity is -72. Find the common ratio and the sum of the first three terms.

7 The second term of a geometric progression is 6.5 and the sum to infinity is 26. Find the common ratio and the first term.

8 The second term of a geometric progression is -96 and the fifth term is $40\dfrac{1}{2}$.

    a Find the common ratio and the first term.

    b Find the sum to infinity.

9 The first three terms of a geometric progression are $175, k$ and 63. Given that all the terms in the progression are positive, find

    a the value of k **b** the sum to infinity.

10 The second term of a geometric progression is 18 and the fourth term is 1.62. Given that the common ratio is positive, find

    a the common ratio and the first term   **b** the sum to infinity.

11 The first three terms of a geometric progression are $k + 15, k$ and $k - 12$ respectively. Find

    a the value of k **b** the sum to infinity.

12 The fourth term of a geometric progression is 48 and the sum to infinity is three times the first term. Find the first term.

13   A geometric progression has first term a and common ratio r. The sum of the first three terms is 62 and the sum to infinity is 62.5. Find the value of a and the value of r.

14   The first term of a geometric progression is 1 and the second term is $2\sin x$ where $-\dfrac{\pi}{2} < x < \dfrac{\pi}{2}$. Find the set of values of x for which this progression is convergent.

15   A ball is dropped from a height of $12\,\text{m}$. After each bounce it rises to $\dfrac{3}{4}$ of the height of the previous bounce. Find the total vertical distance that the ball travels.

16   **CHALLENGE QUESTION**

Starting with an equilateral triangle, a Koch snowflake pattern can be constructed using the following steps:

Step 1:   Divide each line segment into three equal segments.

Step 2:   Draw an equilateral triangle, pointing outwards, which has the middle segment from step 1 as its base.

Step 3:   Remove the line segments that were used as the base of the equilateral triangles in step 2.

These three steps are then repeated to produce the next pattern.

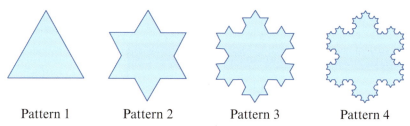

| Pattern 1 | Pattern 2 | Pattern 3 | Pattern 4 |

You are given that the triangle in **Pattern 1** has side length x units.

a   Find, in terms of x, expressions for the perimeter of each of patterns 1, 2, 3 and 4 and explain why this progression for the perimeter of the snowflake diverges to infinity.

b   Show that the area of patterns 1, 2, 3 and 4 can be written as shown in this table:

Pattern	Area
1	$\dfrac{\sqrt{3}\,x^2}{4}$
2	$\dfrac{\sqrt{3}\,x^2}{4} + 3\dfrac{\sqrt{3}\left(\frac{x}{3}\right)^2}{4}$
3	$\dfrac{\sqrt{3}\,x^2}{4} + 3\dfrac{\sqrt{3}\left(\frac{x}{3}\right)^2}{4} + 12\dfrac{\sqrt{3}\left(\frac{x}{9}\right)^2}{4}$
4	$\dfrac{\sqrt{3}\,x^2}{4} + 3\dfrac{\sqrt{3}\left(\frac{x}{3}\right)^2}{4} + 12\dfrac{\sqrt{3}\left(\frac{x}{9}\right)^2}{4} + 48\dfrac{\sqrt{3}\left(\frac{x}{27}\right)^2}{4}$

Hence show that the progression for the area of the snowflake converges to $\dfrac{8}{5}$ times the area of the original triangle.

17 CHALLENGE QUESTION

A circle of radius 1 unit is drawn touching the three edges of an equilateral triangle.

Three smaller circles are then drawn at each corner to touch the original circle and two edges of the triangle.

This process is then repeated an infinite number of times.

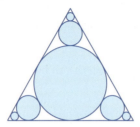

a   Find the sum of the circumferences of all the circles.

b   Find the sum of the areas of all the circles.

11.6 Further arithmetic and geometric series

Some problems may involve more than one progression.

CLASS DISCUSSION

$a, b, c, \ldots$

1   Given that a, b and c are in arithmetic progression, find an equation connecting a, b and c.

2   Given that a, b and c are in geometric progression, find an equation connecting a, b and c.

WORKED EXAMPLE 20

The first, second and third terms of an arithmetic series are x, y and x^2.
The first, second and third terms of a geometric series are x, x^2 and y.
Given that $x < 0$, find

a   the value of x and the value of y

b   the sum to infinity of the geometric series

c   the sum of the first 20 terms of the arithmetic series.

Answers

a   Arithmetic series is: $x + y + x^2 + \ldots$ use common differences

$$y - x = x^2 - y$$

$$2y = x^2 + x \quad (1)$$

CONTINUED

Geometric series is: $x + x^2 + y + \dots$ use common ratios

$$\frac{y}{x^2} = \frac{x^2}{x}$$

$$y = x^3 \quad (2)$$

(1) and (2) give $2x^3 = x^2 + x$ divide by x (since $x \neq 0$) and rearrange

$$2x^2 - x - 1 = 0$$ factorise and solve

$$(2x + 1)(x - 1) = 0$$

$$x = -\frac{1}{2} \text{ or } x = 1$$ $x \neq 1$ since $x < 0$

Hence, $x = -\frac{1}{2}$ and $y = -\frac{1}{8}$

b $S_\infty = \dfrac{a}{1 - r}$ use $a = -\frac{1}{2}$ and $r = -\frac{1}{2}$

$$S_\infty = \frac{-\frac{1}{2}}{1 - \left(-\frac{1}{2}\right)} = -\frac{1}{3}$$

c $S_n = \dfrac{n}{2}[2a + (n - 1)d]$ use $n = 20$, $a = -\frac{1}{2}$, $d = y - x = \frac{3}{8}$

$$S_{20} = \frac{20}{2}\left[-1 + 19\left(\frac{3}{8}\right)\right]$$

$$= 61.25$$

Exercise 11.6

1   The first term of a progression is 8 and the second term is 12. Find the sum of the first six terms, given that the progression is

   a   arithmetic   **b**   geometric.

2   The first term of a progression is 25 and the second term is 20.

   a   Given that the progression is geometric, find the sum to infinity.

   b   Given that the progression is arithmetic, find the number of terms in the progression if the sum of all the terms is -1550.

3   The first, second and third terms of a geometric progression are the first, fifth and 11th terms respectively of an arithmetic progression. Given that the first term in each progression is 48 and the common ratio of the geometric progression is r, where $r \neq 1$, find

   a   the value of r **b**   the sixth term of each progression.

4   A geometric progression has six terms. The first term is 486 and the common ratio is $\frac{2}{3}$. An arithmetic progression has 35 terms and common difference $\frac{3}{2}$. The sum of all the terms in the geometric progression is equal to the sum of all the terms in the arithmetic progression. Find the first term and the last term of the arithmetic progression.

5 The first, second and third terms of a geometric progression are the first, fifth and eighth terms respectively of an arithmetic progression. Given that the first term in each progression is 200 and the common ratio of the geometric progression is r, where $r \neq 1$, find

    a the value of r

    b the fourth term of each progression

    c the sum to infinity of the geometric progression.

6 The first term of an arithmetic progression is 12 and the sum of the first 16 terms is 282.

    a Find the common difference of this progression.

    The first, fifth and nth term of this arithmetic progression are the first, second and third term respectively of a geometric progression.

    b Find the common ratio of the geometric progression and the value of n.

7 The first two terms of a geometric progression are 80 and 64 respectively. The first three terms of this geometric progression are also the first, 11th and nth terms respectively of an arithmetic progression. Find the value of n.

8 The first two terms of a progression are $5x$ and x^2 respectively.

    a For the case where the progression is arithmetic with a common difference of 24, find the two possible values of x and the corresponding values of the third term.

    b For the case where the progression is geometric with a third term of $-\frac{8}{5}$, find the common ratio and the sum to infinity.

SUMMARY

Binomial expansions

If n is a positive integer, then $(a + b)^n$ can be expanded using the formula

$$(a + b)^n = a^n + {}^nC_1 a^{n-1} b + {}^nC_2 a^{n-2} b^2 + {}^nC_3 a^{n-3} b^3 + \ldots + {}^nC_r a^{n-r} b^r + \ldots + b^n$$

or

$$(a + b)^n = a^n + \binom{n}{1} a^{n-1} b + \binom{n}{2} a^{n-2} b^2 + \binom{n}{3} a^{n-3} b^3 + \ldots + \binom{n}{r} a^{n-r} b^r + \ldots + b^n$$

and where ${}^nC_r = \binom{n}{r} = \dfrac{n!}{(n-r)!\,r!}$

In particular,

$$(1 + x)^n = 1 + nx + \frac{n(n-1)}{2!} x^2 + \frac{n(n-1)(n-2)}{3!} x^3 + \frac{n(n-1)(n-2)(n-3)}{4!} x^4 + \ldots + x^n$$

Arithmetic series

For an arithmetic progression with first term a, common difference d and n terms:

- the kth term $= a + (k - 1)d$

- the last term $= l = a + (n - 1)d$

- the sum of the terms $= S_n = \dfrac{n}{2}(a + l) = \dfrac{n}{2}[2a + (n - 1)d]$.

CONTINUED

Geometric series

For a geometric progression with first term a, common ratio r and n terms:

- the kth term $= ar^{k-1}$
- the last term $= ar^{n-1}$
- the sum of the terms $= S_n = \dfrac{a(1-r^n)}{1-r} = \dfrac{a(r^n-1)}{r-1}$.

The condition for a geometric series to converge is $-1 < r < 1$.

When a geometric series converges, $S_\infty = \dfrac{a}{1-r}$.

Past paper questions and practice questions

Worked example

a   Find the first 4 terms in the expansion of $(2 + x^2)^6$ in ascending powers of x.   [3]

b   Find the term independent of x in the expansion of $(2 + x^2)^6\left(1 - \dfrac{3}{x^2}\right)^2$.   [3]

Cambridge IGCSE Additional Mathematics 0606 Paper 11 Q3i,ii Jun 2015

Answers

a   Expanding $(2 + x^2)^6$ using the binomial theorem gives

$2^6 + {}^6C_1 2^5 x^2 + {}^6C_2 2^4 (x^2)^2 + {}^6C_3 2^3 (x^2)^3 = 64 + 192x^2 + 240x^4 + 160x^6 \ldots$

b $(2 + x^2)^6\left(1 - \dfrac{2}{x^2}\right)^2 = (64 + 192x^2 + 240x^4 + 160x^6 \ldots)\left(1 - \dfrac{6}{x^2} + \dfrac{9}{x^4}\right)$

Term independent of $x = (64 \times 1) + \left(192x^2 \times -\dfrac{6}{x^2}\right) + \left(240x^4 \times \dfrac{9}{x^4}\right)$

$$= 64 - 1152 + 2160$$

$$= 1072$$

1   **i**   Write down, in ascending powers of x, the first 3 terms in the expansion of $(3 + 2x)^6$.

Give each term in its simplest form.   [3]

     ii   Hence find the coefficient of x^2 in the expansion of $(2 - x)(3 + 2x)^6$.   [2]

Cambridge IGCSE Additional Mathematics 0606 Paper 12 Q4 Mar 2015

2   **i**   Find the first 4 terms in the expansion of $(2 + x^2)^6$ in ascending powers of x.   [3]

     ii   Find the term independent of x in the expansion of $(2 + x^2)^6\left(1 - \dfrac{3}{x^2}\right)^2$.   [3]

Cambridge IGCSE Additional Mathematics 0606 Paper 11 Q3 Jun 2015

3   a   i   Use the Binomial Theorem to expand $(a + b)^4$, giving each term in its simplest form.   [2]

    ii   Hence find the term independent of x in the expansion of $\left(2x + \dfrac{1}{5x}\right)^4$.   [2]

    b   The coefficient of x^3 in the expansion of $\left(1 + \dfrac{x}{2}\right)^n$ equals $\dfrac{5n}{12}$. Find the value of the positive integer n.   [3]

Cambridge IGCSE Additional Mathematics 0606 Paper 21 Q8 Jun 2016

4   i   Given that a is a constant, expand $(2 + ax)^4$, in ascending powers of x, simplifying each term of your expansion.   [2]

Given also that the coefficient of x^2 is equal to the coefficient of x^3,

    ii   show that $a = 3$,   [1]

    iii   use your expansion to show that the value of 1.97^4 is 15.1 to 1 decimal place.   [2]

Cambridge IGCSE Additional Mathematics 0606 Paper 21 Q5 Jun 2017

5   i   Find the first 3 terms in the expansion of $\left(2x - \dfrac{1}{16x}\right)^8$ in descending powers of x.   [3]

    ii   Hence find the coefficient of x^4 in the expansion of $\left(2x - \dfrac{1}{16x}\right)^8 \left(\dfrac{1}{x^2} + 1\right)^2$.   [3]

Cambridge IGCSE Additional Mathematics 0606 Paper 11 Q9 Jun 2018

6   i   The first 3 terms, in ascending powers of x, in the expansion of $(2 + bx)^8$ can be written as $a + 256x + cx^2$. Find the value of each of the constants a, b and c.   [4]

    ii   Using the values found in **part i**, find the term independent of x in the expansion of $(2 + bx)^8 \left(2x - \dfrac{3}{x}\right)^2$.   [3]

Cambridge IGCSE Additional Mathematics 0606 Paper 11 Q4 Jun 2019

7   i   Find the first 3 terms in the expansion, in ascending powers of x, of $\left(3 - \dfrac{x}{9}\right)^6$.

    Give the terms in their simplest form.   [3]

    ii   Hence find the find the term independent of x in the expansion of $\left(3 - \dfrac{x}{9}\right)^6 \left(x - \dfrac{2}{x}\right)^2$.   [3]

Cambridge IGCSE Additional Mathematics 0606 Paper 12 Q3 Mar 2019

8   The first 3 terms in the expansion of $(a + x)^3 \left(1 - \dfrac{x}{3}\right)^5$, in ascending powers of x, can be written in the form $27 + bx + cx^2$, where a, b and c are integers. Find the values of a, b and c.   [8]

Cambridge IGCSE Additional Mathematics 0606 Paper 11 Q4 Jun 2021

9   The first term of a geometric progression is 35 and the second term is -14.
    a   Find the fourth term.   [3]
    b   Find the sum to infinity.   [2]

Practice question

10   The first three terms of a geometric progression are $2k + 6$, $k + 12$ and k respectively.
    All the terms in the progression are positive.
    a   Find value of k.   [3]
    b   Find the sum to infinity.   [2]

Practice question

11 An arithmetic progression has first term a and common difference d.
Give that the sum of the first 100 terms is 25 times the sum of the first 20 terms.

   a   Find d in terms of a.   [3]

   b   Write down an expression, in terms of a, for the 50th term.   [2]

Practice question

12 The 15th term of an arithmetic progression is 3 and the sum of the first 8 terms is 194.

   a   Find the first term of the progression and the common difference.   [4]

   b   Given that the nth term of the progression is -22, find the value of n.   [2]

Practice question

13 The second term of a geometric progression is -576 and the fifth term is 243. Find

   a   the common ratio   [3]

   b   the first term   [1]

   c   the sum to infinity.   [2]

Practice question

14 **a**   The sixth term of an arithmetic progression is 35 and the sum of the first ten terms is 335.
Find the eighth term.   [4]

   b   A geometric progression has first term 8 and common ratio r. A second geometric progression has first
term 10 and common ratio $\frac{1}{4}r$. The two progressions have the same sum to infinity, S.
Find the values of r and the value of S.   [3]

Practice question

15 **a**   The 10th term of an arithmetic progression is 4 and the sum of the first 7 terms is -28.
Find the first term and the common difference.   [4]

   b   The first term of a geometric progression is 40 and the fourth term is 5.
Find the sum to infinity of the progression.   [3]

Practice question

16 **a**   A geometric progression has first term a, common ratio r and sum to infinity S.
A second geometric progression has first term $3a$, common ratio $2r$ and sum to infinity $4S$.
Find the value of r.   [3]

   b   An arithmetic progression has first term -24. The nth term is -13.8 and the $(2n)$th term is -3.
Find the value of n.   [4]

Practice question

17 **a**   The sum of the first two terms of a geometric progression is 10 and the third term is 9.

      i   Find the possible values of the common ratio and the first term.   [5]

      ii   Find the sum to infinity of the convergent progression.   [1]

   b   In an arithmetic progression, $u_1 = -10$ and $u_4 = 14$.
Find $u_{100} + u_{101} + u_{102} + ... + u_{200}$, the sum of the 100th to the 200th terms of the progression.   [4]

Cambridge IGCSE Additional Mathematics 0606 Paper 22 Q13 Mar 2020

18 a An arithmetic progression has a second term -14 and a sum to 21 terms of 84.
Find the first term and the 21st term of this progression. [5]

b A geometric progression has a second term of $27p^2$ and a fifth term of p^5.
The common ratio, r, is such that $0 < r < 1$.

i Find r in terms of p. [2]

ii Hence find, in terms of p, the sum to infinity of the progression. [3]

iii Given that the sum to infinity is 81, find the value of p. [2]

Cambridge IGCSE Additional Mathematics 0606 Paper 11 Q9 Jun 2020

19 a An arithmetic progression has a second term of 8 and a fourth term of 18.
Find the least number of terms for which the sum of this progression is greater than 1560. [6]

b A geometric progression has a sum to infinity of 72. The sum of the first 3 terms of this

progression is $\dfrac{333}{8}$.

i Find the value of the common ratio. [5]

ii Hence find the value of the first term. [1]

Cambridge IGCSE Additional Mathematics 0606 Paper 11 Q10 Nov 2020

20 a A geometric progression has first term 10 and sum to infinity 6.

i Find the common ratio of this progression. [2]

ii Hence find the sum of the first 7 terms, giving your answer correct to 2 decimal places. [2]

b The first three terms of an arithmetic progression are $\log_x 3$, $\log_x(3^2)$, $\log_x(3^3)$.

i Find the common difference of this progression. [1]

ii Find, in terms of n and $\log_x 3$, the sum to n terms of this progression. Simplify your answer. [2]

iii Given that the sum to n terms is $3081 \log_x 3$, find the value of n. [2]

iv Hence, given that the sum to n terms is also equal to 1027, find the value of x. [2]

Cambridge IGCSE Additional Mathematics 0606 Paper 12 Q6 Mar 2021

21 The 2nd, 8th and 44th terms of an arithmetic progression form the first three terms of a geometric progression.
In the arithmetic progression, the first term is 1 and the common difference is positive.

a i Show that the common difference of the arithmetic progression is 5. [5]

ii Find the sum of the first 20 terms of the arithmetic progression. [2]

b i Find the 5th term of the geometric progression. [2]

ii Explain whether or not the sum to infinity of this geometric progression exists. [1]

Cambridge IGCSE Additional Mathematics 0606 Paper 21 Q11 Jun 2021

› Chapter 12

Calculus – Differentiation 1

THIS SECTION WILL SHOW YOU HOW TO:

- use the notations $f'(x)$, $f''(x)$, $\dfrac{dy}{dx}$, $\dfrac{d^2y}{dx^2}$, $\left[\dfrac{d}{dx}\left(\dfrac{dy}{dx}\right)\right]$

- use the derivative of x^n (for any rational n), together with constant multiples, sums and composite functions of these

- differentiate products and quotients of functions

- apply differentiation to gradients, tangents and normals, stationary points, connected rates of change, small increments and approximations and practical maxima and minima problems

- use the first and second derivative tests to discriminate between maxima and minima.

PRE-REQUISITE KNOWLEDGE

Before you start…

Where it comes from	What you should be able to do	Check your skills
Cambridge IGCSE/O Level Mathematics	Use the rules of indices to simplify expressions to the form ax^n.	**1** Write in the form ax^n: **a** $2x\sqrt{x}$ **b** $7\sqrt[3]{x^2}$ **c** $\dfrac{x}{3\sqrt{x}}$ **d** $\dfrac{1}{2x^2}$
Cambridge IGCSE/O Level Mathematics	Write $\dfrac{k}{(ax+b)^n}$ in the form $k(ax+b)^{-n}$.	**2** Write in the form $k(ax+b)^{-n}$: **a** $\dfrac{3}{(x+1)^2}$ **b** $\dfrac{5}{(2x-3)^4}$
Cambridge IGCSE/O Level Mathematics	Find the gradient of a perpendicular line.	**3** The gradient of a line is $-\dfrac{3}{4}$. Write down the gradient of a line that is perpendicular to it.
Cambridge IGCSE/O Level Mathematics	Find the equation of a line with a given gradient and a given point on the line.	**4** Find the equation of a line with gradient 3 that passes through the point $(5, -1)$.

12.1 The gradient function

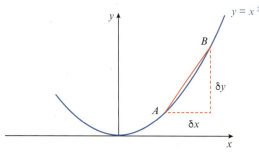

The diagram shows a point $A(x, y)$ on the curve $y = x^2$ and a point B which is close to the point A.

The coordinates of B are $(x + \delta x, y + \delta y)$, where δx is a small increase in the value of x and δy is a small increase in the value of y.

KEY WORD

differentiation

derivative

gradient function

chain rule

product rule

quotient rule

normal

rate of change

first derivative

second derivative

stationary point

turning point

maximum point

minimum point

point of inflexion

The coordinates of A and B can also be written as (x, x^2) and $(x + \delta x, (x + \delta x)^2)$.

gradient of chord $AB = \dfrac{y_2 - y_1}{x_2 - x_1}$

$$= \frac{(x + \delta x)^2 - x^2}{(x + \delta x) - x}$$

$$= \frac{x^2 + 2x\delta x + (\delta x)^2 - x^2}{\delta x}$$

$$= \frac{2x\delta x + (\delta x)^2}{\delta x}$$

$$= 2x + \delta x$$

As $B \to A$, $\delta x \to 0$ and the gradient of the chord $AB \to$ the gradient of the curve at A.

Hence, gradient of curve $= 2x$.

This process of finding the gradient of a curve at any point is called differentiation from first principles.

Notation

There are three different notations that can be used to describe the rule above.

1   If $y = x^2$, then $\dfrac{dy}{dx} = 2x$

2   If $f(x) = x^2$, then $f'(x) = 2x$

3   $\dfrac{d}{dx}(x^2) = 2x$

$\dfrac{dy}{dx}$ is called the **derivative** of y with respect to x.

$f'(x)$ is called the **gradient function** of the curve $y = f(x)$.

$\dfrac{d}{dx}(x^2) = 2x$ means 'if you differentiate x^2 with respect to x, the result is $2x$'.

You do not need to be able to differentiate from first principles, but you are expected to know the rules of differentiation.

Differentiation of power functions

The rule for differentiating power functions is:

$$\text{If } y = x^n, \text{ then } \frac{dy}{dx} = nx^{n-1}$$

It is easier to remember this rule as:

'multiply by the power n and then subtract one from the power'.

So for the earlier example where $y = x^2$,

$$\frac{dy}{dx} = 2 \times x^{2-1}$$

$$= 2x^1$$

$$= 2x$$

WORKED EXAMPLE 1

Differentiate with respect to x.

a x^5 **b** $\dfrac{1}{x^3}$ **c** $\sqrt{x}$ **d** 3

Answers

a $\dfrac{d}{dx}(x^5) = 5x^{5-1}$

$\qquad\qquad = 5x^4$

b $\dfrac{d}{dx}\left(\dfrac{1}{x^3}\right) = \dfrac{d}{dx}(x^{-3})$

$\qquad\qquad\quad = -3x^{-3-1}$

$\qquad\qquad\quad = -3x^{-4}$

$\qquad\qquad\quad = -\dfrac{3}{x^4}$

c $\dfrac{d}{dx}(\sqrt{x}) = \dfrac{d}{dx}\left(x^{\frac{1}{2}}\right)$

$\qquad\qquad = \dfrac{1}{2}x^{\frac{1}{2}-1}$

$\qquad\qquad = \dfrac{1}{2}x^{-\frac{1}{2}}$

$\qquad\qquad = \dfrac{1}{2\sqrt{x}}$

d $\dfrac{d}{dx}(3) = \dfrac{d}{dx}(3x^0)$

$\qquad\qquad = 0x^{0-1}$

$\qquad\qquad = 0$

You need to know and be able to use the following two rules:

Scalar multiple rule:

$$\frac{d}{dx}[k\,f(x)] = k\frac{d}{dx}[f(x)]$$

Addition/subtraction rule:

$$\frac{d}{dx}[f(x) \pm g(x)] = \frac{d}{dx}[f(x)] \pm \frac{d}{dx}[g(x)]$$

WORKED EXAMPLE 2

Differentiate $2x^5 - 3x^2 + \dfrac{1}{2x^3} + \dfrac{5}{\sqrt{x}}$ with respect to x.

Answers

$\dfrac{d}{dx}\left(2x^5 - 3x^2 + \dfrac{1}{2x^3} + \dfrac{5}{\sqrt{x}}\right) = \dfrac{d}{dx}\left(2x^5 - 3x^2 + \dfrac{1}{2}x^{-3} + 5x^{-\frac{1}{2}}\right)$

$\qquad\qquad = 2\dfrac{d}{dx}(x^5) - 3\dfrac{d}{dx}(x^2) + \dfrac{1}{2}\dfrac{d}{dx}(x^{-3}) + 5\dfrac{d}{dx}\left(x^{-\frac{1}{2}}\right)$

$\qquad\qquad = 2(5x^4) - 3(2x) + \dfrac{1}{2}(-3x^{-4}) + 5\left(-\dfrac{1}{2}x^{-\frac{3}{2}}\right)$

$\qquad\qquad = 10x^4 - 6x - \dfrac{3}{2}x^{-4} - \dfrac{5}{2}x^{-\frac{3}{2}}$

WORKED EXAMPLE 3

Find the gradient of the curve $y = (3x + 2)(2x - 1)$ at the point $(-1, 3)$.

Answers

$y = (3x + 2)(2x - 1)$

$y = 6x^2 + x - 2$

$\dfrac{dy}{dx} = 12x + 1$

When $x = -1$, $\dfrac{dy}{dx} = 12(-1) + 1$

$\qquad\qquad\qquad = -11$

Gradient of curve at $(-1, 3)$ is -11.

Exercise 12.1

1 Differentiate with respect to x.

a x^4 b x^9 c x^{-3} d x^{-6} e $\dfrac{1}{x}$

f $\dfrac{1}{x^5}$ g $\sqrt{x}$ h $\sqrt{x^5}$ i $x^{-\frac{1}{5}}$ j $x^{\frac{1}{3}}$

k $\sqrt[3]{x^2}$ l $\dfrac{1}{\sqrt{x}}$ m x n $x^{\frac{3}{2}}$ o $\sqrt[3]{x^5}$

p $x^2 \times x^4$ q $x^2 \times x$ r $\dfrac{x^4}{x^2}$ s $\dfrac{x}{\sqrt{x}}$ t $\dfrac{x\sqrt{x}}{x^3}$

2 Differentiate with respect to x.

a $2x^3 - 5x + 4$ b $8x^5 - 3x^2 - 2$ c $7 - 2x^3 + 4x$

d $3x^2 + \dfrac{2}{x} - \dfrac{1}{x^2}$ e $2x - \dfrac{1}{x} - \dfrac{1}{\sqrt{x}}$ f $\dfrac{x + 5}{\sqrt{x}}$

g $\dfrac{x^2 - 3}{x}$ h $\dfrac{5x^2 - \sqrt{x}}{x}$ i $\dfrac{x^2 - x - 1}{\sqrt{x}}$

j $5x^2(x + 1)$ k $x^{-2}(2x - 5)$ l $\dfrac{1}{x}(x^2 - 2)$

m $(3x + 1)^2$ n $(1 - x^3)^2$ o $(2x - 1)(3x + 4)$

3 Find the value of $\dfrac{dy}{dx}$ at the given point on the curve.

a $y = 3x^2 - 4$ at the point $(1, -1)$

b $y = 4 - 2x^2$ at the point $(-1, 2)$

c $y = 2 + \dfrac{8}{x}$ at the point $(-2, -2)$

d $y = 5x^3 - 2x^2 - 3$ at the point $(0, -3)$

e $y = \dfrac{x + 5}{x}$ at the point $(5, 2)$

f $y = \dfrac{x - 3}{\sqrt{x}}$ at the point $(9, 2)$

4 Find the coordinates of the point on the curve $y = 2x^2 - x - 1$ at which the gradient is 7.

5 Find the gradient of the curve $y = \dfrac{x - 4}{x}$ at the point where the curve crosses the x-axis.

6 Find the gradient of the curve $y = x^3 - 2x^2 + 5x - 3$ at the point where the curve crosses the y-axis.

7 The curve $y = 2x^2 + 7x - 4$ and the line $y = 5$ meet at the points P and Q.

Find the gradient of the curve at the point P and at the point Q.

8 The curve $y = ax^2 + bx$ has gradient 8 when $x = 2$ and has gradient -10 when $x = -1$.

Find the value of a and the value of b.

9 The gradient of the curve $y = ax + \dfrac{b}{x}$ at the point $(-1, -3)$ is -7.

Find the value of a and the value of b.

10 Find the coordinates of the points on this curve $y = \dfrac{x^3}{3} - \dfrac{5x^2}{2} + 6x - 1$ where the gradient is 2.

11 The curve $y = \dfrac{1}{3}x^3 - 2x^2 - 8x + 5$ and the line $y = x + 5$ meet at the points A, B and C.

  a Find the coordinates of the points A, B and C.

  b Find the gradient of the curve at the points A, B and C.

12 $y = 4x^3 + 3x^2 - 6x - 1$

  a Find $\dfrac{dy}{dx}$.

  b Find the range of values of x for which $\dfrac{dy}{dx} \geqslant 0$.

13 $y = x^3 + x^2 - 16x - 16$

  a Find $\dfrac{dy}{dx}$.

  b Find the range of values of x for which $\dfrac{dy}{dx} \leqslant 0$.

14 CHALLENGE QUESTION

A curve has equation $y = x^5 - 5x^3 + 25x^2 + 145x + 10$. Show that the gradient of the curve is never negative.

12.2 The chain rule

To differentiate $y = (2x + 3)^8$, you could expand the brackets and then differentiate each term separately. This would take a long time to do. There is a more efficient method available that allows you to find the derivative without expanding:

Let $u = 2x + 3$. Then $y = (2x + 3)^8$ becomes $y = u^8$.

The derivative of the composite function $y = (2x + 3)^8$ can then be found using the **chain rule**:

$$\frac{dy}{dx} = \frac{dy}{du} \times \frac{du}{dx}$$

WORKED EXAMPLE 4

Find the derivative of $y = (2x + 3)^8$.

Answers

$y = (2x + 3)^8$

Let $u = 2x + 3$ so $y = u^8$

$\dfrac{du}{dx} = 2$ and $\dfrac{dy}{du} = 8u^7$

Using $\dfrac{dy}{dx} = \dfrac{dy}{du} \times \dfrac{du}{dx}$

$\quad = 8u^7 \times 2$

$\quad = 8(2x + 3)^7 \times 2$

$\quad = 16(2x + 3)^7$

With practice you will find that you can do this mentally:

Consider the 'inside' of $(2x + 3)^8$ to be $2x + 3$.

To differentiate $(2x + 3)^8$:

Step 1: Differentiate the 'outside' first: $\longrightarrow$ $8(2x + 3)^7$

Step 2: Then differentiate the 'inside': $\longrightarrow$ 2

Step 3: Multiply these two expressions: $\longrightarrow$ $16(2x + 3)^7$

WORKED EXAMPLE 5

Find the derivative of $y = \dfrac{2}{(5x^2 - 1)^4}$.

Answers

$y = \dfrac{2}{(5x^2 - 1)^4}$

Let $u = 5x^2 - 1$ so $y = 2u^{-4}$

$\dfrac{du}{dx} = 10x$ and $\dfrac{dy}{du} = -8u^{-5}$

Using $\dfrac{dy}{dx} = \dfrac{dy}{du} \times \dfrac{du}{dx}$

$\qquad = -8u^{-5} \times 10x$

$\qquad = -8(5x^2 - 1)^{-5} \times 10x$

$\qquad = \dfrac{-80x}{(5x^2 - 1)^5}$

Alternatively, to differentiate the expression mentally:

Write $\dfrac{2}{(5x^2 - 1)^4}$ as $2(5x^2 - 1)^{-4}$.

Step 1: Differentiate the 'outside' first: $\longrightarrow$ $-8(5x^2 - 1)^{-5}$

Step 2: Then differentiate the 'inside': $\longrightarrow$ $10x$

Step 3: Multiply the two expressions: $\longrightarrow$ $-80x(5x^2 - 1)^{-5} = \dfrac{-80x}{(5x^2 - 1)^5}$

Exercise 12.2

1 Differentiate with respect to x.

   a $(x + 2)^9$ **b** $(3x - 1)^7$ **c** $(1 - 5x)^6$ **d** $\left(\dfrac{1}{2}x - 7\right)^4$

   e $\dfrac{(2x + 1)^6}{3}$ **f** $2(x - 4)^6$ **g** $6(5 - x)^5$ **h** $\dfrac{1}{2}(2x + 5)^8$

   i $(x^2 + 2)^4$ **j** $(1 - 2x^2)^7$ **k** $(x^2 - 3x)^5$ **l** $\left(x^2 + \dfrac{2}{x}\right)^4$

2 Differentiate with respect to x.

   a $\dfrac{1}{(x + 4)}$ **b** $\dfrac{3}{(2x - 1)}$ **c** $\dfrac{5}{(2 - 3x)}$ **d** $\dfrac{16}{(2x^2 - 5)}$

   e $\dfrac{4}{(x^2 - 2x)}$ **f** $\dfrac{1}{(x - 1)^5}$ **g** $\dfrac{2}{(5x - 1)^3}$ **h** $\dfrac{1}{2(3x - 2)^4}$

3 Differentiate with respect to x.

   a $\sqrt{x + 2}$ **b** $\sqrt{5x - 1}$ **c** $\sqrt{2x^2 - 3}$ **d** $\sqrt{x^3 + 2x}$

   e $\sqrt[3]{3 - 2x}$ **f** $4\sqrt{2x - 1}$ **g** $\dfrac{1}{\sqrt{3x - 1}}$ **h** $\dfrac{3}{\sqrt[3]{2 - 5x}}$

4   Find the gradient of the curve $y = (2x - 5)^4$ at the point $(3, 1)$.

5   Find the gradient of the curve $y = \dfrac{8}{(x - 2)^2}$ at the point where the curve crosses the y-axis.

6   Find the gradient of the curve $y = x + \dfrac{4}{x - 5}$ at the points where the curve crosses the x-axis.

7   Find the coordinates of the point on the curve $y = \sqrt{x^2 - 6x + 13}$ where the gradient is 0.

8   The curve $y = \dfrac{a}{\sqrt{bx + 1}}$ passes through the point $(1, 4)$ and has gradient $-\dfrac{3}{2}$ at this point.

Find the value of a and the value of b.

12.3 The product rule

Consider the function $y = x^2(x^5 + 1)$

$$y = x^7 + x^2$$

$$\frac{dy}{dx} = 7x^6 + 2x$$

The function $y = x^2(x^5 + 1)$ can also be considered as the product of two separate functions $y = uv$ where $u = x^2$ and $v = x^5 + 1$.

To differentiate the product of two functions you can use the **product rule**:

$$\frac{d}{dx}(uv) = u\frac{dv}{dx} + v\frac{du}{dx}$$

It is easier to remember this rule as

'(first function × derivative of second function) + (second function × derivative of first function)'

So, for $y = x^2(x^5 + 1)$, $\dfrac{dy}{dx} = \underbrace{(x^2)}_{\text{first}}\ \underbrace{\dfrac{d}{dx}(x^5 + 1)}_{\substack{\text{differentiate}\\\text{second}}} + \underbrace{(x^5 + 1)}_{\text{second}}\ \underbrace{\dfrac{d}{dx}(x^2)}_{\substack{\text{differentiate}\\\text{first}}}$

$$= (x^2)(5x^4) + (x^5 + 1)(2x)$$
$$= 5x^6 + 2x^6 + 2x$$
$$= 7x^6 + 2x$$

WORKED EXAMPLE 6

Find the derivative of $y = (5x + 1)\sqrt{6x - 1}$.

Answers

$$y = (5x + 1)\sqrt{6x - 1}$$

$$= (5x + 1)(6x - 1)^{\frac{1}{2}}$$

$$\frac{dy}{dx} = \underbrace{(5x + 1)}_{\text{first}} \underbrace{\frac{d}{dx}\left[(6x - 1)^{\frac{1}{2}}\right]}_{\substack{\text{differentiate} \\ \text{second}}} + \underbrace{\left[(6x - 1)^{\frac{1}{2}}\right]}_{\text{second}} \underbrace{\frac{d}{dx}(5x + 1)}_{\substack{\text{differentiate} \\ \text{first}}}$$

$$= (5x + 1)\underbrace{\left[\frac{1}{2}(6x - 1)^{-\frac{1}{2}}(6)\right]}_{\text{use the chain rule}} + (\sqrt{6x - 1})(5)$$

$$= \frac{3(5x + 1)}{\sqrt{6x - 1}} + 5\sqrt{6x - 1} \qquad \text{write as a single fraction}$$

$$= \frac{3(5x + 1) + 5(6x - 1)}{\sqrt{6x - 1}} \qquad \text{simplify the denominator}$$

$$= \frac{45x - 2}{\sqrt{6x - 1}}$$

WORKED EXAMPLE 7

Find the x-coordinate of the points on the curve $y = (x + 2)^2(2x - 5)^3$ where the gradient is 0.

Answers

$$y = (x + 2)^2(2x - 5)^3$$

$$\frac{dy}{dx} = \underbrace{(x + 2)^2}_{\text{first}} \underbrace{\frac{d}{dx}\left[(2x - 5)^3\right]}_{\substack{\text{differentiate} \\ \text{second}}} + \underbrace{(2x - 5)^3}_{\text{second}} \underbrace{\frac{d}{dx}\left[(x + 2)^2\right]}_{\substack{\text{differentiate} \\ \text{first}}}$$

$$= (x + 2)^2 \underbrace{\left[3(2x - 5)^2(2)\right]}_{\text{use the chain rule}} + (2x - 5)^3 \underbrace{\left[2(x + 2)^1(1)\right]}_{\text{use the chain rule}}$$

$$= 6(x + 2)^2(2x - 5)^2 + 2(x + 2)(2x - 5)^3 \qquad \text{factorise}$$

$$= 2(x + 2)(2x - 5)^2[3(x + 2) + (2x - 5)] \qquad \text{simplify}$$

$$= 2(x + 2)(2x - 5)^2(5x + 1)$$

$\dfrac{dy}{dx} = 0$ when $2(x + 2)(2x - 5)^2(5x + 1) = 0$

$$x + 2 = 0 \qquad\qquad 2x - 5 = 0 \qquad\qquad 5x + 1 = 0$$
$$x = -2 \qquad\qquad\quad x = 2.5 \qquad\qquad\quad x = -0.2$$

Exercise 12.3

1   Use the product rule to differentiate each of the following with respect to x:

   a $x(x + 4)$ **b** $2x(3x + 5)$

   c $x(x + 2)^3$ **d** $x^2(x - 1)^3$

   e $x\sqrt{x - 5}$ **f** $(x + 2)\sqrt{x}$

   g $x^2\sqrt{x + 3}$ **h** $\sqrt{x}\,(3 - x^2)^3$

   i $(2x + 1)(x^2 + 5)$ **j** $(x + 4)(x - 3)^3$

   k $(x - 1)^2(x + 2)^2$ **l** $(2x + 1)^3(x - 3)^4$

2   Find the gradient of the curve $y = x^2\sqrt{x + 2}$ at the point $(2, 8)$.

3   Find the gradient of the curve $y = (x - 1)^3(x + 3)^2$ at the point where $x = 2$.

4   Find the gradient of the curve $y = (x + 2)(x - 5)^2$ at the points where the curve meets the x-axis.

5   Find the x-coordinate of the points on the curve $y = (2x - 3)^3(x + 2)^4$ where the gradient is zero.

6   Find the x-coordinate of the point on the curve $y = (x + 3)\sqrt{4 - x}$ where the gradient is zero.

12.4 The quotient rule

The function $y = \dfrac{x^3 + 1}{(2x - 3)}$ can be differentiated by writing the function in the form

$y = (x^3 + 1)(2x - 3)^{-1}$ and then by applying the product rule.

Alternatively, $y = \dfrac{x^3 + 1}{(2x - 3)}$ can be considered as the division (quotient) of two separate functions:

$y = \dfrac{u}{v}$ where $u = x^3 + 1$ and $v = 2x - 3$.

To differentiate the quotient of two functions you can use the **quotient rule**:

$$\frac{d}{dx}\left(\frac{u}{v}\right) = \frac{v\dfrac{du}{dx} - u\dfrac{dv}{dx}}{v^2}$$

It is easier to remember this rule as:

$$\frac{(\text{denominator} \times \text{derivative of numerator}) - (\text{numerator} \times \text{derivative of denominator})}{(\text{denominator})^2}$$

WORKED EXAMPLE 8

Use the quotient rule to find the derivative of $y = \dfrac{x^3 + 1}{(2x - 3)}$.

Answers

$$y = \frac{x^3 + 1}{(2x - 3)}$$

$$\frac{dy}{dx} = \frac{\overbrace{(2x - 3)}^{\text{denominator}} \times \overbrace{\dfrac{d}{dx}(x^3 + 1)}^{\substack{\text{differentiate} \\ \text{numerator}}} \overset{-}{} \overbrace{(x^3 + 1)}^{\text{numerator}} \times \overbrace{\dfrac{d}{dx}(2x - 3)}^{\substack{\text{differentiate} \\ \text{denominator}}}}{\underbrace{(2x - 3)^2}_{\text{denominator squared}}}$$

$$= \frac{(2x - 3)(3x^2) - (x^3 + 1)(2)}{(2x - 3)^2}$$

$$= \frac{6x^3 - 9x^2 - 2x^3 - 2}{(2x - 3)^2}$$

$$= \frac{4x^3 - 9x^2 - 2}{(2x - 3)^2}$$

CLASS DISCUSSION

An alternative method for finding $\dfrac{dy}{dx}$ in worked example 8 is to express y in the form

$$y = (x^3 + 1)(2x - 3)^{-1}$$

and to then differentiate using the product rule.

Try this method and then discuss with your classmates which method you prefer.

WORKED EXAMPLE 9

Find the derivative of $y = \dfrac{(x + 1)^2}{\sqrt{x + 2}}$.

Answers

$$y = \frac{(x + 1)^2}{\sqrt{x + 2}}$$

$$\frac{dy}{dx} = \frac{\overbrace{\sqrt{x + 2}}^{\text{denominator}} \times \overbrace{\dfrac{d}{dx}[(x + 1)^2]}^{\substack{\text{differentiate} \\ \text{numerator}}} \overset{-}{} \overbrace{(x + 1)^2}^{\text{numerator}} \times \overbrace{\dfrac{d}{dx}[\sqrt{x + 2}]}^{\substack{\text{differentiate} \\ \text{denominator}}}}{\underbrace{(\sqrt{x + 2})^2}_{\text{denominator squared}}}$$

CONTINUED

$$= \frac{(\sqrt{x+2})\left[2(x+1)^1(1)\right] - (x+1)^2\left[\frac{1}{2}(x+2)^{-\frac{1}{2}}(1)\right]}{x+2}$$

$$= \frac{2(x+1)\sqrt{x+2} - \dfrac{(x+1)^2}{2\sqrt{x+2}}}{x+2} \quad \text{multiply numerator and denominator by } 2\sqrt{x+2}$$

$$= \frac{4(x+1)(x+2) - (x+1)^2}{2(x+2)\sqrt{x+2}} \quad \text{factorise the numerator}$$

$$= \frac{(x+1)[4(x+2) - (x+1)]}{2(x+2)^{\frac{3}{2}}}$$

$$= \frac{(x+1)(3x+7)}{2(x+2)^{\frac{3}{2}}}$$

Exercise 12.4

1 Use the quotient rule to differentiate each of the following with respect to x:

a $\dfrac{1+2x}{5-x}$ b $\dfrac{3x+2}{x+4}$ c $\dfrac{x-1}{3x+4}$ d $\dfrac{5x-2}{3-8x}$

e $\dfrac{x^2}{5x-2}$ f $\dfrac{x}{x^2-1}$ g $\dfrac{5}{3x-1}$ h $\dfrac{x+4}{x^2-2}$

2 Find the gradient of the curve $y = \dfrac{x+3}{x-1}$ at the point $(2, 5)$.

3 Find the coordinates of the points on the curve $y = \dfrac{x^2}{2x-1}$ where $\dfrac{dy}{dx} = 0$.

4 Find the gradient of the curve $y = \dfrac{7x-2}{2x+3}$ at the point where the curve crosses the y-axis.

5 Differentiate with respect to x:

a $\dfrac{\sqrt{x}}{2x+1}$ b $\dfrac{x}{\sqrt{1-2x}}$ c $\dfrac{x^2}{\sqrt{x^2+2}}$ d $\dfrac{5\sqrt{x}}{3+x}$

6 Find the gradient of the curve $y = \dfrac{x-2}{\sqrt{x+5}}$ at the point $(-4, -6)$.

7 Find the coordinates of the point on the curve $y = \dfrac{2(x-5)}{\sqrt{x+1}}$ where the gradient is $\dfrac{5}{4}$.

8 CHALLENGE QUESTION

The line $5x - 5y = 2$ intersects the curve $x^2y - 5x + y + 2 = 0$ at three points.

a Find the coordinates of the points of intersection.

b Find the gradient of the curve at each of the points of intersection.

12.5 Tangents and normals

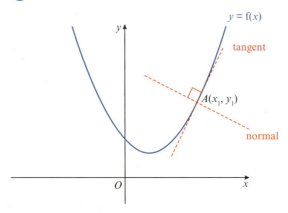

The line perpendicular to the tangent at the point A is called the **normal** at A.

If the value of $\dfrac{dy}{dx}$ at the point (x_1, y_1) is m, then the equation of the tangent is given by:

$$y - y_1 = m(x - x_1)$$

The normal at the point (x_1, y_1) is perpendicular to the tangent, so the gradient of the normal is $-\dfrac{1}{m}$ and the equation of the normal is given by:

$$y - y_1 = -\frac{1}{m}(x - x_1), \ m \neq 0$$

WORKED EXAMPLE 10

Find the equation of the tangent and the normal to the curve $y = 3x^2 - x + \dfrac{8}{x}$ at the point where $x = 2$.

Answers

$y = 3x^2 - x + 8x^{-1}$

$\dfrac{dy}{dx} = 6x - 1 - 8x^{-2}$

When $x = 2$, $y = 3(2)^2 - (2) + 8(2)^{-1} = 14$

and $\dfrac{dy}{dx} = 6(2) - 1 - 8(2)^{-2} = 9$

Tangent: passes through the point $(2, 14)$ and gradient $= 9$

$y - 14 = 9(x - 2)$

$\qquad y = 9x - 4$

Normal: passes through the point $(2, 14)$ and gradient $= -\dfrac{1}{9}$

$y - 14 = -\dfrac{1}{9}(x - 2)$

$9y + x = 128$

WORKED EXAMPLE 11

The normals to the curve $y = x^3 - 5x^2 + 3x + 1$, at the points $A\,(4, -3)$ and $B\,(1, 0)$, meet at the point C.

a Find the coordinates of C.

b Find the area of triangle ABC.

Answers

a
$$\frac{dy}{dx} = 3x^2 - 10x + 3$$

When $x = 4$, $\dfrac{dy}{dx} = 3(4)^2 - 10(4) + 3 = 11$

when $x = 1$, $\dfrac{dy}{dx} = 3(1)^2 - 10(1) + 3 = -4$

Normal at A: passes through the point $(4, -3)$ and gradient $= -\dfrac{1}{11}$

$$y - (-3) = -\frac{1}{11}(x - 4)$$

$$11y = -x - 29 \qquad (1)$$

Normal at B: passes through the point $(1, 0)$ and gradient $= \dfrac{1}{4}$

$$y - 0 = \frac{1}{4}(x - 1)$$

$$4y = x - 1 \qquad (2)$$

Adding equations (1) and (2) gives

$$15y = -30$$

$$y = -2$$

When $y = -2$, $11(-2) = -x - 29$

$$x = -7$$

Hence, C is the point $(-7, -2)$.

b

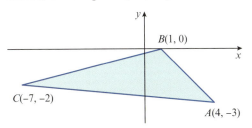

Area of triangle $ABC = \dfrac{1}{2}\begin{vmatrix} 4 & 1 & -7 & 4 \\ -3 & 0 & -2 & -3 \end{vmatrix}$

$$= \frac{1}{2}|0 + (-2) + 21 - (-3) - 0 - (-8)|$$

$$= \frac{1}{2}|30|$$

$$= 15 \text{ units}^2$$

Exercise 12.5

1 Find the equation of the tangent to the curve at the given value of x.

   a $y = x^4 - 3$ at $x = 1$ **b** $y = x^2 + 3x + 2$ at $x = -2$

   c $y = 2x^3 + 5x^2 - 1$ at $x = 1$ **d** $y = 5 + \dfrac{2}{x}$ at $x = -2$

   e $y = (x - 3)(2x - 1)^2$ at $x = 2$ **f** $y = \dfrac{x}{x + 1}$ at $x = -3$

2 Find the equation of the normal to the curve at the given value of x.

   a $y = x^2 + 5x$ at $x = -1$ **b** $y = 3x^2 - 4x + 1$ at $x = 2$

   c $y = 5x^4 - 7x^2 + 2x$ at $x = -1$ **d** $y = 4 - \dfrac{2}{x^2}$ at $x = -2$

   e $y = 2x(x - 3)^3$ at $x = 2$ **f** $y = \dfrac{x + 2}{x - 2}$ at $x = 6$

3 Find the equation of the tangent and the normal to the curve $y = 5x - \dfrac{3}{x}$ at the point where $x = 1$.

4 The normal to the curve $y = x^3 - 2x + 1$ at the point $(2, 5)$ intersects the y-axis at the point P.

   Find the coordinates of P.

5 Find the equation of the tangent and the normal to the curve $y = \dfrac{x - 1}{\sqrt{x + 4}}$ at the point where the curve intersects the y-axis.

6 The tangents to the curve $y = x^2 - 5x + 4$, at the points $(1, 0)$ and $(3, -2)$, meet at the point Q.

   Find the coordinates of Q.

7 The tangent to the curve $y = 3x^2 - 10x - 8$ at the point P is parallel to the line $y = 2x - 5$.

   Find the equation of the tangent at P.

8 A curve has equation $y = x^3 - x + 6$.

   a Find the equation of the tangent to this curve at the point $P(-1, 6)$.

   The tangent at the point Q is parallel to the tangent at P.

   b Find the coordinates of Q.

   c Find the equation of the normal at Q.

9 A curve has equation $y = 4 + (x - 1)^4$.

   The normal at the point $P(1, 4)$ and the normal at the point $Q(2, 5)$ intersect at the point R.

   Find the coordinates of R.

10 A curve has equation $y = (2 - \sqrt{x})^4$.

   The normal at the point $P(1, 1)$ and the normal at the point $Q(9, 1)$ intersect at the point R.

   a Find the coordinates of R.

   b Find the area of triangle PQR.

11 A curve has equation $y = \sqrt{x}\,(x - 2)^3$.

The tangent at the point $P(3, \sqrt{3})$ and the normal at the point $Q(9, 1)$ intersect at the point R.

   a   Show that the equation of the tangent at the point $P(3, \sqrt{3})$ is

$$y = \frac{19\sqrt{3}}{6}x - \frac{17\sqrt{3}}{2}.$$

   b   Find the equation of the normal at the point $Q(1, -1)$.

12 The equation of a curve is $y = \dfrac{x^2}{x + 2}$.

The tangent to the curve at the point where $x = -3$ meets the y-axis at M.

The normal to the curve at the point where $x = -3$ meets the x-axis at N.

Find the area of the triangle MNO, where O is the origin.

13 The equation of a curve is $y = \dfrac{x - 3}{x + 2}$.

The curve intersects the x-axis at the point P.

The normal to the curve at P meets the y-axis at the point Q.

Find the area of the triangle POQ, where O is the origin.

12.6 Small increments and approximations

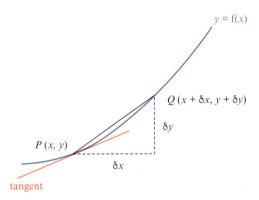

The diagram shows the tangent to the curve $y = \text{f}(x)$ at the point $P(x, y)$.

The gradient of the tangent at the point P is $\dfrac{\mathrm{d}y}{\mathrm{d}x}$.

The point $Q(x + \delta x, y + \delta y)$ is a point on the curve close to the point P.

The gradient of the chord PQ is $\dfrac{\delta y}{\delta x}$.

If P and Q are sufficiently close then:

$$\frac{\delta y}{\delta x} \approx \frac{\mathrm{d}y}{\mathrm{d}x}$$

WORKED EXAMPLE 12

Variables x and y are connected by the equation $y = x^3 + x^2$.

Find the approximate increase in y as x increases from 2 to 2.05.

Answers

$y = x^3 + x^2$

$\dfrac{dy}{dx} = 3x^2 + 2x$

When $x = 2$, $\dfrac{dy}{dx} = 3(2)^2 + 2(2) = 16$

Using $\dfrac{\delta y}{\delta x} \approx \dfrac{dy}{dx}$

$\dfrac{\delta y}{0.05} \approx 16$

$\delta y \approx 16 \times 0.05$

$\delta y \approx 0.8$

WORKED EXAMPLE 13

The volume, $V\,\text{cm}^3$, of a sphere with radius r cm is $V = \dfrac{4}{3}\pi r^3$.

Find, in terms of p, the approximate change in V as r increases from 10 to $10 + p$, where p is small.

Answers

$V = \dfrac{4}{3}\pi r^3$

$\dfrac{dV}{dr} = 4\pi r^2$

When $r = 10$, $\dfrac{dV}{dr} = 4\pi (10)^2 = 400\pi$

Using $\dfrac{\delta V}{\delta r} \approx \dfrac{dV}{dr}$

$\dfrac{\delta V}{p} \approx 400\pi$

$\delta V \approx 400\pi p$

Exercise 12.6

1   Variables x and y are connected by the equation $y = 2x^3 - 3x$.
    Find the approximate change in y as x increases from 2 to 2.01.

2   Variables x and y are connected by the equation $y = 5x^2 - \dfrac{8}{x^3}$.

    Find the approximate change in y as x increases from 1 to 1.02.

3   Variables x and y are connected by the equation $x^2 y = 400$.
    Find, in terms of p, the approximate change in y as x increases from 10 to $10 + p$,
    where p is small.

4   Variables x and y are connected by the equation $y = \left(\dfrac{1}{3}x - 2\right)^6$.

    Find, in terms of p, the approximate change in y as x increases from 9 to $9 + p$,
    where p is small.

5   A curve has equation $y = (x + 1)(2x - 3)^4$.
    Find, in terms of p, the approximate change in y as x increases from 2 to $2 + p$,
    where p is small.

6   A curve has equation $y = (x - 2)\sqrt{2x + 1}$.
    Find, in terms of p, the approximate change in y as x increases from 4 to $4 + p$,
    where p is small.

7   The periodic time, T seconds, for a pendulum of length L cm is $T = 2\pi\sqrt{\dfrac{L}{10}}$.
    Find the approximate increase in T as L increases from 40 to 41.

8   The volume of the solid cuboid is $360\,\text{cm}^3$ and the surface area is $A\,\text{cm}^2$.

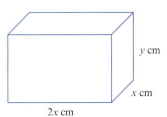

    a   Express y in terms of x.

    b   Show that $A = 4x^2 + \dfrac{1080}{x}$.

    c   Find, in terms of p, the approximate change in A as x increases
        from 2 to $2 + p$, where p is small.
        State whether the change is an increase or decrease.

12.7 Rates of change

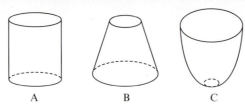

Consider pouring water at a constant rate of $10\,\mathrm{cm^3\,s^{-1}}$ into each of these three large containers.

1   Discuss how the height of water in container A changes with time.

2   Discuss how the height of water in container B changes with time.

3   Discuss how the height of water in container C changes with time.

On copies of the axes below, sketch graphs to show how the height (h cm) varies with time (t seconds) for each container.

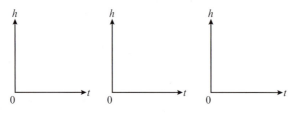

You already know that $\dfrac{\mathrm{d}y}{\mathrm{d}x}$ represents the **rate of change** of y with respect to x.

There are many situations where the rate of change of one quantity depends on the changing value of a second quantity.

In the class discussion, the rate of change of the height of water at a particular time, t, can be found by finding the value of $\dfrac{\mathrm{d}h}{\mathrm{d}t}$ at time t.

(The gradient of the tangent at time t.)

WORKED EXAMPLE 14

Variables V and t are connected by the equation $V = 5t^2 - 8t + 3$.
Find the rate of change of V with respect to t when $t = 4$.

Answers

$V = 5t^2 - 8t + 3$

$\dfrac{\mathrm{d}V}{\mathrm{d}t} = 10t - 8$

When $t = 4$, $\dfrac{\mathrm{d}V}{\mathrm{d}t} = 10(4) - 8 = 32$

Connected rates of change

When two variables x and y both vary with a third variable t, the three variables can be connected using the chain rule:

$$\frac{dy}{dt} = \frac{dy}{dx} \times \frac{dx}{dt}$$

You may also need to use the rule that:

$$\frac{dx}{dy} = \frac{1}{\left(\dfrac{dy}{dx}\right)}$$

WORKED EXAMPLE 15

Variables x and y are connected by the equation $y = x^3 - 5x^2 + 15$.

Given that x increases at a rate of 0.1 units per second, find the rate of change of y when $x = 4$.

Answers

$y = x^3 - 5x^2 + 15$ and $\dfrac{dx}{dt} = 0.1$

$\dfrac{dy}{dx} = 3x^2 - 10x$

When $x = 4$, $\dfrac{dy}{dx} = 3(4)^2 - 10(4)$

$\qquad\qquad = 8$

Using the chain rule, $\dfrac{dy}{dt} = \dfrac{dy}{dx} \times \dfrac{dx}{dt}$

$\qquad\qquad\qquad = 8 \times 0.1$

$\qquad\qquad\qquad = 0.8$

Rate of change of y is 0.8 units per second.

WORKED EXAMPLE 16

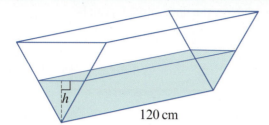

120 cm

The diagram shows a water container in the shape of a triangular prism of length 120 cm.

The vertical cross-section is an equilateral triangle.

Water is poured into the container at a rate of $24\,\text{cm}^3\,\text{s}^{-1}$.

a Show that the volume of water in the container, $V\,\text{cm}^3$, is given by $V = 40\sqrt{3}\,h^2$, where $h\,\text{cm}$ is the height of the water in the container.

b Find the rate of change of h when $h = 12$.

Answers

a Length of side of triangle $= \dfrac{h}{\sin 60°}$

$$= \dfrac{2\sqrt{3}\,h}{3}$$

Area of triangle $= \dfrac{1}{2} \times \dfrac{2\sqrt{3}\,h}{3} \times h$

$$= \dfrac{\sqrt{3}\,h^2}{3}$$

V = area of triangle × 120

$$= \dfrac{\sqrt{3}\,h^2}{3} \times 120$$

$$= 40\sqrt{3}\,h^2$$

b $\dfrac{dV}{dh} = 80\sqrt{3}\,h$ and $\dfrac{dV}{dt} = 24$

When $h = 12$, $\dfrac{dV}{dh} = 80\sqrt{3}(12)$

$$= 960\sqrt{3}$$

Using the chain rule, $\dfrac{dh}{dt} = \dfrac{dh}{dV} \times \dfrac{dV}{dt}$

$$= \dfrac{1}{960\sqrt{3}} \times 24$$

$$= \dfrac{\sqrt{3}}{120}$$

Rate of change of h is $\dfrac{\sqrt{3}}{120}$ cm per second.

Exercise 12.7

1. Variables x and y are connected by the equation $y = x^2 - 5x$.
    Given that x increases at a rate of 0.05 units per second, find the rate of change of y when $x = 4$.

2. Variables x and y are connected by the equation $y = x + \sqrt{x - 5}$.
    Given that x increases at a rate of 0.1 units per second, find the rate of change of y when $x = 9$.

3. Variables x and y are connected by the equation $y = (x - 3)(x + 5)^3$.
    Given that x increases at a rate of 0.2 units per second, find the rate of change of y when $x = -4$.

4. Variables x and y are connected by the equation $y = \dfrac{5}{2x - 1}$.
    Given that y increases at a rate of 0.1 units per second, find the rate of change of x when $x = -2$.

5. Variables x and y are connected by the equation $y = \dfrac{2x}{x^2 + 3}$.
    Given that x increases at a rate of 2 units per second, find the rate of increase of y when $x = 1$.

6. Variables x and y are connected by the equation $y = \dfrac{2x - 5}{x - 1}$.
    Given that x increases at a rate of 0.02 units per second, find the rate of change of y when $y = 1$.

7. Variables x and y are connected by the equation $\dfrac{1}{y} = \dfrac{1}{8} - \dfrac{2}{x}$.
    Given that x increases at a rate of 0.01 units per second, find the rate of change of y when $x = 8$.

8. A square has sides of length $x\,\text{cm}$ and area $A\,\text{cm}^2$.
    The area is increasing at a constant rate of $0.2\,\text{cm}^2\,\text{s}^{-1}$.
    Find the rate of increase of x when $A = 16$.

9. A cube has sides of length $x\,\text{cm}$ and volume $V\,\text{cm}^3$.
    The volume is increasing at a rate of $2\,\text{cm}^3\,\text{s}^{-1}$.
    Find the rate of increase of x when $V = 512$.

10. A sphere has radius $r\,\text{cm}$ and volume $V\,\text{cm}^3$.
    The radius is increasing at a rate of $\dfrac{1}{\pi}\,\text{cm s}^{-1}$.
    Find the rate of increase of the volume when $V = 972\pi$.

11. A solid metal cuboid has dimensions $x\,\text{cm}$ by $x\,\text{cm}$ by $5x\,\text{cm}$.
    The cuboid is heated and the volume increases at a rate of $0.5\,\text{cm}^3\,\text{s}^{-1}$.
    Find the rate of increase of x when $x = 4$.

12. A cone has base radius $r\,\text{cm}$ and a fixed height $18\,\text{cm}$.
    The radius of the base is increasing at a rate of $0.1\,\text{cm s}^{-1}$.
    Find the rate of change of the volume when $r = 10$.

13 Water is poured into the conical container at a rate of $5\,\text{cm}^3\,\text{s}^{-1}$.
After t seconds, the volume of water in the container, $V\,\text{cm}^3$, is given by
$V = \dfrac{1}{12}\pi h^3$, where $h\,\text{cm}$ is the height of the water in the container.

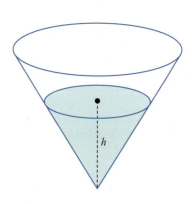

   a   Find the rate of change of h when $h = 5$.

   b   Find the rate of change of h when $h = 10$.

14 Water is poured into the hemispherical bowl at a rate of $4\pi\,\text{cm}^3\,\text{s}^{-1}$.
After t seconds, the volume of water in the bowl, $V\,\text{cm}^3$, is given by
$V = 8\pi h^2 - \dfrac{1}{3}\pi h^3$, where $h\,\text{cm}$ is the height of the water in the bowl.

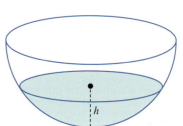

   a   Find the rate of change of h when $h = 2$.

   b   Find the rate of change of h when $h = 4$.

12.8 Second derivatives

If you differentiate y with respect to x you obtain $\dfrac{dy}{dx}$.

$\dfrac{dy}{dx}$ is called the **first derivative** of y with respect to x.

If you differentiate $\dfrac{dy}{dx}$ with respect to x you obtain $\dfrac{d}{dx}\left(\dfrac{dy}{dx}\right)$ which can also be written as $\dfrac{d^2y}{dx^2}$.

$\dfrac{d^2y}{dx^2}$ is called the **second derivative** of y with respect to x.

So for $y = x^3 - 7x^2 + 2x + 1$ or $f(x) = x^3 - 7x^2 + 2x + 1$

$\qquad \dfrac{dy}{dx} = 3x^2 - 14x + 2 \qquad$ or $\quad f'(x) = 3x^2 - 14x + 2$

$\qquad \dfrac{d^2y}{dx^2} = 6x - 14 \qquad\qquad$ or $\quad f''(x) = 6x - 14$

WORKED EXAMPLE 17

Given that $y = 3x^2 - \dfrac{5}{x}$, find $\dfrac{d^2y}{dx^2}$.

Answers

$y = 3x^2 - 5x^{-1}$

$\dfrac{dy}{dx} = 6x + 5x^{-2}$

$\dfrac{d^2y}{dx^2} = 6 - 10x^{-3}$

$\qquad = 6 - \dfrac{10}{x^3}$

Exercise 12.8

1   Find $\dfrac{d^2y}{dx^2}$ for each of the following functions.

  a $y = 5x^2 - 7x + 3$ b $y = 2x^3 + 3x^2 - 1$ c $y = 4 - \dfrac{3}{x^2}$

  d $y = (4x + 1)^5$ e $y = \sqrt{2x + 1}$ f $y = \dfrac{4}{\sqrt{x + 3}}$

2   Find $\dfrac{d^2y}{dx^2}$ for each of the following functions.

  a $y = x(x - 4)^3$ b $y = \dfrac{4x - 1}{x^2}$ c $y = \dfrac{x + 1}{x - 3}$

  d $y = \dfrac{x + 2}{x^2 - 1}$ e $y = \dfrac{x^2}{x - 5}$ f $y = \dfrac{2x + 5}{3x - 1}$

3   Given that $f(x) = x^3 - 7x^2 + 2x + 1$, find

  a $f(1)$ b $f'(1)$ c $f''(1)$.

4   A curve has equation $y = 4x^3 + 3x^2 - 6x - 1$.

  a   Show that $\dfrac{dy}{dx} = 0$ when $x = -1$ and when $x = 0.5$.

  b   Find the value of $\dfrac{d^2y}{dx^2}$ when $x = -1$ and when $x = 0.5$.

5 A curve has equation $y = 2x^3 - 15x^2 + 24x + 6$.

Copy and complete the table to show whether $\dfrac{dy}{dx}$ and $\dfrac{d^2y}{dx^2}$ are positive (+), negative (−) or zero (0) for the given values of x.

x	0	1	2	3	4	5
$\dfrac{dy}{dx}$						
$\dfrac{d^2y}{dx^2}$						

6 A curve has equation $y = 2x^3 + 3x^2 - 36x + 5$. Find the range of values of x for which both $\dfrac{dy}{dx}$ and $\dfrac{d^2y}{dx^2}$ are both positive.

7 Given that $y = x^2 - 2x + 5$, show that $4\dfrac{d^2y}{dx^2} + (x - 1)\dfrac{dy}{dx} = 2y$.

8 **CHALLENGE QUESTION**

Given that $y = 8\sqrt{x}$, show that $4x^2\dfrac{d^2y}{dx^2} + 4x\dfrac{dy}{dx} = y$.

12.9 Stationary points

Consider the graph of the function $y = f(x)$ shown below.

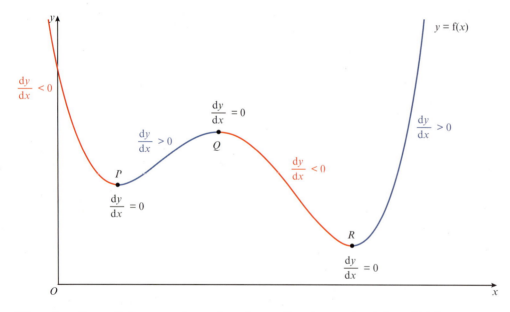

The red sections of the curve show where the gradient is negative (where $f(x)$ is a decreasing function) and the blue sections show where the gradient is positive (where $f(x)$ is an increasing function). The gradient of the curve is zero at the points P, Q and R.

A point where the gradient is zero is called a **stationary point** or a **turning point**.

Maximum points

The stationary point Q is called a **maximum point** because the value of y at this point is greater than the *value of y at other points close to Q.

At a maximum point:

- $\dfrac{dy}{dx} = 0$

- the gradient is positive to the left of the maximum and negative to the right

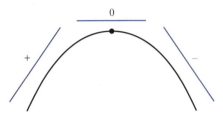

Minimum points

The stationary points P and R are called **minimum points**.

At a minimum point:

- $\dfrac{dy}{dx} = 0$

- the gradient is negative to the left of the minimum and positive to the right

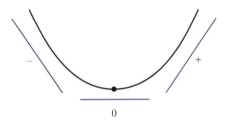

Stationary points of inflexion

There is a third type of stationary point (turning point) called a **point of inflexion**.

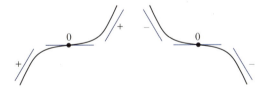

At a stationary point of inflexion:

- $\dfrac{dy}{dx} = 0$

- the gradient changes $\begin{cases} \text{from positive to zero and then to positive again} \\ \text{or} \\ \text{from negative to zero and then to negative again} \end{cases}$

Points of inflexion are not part of the syllabus. They have been included here for completeness.

WORKED EXAMPLE 18

Find the coordinates of the stationary points on the curve $y = x^3 - 3x + 1$ and determine the nature of these points. Sketch the graph of $y = x^3 - 3x + 1$.

Answers

$y = x^3 - 3x + 1$

$\dfrac{dy}{dx} = 3x^2 - 3$

For stationary points, $\quad\quad\quad\quad\dfrac{dy}{dx} = 0$

$$3x^2 - 3 = 0$$
$$x^2 - 1 = 0$$
$$(x + 1)(x - 1) = 0$$
$$x = -1 \text{ or } x = 1$$

When $x = -1$, $y = (-1)^3 - 3(-1) + 1 = 3$

When $x = 1$, $y = (1)^3 - 3(1) + 1 = -1$

The stationary points are $(-1, 3)$ and $(1, -1)$.

Now consider the gradient on either side of the points $(-1, 3)$ and $(1, -1)$:

x	-1.1	-1	-0.9
$\dfrac{dy}{dx}$	$3(-1.1)^2 - 3 = $ positive	0	$3(-0.9)^2 - 3 = $ negative
direction of tangent	 ╱	 —	 ╲
shape of curve	 ⌢		

x	0.9	1	1.1
$\dfrac{dy}{dx}$	$3(0.9)^2 - 3 = $ negative	0	$3(1.1)^2 - 3 = $ positive
direction of tangent	╲	—	╱
shape of curve	 ⌣		

CONTINUED

So $(-1, 3)$ is a maximum point and $(1, -1)$ is a minimum point.

The sketch graph of $y = x^3 - 3x + 1$ is:

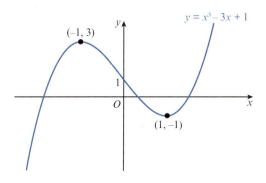

Second derivatives and stationary points

Consider moving from left to right along a curve, passing through a maximum point:

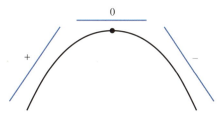

The gradient, $\dfrac{dy}{dx}$, starts as a positive value, decreases to zero at the maximum point and then decreases to a negative value.

Since $\dfrac{dy}{dx}$ decreases as x increases, then the rate of change of $\dfrac{dy}{dx}$ is negative.

[The rate of change of $\dfrac{dy}{dx}$ is written as $\dfrac{d}{dx}\left(\dfrac{dy}{dx}\right) = \dfrac{d^2y}{dx^2}$.]

This leads to the rule:

If $\dfrac{dy}{dx} = 0$ and $\dfrac{d^2y}{dx^2} < 0$, then the point is a maximum point.

Now, consider moving from left to right along a curve, passing through a minimum point:

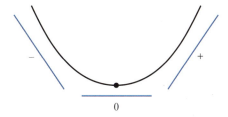

The gradient, $\dfrac{dy}{dx}$, starts as a negative value, increases to zero at the minimum point and then increases to a positive value.

Since $\dfrac{dy}{dx}$ increases as x increases, then the rate of change of $\dfrac{dy}{dx}$ is positive.

This leads to the rule:

> If $\dfrac{dy}{dx} = 0$ and $\dfrac{d^2y}{dx^2} > 0$, then the point is a minimum point.
>
> If $\dfrac{dy}{dx} = 0$ and $\dfrac{d^2y}{dx^2} = 0$, then the nature of the stationary point can be found using the First Derivative Test.

WORKED EXAMPLE 19

Find the coordinates of the stationary points on the curve $y = 2x^3 - 15x^2 + 24x + 6$ and determine the nature of these points.

Sketch the graph of $y = 2x^3 - 15x^2 + 24x + 6$.

Answers

$y = 2x^3 - 15x^2 + 24x + 6$

$\dfrac{dy}{dx} = 6x^2 - 30x + 24$

For stationary points, $\qquad\qquad \dfrac{dy}{dx} = 0$

$$6x^2 - 30x + 24 = 0$$

$$x^2 - 5x + 4 = 0$$

$$(x - 1)(x - 4) = 0$$

$$x = 1 \text{ or } x = 4$$

When $x = 1$, $y = 2(1)^3 - 15(1)^2 + 24(1) + 6 = 17$

When $x = 4$, $y = 2(4)^3 - 15(4)^2 + 24(4) + 6 = -10$

The stationary points are $(1, 17)$ and $(4, -10)$.

$$\dfrac{d^2y}{dx^2} = 12x - 30$$

When $x = 1$, $\dfrac{d^2y}{dx^2} = -18$ which is < 0

When $x = 4$, $\dfrac{d^2y}{dx^2} = 18$ which is > 0

So $(1, 17)$ is a maximum point and $(4, -10)$ is a minimum point.

CONTINUED

The sketch graph of $y = 2x^3 - 15x^2 + 24x + 6$ is:

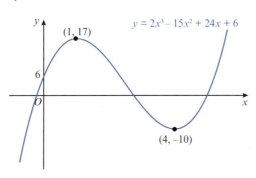

REFLECTION

In this section you have been shown two methods for determining the nature of stationary points. Explain these two methods to a friend. Which method do you prefer and why?

Exercise 12.9

1 Find the coordinates of the stationary points on each of the following curves and determine the nature of each of the stationary points.

   a $y = x^2 - 12x + 8$ b $y = (5 + x)(1 - x)$

   c $y = x^3 - 12x + 2$ d $y = x^3 + x^2 - 16x - 16$

   e $y = x(3 - 4x - x^2)$ f $y = (x - 1)(x^2 - 6x + 2)$

2 Find the coordinates of the stationary points on each of the following curves and determine the nature of each of the stationary points.

   a $y = \sqrt{x} + \dfrac{4}{\sqrt{x}}$ b $y = x^2 - \dfrac{2}{x}$ c $y = \dfrac{4}{x} + \sqrt{x}$

   d $y = \dfrac{2x}{x^2 + 9}$ e $y = \dfrac{x^2}{x + 1}$ f $y = \dfrac{x^2 - 5x + 3}{x + 1}$

3 The equation of a curve is $y = \dfrac{2x + 5}{x + 1}$.

   Find $\dfrac{dy}{dx}$ and hence explain why the curve has no turning points.

4 The curve $y = 2x^3 + ax^2 - 12x + 7$ has a maximum point at $x = -2$.
   Find the value of a.

5 The curve $y = x^3 + ax + b$ has a stationary point at $(1, 3)$.

   a   Find the value of a and the value of b.

   b   Determine the nature of the stationary point $(1, 3)$.

   c   Find the coordinates of the other stationary point on the curve and determine the nature of this stationary point.

6 The curve $y = x^2 + \dfrac{a}{x} + b$ has a stationary point at $(1, -1)$.

   a   Find the value of a and the value of b.

   b   Determine the nature of the stationary point $(1, -1)$.

7 The curve $y = ax + \dfrac{b}{x^2}$ has a stationary point at $(-1, -12)$.

    a   Find the value of a and the value of b.

    b   Determine the nature of the stationary point $(-1, -12)$.

8   **CHALLENGE QUESTION**

    The curve $y = 2x^3 - 3x^2 + ax + b$ has a stationary point at the point $(3, -77)$.

    a   Find the value of a and the value of b.

    b   Find the coordinates of the second stationary point on the curve.

    c   Determine the nature of the two stationary points.

    d   Find the coordinates of the point on the curve where the gradient is minimum and state the value of the minimum gradient.

12.10 Practical maximum and minimum problems

There are many problems for which you need to find the maximum or minimum value of a function, such as the maximum area that can be enclosed within a shape or the minimum amount of material that can be used to make a container.

WORKED EXAMPLE 20

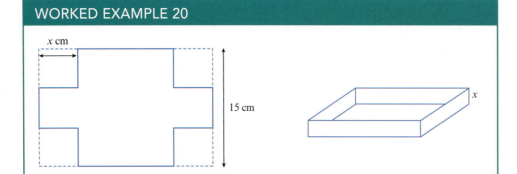

The diagram shows a 24 cm by 15 cm sheet of metal with a square of side x cm removed from each corner. The metal is then folded to make an open rectangular box of depth x cm and volume V cm³.

    a   Show that $V = 4x^3 - 78x^2 + 360x$.

    b   Find the stationary value of V and the value of x for which this occurs.

    c   Determine the nature of this stationary value.

CONTINUED

Answers

a $V = $ length $\times$ breadth $\times$ height

$$= (24 - 2x)(15 - 2x)x$$
$$= (360 - 78x + 4x^2)x$$
$$= 4x^3 - 78x^2 + 360x$$

b $\dfrac{\mathrm{d}V}{\mathrm{d}x} = 12x^2 - 156x + 360$

Stationary values occur when $\dfrac{\mathrm{d}y}{\mathrm{d}x} = 0$

$$12x^2 - 156x + 360 = 0$$
$$x^2 - 13x + 30 = 0$$
$$(x - 10)(x - 3) = 0$$
$$x = 10 \text{ or } x = 3$$

The dimensions of the box must be positive so $x = 3$. (One side of the metal sheet is 15 cm in length, so you couldn't cut two squares of side length 10 cm each from that side.)

When $x = 3$, $V = 4(3)^3 - 78(3)^2 + 360(3) = 486$.

The stationary value of V is 486 and occurs when $x = 3$.

c $\dfrac{\mathrm{d}^2V}{\mathrm{d}x^2} = 24x - 156$

When $x = 3$, $\dfrac{\mathrm{d}^2V}{\mathrm{d}x^2} = 24(3) - 156 = -84$ which is < 0

The stationary value is a maximum value.

WORKED EXAMPLE 21

A piece of wire, of length 2 m, is bent to form the shape $PQRST$.

$PQST$ is a rectangle and QRS is a semi-circle with diameter SQ.

$PT = x$ m and $PQ = ST = y$ m.

The total area enclosed by the shape is A m^2.

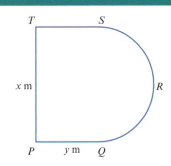

a   Express y in terms of x.

b   Show that $A = x - \dfrac{1}{2}x^2 - \dfrac{1}{8}\pi x^2$.

c   Find $\dfrac{\mathrm{d}A}{\mathrm{d}x}$ and $\dfrac{\mathrm{d}^2A}{\mathrm{d}x^2}$.

d   Find the value for x for which there is a stationary value of A.

e   Determine the magnitude and nature of this stationary value.

Answers

a   Perimeter = $PQ + \text{arc } QRS + ST + TP$

$$2 = y + \frac{1}{2} \times 2\pi\left(\frac{x}{2}\right) + y + x$$

$$2 = 2y + \frac{1}{2}\pi x + x$$

$$2y = 2 - x - \frac{1}{2}\pi x$$

$$y = 1 - \frac{1}{2}x - \frac{1}{4}\pi x$$

b A = area of rectangle + area of semi-circle

$$= xy + \frac{1}{2}\pi\left(\frac{x}{2}\right)^2$$

$$= x\left(1 - \frac{1}{2}x - \frac{1}{4}\pi x\right) + \frac{1}{8}\pi x^2$$

$$= x - \frac{1}{2}x^2 - \frac{1}{4}\pi x^2 + \frac{1}{8}\pi x^2$$

$$= x - \frac{1}{2}x^2 - \frac{1}{8}\pi x^2$$

c $\frac{dA}{dx} = 1 - x - \frac{1}{4}\pi x$

$\frac{d^2A}{dx^2} = -1 - \frac{1}{4}\pi$

d   Stationary values occur when $\frac{dA}{dx} = 0$.

$$1 - x - \frac{1}{4}\pi x = 0$$

$$4 - 4x - \pi x = 0$$

$$x(4 + \pi) = 4$$

Stationary value occurs when $x = \frac{4}{(4 + \pi)}$.

e   When $x = \frac{4}{(4 + \pi)}$, $A = \frac{4}{(4 + \pi)} - \frac{1}{2}\left[\frac{4}{(4 + \pi)}\right]^2 - \frac{1}{8}\pi\left[\frac{4}{(4 + \pi)}\right]^2$

$$= \frac{4(4 + \pi) - 8 - 2\pi}{(4 + \pi)^2}$$

$$= \frac{8 + 2\pi}{(4 + \pi)^2}$$

$$= \frac{2(4 + \pi)}{(4 + \pi)^2}$$

$$= \frac{2}{(4 + \pi)}$$

When $x = \frac{4}{(4 + \pi)}$, $\frac{d^2A}{dx^2} = -1 - \frac{1}{4}\pi$ which is < 0

The stationary value of A is $\frac{2}{(4 + \pi)}$ m^2 and it is a maximum value.

Exercise 12.10

1 The sum of two numbers x and y is 8.

    a Express y in terms of x.

    b **i** Given that $P = xy$, write down an expression for P in terms of x.

        ii Find the maximum value of P.

    c **i** Given that $S = x^2 + y^2$, write down an expression for S, in terms of x.

        ii Find the minimum value of S.

2 The diagram shows a rectangular garden with a fence on three of its sides and a wall on its fourth side. The total length of the fence is 100 m and the area enclosed is $A\,\text{m}^2$.

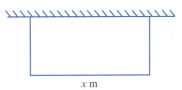

    a Show that $A = \dfrac{1}{2}x(100 - x)$.

    b Find the maximum area of the garden enclosed and the value of x for which this occurs.

3 The volume of the solid cuboid is $576\,\text{cm}^3$ and the surface area is $A\,\text{cm}^2$.

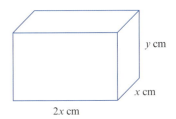

    a Express y in terms of x.

    b Show that $A = 4x^2 + \dfrac{1728}{x}$.

    c Find the maximum value of A and state the dimensions of the cuboid for which this occurs.

4 A cuboid has a total surface area of $400\,\text{cm}^2$ and a volume of $V\,\text{cm}^3$. The dimensions of the cuboid are $4x$ cm by x cm by h cm.

    a Express h in terms of V and x.

    b Show that $V = 160x - \dfrac{16}{5}x^3$.

    c Find the value of x when V is a maximum.

5

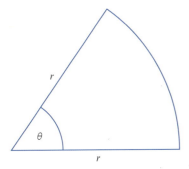

A piece of wire, of length 60 cm, is bent to form a sector of a circle with radius r cm and sector angle θ radians. The total area enclosed by the shape is $A\,\text{cm}^2$.

    a Express θ in terms of r.

    b Show that $A = 30r - r^2$.

    c Find $\dfrac{\mathrm{d}A}{\mathrm{d}r}$ and $\dfrac{\mathrm{d}^2 A}{\mathrm{d}r^2}$.

    d Find the value for r for which there is a stationary value of A.

    e Determine the magnitude and nature of this stationary value.

6   The diagram shows a window made from a rectangle with base $2r\,$m and height $h\,$m and a semicircle of radius $r\,$m. The perimeter of the window is $6\,$m and the surface area is $A\,$m^2.

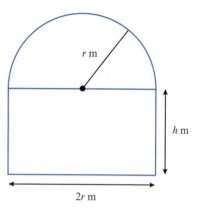

   a   Express h in terms of r.

   b   Show that $A = 6r - 2r^2 - \dfrac{1}{2}\pi r^2$.

   c   Find $\dfrac{\mathrm{d}A}{\mathrm{d}r}$ and $\dfrac{\mathrm{d}^2A}{\mathrm{d}r^2}$.

   d   Find the value for r for which there is a stationary value of A.

   e   Determine the magnitude and nature of this stationary value.

7   $ABCD$ is a rectangle with base length $2p$ units, and area A units^2.
    The points A and B lie on the x-axis and the points C and D lie on the curve $y = 4 - x^2$.

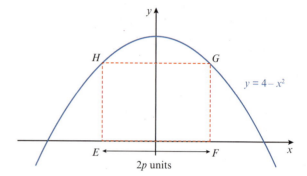

   a   Express BC in terms of p.

   b   Show that $A = 2p(4 - p^2)$.

   c   Find the value of p for which A has a stationary value.

   d   Find this stationary value and determine its nature.

8   A solid cylinder has radius $r\,$cm and height $h\,$cm.
    The volume of this cylinder is $250\pi\,$cm^3 and the surface area is $A\,$cm^2.

   a   Express h in terms of r.

   b   Show that $A = 2\pi r^2 + \dfrac{500\pi}{r}$.

   c   Find $\dfrac{\mathrm{d}A}{\mathrm{d}r}$ and $\dfrac{\mathrm{d}^2A}{\mathrm{d}r^2}$.

   d   Find the value for r for which there is a stationary value of A.

   e   Determine the magnitude and nature of this stationary value.

9   The diagram shows a solid formed by joining a hemisphere of radius $r\,$m to a cylinder of radius $r\,$cm and height $h\,$cm. The surface area of the solid is $288\pi\,$cm^2 and the volume is $V\,$cm^3.

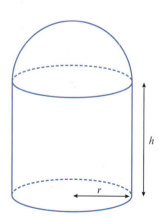

   a   Express h in terms of r.

   b   Show that $V = 144\pi r - \dfrac{5}{6}\pi r^3$.

   c   Find the exact value of r such that V is a maximum.

10 A piece of wire, of length 50 cm, is cut into two pieces.
One piece is bent to make a square of side x cm and the other is bent to make a
circle of radius r cm. The total area enclosed by the two shapes is A cm^2.

  a Express r in terms of x.

  b Show that $A = \dfrac{(\pi + 4)\,x^2 - 100x + 625}{\pi}$.

  c Find the stationary value of A and the value of x for which this occurs.
Give your answers correct to 3 sf.

11 The diagram shows a solid cylinder of radius r cm and height $2h$ cm cut from a
solid sphere of radius 5 cm. The volume of the cylinder is V cm^3.

  a Express r in terms of h.

  b Show that $V = 2\pi h (25 - h^2)$.

  c Find the value for h for which there is a stationary value of V.

  d Determine the nature of this stationary value.

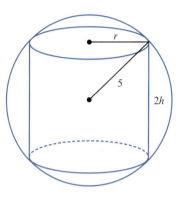

12 CHALLENGE QUESTION

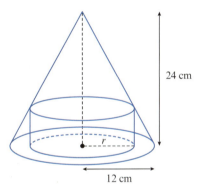

The diagram shows a hollow cone with base radius 12 cm and height 24 cm.
A solid cylinder stands on the base of the cone and the upper edge touches the
inside of the cone. The cylinder has base radius r cm, height h cm and
volume V cm^3.

  a Express h in terms of r.

  b Show that $V = 2\pi r^2 (12 - r)$.

  c Find the volume of the largest cylinder which can stand inside the cone.

13 CHALLENGE QUESTION

The diagram shows a right circular cone of base radius r cm and height h cm cut from a solid sphere of radius 10 cm. The volume of the cone is V cm^3.

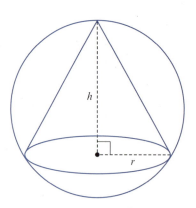

a Express r in terms of h.

b Show that $V = \dfrac{1}{3}\pi h^2(20 - h)$

c Find the value for h for which there is a stationary value of V.

d Determine the magnitude and nature of this stationary value.

SUMMARY

Rules of differentiation

Power rule:	If $y = x^n$, then $\dfrac{dy}{dx} = n x^{n-1}$
Scalar multiple rule:	$\dfrac{d}{dx}[kf(x)] = k\dfrac{d}{dx}[f(x)]$
Addition/subtraction rule:	$\dfrac{d}{dx}[f(x) \pm g(x)] = \dfrac{d}{dx}[f(x)] \pm \dfrac{d}{dx}[g(x)]$
Chain rule:	$\dfrac{dy}{dx} = \dfrac{dy}{du} \times \dfrac{du}{dx}$
Product rule:	$\dfrac{d}{dx}(uv) = u\dfrac{dv}{dx} + v\dfrac{du}{dx}$
Quotient rule:	$\dfrac{d}{dx}\left(\dfrac{u}{v}\right) = \dfrac{v\dfrac{du}{dx} - u\dfrac{dv}{dx}}{v^2}$

Tangents and normals

If the value of $\dfrac{dy}{dx}$ at the point (x_1, y_1) is m, then:

- the equation of the tangent is given by $y - y_1 = m(x - x_1)$
- the equation of the normal is given by $y - y_1 = -\dfrac{1}{m}(x - x_1)$

Small increments and approximations

If δx and δy are sufficiently small then $\dfrac{\delta y}{\delta x} \approx \dfrac{dy}{dx}$.

Stationary points

Stationary points (turning points) of a function $y = f(x)$ occur when $\dfrac{dy}{dx} = 0$.

CONTINUED

First derivative test for maximum and minimum points

At a maximum point:

- $\dfrac{\mathrm{d}y}{\mathrm{d}x} = 0$

- the gradient is positive to the left of the maximum and negative to the right

At a minimum point:

- $\dfrac{\mathrm{d}y}{\mathrm{d}x} = 0$

- the gradient is negative to the left of the minimum and positive to the right

Second derivative test for maximum and minimum points

If $\dfrac{\mathrm{d}y}{\mathrm{d}x} = 0$ and $\begin{cases} \dfrac{\mathrm{d}^2 y}{\mathrm{d}x^2} < 0, \text{ then the point is a maximum point} \\ \dfrac{\mathrm{d}^2 y}{\mathrm{d}x^2} > 0, \text{ then the point is a minimum point} \end{cases}$

Past paper questions

Worked example

The diagram shows a shape made by cutting an equilateral triangle out of a rectangle of width x cm.

x cm

The perimeter of the shape is 20 cm.

i   Show that the area, A cm^2, of the shape is given by

$$A = 10x - \left(\frac{6 + \sqrt{3}}{4}\right)x^2.$$ [3]

ii Given that x can vary, find the value of x which produces the maximum area and calculate this maximum area. Give your answers to 2 significant figures.   [4]

Cambridge IGCSE Additional Mathematics 0606 Paper 22 Q12 Mar 2017

Answers

i Area of triangle $= \dfrac{1}{2}ab\sin C = \dfrac{1}{2}x^2\sin 60° = \dfrac{\sqrt{3}}{4}x^2$

Let height of rectangle $= h$

Perimeter of shaded shape $= 3x + 2h$

$$20 = 3x + 2h$$

$$h = \frac{20 - 3x}{2}$$

Area of rectangle $= xh = x\left(\dfrac{20 - 3x}{2}\right)$

Shaded area = area of rectangle − area of triangle

$$= x\left(\frac{20 - 3x}{2}\right) - \frac{\sqrt{3}}{4}x^2$$

$$= 10x - \frac{3}{2}x^2 - \frac{\sqrt{3}}{4}x^2$$

$$= 10x - x^2\left(\frac{3}{2} + \frac{\sqrt{3}}{4}\right)$$

$$A = 10x - \left(\frac{6 + \sqrt{3}}{4}\right)x^2$$

ii $A = 10x - \left(\dfrac{6 + \sqrt{3}}{4}\right)x^2$

$$\frac{\mathrm{d}A}{\mathrm{d}x} = 10 - 2\left(\frac{6 + \sqrt{3}}{4}\right)x$$

$$\frac{\mathrm{d}A}{\mathrm{d}x} = 0 \text{ when } 2\left(\frac{6 + \sqrt{3}}{4}\right)x = 10$$

$$x = \frac{20}{6 + \sqrt{3}}$$

$$x = 2.58663...$$

$\dfrac{\mathrm{d}^2 A}{\mathrm{d}x^2} = -2\left(\dfrac{6 + \sqrt{3}}{4}\right)$ which is < 0

When $x = 2.58663...$, $A = 12.933...$ and is a maximum

Hence $x = 2.6$ and $A = 13$ to 2 significant figures

1

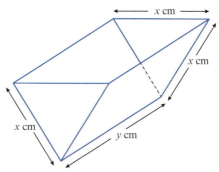

The diagram shows an empty container in the form of an open triangular prism. The triangular faces are equilateral with a side of x cm and the length of each rectangular face is y cm. The container is made from thin sheet metal. When full, the container holds $200\sqrt{3}$ cm^3.

i   Show that A cm^2, the total area of the thin sheet metal used, is given by $A = \dfrac{\sqrt{3}x^2}{2} + \dfrac{1600}{x}$.   [5]

ii   Given that x and y can vary, find the stationary value of A and determine its nature.   [6]

Cambridge IGCSE Additional Mathematics 0606 Paper 12 Q9 Mar 2015

2   A cube of side x cm has surface area S cm^2. The volume, V cm^3, of the cube is increasing at a rate of 480 cm^3 s^{-1}. Find, at the instant when $V = 512$,

a   the rate of increase of x,   [4]

b   the rate of increase of S.   [2]

Cambridge IGCSE Additional Mathematics 0606 Paper 22 Q5 Mar 2021

3

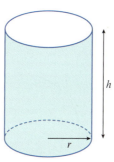

A container is a circular cylinder, open at one end, with a base radius of r cm and a height of h cm. The volume of the container is 1000 cm^3. Given that r and h can vary and that the total outer surface area of the container has a minimum value, find this value.   [8]

Cambridge IGCSE Additional Mathematics 0606 Paper 22 Q11 Mar 2020

4   i   Given that $y = x\sqrt{x^2 + 1}$, show that $\dfrac{\mathrm{d}y}{\mathrm{d}x} = \dfrac{ax^2 + b}{(x^2 + 1)^p}$, where a, b and p are positive constants.   [4]

ii   Explain why the graph of $y = x\sqrt{x^2 + 1}$ has no stationary points.   [2]

Cambridge IGCSE Additional Mathematics 0606 Paper 22 Q7 Mar 2019

5   The area of a sector of a circle of radius r cm is 36 cm^2.

i   Show that the perimeter, P cm, of the sector is such that $P = 2r + \dfrac{72}{r}$.   [3]

ii   Hence, given that r can vary, find the stationary value of P and determine its nature.   [4]

Cambridge IGCSE Additional Mathematics 0606 Paper 12 Q9 Mar 2019

6   The volume, V, and surface area, S, of a sphere of radius r are given by $V = \frac{4}{3}\pi r^3$ and $S = 4\pi r^2$ respectively.
    The volume of a sphere increases at a rate of $200\,\text{cm}^3$ per second. At the instant when the radius of the sphere is $10\,\text{cm}$, find

    i   the rate of increase of the radius of the sphere,   [4]

    ii   the rate of increase of the surface area of the sphere.   [3]

    Cambridge IGCSE Additional Mathematics 0606 Paper 22 Q12 Mar 2018

7   Two variables x and y are such that $y = \dfrac{5}{\sqrt{x - 9}}$ for $x > 9$.

    i   Find an expression for $\dfrac{dy}{dx}$.   [2]

    ii   Hence, find the approximate change in y as x increases from 13 to $13 + h$, where h is small.   [2]

    Cambridge IGCSE Additional Mathematics 0606 Paper 22 Q1 Mar 2016

8   A curve has equation $y = \dfrac{x}{x^2 + 1}$.

    i   Find the coordinates of the stationary points of the curve.   [5]

    ii   Show that $\dfrac{d^2y}{dx^2} = \dfrac{px^3 + qx}{(x^2 + 1)^3}$, where p and q are integers to be found, and determine the nature of the
         stationary points of the curve.   [5]

    Cambridge IGCSE Additional Mathematics 0606 Paper 22 Q11 Mar 2016

9   Find the equation of the normal to the curve $y = \sqrt{8x + 5}$ at the point where $x = \dfrac{1}{2}$, giving your answer in
    the form $ax + by + c = 0$, where a, b and c are integers.   [5]

    Cambridge IGCSE Additional Mathematics 0606 Paper 11 Q6 Nov 2019

10 In this question, all lengths are in metres.

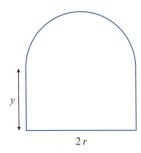

$2r$

    The diagram shows a window formed by a semi-circle of radius r on top of a rectangle with dimensions $2r$ by y.
    The total perimeter of the window is 5.

    i   Find y in terms of r.   [2]

    ii   Show that the total area of the window is $A = 5r - \dfrac{\pi r^2}{2} - 2r^2$.   [2]

    iii Given that r can vary, find the value of r which gives a maximum area of the window and find this area.
         (You are not required to show that this area is a maximum.)   [5]

    Cambridge IGCSE Additional Mathematics 0606 Paper 21 Q9 Nov 2018

11

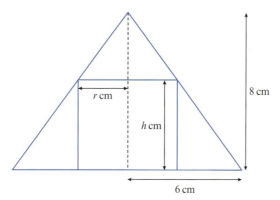

A cone, of height 8 cm and base radius 6 cm, is placed over a cylinder of radius r cm and height h cm and is in contact with the cylinder along the cylinder's upper rim. The arrangement is symmetrical and the diagram shows a vertical cross-section through the vertex of the cone.

  i   Use similar triangles to express h in terms of r. [2]

  ii   Hence show that the volume, V cm^3, of the cylinder is given by $V = 8\pi r^2 - \frac{4}{3}\pi r^3$. [1]

  iii   Given that r can vary, find the value of r which gives a stationary value of V. Find this stationary value of V in terms of π and determine its nature. [6]

Cambridge IGCSE Additional Mathematics 0606 Paper 21 Q7 Nov 2015

12   Find the equation of the tangent to the curve $y = \dfrac{2x - 1}{\sqrt{x^2 + 5}}$ at the point where $x = 2$. [7]

Cambridge IGCSE Additional Mathematics 0606 Paper 11 Q8 Nov 2015

13   Show that the curve $y = (3x^2 + 8)^{\frac{5}{3}}$ has only one stationary point. Find the coordinates of this stationary point and determine its nature. [8]

Cambridge IGCSE Additional Mathematics 0606 Paper 11 Q7 Jun 2017

14   In this question all lengths are in metres.

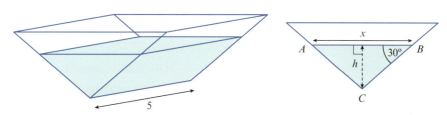

A water container is in the shape of a triangular prism. The diagrams show the container and its cross-section. The cross-section of the water in the container is an isosceles triangle ABC, with angle ABC = angle BAC = 30°. The length of AB is x and the depth of water is h. The length of the container is 5.

  i   Show that $x = 2\sqrt{3}\,h$ and hence find the volume of water in the container in terms of h. [3]

  ii   The container is filled at a rate of 0.5 m^3 per minute.

   At the instant when h is 0.25 m, find

   a   the rate at which h is increasing [4]

   b   the rate at which x is increasing. [2]

Cambridge IGCSE Additional Mathematics 0606 Paper 21 Q12 Jun 2018

15 It is given that $y = (x^2 + 1)(2x - 3)^{\frac{1}{2}}$.

   i   Show that $\dfrac{\mathrm{d}y}{\mathrm{d}x} = \dfrac{Px^2 + Qx + 1}{(2x - 3)^{\frac{1}{2}}}$, where P and Q are integers. [5]

   ii   Hence find the equation of the normal to the curve $y = (x^2 + 1)(2x - 3)^{\frac{1}{2}}$ at the point where $x = 2$, giving your answer in the form $ax + by + c = 0$, where a, b and c are integers. [4]

Cambridge IGCSE Additional Mathematics 0606 Paper 11 Q11 Jun 2019

16 The radius, r cm, of a circle is increasing at the rate of $5\,\mathrm{cm\,s^{-1}}$. Find, in terms of π, the rate at which the area of the circle is increasing when $r = 3$. [4]

Cambridge IGCSE Additional Mathematics 0606 Paper 11 Q3 Jun 2020

Chapter 13
Vectors

THIS SECTION WILL SHOW YOU HOW TO:

- use vectors in any form, e.g. $\begin{pmatrix} a \\ b \end{pmatrix}$, $\overrightarrow{AB}$, $\mathbf{p}$, $a\mathbf{i} + b\mathbf{j}$
- use position vectors and unit vectors
- find the magnitude of a vector; add and subtract vectors and multiply vectors by scalars
- compose and resolve velocities.

PRE-REQUISITE KNOWLEDGE

Before you start…

Where it comes from	What you should be able to do	Check your skills
Cambridge IGCSE/O Level Mathematics	Draw vector diagrams to represent the addition and subtraction of vectors.	1 $\mathbf{a} = \begin{pmatrix} 2 \\ 3 \end{pmatrix}$ $\mathbf{b} = \begin{pmatrix} 4 \\ 1 \end{pmatrix}$ Draw vector diagrams to represent: a $\mathbf{a} + \mathbf{b}$ b $\mathbf{a} - \mathbf{b}$
Cambridge IGCSE/O Level Mathematics	Add and subtract column vectors without using a diagram.	2 $\mathbf{f} = \begin{pmatrix} -1 \\ 3 \end{pmatrix}$ $\mathbf{g} = \begin{pmatrix} -5 \\ -2 \end{pmatrix}$ Write these vectors as column vectors: a $\mathbf{f} + \mathbf{g}$ b $\mathbf{f} - \mathbf{g}$
Cambridge IGCSE/O Level Mathematics	Multiply a vector by a **scalar**.	3 $\mathbf{p} = \begin{pmatrix} -1 \\ 2 \end{pmatrix}$ Write these vectors as column vectors: a $2\mathbf{p}$ b $-3\mathbf{p}$
Cambridge IGCSE/O Level Mathematics	Recognise parallel vectors.	4 Which of these vectors are parallel? $\begin{pmatrix} -2 \\ 4 \end{pmatrix}$ $\begin{pmatrix} -3 \\ 5 \end{pmatrix}$ $\begin{pmatrix} -1 \\ 2 \end{pmatrix}$ $\begin{pmatrix} 4 \\ -8 \end{pmatrix}$ $\begin{pmatrix} -2 \\ -4 \end{pmatrix}$
Cambridge IGCSE/O Level Mathematics	Calculate the magnitude (**modulus**) of a vector.	5 $\mathbf{q} = \begin{pmatrix} 6 \\ -8 \end{pmatrix}$ Find $|\mathbf{q}|$.
Cambridge IGCSE/O Level Mathematics	Use the sum and difference of two vectors to express given vectors in terms of two coplanar vectors.	6 ABC is a triangle. M is the midpoint of AB. $\overrightarrow{CA} = \mathbf{p}$ and $\overrightarrow{CB} = \mathbf{q}$ Find in terms of $\mathbf{p}$ and $\mathbf{q}$: a $\overrightarrow{AB}$ b $\overrightarrow{AM}$ c $\overrightarrow{BM}$ d $\overrightarrow{CM}$

13.1 Further vector notation

The **vector** $\overrightarrow{AB}$ in the diagram can be written in component form as $\begin{pmatrix} 4 \\ 3 \end{pmatrix}$

$\overrightarrow{AB}$ can also be written as $4\mathbf{i} + 3\mathbf{j}$, where:

   $\mathbf{i}$ is a vector of length 1 unit in the positive x-direction

and $\mathbf{j}$ is a vector of length 1 unit in the positive y-direction.

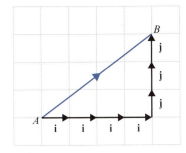

KEY WORDS

scalar

modulus

vector

unit vector

collinear

> **Note:**
> A vector of length 1 unit is called a **unit vector**.

WORKED EXAMPLE 1

a   Write $\overrightarrow{PQ}$ in the form $a\mathbf{i} + b\mathbf{j}$.

b   Find $|\overrightarrow{PQ}|$.

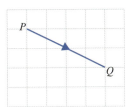

Answers

a $PQ = 4\mathbf{i} - 2\mathbf{j}$

b   Using Pythagoras, $|\overrightarrow{PQ}| = \sqrt{(4)^2 + (-2)^2} = \sqrt{20} = 2\sqrt{5}$.

You could be asked to find the unit vector in the direction of a given vector.

The method is outlined in the following example.

WORKED EXAMPLE 2

$\overrightarrow{EF} = 4\mathbf{i} + 3\mathbf{j}$

Find the unit vector in the direction of the vector $\overrightarrow{EF}$.

Answers

First find the length of the vector $\overrightarrow{EF}$:

$EF^2 = 4^2 + 3^2$ using Pythagoras

$EF = 5$

Hence the unit vector in the direction of $\overrightarrow{EF}$ is:

$\frac{1}{5}(4\mathbf{i} + 3\mathbf{j})$

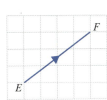

WORKED EXAMPLE 3

$\mathbf{a} = -2\mathbf{i} + 3\mathbf{j}$, $\mathbf{b} = 4\mathbf{i} - \mathbf{j}$ and $\mathbf{c} = -22\mathbf{i} + 18\mathbf{j}$.

Find λ and μ such that $\lambda\mathbf{a} + \mu\mathbf{b} = \mathbf{c}$.

Answers

$\lambda\mathbf{a} + \mu\mathbf{b} = \mathbf{c}$

$\lambda(-2\mathbf{i} + 3\mathbf{j}) + \mu(4\mathbf{i} - \mathbf{j}) = -22\mathbf{i} + 18\mathbf{j}$

Equating the **i**'s gives

$\qquad -2\lambda + 4\mu = -22$

$\qquad -\lambda + 2\mu = -11$ (1)

Equating the **j**'s gives

$\qquad 3\lambda - \mu = 18$

$\qquad 6\lambda - 2\mu = 36$ (2)

Adding equations (1) and (2) gives

$\qquad 5\lambda = 25$

$\qquad \lambda = 5$

Substituting for λ in equation (1) gives

$\qquad -5 + 2\mu = -11$

$\qquad 2\mu = -6$

$\qquad \mu = -3$

So $\lambda = 5$, $\mu = -3$.

Exercise 13.1

1   Write each vector in the form $a\mathbf{i} + b\mathbf{j}$.

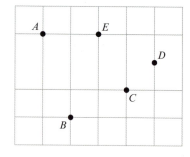

a $\overrightarrow{AB}$ b $\overrightarrow{AC}$ c $\overrightarrow{AD}$

d $\overrightarrow{AE}$ e $\overrightarrow{BE}$ f $\overrightarrow{DE}$

g $\overrightarrow{EA}$ h $\overrightarrow{DB}$ i $\overrightarrow{DC}$

2   Find the magnitude of each of these vectors.

a $-2\mathbf{i}$ b $4\mathbf{i} + 3\mathbf{j}$ c $5\mathbf{i} - 12\mathbf{j}$ d $-8\mathbf{i} - 6\mathbf{j}$

e $7\mathbf{i} + 24\mathbf{j}$ f $15\mathbf{i} - 8\mathbf{j}$ g $-4\mathbf{i} + 4\mathbf{j}$ h $5\mathbf{i} - 10\mathbf{j}$

3   The vector $\overrightarrow{AB}$ has a magnitude of 20 units and is parallel to the vector $4\mathbf{i} + 3\mathbf{j}$.

Find $\overrightarrow{AB}$.

4   The vector $\overrightarrow{PQ}$ has a magnitude of 39 units and is parallel to the vector $12\mathbf{i} - 5\mathbf{j}$.
     Find $\overrightarrow{PQ}$.

5   Find the unit vector in the direction of each of these vectors.
     a $6\mathbf{i} + 8\mathbf{j}$ **b** $5\mathbf{i} + 12\mathbf{j}$ **c** $-4\mathbf{i} - 3\mathbf{j}$ **d** $8\mathbf{i} - 15\mathbf{j}$ **e** $3\mathbf{i} + 3\mathbf{j}$

6    $\mathbf{p} = 8\mathbf{i} - 6\mathbf{j}, \mathbf{q} = -2\mathbf{i} + 3\mathbf{j}$ and $\mathbf{r} = 10\mathbf{i}$
     Find
     a $2\mathbf{q}$ **b** $2\mathbf{p} + \mathbf{q}$ **c** $\dfrac{1}{2}\mathbf{p} - 3\mathbf{r}$ **d** $\dfrac{1}{2}\mathbf{r} - \mathbf{p} - \mathbf{q}$.

7    $\mathbf{p} = 9\mathbf{i} + 12\mathbf{j}, \mathbf{q} = 3\mathbf{i} - 3\mathbf{j}$ and $\mathbf{r} = 7\mathbf{i} + \mathbf{j}$
     Find
     a $|\mathbf{p} + \mathbf{q}|$ **b** $|\mathbf{p} + \mathbf{q} + \mathbf{r}|$.

8    $\mathbf{p} = 7\mathbf{i} - 2\mathbf{j}$ and $\mathbf{q} = \mathbf{i} + \mu\mathbf{j}$.
     Find λ and μ such that $\lambda\mathbf{p} + \mathbf{q} = 36\mathbf{i} - 13\mathbf{j}$.

9    $\mathbf{a} = 5\mathbf{i} - 6\mathbf{j}, \mathbf{b} = -\mathbf{i} + 2\mathbf{j}$ and $\mathbf{c} = -13\mathbf{i} + 18\mathbf{j}$.
     Find λ and μ such that $\lambda\mathbf{a} + \mu\mathbf{b} = \mathbf{c}$.

13.2 Position vectors

The position vector of a point P relative to an origin, O,
means the displacement of the point P from O.

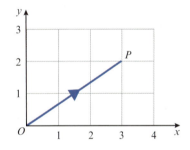

For this diagram, the position vector of P is

$$\overrightarrow{OP} = \begin{pmatrix} 3 \\ 2 \end{pmatrix} \quad \text{or} \quad \overrightarrow{OP} = 3\mathbf{i} + 2\mathbf{j}$$

Now consider two points A and B with position vectors $\mathbf{a}$ and $\mathbf{b}$.

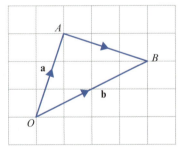

$\overrightarrow{AB}$ means the position vector of B relative to A.

$\overrightarrow{AB} = \overrightarrow{AO} + \overrightarrow{OB}$

$\qquad = -\overrightarrow{OA} + \overrightarrow{OB}$

Hence:

$$\overrightarrow{AB} = \overrightarrow{OB} - \overrightarrow{OA} \quad \text{or} \quad \overrightarrow{AB} = \mathbf{b} - \mathbf{a}$$

WORKED EXAMPLE 4

Relative to an origin O, the position vector of P is $4\mathbf{i} + 5\mathbf{j}$ and the position vector of Q is $10\mathbf{i} - 3\mathbf{j}$.

a Find $\overrightarrow{PQ}$.

The point R lies on PQ such that $\overrightarrow{PR} = \dfrac{1}{4}\overrightarrow{PQ}$.

b Find the position vector of R.

Answers

a $\overrightarrow{PQ} = \overrightarrow{OQ} - \overrightarrow{OP}$ $\overrightarrow{OQ} = 10\mathbf{i} - 3\mathbf{j}$ and $\overrightarrow{OP} = 4\mathbf{i} + 5\mathbf{j}$

   $= (10\mathbf{i} - 3\mathbf{j}) - (4\mathbf{i} + 5\mathbf{j})$ collect $\mathbf{i}$'s and $\mathbf{j}$'s

   $\overrightarrow{PQ} = 6\mathbf{i} - 8\mathbf{j}$

b $\overrightarrow{PR} = \dfrac{1}{4}\overrightarrow{PQ}$

   $= \dfrac{1}{4}(6\mathbf{i} - 8\mathbf{j})$

   $= 1.5\mathbf{i} - 2\mathbf{j}$

   $\overrightarrow{OR} = \overrightarrow{OP} + \overrightarrow{PR}$

   $= (4\mathbf{i} + 5\mathbf{j}) + (1.5\mathbf{i} - 2\mathbf{j})$

   $\overrightarrow{OR} = 5.5\mathbf{i} + 3\mathbf{j}$

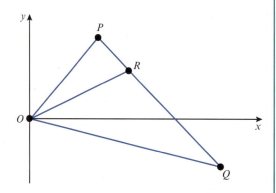

WORKED EXAMPLE 5

Relative to an origin O, the position vectors of points A, B and C are $-2\mathbf{i} + 5\mathbf{j}$, $10\mathbf{i} - \mathbf{j}$ and $\lambda(2\mathbf{i} + \mathbf{j})$ respectively. Given that C lies on the line AB, find the value of λ.

Answers

$\overrightarrow{AB} = \overrightarrow{OB} - \overrightarrow{OA}$ $\overrightarrow{OB} = 10\mathbf{i} - \mathbf{j}$ and $\overrightarrow{OA} = -2\mathbf{i} + 5\mathbf{j}$

   $= (10\mathbf{i} - \mathbf{j}) - (-2\mathbf{i} + 5\mathbf{j})$ collect $\mathbf{i}$'s and $\mathbf{j}$'s

   $= 12\mathbf{i} - 6\mathbf{j}$

If C lies on the line AB, then $\overrightarrow{AC} = k\overrightarrow{AB}$.

$\overrightarrow{AC} = \overrightarrow{OC} - \overrightarrow{OA}$ $\overrightarrow{OC} = \lambda(2\mathbf{i} + \mathbf{j})$ and $\overrightarrow{OA} = -2\mathbf{i} + 5\mathbf{j}$

   $= \lambda(2\mathbf{i} + \mathbf{j}) - (-2\mathbf{i} + 5\mathbf{j})$ collect $\mathbf{i}$'s and $\mathbf{j}$'s

   $= (2\lambda + 2)\mathbf{i} - (5 - \lambda)\mathbf{j}$

$k\overrightarrow{AB} = k(12\mathbf{i} - 6\mathbf{j})$

   $= 12k\mathbf{i} - 6k\mathbf{j}$

Hence, $(2\lambda + 2)\mathbf{i} - (5 - \lambda)\mathbf{j} = 12k\mathbf{i} - 6k\mathbf{j}$

CONTINUED

Equating the **i**'s gives: $2\lambda + 2 = 12k$ (1)

Equating the **j**'s gives: $5 - \lambda = 6k$ multiply both sides by 2

$$10 - 2\lambda = 12k \quad (2)$$

Using equation (1) and equation (2) gives

$$2\lambda + 2 = 10 - 2\lambda$$
$$4\lambda = 8$$
$$\lambda = 2$$

Exercise 13.2

1 Find $\overrightarrow{AB}$, in the form $a\mathbf{i} + b\mathbf{j}$, for each of the following.

a $A(4, 7)$ and $B(3, 4)$ b $A(0, 6)$ and $B(2, -4)$ c $A(3, -3)$ and $B(6, -2)$

d $A(7, 0)$ and $B(-5, 3)$ e $A(-4, -2)$ and $B(-3, 5)$ f $A(5, -6)$ and $B(-1, -7)$.

2   a O is the origin, P is the point $(1, 5)$ and $\overrightarrow{PQ} = \begin{pmatrix} 3 \\ 5 \end{pmatrix}$. Find $\overrightarrow{OQ}$.

   b O is the origin, E is the point $(-3, 4)$ and $\overrightarrow{EF} = \begin{pmatrix} -2 \\ 7 \end{pmatrix}$. Find the position vector of F.

   c O is the origin, M is the point $(4, -2)$ and $\overrightarrow{NM} = \begin{pmatrix} 3 \\ -5 \end{pmatrix}$. Find the position vector of N.

3   The vector $\overrightarrow{OA}$ has a magnitude of 25 units and is parallel to the vector $-3\mathbf{i} + 4\mathbf{j}$.

The vector $\overrightarrow{OB}$ has a magnitude of 26 units and is parallel to the vector $12\mathbf{i} + 5\mathbf{j}$.

Find

a $\overrightarrow{OA}$ b $\overrightarrow{OB}$ c $\overrightarrow{AB}$ d $|\overrightarrow{AB}|$.

4   Relative to an origin O, the position vector of A is $-7\mathbf{i} - 7\mathbf{j}$ and the position vector of B is $9\mathbf{i} + 5\mathbf{j}$. The point C lies on AB such that $\overrightarrow{AC} = 3\overrightarrow{CB}$.

a   Find $\overrightarrow{AB}$.

b   Find the unit vector in the direction of $\overrightarrow{AB}$.

c   Find the position vector of C.

5   Relative to an origin O, the position vector of P is $-2\mathbf{i} - 4\mathbf{j}$ and the position vector of Q is $8\mathbf{i} + 20\mathbf{j}$.

a   Find $\overrightarrow{PQ}$.

b   Find $|\overrightarrow{PQ}|$.

c   Find the unit vector in the direction of $\overrightarrow{PQ}$.

d   Find the position vector of M, the midpoint of PQ.

6   Relative to an origin O, the position vector of A is $4\mathbf{i} - 2\mathbf{j}$ and the position vector of B is $\lambda\mathbf{i} + 2\mathbf{j}$. The unit vector in the direction of $\overrightarrow{AB}$ is $0.3\mathbf{i} + 0.4\mathbf{j}$. Find the value of λ.

7 Relative to an origin O, the position vector of A is $\begin{pmatrix} -10 \\ 10 \end{pmatrix}$ and the position vectors of B is $\begin{pmatrix} 10 \\ -11 \end{pmatrix}$.

   a Find $\overrightarrow{AB}$.

   The points A, B and C lie on a straight line such that $\overrightarrow{AC} = 2\overrightarrow{AB}$.

   b Find the position vector of the point C.

8 Relative to an origin O, the position vector of A is $\begin{pmatrix} 21 \\ -20 \end{pmatrix}$ and the position vector of B is $\begin{pmatrix} 24 \\ 18 \end{pmatrix}$.

   a Find:

      i $|\overrightarrow{OA}|$ **ii** $|\overrightarrow{OB}|$ **iii** $|\overrightarrow{AB}|$.

   The points A, B and C lie on a straight line such that $\overrightarrow{AC} = \overrightarrow{CB}$.

   b Find the position vector of the point C.

9 Relative to an origin O, the position vector of A is $3\mathbf{i} - 2\mathbf{j}$ and the position vector of B is $15\mathbf{i} + 7\mathbf{j}$.

   a Find $\overrightarrow{AB}$.

   The point C lies on AB such that $\overrightarrow{AC} = \dfrac{1}{3} \overrightarrow{AB}$.

   b Find the position vector of C.

10 Relative to an origin O, the position vector of A is $6\mathbf{i} + 6\mathbf{j}$ and the position vector of B is $12\mathbf{i} - 2\mathbf{j}$.

   a Find $\overrightarrow{AB}$

   The point C lies on AB such that $\overrightarrow{AC} = \dfrac{3}{4} \overrightarrow{AB}$.

   b Find the position vector of C.

11 Relative to an origin O, the position vector of A is $\begin{pmatrix} 3 \\ 4 \end{pmatrix}$ and the position vector of B is $\begin{pmatrix} 5 \\ 5 \end{pmatrix}$.

   The points A, B and C are such that $\overrightarrow{BC} = 2\overrightarrow{AB}$. Find the position vector of C.

12 Relative to an origin O, the position vectors of points A, B and C are $-5\mathbf{i} - 11\mathbf{j}$, $23\mathbf{i} - 4\mathbf{j}$ and $\lambda(\mathbf{i} - 3\mathbf{j})$ respectively. Given that C lies on the line AB, find the value of λ.

13 Relative to an origin O, the position vectors of A, B and C are $-2\mathbf{i} + 7\mathbf{j}$, $2\mathbf{i} - \mathbf{j}$ and $6\mathbf{i} + \lambda\mathbf{j}$ respectively.

   a Find the value of λ when $AC = 17$.

   b Find the value of λ when ABC is a straight line.

   c Find the value of λ when ABC is a right-angle.

14 Relative to an origin O, the position vector of A is $-6\mathbf{i} + 4\mathbf{j}$ and the position vector of B is $18\mathbf{i} + 6\mathbf{j}$. C lies on the y-axis and $\overrightarrow{OC} = \overrightarrow{OA} + \lambda\overrightarrow{OB}$.

   Find $\overrightarrow{OC}$.

15 Relative to an origin O, the position vector of P is $8\mathbf{i} + 3\mathbf{j}$ and the position vector of Q is $-12\mathbf{i} - 7\mathbf{j}$. R lies on the x-axis and $\overrightarrow{OR} = \overrightarrow{OP} + \mu\overrightarrow{OQ}$.

   Find $\overrightarrow{OR}$.

16 CHALLENGE QUESTION

Relative to an origin O, the position vectors of points P, Q and R are $-6\mathbf{i} + 8\mathbf{j}$, $-4\mathbf{i} + 2\mathbf{j}$ and $5\mathbf{i} + 5\mathbf{j}$ respectively.

a Find the magnitude of:

    i $\overrightarrow{PQ}$ **ii** $\overrightarrow{PR}$ **iii** $\overrightarrow{QR}$

b Show that angle PQR is 90°.

c If $\overrightarrow{OP} = \lambda\,\overrightarrow{OQ} + \mu\,\overrightarrow{OR}$, find the value of λ and the value of μ.

13.3 Vector geometry

WORKED EXAMPLE 6

$\overrightarrow{OA} = \mathbf{a}$, $\overrightarrow{OB} = \mathbf{b}$, $\overrightarrow{BX} = \dfrac{3}{5}\overrightarrow{BA}$ and $\overrightarrow{OY} = \dfrac{3}{4}\overrightarrow{OA}$.

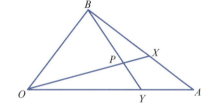

a Find in terms of $\mathbf{a}$ and $\mathbf{b}$:

    i $\overrightarrow{BA}$ **ii** $\overrightarrow{BX}$

    iii $\overrightarrow{OX}$ **iv** $\overrightarrow{BY}$

b Given that $\overrightarrow{OP} = \lambda\overrightarrow{OX}$, find $\overrightarrow{OP}$ in terms of λ, $\mathbf{a}$ and $\mathbf{b}$.

c Given that $\overrightarrow{BP} = \mu\overrightarrow{BY}$, find $\overrightarrow{OP}$ in terms of μ, $\mathbf{a}$ and $\mathbf{b}$.

d Find the value of λ and the value of μ.

Answers

a **i** $\overrightarrow{BA} = \overrightarrow{OA} - \overrightarrow{OB} = \mathbf{a} - \mathbf{b}$

    ii $\overrightarrow{BX} = \dfrac{3}{5}\overrightarrow{BA} = \dfrac{3}{5}(\mathbf{a} - \mathbf{b})$

    iii $\overrightarrow{OX} = \overrightarrow{OB} + \overrightarrow{BX} = \mathbf{b} + \dfrac{3}{5}(\mathbf{a} - \mathbf{b}) = \dfrac{3}{5}\mathbf{a} + \dfrac{2}{5}\mathbf{b}$

    iv $\overrightarrow{BY} = \overrightarrow{BO} + \overrightarrow{OY} = -\mathbf{b} + \dfrac{3}{4}\overrightarrow{OA} = \dfrac{3}{4}\mathbf{a} - \mathbf{b}$

b $\overrightarrow{OP} = \lambda\overrightarrow{OX}$

    $= \lambda\left(\dfrac{3}{5}\mathbf{a} + \dfrac{2}{5}\mathbf{b}\right)$

    $= \dfrac{3\lambda}{5}\mathbf{a} + \dfrac{2\lambda}{5}\mathbf{b}$

c $\overrightarrow{OP} = \overrightarrow{OB} + \overrightarrow{BP}$

    $= \mathbf{b} + \mu\overrightarrow{BY}$

    $= \mathbf{b} + \mu\left(\dfrac{3}{4}\mathbf{a} - \mathbf{b}\right)$

    $= \dfrac{3\mu}{4}\mathbf{a} + (1 - \mu)\mathbf{b}$

CONTINUED

d   Equating the coefficients of **a** for $\overrightarrow{OP}$ gives:

$\dfrac{3\lambda}{5} = \dfrac{3\mu}{4}$ divide both sides by 3

$\dfrac{\lambda}{5} = \dfrac{\mu}{4}$ multiply both sides by 20

$4\lambda = 5\mu$ (1)

Equating the coefficients of **b** for $\overrightarrow{OP}$ gives:

$\dfrac{2\lambda}{5} = 1 - \mu$ multiply both sides by 5

$2\lambda = 5 - 5\mu$ (2)

Adding equation (1) and equation (2) gives:

$6\lambda = 5$

$\lambda = \dfrac{5}{6}$

Substituting $\lambda = \dfrac{5}{6}$ in equation (1) gives $\mu = \dfrac{2}{3}$.

Hence, $\lambda = \dfrac{5}{6}$ and $\mu = \dfrac{2}{3}$.

WORKED EXAMPLE 7

$\overrightarrow{OA} = 3\mathbf{a}$, $\overrightarrow{OB} = 4\mathbf{b}$ and M is the midpoint of OB.

$OP : PA = 4 : 3$ and $\overrightarrow{BX} = \lambda\overrightarrow{BP}$.

a   Find in terms of **a** and **b**:

   i $\overrightarrow{AB}$ **ii** $\overrightarrow{MA}$.

b   Find in terms of λ, **a** and **b**:

   i $\overrightarrow{BX}$ **ii** $\overrightarrow{MX}$.

c   If M, X and A are **collinear**, find the value of λ.

Answers

a   **i** $\overrightarrow{AB} = \overrightarrow{AO} + \overrightarrow{OB}$

      $= -3\mathbf{a} + 4\mathbf{b}$

   ii $\overrightarrow{MA} = \overrightarrow{MO} + \overrightarrow{OA}$

      $= -2\mathbf{b} + 3\mathbf{a}$

      $= 3\mathbf{a} - 2\mathbf{b}$

CONTINUED

b **i** $\overrightarrow{BX} = \lambda\overrightarrow{BP}$

$\qquad\qquad = \lambda(\overrightarrow{BO} + \overrightarrow{OP})$

$\qquad\qquad = \lambda\left(-4\mathbf{b} + \dfrac{4}{7}\overrightarrow{OA}\right)\qquad$ use $\overrightarrow{OA} = 3\mathbf{a}$

$\qquad\qquad = \dfrac{12\lambda}{7}\mathbf{a} - 4\lambda\mathbf{b}$

ii $\overrightarrow{MX} = \overrightarrow{MB} + \overrightarrow{BX}\qquad$ use $\overrightarrow{BX} = \dfrac{12\lambda}{7}\mathbf{a} - 4\lambda\mathbf{b}$

$\qquad\qquad = 2\mathbf{b} + \dfrac{12\lambda}{7}\mathbf{a} - 4\lambda\mathbf{b}\qquad$ collect **a**'s and **b**'s

$\qquad\qquad = \dfrac{12\lambda}{7}\mathbf{a} + (2 - 4\lambda)\mathbf{b}$

c If M, X and A are collinear, then $\overrightarrow{MX} = k\overrightarrow{MA}$.

$\dfrac{12\lambda}{7}\mathbf{a} + (2 - 4\lambda)\mathbf{b} = k(3\mathbf{a} - 2\mathbf{b})$

Equating the coefficients of **a** gives:

$\qquad 3k = \dfrac{12\lambda}{7}\qquad$ divide both sides by 3

$\qquad k = \dfrac{4\lambda}{7}\qquad$ (1)

Equating the coefficients of **b** gives:

$\qquad -2k = 2 - 4\lambda\qquad$ divide both sides by -2

$\qquad k = 2\lambda - 1\qquad$ (2)

Using equation (1) and equation (2) gives:

$\qquad 2\lambda - 1 = \dfrac{4\lambda}{7}$

$\qquad 14\lambda - 7 = 4\lambda$

$\qquad 10\lambda = 7$

$\qquad \lambda = 0.7$

Exercise 13.3

1 $\overrightarrow{OA} = \mathbf{a}$, $\overrightarrow{OB} = \mathbf{b}$

R is the midpoint of OA and $\overrightarrow{OP} = 3\overrightarrow{OB}$.

$\overrightarrow{AQ} = \lambda\overrightarrow{AB}$ and $\overrightarrow{RQ} = \mu\overrightarrow{RP}$.

a Find $\overrightarrow{OQ}$ in terms of λ, **a** and **b**.

b Find $\overrightarrow{OQ}$ in terms of μ, **a** and **b**.

c Find the value of λ and the value of μ.

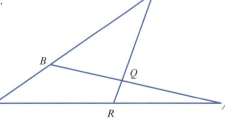

2 $\overrightarrow{AB} = 5\mathbf{a}$, $\overrightarrow{DC} = 3\mathbf{a}$ and $\overrightarrow{CB} = \mathbf{b}$

$\overrightarrow{AX} = \lambda\overrightarrow{AC}$ and $\overrightarrow{DX} = \mu\overrightarrow{DB}$.

a Find in terms of **a** and **b**,

   i $\overrightarrow{AD}$,   **ii** $\overrightarrow{DB}$.

b Find in terms of λ, μ, **a** and/or **b**,

   i $\overrightarrow{AX}$,   **ii** $\overrightarrow{DX}$.

c Find the value of λ and the value of μ.

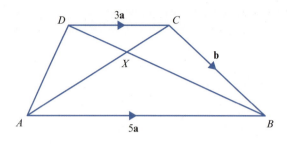

3 $\overrightarrow{OP} = \mathbf{a}$, $\overrightarrow{PY} = 2\mathbf{b}$, and $\overrightarrow{OQ} = 3\mathbf{b}$

$\overrightarrow{OX} = \lambda\overrightarrow{OY}$ and $\overrightarrow{QX} = \mu\overrightarrow{QP}$.

a Find $\overrightarrow{OX}$ in terms of λ, **a** and **b**.

b Find $\overrightarrow{OX}$ in terms of μ, **a** and **b**.

c Find the value of λ and the value of μ.

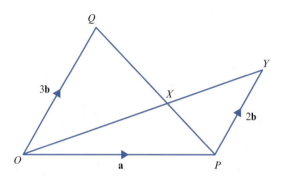

4 $\overrightarrow{OA} = \mathbf{a}$, $\overrightarrow{OB} = \mathbf{b}$

B is the midpoint of OD and $\overrightarrow{AC} = \dfrac{2}{3}\overrightarrow{OA}$.

$\overrightarrow{AX} = \lambda\overrightarrow{AD}$ and $\overrightarrow{BX} = \mu\overrightarrow{BC}$.

a Find $\overrightarrow{OX}$ in terms of λ, **a** and **b**.

b Find $\overrightarrow{OX}$ in terms of μ, **a** and **b**.

c Find the value of λ and the value of μ.

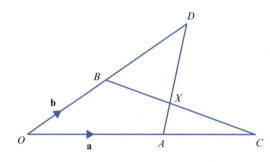

5 $\overrightarrow{OA} = \mathbf{a}$, $\overrightarrow{OB} = \mathbf{b}$

M is the midpoint of AB and $\overrightarrow{OY} = \dfrac{3}{4}\overrightarrow{OA}$.

$\overrightarrow{OX} = \lambda\overrightarrow{OM}$ and $\overrightarrow{BX} = \mu\overrightarrow{BY}$.

a Find in terms of **a** and **b**

   i $\overrightarrow{AB}$ **ii** $\overrightarrow{OM}$.

b Find $\overrightarrow{OX}$ in terms of λ, **a** and **b**.

c Find $\overrightarrow{OX}$ in terms of μ, **a** and **b**.

d Find the value of λ and the value of μ.

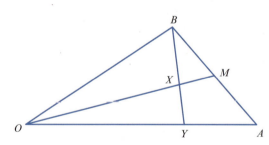

6 $\overrightarrow{OA} = \mathbf{a}$, $\overrightarrow{OB} = \mathbf{b}$, $\overrightarrow{BX} = \dfrac{3}{5}\overrightarrow{BA}$ and $\overrightarrow{OY} = \dfrac{5}{7}\overrightarrow{OA}$.

$\overrightarrow{OP} = \lambda\overrightarrow{OX}$ and $\overrightarrow{BP} = \mu\overrightarrow{BY}$.

a Find $\overrightarrow{OP}$ in terms of λ, **a** and **b**.

b Find $\overrightarrow{OP}$ in terms of μ, **a** and **b**.

c Find the value of λ and the value of μ.

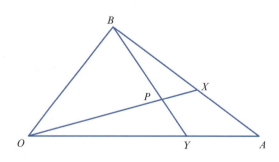

7   $\overrightarrow{OA} = \mathbf{a}$, $\overrightarrow{OB} = \mathbf{b}$ and O is the origin.

$\overrightarrow{OX} = \lambda\overrightarrow{OA}$ and $\overrightarrow{OY} = \mu\overrightarrow{OB}$.

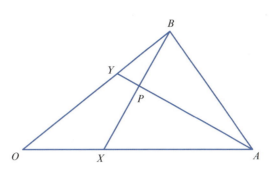

   a   i   Find $\overrightarrow{BX}$ in terms of λ, $\mathbf{a}$ and $\mathbf{b}$.

       ii   Find $\overrightarrow{AY}$ in terms of μ, $\mathbf{a}$ and $\mathbf{b}$.

   b $5\overrightarrow{BP} = 2\overrightarrow{BX}$ and $\overrightarrow{AY} = 4\overrightarrow{PY}$.

       i   Find $\overrightarrow{OP}$ in terms of λ, $\mathbf{a}$ and $\mathbf{b}$.

       ii   Find $\overrightarrow{OP}$ in terms of μ, $\mathbf{a}$ and $\mathbf{b}$.

       iii   Find the value of λ and the value of μ.

8   O, A, B and C are four points such that

$\overrightarrow{OA} = 7\mathbf{a} - 5\mathbf{b}$, $\overrightarrow{OB} = 2\mathbf{a} + 5\mathbf{b}$ and $\overrightarrow{OC} = -2\mathbf{a} + 13\mathbf{b}$.

   a   Find   i $\overrightarrow{AC}$ ii $\overrightarrow{AB}$.

   b   Use your answers to **part a** to explain why B lies on the line AC.

9   **CHALLENGE QUESTION**

$\overrightarrow{OA} = \mathbf{a}$ and $\overrightarrow{OB} = \mathbf{b}$

$OA : AE = 1 : 3$ and $AB : BC = 1 : 2$

$OB = BD$

   a   Find, in terms of $\mathbf{a}$ and/or $\mathbf{b}$,

       i $\overrightarrow{OE}$ ii $\overrightarrow{OD}$ iii $\overrightarrow{OC}$.

   b   Find, in terms of $\mathbf{a}$ and/or $\mathbf{b}$,

       i $\overrightarrow{CE}$ ii $\overrightarrow{CD}$ iii $\overrightarrow{DE}$.

   c   Use your answers to **part b** to explain why C, D and E are collinear.

   d   Find the ratio $CD : DE$.

13.4 Constant velocity problems

If an object moves with a constant velocity, $\mathbf{v}$, where $\mathbf{v} = (4\mathbf{i} - 2\mathbf{j})\,\mathrm{m\,s}^{-1}$, the velocity can be represented on a diagram, as shown here.

Velocity is a quantity that has both magnitude and direction.

The magnitude of the velocity is the speed.

If $\mathbf{v} = (4\mathbf{i} - 2\mathbf{j})\,\mathrm{m\,s}^{-1}$ then,

$\text{speed} = \sqrt{(4)^2 + (-2)^2}$

$\qquad\quad = \sqrt{20}$

$\qquad\quad = 2\sqrt{5}\,\mathrm{m\,s}^{-1}$

You should already know the formula for an object moving with constant speed:

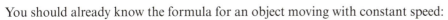

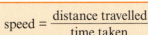

$$\text{speed} = \frac{\text{distance travelled}}{\text{time taken}}$$

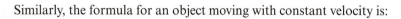
Similarly, the formula for an object moving with constant velocity is:

$$\text{velocity} = \frac{\text{displacement}}{\text{time taken}}$$

Splitting a velocity into its components

The velocity of a particle travelling north-east at $4\sqrt{2}\,\mathrm{m\,s}^{-1}$ can be written in the form $(a\mathbf{i} + b\mathbf{j})\,\mathrm{m\,s}^{-1}$:

$\cos 45° = \dfrac{a}{4\sqrt{2}}$ and $\sin 45° = \dfrac{b}{4\sqrt{2}}$

$\qquad a = 4\sqrt{2} \times \cos 45° \qquad\qquad b = 4\sqrt{2} \times \sin 45°$

$\qquad a = 4 \qquad\qquad\qquad\qquad\quad b = 4$

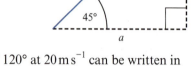

Hence the velocity vector is $(4\mathbf{i} + 4\mathbf{j})\,\mathrm{m\,s}^{-1}$.

The velocity of a particle travelling on a bearing of $120°$ at $20\,\mathrm{m\,s}^{-1}$ can be written in the form $(x\mathbf{i} + y\mathbf{j})\,\mathrm{m\,s}^{-1}$:

$\sin 60° = \dfrac{x}{20}$ and $\cos 60° = \dfrac{y}{20}$

$\qquad x = 20 \times \sin 60° \qquad\qquad y = 20 \times \cos 60°$

$\qquad x = 10\sqrt{3} \qquad\qquad\qquad\quad y = 10$

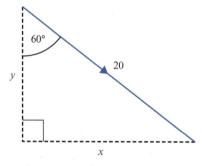

Hence the velocity vector is $\left(10\sqrt{3}\,\mathbf{i} - 10\mathbf{j}\right)\mathrm{m\,s}^{-1}$.

WORKED EXAMPLE 8

An object travels at a constant velocity from point A to point B.

$\overrightarrow{AB} = (32\mathbf{i} - 24\mathbf{j})\,\mathrm{m}$ and the time taken is $4\,\mathrm{s}$. Find

a the velocity **b** the speed.

Answers

a $\text{velocity} = \dfrac{\text{displacement}}{\text{time taken}} = \dfrac{32\mathbf{i} - 24\mathbf{j}}{4} = (8\mathbf{i} - 6\mathbf{j})\,\mathrm{m\,s}^{-1}$

b $\text{speed} = \sqrt{(8)^2 + (-6)^2} = 10\,\mathrm{m\,s}^{-1}$

Consider a boat sailing with velocity $\begin{pmatrix} 3 \\ -2 \end{pmatrix}\mathrm{km\,h}^{-1}$.

At $12\,00$ hours the boat is at the point A with position vector $\begin{pmatrix} 3 \\ 13 \end{pmatrix}\mathrm{km}$ relative to an origin O.

The diagram shows the positions of the boat at 12 00 hours, 1 pm, 2 pm, 3 pm, 4 pm …

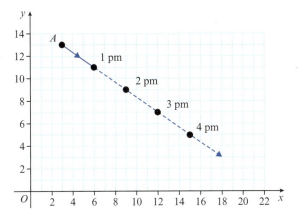

The position at 1 pm $= \begin{pmatrix} 3 \\ 13 \end{pmatrix} + 1\begin{pmatrix} 3 \\ -2 \end{pmatrix} = \begin{pmatrix} 6 \\ 11 \end{pmatrix}$

The position at 2 pm $= \begin{pmatrix} 3 \\ 13 \end{pmatrix} + 2\begin{pmatrix} 3 \\ -2 \end{pmatrix} = \begin{pmatrix} 9 \\ 9 \end{pmatrix}$

The position at 3 pm $= \begin{pmatrix} 3 \\ 13 \end{pmatrix} + 3\begin{pmatrix} 3 \\ -2 \end{pmatrix} = \begin{pmatrix} 12 \\ 7 \end{pmatrix}$

The position at 4 pm $= \begin{pmatrix} 3 \\ 13 \end{pmatrix} + 4\begin{pmatrix} 3 \\ -2 \end{pmatrix} = \begin{pmatrix} 15 \\ 5 \end{pmatrix}$

Hence the position vector, **r**, of the boat t hours after 12 00 hours is given by the expression:

$\mathbf{r} = \begin{pmatrix} 3 \\ 13 \end{pmatrix} + t\begin{pmatrix} 3 \\ -2 \end{pmatrix}.$

This leads to the general rule:

> If an object has initial position **a** and moves with a constant velocity **v**, the position vector **r**, at time t, is given by the formula: $\mathbf{r} = \mathbf{a} + t\mathbf{v}$.

WORKED EXAMPLE 9

Particle A starts moving at time $t = 0$ from the point with position vector $52\mathbf{i} + \mathbf{j}$ with a speed of $15\,\mathrm{m\,s^{-1}}$ in the direction $-3\mathbf{i} + 4\mathbf{j}$.

a   Find the velocity vector, $\mathbf{v}_A$, of A.

b   Find the position vector, $\mathbf{r}_A$, of A after t seconds.

Particle B starts moving at time $t = 0$ from the point with position vector $-11\mathbf{i} - 8\mathbf{j}$ with velocity $\mathbf{v}_B = 12\mathbf{i} + 15\mathbf{j}\,\mathrm{m\,s^{-1}}$.

c   Show that A and B collide, and find the value of t when they collide and the position vector of the point of collision.

CONTINUED

Answers

a $\quad \sqrt{(-3)^2 + 4^2} = 5$

$$\mathbf{v}_A = \frac{15}{5}(-3\mathbf{i} + 4\mathbf{j})$$

$$\mathbf{v}_A = -9\mathbf{i} + 12\mathbf{j}\,\mathrm{m\,s^{-1}}$$

b $\quad \mathbf{r}_A = 52\mathbf{i} + \mathbf{j} + t(-9\mathbf{i} + 12\mathbf{j})$

c $\quad \mathbf{r}_B = -11\mathbf{i} - 8\mathbf{j} + t(12\mathbf{i} + 15\mathbf{j})$

Particles A and B collide if $\mathbf{r}_A = \mathbf{r}_B$ at a particular time.

$$52\mathbf{i} + \mathbf{j} + t(-9\mathbf{i} + 12\mathbf{j}) = -11\mathbf{i} - 8\mathbf{j} + t(12\mathbf{i} + 15\mathbf{j})$$

Equating the $\mathbf{i}$'s: $\qquad 52 - 9t = -11 + 12t$

$$21t = 63$$

$$t = 3$$

Equating the $\mathbf{j}$'s: $\qquad 1 + 12t = -8 + 15t$

$$3t = 9$$

$$t = 3$$

Particles A and B collide when $t = 3$.

When $t = 3$, $\mathbf{r}_A = 52\mathbf{i} + \mathbf{j} + 3(-9\mathbf{i} + 12\mathbf{j}) = 25\mathbf{i} + 37\mathbf{j}$

$$\mathbf{r}_B = -11\mathbf{i} - 8\mathbf{j} + 3(12\mathbf{i} + 15\mathbf{j}) = 25\mathbf{i} + 37\mathbf{j}$$

Hence particles A and B collide when $t = 3$ at the point with position vector $25\mathbf{i} + 37\mathbf{j}$.

Exercise 13.4

1 a Displacement $= (21\mathbf{i} + 54\mathbf{j})\,\mathrm{m}$, time taken $= 6$ seconds. Find the velocity.

   b Velocity $= (5\mathbf{i} - 6\mathbf{j})\,\mathrm{m\,s^{-1}}$, time taken $= 6$ seconds. Find the displacement.

   c Velocity $= (-4\mathbf{i} + 4\mathbf{j})\,\mathrm{km\,h^{-1}}$, displacement $= (-50\mathbf{i} + 50\mathbf{j})\,\mathrm{km}$.
      Find the time taken.

2 A car travels from a point A with position vector $(60\mathbf{i} - 40\mathbf{j})\,\mathrm{km}$ to a point B with position vector $(-50\mathbf{i} + 18\mathbf{j})\,\mathrm{km}$.

   The car travels with constant velocity and takes 5 hours to complete the journey.

   Find the velocity vector.

3 A helicopter flies from a point P with position vector $(50\mathbf{i} + 100\mathbf{j})\,\mathrm{km}$ to a point Q.

   The helicopter flies with a constant velocity of $(30\mathbf{i} - 40\mathbf{j})\,\mathrm{km\,h^{-1}}$ and takes 2.5 hours to complete the journey. Find the position vector of the point Q.

4 a A car travels north-east with a speed of $18\sqrt{2}\,\mathrm{km\,h^{-1}}$.

Find the velocity vector of the car.

b A boat sails on a bearing of 030° with a speed of $20\,\mathrm{km\,h^{-1}}$.

Find the velocity vector of the boat.

c A plane flies on a bearing of 240° with a speed of $100\,\mathrm{m\,s^{-1}}$.

Find the velocity vector of the plane.

5 A particle starts at a point P with position vector $(-80\mathbf{i} + 60\mathbf{j})\,\mathrm{m}$ relative to an origin O.
The particle travels with velocity $(12\mathbf{i} - 16\mathbf{j})\,\mathrm{m\,s^{-1}}$.

a Find the speed of the particle.

b Find the position vector of the particle after

i 1 second   ii 2 seconds   iii 3 seconds.

c Find the position vector of the particle t seconds after leaving P.

6 At 12 00 hours, a ship leaves a point Q with position vector $(10\mathbf{i} + 38\mathbf{j})\,\mathrm{km}$ relative
to an origin O. The ship travels with velocity $(6\mathbf{i} - 8\mathbf{j})\,\mathrm{km\,h^{-1}}$.

a Find the speed of the ship.

b Find the position vector of the ship at 3 pm.

c Find the position vector of the ship t hours after leaving Q.

d Find the time when the ship is at the point with position vector $(61\mathbf{i} - 30\mathbf{j})\,\mathrm{km}$.

7 At 12 00 hours, a tanker sails from a point P with position vector $(5\mathbf{i} + 12\mathbf{j})\,\mathrm{km}$
relative to an origin O. The tanker sails south-east with a speed of $12\sqrt{2}\,\mathrm{km\,h^{-1}}$.

a Find the velocity vector of the tanker.

b Find the position vector of the tanker at

i 14 00 hours   ii 12 45 hours.

c Find the position vector of the tanker t hours after leaving P.

8 At 12 00 hours, a boat sails from a point P.

The position vector, $\mathbf{r}\,\mathrm{km}$, of the boat relative to an origin O, t hours after 12 00 is

given by $\mathbf{r} = \begin{pmatrix} 10 \\ 6 \end{pmatrix} + t\begin{pmatrix} 5 \\ 12 \end{pmatrix}$.

a Write down the position vector of the point P.

b Write down the velocity vector of the boat.

c Find the speed of the boat.

d Find the distance of the boat from P after 4 hours.

9 At 15 00 hours, a submarine departs from point A and travels a distance of
120 km to a point B.

The position vector, $\mathbf{r}\,\mathrm{km}$, of the submarine relative to an origin O, t hours after

15 00 is given by $\mathbf{r} = \begin{pmatrix} 15 + 8t \\ 20 + 6t \end{pmatrix}$.

a Write down the position vector of the point A.

b Write down the velocity vector of the submarine.

c Find the position vector of the point B.

10 At 12 00 hours, boats A and B have position vectors $(-10\mathbf{i} + 40\mathbf{j})\,\text{km}$ and $(70\mathbf{i} + 10\mathbf{j})\,\text{km}$ and are moving with velocities $(20\mathbf{i} + 10\mathbf{j})\,\text{km h}^{-1}$ and $(-10\mathbf{i} + 30\mathbf{j})\,\text{km h}^{-1}$ respectively.

    a Find the position vectors of A and B at 15 00 hours.

    b Find the distance between A and B at 15 00 hours.

11 At time $t = 0$, boat P leaves the origin and travels with velocity $(3\mathbf{i} + 4\mathbf{j})\,\text{km h}^{-1}$. Also at time $t = 0$, boat Q leaves the point with position vector $(-10\mathbf{i} + 17\mathbf{j})\,\text{km}$ and travels with velocity $(5\mathbf{i} + 2\mathbf{j})\,\text{km h}^{-1}$.

    a Write down the position vectors of boats A and B after 2 hours.

    b Find the distance between boats P and Q when $t = 2$.

> **REFLECTION**
>
> Explain to a friend the difference between speed and velocity.

SUMMARY

Position vectors

$\overrightarrow{AB}$ means the position vector of B relative to A.

$\overrightarrow{AB} = \overrightarrow{OB} - \overrightarrow{OA}$ or $\overrightarrow{AB} = \mathbf{b} - \mathbf{a}$

If an object has initial position $\mathbf{a}$ and moves with a constant velocity $\mathbf{v}$, the position vector $\mathbf{r}$, at time t, is given by the formula: $\mathbf{r} = \mathbf{a} + t\mathbf{v}$.

Velocity

$$\text{Velocity} = \frac{\text{displacement}}{\text{time taken}}$$

Past paper questions

Worked example

Particle A is at the point with position vector $\begin{pmatrix} 2 \\ -5 \end{pmatrix}$ at time $t = 0$ and moves with a speed of $10\,\text{m s}^{-1}$ in the same direction as $\begin{pmatrix} 3 \\ 4 \end{pmatrix}$.

i Given that A is at the point with position vector $\begin{pmatrix} 38 \\ a \end{pmatrix}$ when $t = 6\,\text{s}$, find the value of the constant a.   [3]

Particle B is at the point with position vector $\begin{pmatrix} 16 \\ 37 \end{pmatrix}$ at time $t = 0$ and moves with velocity $\begin{pmatrix} 4 \\ 2 \end{pmatrix}\,\text{m s}^{-1}$.

ii Write down, in terms of t, the position vector of B at time $t\,\text{s}$.   [1]

iii Verify that particles A and B collide.   [4]

iv Write down the position vector of the point of collision.   [1]

Cambridge IGCSE Additional Mathematics 0606 Paper 11 Q10 Nov 2018

Answers

i Magnitude of $\begin{pmatrix} 3 \\ 4 \end{pmatrix} = \sqrt{3^2 + 4^2} = 5$

Velocity vector of A is: $\mathbf{v}_A = 2\begin{pmatrix} 3 \\ 4 \end{pmatrix} = \begin{pmatrix} 6 \\ 8 \end{pmatrix}$

Position vector of A at time t is: $\mathbf{r}_A = \begin{pmatrix} 2 \\ -5 \end{pmatrix} + t\begin{pmatrix} 6 \\ 8 \end{pmatrix}$

When $t = 6$, $\mathbf{r}_A = \begin{pmatrix} 38 \\ a \end{pmatrix}$

$$\begin{pmatrix} 38 \\ a \end{pmatrix} = \begin{pmatrix} 2 \\ -5 \end{pmatrix} + 6\begin{pmatrix} 6 \\ 8 \end{pmatrix}$$

$$\begin{pmatrix} 38 \\ a \end{pmatrix} = \begin{pmatrix} 38 \\ 43 \end{pmatrix}$$

Hence $a = 43$

ii Position vector of B at time t is: $\mathbf{r}_B = \begin{pmatrix} 16 \\ 37 \end{pmatrix} + t\begin{pmatrix} 4 \\ 2 \end{pmatrix}$

iii Particles A and B collide if $\mathbf{r}_A = \mathbf{r}_B$ at a particular time.

$$\begin{pmatrix} 2 \\ -5 \end{pmatrix} + t\begin{pmatrix} 6 \\ 8 \end{pmatrix} = \begin{pmatrix} 16 \\ 37 \end{pmatrix} + t\begin{pmatrix} 4 \\ 2 \end{pmatrix}$$

$2 + 6t = 16 + 4t$ and $-5 + 8t = 37 + 2t$

$\quad 2t = 14 \qquad\qquad\qquad 6t = 42$

$\quad\quad t = 7 \qquad\qquad\qquad\quad t = 7$

Hence particles A and B collide when $t = 7$.

iv When $t = 7$:

$\mathbf{r}_A = \begin{pmatrix} 2 \\ -5 \end{pmatrix} + 7\begin{pmatrix} 6 \\ 8 \end{pmatrix}$ and $\mathbf{r}_B = \begin{pmatrix} 16 \\ 37 \end{pmatrix} + 7\begin{pmatrix} 4 \\ 2 \end{pmatrix}$

$\mathbf{r}_A = \begin{pmatrix} 44 \\ 51 \end{pmatrix}$ and $\mathbf{r}_B = \begin{pmatrix} 44 \\ 51 \end{pmatrix}$

Position vector of the point of collision is $\begin{pmatrix} 44 \\ 51 \end{pmatrix}$.

1 The position vectors of the points A and B relative to an origin O are $-2\mathbf{i} + 17\mathbf{j}$ and $6\mathbf{i} + 2\mathbf{j}$ respectively.

   i Find the vector $\overrightarrow{AB}$. [1]

   ii Find the unit vector in the direction of $\overrightarrow{AB}$. [2]

   iii The position vector of the point C relative to the origin O is such that $\overrightarrow{OC} = \overrightarrow{OA} + m\overrightarrow{OB}$, where m is a constant. Given that C lies on the x-axis, find the vector $\overrightarrow{OC}$. [3]

Cambridge IGCSE Additional Mathematics 0606 Paper 22 Q5 Mar 2015

2 **a** The four points O, A, B and C are such that $\overrightarrow{OA} = 5\mathbf{a}$, $\overrightarrow{OB} = 15\mathbf{b}$, $\overrightarrow{OC} = 24\mathbf{b} - 3\mathbf{a}$.
     Show that B lies on the line AC. [3]

   b Relative to an origin O, the position vector of the point P is $\mathbf{i} - 4\mathbf{j}$ and the position vector of the point Q is $3\mathbf{i} + 7\mathbf{j}$. Find

     i $|\overrightarrow{PQ}|$, [2]

     ii the unit vector in the direction $\overrightarrow{PQ}$, [1]

     iii the position vector of M, the mid-point of PQ. [2]

Cambridge IGCSE Additional Mathematics 0606 Paper 21 Q7 Jun 2015

3 Relative to an origin O, points A, B and C have position vectors $\begin{pmatrix} 5 \\ 4 \end{pmatrix}$, $\begin{pmatrix} -10 \\ 12 \end{pmatrix}$ and $\begin{pmatrix} 6 \\ -18 \end{pmatrix}$ respectively.

All distances are measured in kilometres. A man drives at a constant speed directly from A to B in 20 minutes.

i Calculate the speed in km h^{-1} at which the man drives from A to B. [3]

He now drives directly from B to C at the same speed.

ii Find how long it takes him to drive from B to C. [3]

Cambridge IGCSE Additional Mathematics 0606 Paper 21 Q3 Nov 2015

4

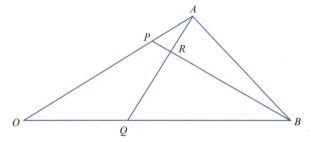

The position vectors of points A and B relative to an origin O are **a** and **b** respectively. The point P is such that $\overrightarrow{OP} = \mu\overrightarrow{OA}$. The point Q is such that $\overrightarrow{OQ} = \lambda\overrightarrow{OB}$. The lines AQ and BP intersect at the point R.

i Express $\overrightarrow{AQ}$ in terms of λ, **a** and **b**. [1]

ii Express $\overrightarrow{BP}$ in terms of μ, **a** and **b**. [1]

It is given that $3\overrightarrow{AR} = \overrightarrow{AQ}$ and $8\overrightarrow{BR} = 7\overrightarrow{BP}$.

iii Express $\overrightarrow{OR}$ in terms of λ, **a** and **b**. [2]

iv Express $\overrightarrow{OR}$ in terms of μ, **a** and **b**. [2]

v Hence find the value of μ and of λ. [3]

Cambridge IGCSE Additional Mathematics 0606 Paper 11 Q12 Nov 2014

5 a The vectors **p** and **q** are such that $\mathbf{p} = 11\mathbf{i} - 24\mathbf{j}$ and $\mathbf{q} = 2\mathbf{i} + \alpha\mathbf{j}$.

    i Find the value of each of the constants α and β such that $\mathbf{p} + 2\mathbf{q} = (\alpha + \beta)\mathbf{i} - 20\mathbf{j}$. [3]

    ii Using the values of α and β found in part **i**, find the unit vector in the direction $\mathbf{p} + 2\mathbf{q}$. [2]

b

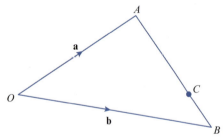

The points A and B have position vectors **a** and **b** with respect to an origin O.

The point C lies on AB and is such that $AB:AC$ is $1:\lambda$. Find an expression for $\overrightarrow{OC}$ in terms of **a**, **b** and λ. [3]

c The points S and T have position vectors **s** and **t** with respect to an origin O.

The points O, S and T do not lie in a straight line. Given that the vector $2\mathbf{s} + \mu\mathbf{t}$ is parallel to the vector $(\mu + 3)\mathbf{s} + 9\mathbf{t}$, where μ is a positive constant, find the value of μ. [3]

Cambridge IGCSE Additional Mathematics 0606 Paper 22 Q10 Mar 2016

6   **a**   A vector **v** has a magnitude of 102 units and has the same direction as $\begin{pmatrix} 8 \\ -15 \end{pmatrix}$.

Find **v** in the form $\begin{pmatrix} a \\ b \end{pmatrix}$, where a and b are integers.   [2]

b   Vectors $\mathbf{c} = \begin{pmatrix} 4 \\ 3 \end{pmatrix}$ and $\mathbf{d} = \begin{pmatrix} p - q \\ 5p + q \end{pmatrix}$ are such that $\mathbf{c} + 2\mathbf{d} = \begin{pmatrix} p^2 \\ 27 \end{pmatrix}$.

Find the possible values of the constants p and q.   [6]

Cambridge IGCSE Additional Mathematics 0606 Paper 12 Q7 Mar 2017

7   The diagram shows the quadrilateral $OABC$
such that $\overrightarrow{OA} = \mathbf{a}$, $\overrightarrow{OB} = \mathbf{b}$ and $\overrightarrow{OC} = \mathbf{c}$.

It is given that $AM:MC = 2:1$ and $OM:MB = 3:2$.

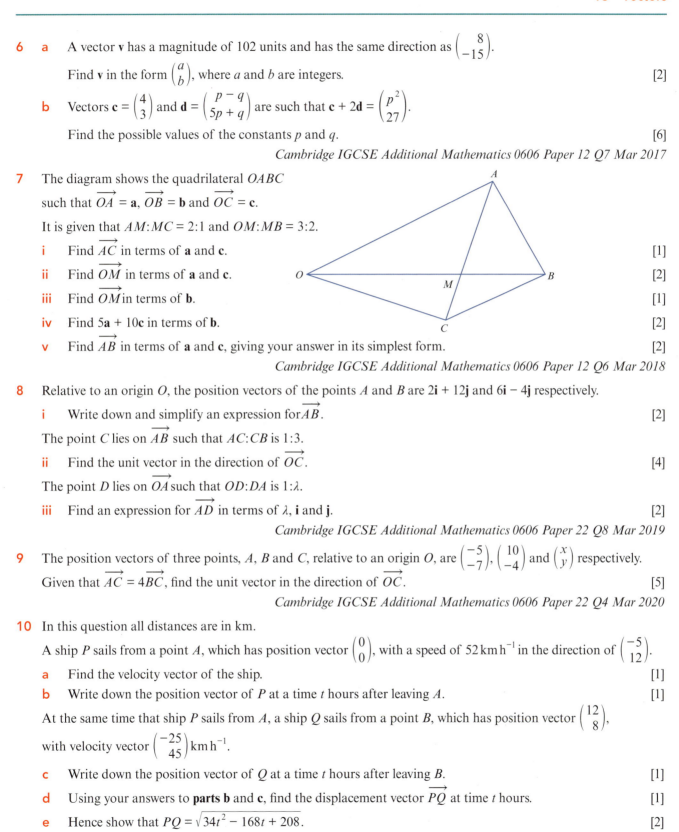

i   Find $\overrightarrow{AC}$ in terms of **a** and **c**.   [1]

ii   Find $\overrightarrow{OM}$ in terms of **a** and **c**.   [2]

iii   Find $\overrightarrow{OM}$ in terms of **b**.   [1]

iv   Find $5\mathbf{a} + 10\mathbf{c}$ in terms of **b**.   [2]

v   Find $\overrightarrow{AB}$ in terms of **a** and **c**, giving your answer in its simplest form.   [2]

Cambridge IGCSE Additional Mathematics 0606 Paper 12 Q6 Mar 2018

8   Relative to an origin O, the position vectors of the points A and B are $2\mathbf{i} + 12\mathbf{j}$ and $6\mathbf{i} - 4\mathbf{j}$ respectively.

i   Write down and simplify an expression for $\overrightarrow{AB}$.   [2]

The point C lies on $\overrightarrow{AB}$ such that $AC:CB$ is $1:3$.

ii   Find the unit vector in the direction of $\overrightarrow{OC}$.   [4]

The point D lies on $\overrightarrow{OA}$ such that $OD:DA$ is $1:\lambda$.

iii   Find an expression for $\overrightarrow{AD}$ in terms of λ, **i** and **j**.   [2]

Cambridge IGCSE Additional Mathematics 0606 Paper 22 Q8 Mar 2019

9   The position vectors of three points, A, B and C, relative to an origin O, are $\begin{pmatrix} -5 \\ -7 \end{pmatrix}$, $\begin{pmatrix} 10 \\ -4 \end{pmatrix}$ and $\begin{pmatrix} x \\ y \end{pmatrix}$ respectively.
Given that $\overrightarrow{AC} = 4\overrightarrow{BC}$, find the unit vector in the direction of $\overrightarrow{OC}$.   [5]

Cambridge IGCSE Additional Mathematics 0606 Paper 22 Q4 Mar 2020

10   In this question all distances are in km.

A ship P sails from a point A, which has position vector $\begin{pmatrix} 0 \\ 0 \end{pmatrix}$, with a speed of $52\,\text{km h}^{-1}$ in the direction of $\begin{pmatrix} -5 \\ 12 \end{pmatrix}$.

a   Find the velocity vector of the ship.   [1]

b   Write down the position vector of P at a time t hours after leaving A.   [1]

At the same time that ship P sails from A, a ship Q sails from a point B, which has position vector $\begin{pmatrix} 12 \\ 8 \end{pmatrix}$,

with velocity vector $\begin{pmatrix} -25 \\ 45 \end{pmatrix} \text{km h}^{-1}$.

c   Write down the position vector of Q at a time t hours after leaving B.   [1]

d   Using your answers to **parts b** and **c**, find the displacement vector $\overrightarrow{PQ}$ at time t hours.   [1]

e   Hence show that $PQ = \sqrt{34t^2 - 168t + 208}$.   [2]

f   Find the value of t when P and Q are first $2\,\text{km}$ apart.   [2]

Cambridge IGCSE Additional Mathematics 0606 Paper 12 Q8 Mar 2020

11 A particle P is moving with a velocity of $20\,\text{m}\,\text{s}^{-1}$ in the same direction as $\begin{pmatrix} 3 \\ 4 \end{pmatrix}$.

    i   Find the velocity vector of P.   [2]

At time $t = 0\,\text{s}$, P has position vector $\begin{pmatrix} 1 \\ 2 \end{pmatrix}$ relative to a fixed point O.

    ii   Write down the position vector of P after $t\,\text{s}$.   [2]

A particle Q has position vector $\begin{pmatrix} 17 \\ 18 \end{pmatrix}$ relative to O at time $t = 0\,\text{s}$ and has a velocity vector $\begin{pmatrix} 8 \\ 12 \end{pmatrix}\text{m}\,\text{s}^{-1}$.

    iii   Given that P and Q collide, find the value of t when they collide and the position vector of the point
           of collision.   [3]

Cambridge IGCSE Additional Mathematics 0606 Paper 11 Q5 Jun 2019

12 a

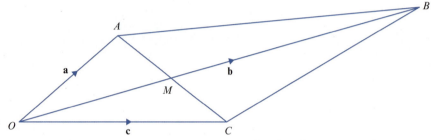

The diagram shows a figure $OABC$, where $\overrightarrow{OA} = \mathbf{a}$, $\overrightarrow{OB} = \mathbf{b}$ and $\overrightarrow{OC} = \mathbf{c}$.
The lines AC and OB intersect at the point M where M is the midpoint of the line AC.

    i   Find, in terms of $\mathbf{a}$ and $\mathbf{c}$, the vector $\overrightarrow{OM}$.   [2]

    ii   Given that $OM:MB = 2:3$, find $\mathbf{b}$ in terms of $\mathbf{a}$ and $\mathbf{c}$.   [2]

  b   Vectors $\mathbf{i}$ and $\mathbf{j}$ are unit vectors parallel to the x-axis and y-axis respectively.

       The vector $\mathbf{p}$ has a magnitude of 39 units and has the same direction as $-10\mathbf{i} + 24\mathbf{j}$.

    i   Find $\mathbf{p}$ in terms of $\mathbf{i}$ and $\mathbf{j}$.   [2]

    ii   Hence find the vector $\mathbf{q}$ such that $2\mathbf{p} + \mathbf{q}$ is parallel to the positive y-axis and has a magnitude of
          12 units.   [3]

    iii   Hence show that $|\mathbf{q}| = k\sqrt{5}$, where k is an integer to be found.   [2]

Cambridge IGCSE Additional Mathematics 0606 Paper 11 Q5 Jun 2017

13 A particle P is initially at the point with position vector $\begin{pmatrix} 30 \\ 10 \end{pmatrix}$ and moves with a constant speed of $10\,\text{m}\,\text{s}^{-1}$

in the same direction as $\begin{pmatrix} -4 \\ 3 \end{pmatrix}$.

  a   Find the position vector of P after $t\,\text{s}$.   [3]

As P starts moving, a particle Q starts to move such that its position vector after $t\,\text{s}$ is given by $\begin{pmatrix} -80 \\ 90 \end{pmatrix} + t \begin{pmatrix} 5 \\ 12 \end{pmatrix}$.

  b   Write down the speed of Q.   [1]

  c   Find the exact distance between P and Q when $t = 10$, giving your answer in its simplest surd form.   [3]

Cambridge IGCSE Additional Mathematics 0606 Paper 11 Q6 Nov 2020

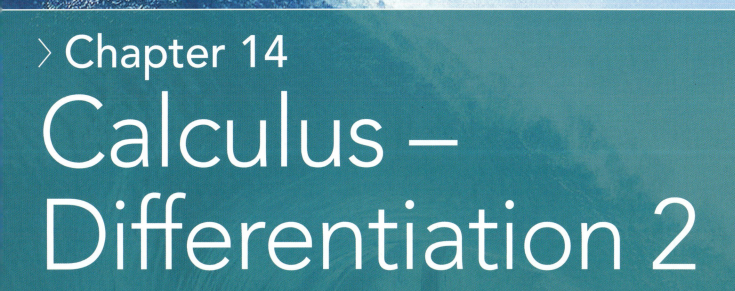

> Chapter 14
Calculus – Differentiation 2

THIS SECTION WILL SHOW YOU HOW TO:

- differentiate $\sin x$, $\cos x$, $\tan x$, e^x and $\ln x$ together with constant multiples, sums and composite functions of these.

PRE-REQUISITE KNOWLEDGE

Before you start…

Where it comes from	What you should be able to do	Check your skills
Chapter 12	Differentiate x^n together with constant multiples, sums and differences.	**1** Given that $y = 3x^2 - \dfrac{4}{x^3} + 2\sqrt{x} - 5$, find $\dfrac{dy}{dx}$.
	Differentiate composite functions using the chain rule.	**2** Given that $y = (2x - 3)^4$, find $\dfrac{dy}{dx}$.
	Differentiate products and quotients.	**3** Differentiate with respect to x. **a** $(x - 3)^4 (2x + 1)^5$ **b** $\dfrac{x^2 - 5}{2x + 1}$
	Find tangents and normal to curves.	**4** Find the equation of the normal to the curve $y = 3x^3 + x^2 - 4x + 1$ at the point $(0, 1)$.
	Find stationary points on curves and determine their nature.	**5** Find the coordinates of the stationary points on the curve $y = x^3 - 12x + 5$ and determine their nature.
	Apply differentiation to connected rates of change, small increments and approximations.	**6** Variables x and y and are related by the equation $y = 2x^2 + 5x$. Find the approximate change in as increases from 1 to 1.01.

14.1 Derivatives of exponential functions

CLASS DISCUSSION

Graphing software has been used to draw the graphs of $y = 2^x$ and $y = 3^x$ together with their gradient (derived) functions.

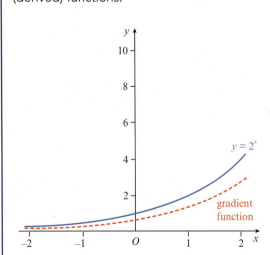

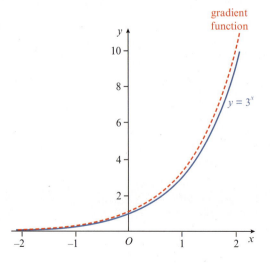

Discuss with your classmates what conclusions can be made from these two graphs.

In Chapter 6 you learned about the exponential function $y = e^x$ where $e \approx 2.718$.

This function has the very special property that the gradient function is identical to the original function. This leads to the rule:

$$\frac{d}{dx}(e^x) = e^x$$

The derivative of $e^{f(x)}$

Consider the function $y = e^{f(x)}$.

Let $y = e^u$ where $u = f(x)$

$$\frac{dy}{du} = e^u \qquad\qquad \frac{du}{dx} = f'(x)$$

Using the chain rule:
$$\frac{dy}{dx} = \frac{dy}{du} \times \frac{du}{dx}$$
$$= e^u \times f'(x)$$
$$= f'(x) \times e^{f(x)}$$

$$\frac{d}{dx}[e^{f(x)}] = f'(x) \times e^{f(x)}$$

In particular:

$$\frac{d}{dx}[e^{ax+b}] = a \times e^{ax+b}$$

WORKED EXAMPLE 1

Differentiate with respect to x.

a e^{5x} **b** e^{x^2-3x} **c** $x^2 e^{-3x}$ **d** $\dfrac{e^{2x}}{x}$

Answers

a $\dfrac{d}{dx}(e^{5x}) = \underset{\substack{\text{differentiate}\\\text{index}}}{5} \times \underset{\substack{\text{original}\\\text{function}}}{e^{5x}} = 5e^{5x}$

b $\dfrac{d}{dx}(e^{x^2-3x}) = \underset{\substack{\text{differentiate}\\\text{index}}}{(2x-3)} \times \underset{\substack{\text{original}\\\text{function}}}{e^{x^2-3x}} = (2x-3)e^{x^2-3x}$

c $\dfrac{d}{dx}(x^2 e^{-3x}) = x^2 \times \dfrac{d}{dx}(e^{-3x}) + e^{-3x} \times \dfrac{d}{dx}(x^2)$ product rule

$\qquad\qquad\qquad = x^2 \times (-3e^{-3x}) + e^{-3x} \times 2x$

$\qquad\qquad\qquad = -3x^2 e^{-3x} + 2x e^{-3x}$

$\qquad\qquad\qquad = x e^{-3x}(2-3x)$

CONTINUED

d $\dfrac{d}{dx}\left(\dfrac{e^{2x}}{x}\right) = \dfrac{x \times \dfrac{d}{dx}(e^{2x}) - e^{2x} \times \dfrac{d}{dx}(x)}{x^2}$ quotient rule

$= \dfrac{x \times 2e^{2x} - e^{2x} \times 1}{x^2}$

$= \dfrac{e^{2x}(2x - 1)}{x^2}$

WORKED EXAMPLE 2

A curve has equation $y = (e^{2x} + e^{3x})^5$.

Find the value of $\dfrac{dy}{dx}$ when $x = 0$.

Answers

$y = \left(e^{2x} + e^{3x}\right)^5$

$\dfrac{dy}{dx} = 5\left(e^{2x} + e^{3x}\right)^4 \times \left(2e^{2x} + 3e^{3x}\right)$ chain rule

When $x = 0$, $\dfrac{dy}{dx} = 5\left(e^0 + e^0\right)^4 \times \left(2e^0 + 3e^0\right)$

$= 5 \times 2^4 \times 5$

$= 400$

CLASS DISCUSSION

By writing 2 as $e^{\ln 2}$, find an expression for $\dfrac{d}{dx}(2^x)$.

Discuss with your classmates whether you can find similar expressions for $\dfrac{d}{dx}(3^x)$ and $\dfrac{d}{dx}(4^x)$.

Exercise 14.1

1   Differentiate with respect to x.

   a e^{7x} **b** e^{3x} **c** $3e^{5x}$ **d** $2e^{-4x}$

   e $6e^{-\frac{x}{2}}$ **f** e^{3x+1} **g** e^{x^2+1} **h** $5x - 3e^{\sqrt{x}}$

   i $2 + \dfrac{1}{e^{3x}}$ **j** $2(3 - e^{2x})$ **k** $\dfrac{e^x + e^{-x}}{2}$ **l** $5\left(x^2 + e^{x^2}\right)$

2   Differentiate with respect to x.

   a xe^x **b** x^2e^{2x} **c** $3xe^{-x}$ **d** $\sqrt{x}\,e^x$ **e** $\dfrac{e^x}{x}$

   f $\dfrac{e^{2x}}{\sqrt{x}}$ **g** $\dfrac{e^x + 1}{e^x - 1}$ **h** $xe^{2x} - \dfrac{e^{2x}}{2}$ **i** $\dfrac{x^2e^x - 5}{e^x + 1}$

3 Find the equation of the tangent to

    a $y = \dfrac{5}{e^{2x} + 3}$ at $x = 0$ **b** $y = \sqrt{e^{2x} + 1}$ at $x = \ln 5$ **c** $y = x^2(1 + e^x)$ at $x = 1$.

4 A curve has equation $y = 5e^{2x} - 4x - 3$.

     The tangent to the curve at the point $(0, 2)$ meets the x-axis at the point A.

     Find the coordinates of A.

5 A curve has equation $y = xe^x$.

    a   Find, in terms of e, the coordinates of the stationary point on this curve and determine its nature.

    b   Find, in terms of e, the equation of the normal to the curve at the point $P(1, e)$.

    c   The normal at P meets the x-axis at A and the y-axis at B. Find, in terms of e, the area of triangle OAB, where O is the origin.

14.2 Derivatives of logarithmic functions

CLASS DISCUSSION

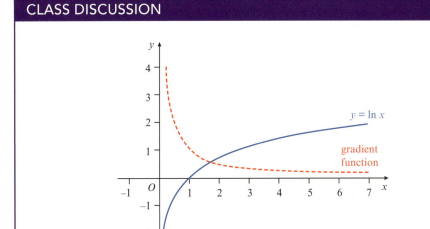

Graphing software has been used to draw the graph of $y = \ln x$ together with its gradient (derived) function.

The gradient function passes through the points $\left(\dfrac{1}{2}, 2\right)$, $(1, 1)$, $\left(2, \dfrac{1}{2}\right)$, and $\left(5, \dfrac{1}{5}\right)$.

Discuss with your classmates what conclusions you can make about the gradient function.

From the class discussion you should have concluded that:

$$\frac{\mathrm{d}}{\mathrm{d}x}(\ln x) = \frac{1}{x}$$

The derivative of $\ln f(x)$

Consider the function $y = \ln f(x)$.

Let $\quad y = \ln u \quad$ where $\quad u = f(x)$

$$\frac{dy}{du} = \frac{1}{u} \qquad\qquad \frac{du}{dx} = f'(x)$$

Using the chain rule: $\quad \dfrac{dy}{dx} = \dfrac{dy}{du} \times \dfrac{du}{dx}$

$$= \frac{1}{u} \times f'(x)$$

$$= \frac{f'(x)}{f(x)}$$

$$\frac{d}{dx}[\ln f(x)] = \frac{f'(x)}{f(x)}$$

In particular:

$$\frac{d}{dx}[\ln(ax + b)] = \frac{a}{ax + b}$$

WORKED EXAMPLE 3

Differentiate with respect to x.

a $\quad \ln 8x \qquad$ **b** $\quad \ln(5x - 7) \qquad$ **c** $\quad \ln(2x^2 + 5) \qquad$ **d** $\quad \ln\sqrt{x - 10}$

Answers

a $\quad \dfrac{d}{dx}(\ln 8x) = \dfrac{8}{8x} \qquad\qquad$ ⟵ 'inside' differentiated
⟵ 'inside'

$$= \frac{1}{x}$$

b $\quad \dfrac{d}{dx}\ln[5x - 7] = \dfrac{5}{5x - 7} \qquad\qquad$ ⟵ 'inside' differentiated
⟵ 'inside'

c $\quad \dfrac{d}{dx}\ln[2x^2 + 5] = \dfrac{4x}{2x^2 + 5} \qquad\qquad$ ⟵ 'inside' differentiated
⟵ 'inside'

d **Method 1:**

$$\frac{d}{dx}[\ln\sqrt{x - 10}] = \frac{\frac{1}{2}(x - 10)^{-\frac{1}{2}}(1)}{\sqrt{x - 10}} \qquad\qquad$$ ⟵ 'inside' differentiated
⟵ 'inside'

$$= \frac{1}{2(x - 10)}$$

CONTINUED

Method 2: using the rules of logarithms before differentiating.

$$\frac{d}{dx}[\ln\sqrt{x-10}] = \frac{d}{dx}\left[\ln(x-10)^{\frac{1}{2}}\right] \qquad \text{use } \ln a^m = m\ln a$$

$$= \frac{d}{dx}\left[\frac{1}{2}\ln(x-10)\right]$$

$$= \frac{1}{2} \times \frac{d}{dx}[\ln(x-10)]$$

$$= \frac{1}{2} \times \frac{1}{x-10} \qquad \longleftarrow \quad \text{'inside' differentiated}$$
$$\qquad\qquad\qquad\qquad \longleftarrow \quad \text{'inside'}$$

$$= \frac{1}{2(x-10)}$$

WORKED EXAMPLE 4

Differentiate with respect to x.

a $5x^6 \ln 2x$

b $\dfrac{\ln x}{3x}$

Answers

a $\dfrac{d}{dx}(5x^6 \ln 2x) = 5x^6 \times \dfrac{d}{dx}(\ln 2x) + \ln 2x \times \dfrac{d}{dx}(5x^6)$ product rule

$$= 5x^6 \times \frac{2}{2x} + \ln 2x \times 30x^5$$

$$= 5x^5 + 30x^5 \ln 2x$$

b $\dfrac{d}{dx}\left(\dfrac{\ln x}{3x}\right) = \dfrac{3x \times \dfrac{d}{dx}(\ln x) - \ln x \times \dfrac{d}{dx}(3x)}{(3x)^2}$ quotient rule

$$= \frac{3x \times \dfrac{1}{x} - \ln x \times 3}{9x^2}$$

$$= \frac{1 - \ln x}{3x^2}$$

WORKED EXAMPLE 5

A curve has equation $y = \ln\left[\dfrac{\sqrt{2x-1}}{x^2+1}\right]$.

a   Show that $y = \dfrac{1}{2}\ln(2x-1) - \ln(x^2+1)$.

b   Hence, find the value of $\dfrac{dy}{dx}$ when $x = 1$.

CONTINUED

Answers

a $y = \ln\left[\dfrac{\sqrt{2x-1}}{x^2+1}\right]$ $\qquad$ use $\log_a\left(\dfrac{x}{y}\right) = \log_a x - \log_a y$

$\quad = \ln(2x-1)^{\frac{1}{2}} - \ln(x^2+1)$ $\qquad$ use $\log_a(x)^m = m\log_a x$

$\quad = \dfrac{1}{2}\ln(2x-1) - \ln(x^2+1)$

b $\dfrac{dy}{dx} = \dfrac{d}{dx}\left[\dfrac{1}{2}\ln(2x-1)\right] - \dfrac{d}{dx}\left[\ln(x^2+1)\right]$

$\quad = \dfrac{1}{2}\times\dfrac{d}{dx}[\ln(2x-1)] - \dfrac{d}{dx}[\ln(x^2+1)]$

$\quad = \dfrac{1}{2}\times\dfrac{2}{2x-1} - \dfrac{2x}{x^2+1}$

$\quad = \dfrac{1}{2x-1} - \dfrac{2x}{x^2+1}$

$\quad = \dfrac{x^2+1-2x(2x-1)}{(2x-1)(x^2+1)}$

$\quad = \dfrac{-3x^2+2x+1}{(2x-1)(x^2+1)}$

When $x = 1$, $\dfrac{dy}{dx} = 0$.

WORKED EXAMPLE 6

A curve has equation $y = \log_2(5x-2)$.

a Show that $y = \dfrac{1}{\ln 2}[\ln(5x-2)]$.

b Hence, find the value of $\dfrac{dy}{dx}$ when $x = 2$.

Answers

a $y = \log_2(5x-2)$ $\qquad$ use $\log_b a = \dfrac{\log_c a}{\log_c b}$

$\quad = \dfrac{\ln(5x-2)}{\ln 2}$

$\quad = \dfrac{1}{\ln 2}[\ln(5x-2)]$

b $\dfrac{dy}{dx} = \dfrac{1}{\ln 2}\times\dfrac{d}{dx}[\ln(5x-2)]$

$\quad = \dfrac{1}{\ln 2}\times\dfrac{5}{5x-2}$

$\quad = \dfrac{5}{(5x-2)\ln 2}$

When $x = 2$, $\dfrac{dy}{dx} = \dfrac{5}{8\ln 2}$.

Exercise 14.2

1   Differentiate with respect to x.

   a $\ln 5x$ b $\ln 12x$ c $\ln(2x + 3)$

   d $2 + \ln(1 - x^2)$ e $\ln(3x + 1)^2$ f $\ln\sqrt{x + 2}$

   g $\ln(2 - 5x)^4$ h $2x + \ln\left(\dfrac{4}{x}\right)$ i $5 - \ln\dfrac{3}{(2 - 3x)}$

   j $\ln(\ln x)$ k $\ln(\sqrt{x} + 1)^2$ l $\ln(x^2 + \ln x)$

2   Differentiate with respect to x.

   a $x\ln x$ b $2x^2\ln x$ c $(x - 1)\ln x$

   d $5x\ln x^2$ e $x^2\ln(\ln x)$ f $\dfrac{\ln 2x}{x}$

   g $\dfrac{4}{\ln x}$ h $\dfrac{\ln(2x + 1)}{x^2}$ i $\dfrac{\ln(x^3 - 1)}{2x + 3}$

3   A curve has equation $y = x^2\ln 3x$.

   Find the value of $\dfrac{dy}{dx}$ and $\dfrac{d^2y}{dx^2}$ at the point where $x = 2$.

4   Use the laws of logarithms to help differentiate these expressions with respect to x.

   a $\ln\sqrt{3x + 1}$ b $\ln\dfrac{1}{(2x - 5)}$ c $\ln\left[x(x - 5)^4\right]$

   d $\ln\left(\dfrac{2x + 1}{x - 1}\right)$ e $\ln\left(\dfrac{2 - x}{x^2}\right)$ f $\ln\left[\dfrac{x(x + 1)}{x + 2}\right]$

   g $\ln\left[\dfrac{2x + 3}{(x - 5)(x + 1)}\right]$ h $\ln\left[\dfrac{2}{(x + 3)^2(x - 1)}\right]$ i $\ln\left[\dfrac{(x + 1)(2x - 3)}{x(x - 1)}\right]$

5   Find $\dfrac{dy}{dx}$ for each of the following.

   a $y = \log_3 x$

   b $y = \log_2 x^2$

   c $y = \log_4(5x - 1)$

> **TIP**
>
> Use change of base of logarithms before differentiating.

6   Find $\dfrac{dy}{dx}$ for each of the following.

   a $e^y = 4x^2 - 1$

   b $e^y = 5x^3 - 2x$

   c $e^y = (x + 3)(x - 4)$

> **TIP**
>
> Take the natural logarithm of both sides of the equation before differentiating.

7   A curve has equation $x = \dfrac{1}{2}\left[e^{y(3x+7)} + 1\right]$.

   Find the value of $\dfrac{dy}{dx}$ when $x = 1$.

14.3 Derivatives of trigonometric functions

Graphing software has been used to draw the graphs of $y = \sin x$ and $y = \cos x$ together with their gradient (derived) functions.

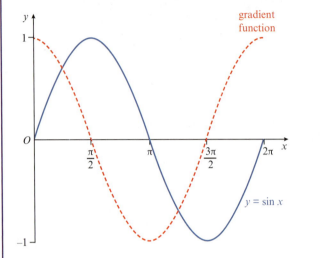 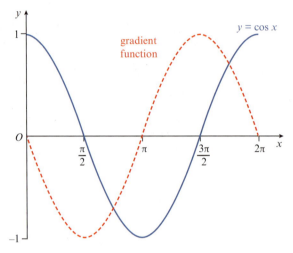

Discuss with your classmates what conclusions you can make from these two graphs.

From the class discussion you should have concluded that if x is measured in radians then:

$$\frac{\mathrm{d}}{\mathrm{d}x}(\sin x) = \cos x \qquad\qquad \frac{\mathrm{d}}{\mathrm{d}x}(\cos x) = -\sin x$$

The derivative of $\tan x$ can be found using these two results together with the quotient rule.

$$\frac{\mathrm{d}}{\mathrm{d}x}(\tan x) = \frac{\mathrm{d}}{\mathrm{d}x}\left(\frac{\sin x}{\cos x}\right) \qquad\qquad \text{use the quotient rule}$$

$$= \frac{\cos x \times \dfrac{\mathrm{d}}{\mathrm{d}x}(\sin x) - \sin x \times \dfrac{\mathrm{d}}{\mathrm{d}x}(\cos x)}{(\cos x)^2}$$

$$= \frac{\cos x \times \cos x - \sin x \times (-\sin x)}{(\cos x)^2}$$

$$= \frac{\cos^2 x + \sin^2 x}{\cos^2 x} \qquad\qquad \text{use } \cos^2 x + \sin^2 x = 1$$

$$= \frac{1}{\cos^2 x} \qquad\qquad \text{use } \frac{1}{\cos x} = \sec x$$

$$= \sec^2 x$$

$$\frac{\mathrm{d}}{\mathrm{d}x}(\tan x) = \sec^2 x$$

WORKED EXAMPLE 7

Differentiate with respect to x.

a $7\cos x$ **b** $x^2\sin x$ **c** $\dfrac{3\tan x}{x}$ **d** $(5-3\cos x)^8$

Answers

a $\dfrac{d}{dx}(7\cos x)=7\dfrac{d}{dx}(\cos x)$

$\qquad = -7\sin x$

b $\dfrac{d}{dx}(x^2\sin x)=x^2\times\dfrac{d}{dx}(\sin x)+\sin x\times\dfrac{d}{dx}(x^2)$ product rule

$\qquad = x^2\cos x + 2x\sin x$

c $\dfrac{d}{dx}\left(\dfrac{3\tan x}{x}\right)=\dfrac{x\times\dfrac{d}{dx}(3\tan x)-3\tan x\times\dfrac{d}{dx}(x)}{x^2}$ quotient rule

$\qquad = \dfrac{x\times 3\sec^2 x - 3\tan x\times 1}{x^2}$

$\qquad = \dfrac{3x\sec^2 x - 3\tan x}{x^2}$

d $\dfrac{d}{dx}\left[(5-3\cos x)^8\right]=8(5-3\cos x)^7\times 3\sin x$ chain rule

$\qquad = 24\sin x(5-3\cos x)^7$

Derivatives of $\sin(ax+b)$, $\cos(ax+b)$ and $\tan(ax+b)$

Consider the function $y=\sin(ax+b)$, where x is measured in radians.

Let $y=\sin u$ where $u=ax+b$

$\dfrac{dy}{du}=\cos u$ $\qquad \dfrac{du}{dx}=a$

Using the chain rule: $\dfrac{dy}{dx}=\dfrac{dy}{du}\times\dfrac{du}{dx}$

$\qquad = \cos u\times a$

$\qquad = a\cos(ax+b)$

$$\dfrac{d}{dx}[\sin(ax+b)]=a\cos(ax+b)$$

Similarly, it can be shown that:

$$\dfrac{d}{dx}[\cos(ax+b)]=-a\sin(ax+b) \qquad \dfrac{d}{dx}[\tan(ax+b)]=a\sec^2(ax+b)$$

> **TIP**
>
> It is important to remember that, in calculus, all angles are measured in radians unless a question tells you otherwise.

WORKED EXAMPLE 8

Differentiate with respect to x.

a $\quad 5\sin\left(\dfrac{\pi}{3} - 2x\right)$ **b** $\quad x\cos 3x$ **c** $\quad \dfrac{x^2}{\sin\left(2x + \dfrac{\pi}{4}\right)}$ **d** $\quad (1 + 3\tan 2x)^5$

Answers

a $\quad \dfrac{d}{dx}\left[5\sin\left(\dfrac{\pi}{3} - 2x\right)\right] = 5\dfrac{d}{dx}\left[\sin\left(\dfrac{\pi}{3} - 2x\right)\right]$

$$= 5 \times \cos\left(\dfrac{\pi}{3} - 2x\right) \times (-2)$$

$$= -10\cos\left(\dfrac{\pi}{3} - 2x\right)$$

b $\quad \dfrac{d}{dx}(x\cos 3x) = x \times \dfrac{d}{dx}(\cos 3x) + \cos 3x \times \dfrac{d}{dx}(x)$ $\qquad$ product rule

$$= x \times (-3\sin 3x) + \cos 3x \times (1)$$

$$= \cos 3x - 3x\sin 3x$$

c $\quad \dfrac{d}{dx}\left[\dfrac{x^2}{\sin\left(2x + \dfrac{\pi}{4}\right)}\right] = \dfrac{\sin\left(2x + \dfrac{\pi}{4}\right) \times \dfrac{d}{dx}(x^2) - x^2 \times \dfrac{d}{dx}\left[\sin\left(2x + \dfrac{\pi}{4}\right)\right]}{\left[\sin\left(2x + \dfrac{\pi}{4}\right)\right]^2}$ $\qquad$ quotient rule

$$= \dfrac{\sin\left(2x + \dfrac{\pi}{4}\right) \times (2x) - x^2 \times \left[2\cos\left(2x + \dfrac{\pi}{4}\right)\right]}{\sin^2\left(2x + \dfrac{\pi}{4}\right)}$$

$$= \dfrac{2x\sin\left(2x + \dfrac{\pi}{4}\right) - 2x^2\cos\left(2x + \dfrac{\pi}{4}\right)}{\sin^2\left(2x + \dfrac{\pi}{4}\right)}$$

d $\quad \dfrac{d}{dx}\left[(1 + 3\tan 2x)^5\right] = 5(1 + 3\tan 2x)^4 \times 6\sec^2 2x$ $\qquad$ chain rule

$$= 30\sec^2 2x(1 + 3\tan 2x)^4$$

Exercise 14.3

1 Differentiate with respect to x.

  a $2 + \sin x$ **b** $2\sin x + 3\cos x$ **c** $2\cos x - \tan x$

  d $3\sin 2x$ **e** $4\tan 5x$ **f** $2\cos 3x - \sin 2x$

  g $\tan(3x + 2)$ **h** $\sin\left(2x + \dfrac{\pi}{3}\right)$ **i** $2\cos\left(3x - \dfrac{\pi}{6}\right)$

2 Differentiate with respect to x.

  a $\sin^3 x$ **b** $5\cos^2(3x)$ **c** $\sin^2 x - 2\cos x$

  d $(3 - \cos x)^4$ **e** $2\sin^3\left(2x + \dfrac{\pi}{6}\right)$ **f** $3\cos^4 x + 2\tan^2\left(2x - \dfrac{\pi}{4}\right)$

3 Differentiate with respect to x.

  a $x\sin x$ **b** $2\sin 2x\cos 3x$ **c** $x^2\tan x$

  d $x\tan^3\left(\dfrac{x}{2}\right)$ **e** $\dfrac{5}{\cos 3x}$ **f** $\dfrac{x}{\cos x}$

  g $\dfrac{\tan x}{x}$ **h** $\dfrac{\sin x}{2 + \cos x}$ **i** $\dfrac{\sin x}{3x - 1}$

  j $\dfrac{1}{\sin^3 2x}$ **k** $\dfrac{3x}{\sin 2x}$ **l** $\dfrac{\sin x + \cos x}{\sin x - \cos x}$

4 Differentiate with respect to x.

  a $e^{\cos x}$ **b** $e^{\cos 5x}$ **c** $e^{\tan x}$ **d** $e^{(\sin x + \cos x)}$

  e $e^x\sin x$ **f** $e^x\cos\dfrac{1}{2}x$ **g** $e^x(\cos x + \sin x)$ **h** $x^2 e^{\cos x}$

  l $\ln(\sin x)$ **j** $x^2\ln(\cos x)$ **k** $\dfrac{\sin 3x}{e^{2x-1}}$ **l** $\dfrac{x\sin x}{e^x}$

5 Find the gradient of the tangent to

  a $y = 2x\cos 3x$ when $x = \dfrac{\pi}{3}$ **b** $y = \dfrac{2 - \cos x}{3\tan x}$ when $x = \dfrac{\pi}{4}$.

6 a By writing $\sec x$ as $\dfrac{1}{\cos x}$, find $\dfrac{d}{dx}(\sec x)$.

  b By writing $\operatorname{cosec} x$ as $\dfrac{1}{\sin x}$, find $\dfrac{d}{dx}(\operatorname{cosec} x)$.

  c By writing $\cot x$ as $\dfrac{\cos x}{\sin x}$, find $\dfrac{d}{dx}(\cot x)$.

7 Find $\dfrac{dy}{dx}$ for each of the following.

  a $e^y = \sin 3x$ **b** $e^y = 3\cos 2x$

8 CHALLENGE QUESTION

A curve has equation $y = A\sin x + B\sin 2x$.

The curve passes through the point $P\left(\dfrac{\pi}{2}, 3\right)$ and has a gradient of $\dfrac{3\sqrt{2}}{2}$ when $x = \dfrac{\pi}{4}$.

Find the value of A and the value of B.

> **TIP**
>
> Take the natural logarithm of both sides of the equation before differentiating.

9 CHALLENGE QUESTION

A curve has equation $y = A \sin x + B \cos 2x$.

The curve has a gradient of $5\sqrt{3}$ when $x = \dfrac{\pi}{6}$ and has a gradient of $6 + 2\sqrt{2}$ when $x = \dfrac{\pi}{4}$.

Find the value of A and the value of B.

14.4 Further applications of differentiation

You need to be able to answer questions that involve the differentiation of exponential, logarithmic and trigonometric functions.

REFLECTION

Without looking back in this chapter, can you prove that
$\dfrac{d}{dx}(\tan x) = \sec^2 x$?

Try using a similar method to find
$\dfrac{d}{dx}(\cot x)$.

WORKED EXAMPLE 9

A curve has equation $y = 3x \sin x + \dfrac{\pi}{6}$.

The curve passes through the point $P\left(\dfrac{\pi}{2}, a\right)$.

a Find the value of a.

b Find the equation of the normal to the curve at P.

Answers

a When $x = \dfrac{\pi}{2}$, $y = 3 \times \dfrac{\pi}{2} \times \sin\left(\dfrac{\pi}{2}\right) + \dfrac{\pi}{6}$

$$y = \dfrac{3\pi}{2} + \dfrac{\pi}{6} = \dfrac{5\pi}{3}$$

Hence, $a = \dfrac{5\pi}{3}$.

b $y = 3x \sin x + \dfrac{\pi}{6}$

$\dfrac{dy}{dx} = 3x \cos x + 3 \sin x$

When $x = \dfrac{\pi}{2}$, $\dfrac{dy}{dx} = 3\left(\dfrac{\pi}{2}\right)\cos\left(\dfrac{\pi}{2}\right) + 3\sin\left(\dfrac{\pi}{2}\right) = 3$

Normal: passes through the point $\left(\dfrac{\pi}{2}, \dfrac{5\pi}{3}\right)$ and gradient $= -\dfrac{1}{3}$

$$y - \dfrac{5\pi}{3} = -\dfrac{1}{3}\left(x - \dfrac{\pi}{2}\right)$$

$$y - \dfrac{5\pi}{3} = -\dfrac{1}{3}x + \dfrac{\pi}{6}$$

$$y = -\dfrac{1}{3}x + \dfrac{11\pi}{6}$$

WORKED EXAMPLE 10

A curve has equation $y = x^2 \ln x$.
Find the approximate increase in y as x increases from e to e $+ p$, where p is small.

Answers

$y = x^2 \ln x$

$\dfrac{\mathrm{d}y}{\mathrm{d}x} = x^2 \times \dfrac{1}{x} + \ln x \times 2x$ product rule

$\phantom{\dfrac{\mathrm{d}y}{\mathrm{d}x}} = x + 2x \ln x$

When $x = \mathrm{e}$, $\dfrac{\mathrm{d}y}{\mathrm{d}x} = \mathrm{e} + 2\mathrm{e} \ln \mathrm{e}$

$\phantom{When x = e, \dfrac{\mathrm{d}y}{\mathrm{d}x}} = 3\mathrm{e}$

Using

$\dfrac{\delta y}{\delta x} \approx \dfrac{\mathrm{d}y}{\mathrm{d}x}$

$\dfrac{\delta y}{p} \approx 3\mathrm{e}$

$\delta y \approx 3\mathrm{e}p$

WORKED EXAMPLE 11

Variables x and y are connected by the equation $y = \dfrac{\ln x}{2x + 5}$.

Given that y increases at a rate of 0.1 units per second, find the rate of change of x when $x = 2$.

Answers

$y = \dfrac{\ln x}{2x + 5}$ and $\dfrac{\mathrm{d}y}{\mathrm{d}t} = 0.1$

$\dfrac{\mathrm{d}y}{\mathrm{d}x} = \dfrac{(2x + 5)\dfrac{1}{x} - 2\ln x}{(2x + 5)^2}$ quotient rule

$\phantom{\dfrac{\mathrm{d}y}{\mathrm{d}x}} = \dfrac{(2x + 5) - 2x \ln x}{x(2x + 5)^2}$

When $x = 2$, $\dfrac{\mathrm{d}y}{\mathrm{d}x} = \dfrac{(4 + 5) - 4\ln 2}{2(4 + 5)^2}$

$\phantom{When x = 2, \dfrac{\mathrm{d}y}{\mathrm{d}x}} = \dfrac{9 - 4\ln 2}{162}$

Using the chain rule, $\dfrac{\mathrm{d}x}{\mathrm{d}t} = \dfrac{\mathrm{d}x}{\mathrm{d}y} \times \dfrac{\mathrm{d}y}{\mathrm{d}t}$

$ = \dfrac{162}{9 - 4\ln 2} \times 0.1$

$ = 2.6014...$

Rate of change of x is 2.60 units per second correct to 3 sf.

WORKED EXAMPLE 12

A curve has equation $y = e^{-x}(2\sin 2x - 3\cos 2x)$ for $0 < x < \dfrac{\pi}{2}$ radians.

Find the x-coordinate of the stationary point on the curve and determine the nature of this point.

Answers

$$y = e^{-x}(2\sin 2x - 3\cos 2x)$$

$$\frac{dy}{dx} = e^{-x}(4\cos 2x + 6\sin 2x) - e^{-x}(2\sin 2x - 3\cos 2x) \qquad \text{product rule}$$

$$= e^{-x}(7\cos 2x + 4\sin 2x)$$

Stationary points occur when $\dfrac{dy}{dx} = 0$.

$$e^{-x}(7\cos 2x + 4\sin 2x) = 0$$

$$7\cos 2x + 4\sin 2x = 0 \quad \text{or} \quad e^{-x} = 0$$

$$\tan 2x = -\frac{7}{4} \qquad \text{no solution}$$

$$2x = 2.0899 \qquad \text{There are other values of } x \text{ for}$$

$$x = 1.045 \qquad \text{which } \tan 2x = -\frac{7}{4} \text{ but they are}$$

$$\text{outside the range } 0 < x < \frac{\pi}{2}.$$

$$\frac{d^2y}{dx^2} = e^{-x}(-14\sin 2x + 8\cos 2x) - e^{-x}(7\cos 2x + 4\sin 2x)$$

$$= e^{-x}(\cos 2x - 18\sin 2x)$$

When $x = 1.045$, $\dfrac{d^2y}{dx^2} < 0$.

Hence the stationary point is a maximum point.

WORKED EXAMPLE 13

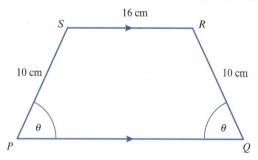

The diagram shows an isosceles trapezium $PQRS$ with area $A\,\text{cm}^2$.

Angle $SPQ = $ angle $PQR = \theta$ radians.

$PS = QR = 10\,\text{cm}$ and $SR = 16\,\text{cm}$.

a Show that $A = 160\sin\theta + 100\sin\theta\cos\theta$.

b Find the value of θ for which A has a stationary value.

c Determine the nature of this stationary value.

CONTINUED

Answers

a $A = \frac{1}{2}(a+b)h$ use $h = 10\sin\theta$

$\quad\quad = \frac{1}{2}(PQ + SR) \times 10\sin\theta$ use $PQ = 10\cos\theta + 16 + 10\cos\theta = 16 + 20\cos\theta$

$\quad\quad = \frac{1}{2}[(16 + 20\cos\theta) + 16] \times 10\sin\theta$

$\quad\quad = 5\sin\theta(32 + 20\cos\theta)$

$\quad\quad = 160\sin\theta + 100\sin\theta\cos\theta$

b $\frac{dA}{dx} = 160\cos\theta + [100\sin\theta(-\sin\theta) + 100\cos^2\theta]$ use the product rule on $100\sin\theta\cos\theta$

$\quad\quad = 160\cos\theta - 100\sin^2\theta + 100\cos^2\theta$ use $\sin^2\theta = 1 - \cos^2\theta$

$\quad\quad = 160\cos\theta - 100(1 - \cos^2\theta) + 100\cos^2\theta$

$\quad\quad = 200\cos^2\theta + 160\cos\theta - 100$

Stationary values occur when $\frac{dA}{dx} = 0$.

$\quad\quad 200\cos^2\theta + 160\cos\theta - 100 = 0$

$\quad\quad\quad 10\cos^2\theta + 8\cos\theta - 5 = 0$ use the quadratic formula

$\cos\theta = 0.412$ or $\cos\theta = -1.212$

$\quad\quad = 1.146$ radians   no solution

c $\frac{d^2A}{dx^2} = -400\cos\theta\sin\theta - 160\sin\theta$

When $\theta = 1.146$, $\frac{d^2A}{dx^2} < 0$.

Hence the stationary value is a maximum value.

Exercise 14.4

1   A curve has equation $y = 3\sin\left(2x + \frac{\pi}{2}\right)$.

Find the equation of the normal to the curve at the point on the curve where $x = \frac{\pi}{4}$.

2   A curve has equation $y = x\sin 2x$ for $0 \leqslant x \leqslant \pi$ radians.

a   Find the equation of the normal to the curve at the point $P\left(\frac{\pi}{4}, \frac{\pi}{4}\right)$.

b   The normal at P intersects the x-axis at Q and the y-axis at R.
Find the coordinates of Q and R.

c   Find the area of triangle OQR where O is the origin.

3   A curve has equation $y = e^{\frac{1}{2}x} + 1$.
The curve crosses the y-axis at P.
The normal to the curve at P meets the x-axis at Q.
Find the coordinates of Q.

4 A curve has equation $y = 5 - e^{2x}$.

The curve crosses the x-axis at A and the y-axis at B.

   a   Find the coordinates of A and B.

   b   The normal to the curve at B meets the x-axis at the point C.

      Find the coordinates of C.

5 A curve has equation $y = xe^x$.

The tangent to the curve at the point $P(1, e)$ meets the y-axis at the point A.

The normal to the curve at P meets the x-axis at the point B.

Find the area of triangle OAB, where O is the origin.

6 Variables x and y are connected by the equation $y = \sin 2x$.

Find the approximate increase in y as x increases from $\dfrac{\pi}{8}$ to $\dfrac{\pi}{8} + p$, where p is small.

7 Variables x and y are connected by the equation $y = 3 + \ln(2x - 5)$

Find the approximate change in y as x increases from 4 to $4 + p$, where p is small.

8 Variables x and y are connected by the equation $y = \dfrac{\ln x}{x^2 + 3}$.

Find the approximate change in y as x increases from 1 to $1 + p$, where p is small.

9 Variables x and y are connected by the equation $y = 3 + 2x - 5e^{-x}$.

Find the approximate change in y as x increases from $\ln 2$ to $\ln 2 + p$, where p is small.

10 A curve has equation $y = \dfrac{\ln(x^2 - 2)}{x^2 - 2}$.

Find the approximate change in y as x increases from $\sqrt{3}$ to $\sqrt{3} + p$, where p is small.

11 Find the coordinates of the stationary points on these curves and determine their nature.

   a $y = xe^{\frac{x}{2}}$ b $y = x^2 e^{2x}$ c $y = e^x - 7x + 2$

   d $y = 5e^{2x} - 10x - 1$ e $y = (x^2 - 8)e^{-x}$ f $y = x^2 \ln x$

   g $y = \dfrac{\ln x}{x^2}$ h $y = \dfrac{\ln(x^2 + 1)}{x^2 + 1}$

12 Find the coordinates of the stationary points on these curves and determine their nature.

   a $y = 4\sin x + 3\cos x$ for $0 \le x \le \dfrac{\pi}{2}$ b $y = 6\cos\dfrac{x}{2} + 8\sin\dfrac{x}{2}$ for $0 \le x \le 2\pi$

   c $y = 5\sin\left(2x + \dfrac{\pi}{2}\right)$ for $-\dfrac{\pi}{6} \le x \le \dfrac{5x}{6}$ d $y = \dfrac{e^x}{\sin x}$ for $0 < x < \pi$

   e $y = 2\sin x \cos x + 2\cos x$ for $0 \le x \le \pi$

13 A curve has equation $y = Ae^{2x} + Be^{-2x}$.

The gradient of the tangent at the point $(0, 10)$ is -12.

   a   Find the value of A and the value of B.

   b   Find the coordinates of the turning point on the curve and determine its nature.

14 A curve has equation $y = x\ln x$.

The curve crosses the x-axis at the point A and has a minimum point at B.

Find the coordinates of A and the coordinates of B.

15 A curve has equation $y = x^2 e^x$.

The curve has a minimum point at P and a maximum point at Q.

  a Find the coordinates of P and the coordinates of Q.

  b The tangent to the curve at the point $A(1, e)$ meets the y-axis at the point B.

  The normal to the curve at the point $A(1, e)$ meets the y-axis at the point C.

  Find the coordinates of B and the coordinates of C.

  c Find the area of triangle ABC.

16

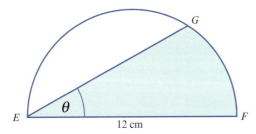

The diagram shows a semi-circle with diameter EF of length $12\,$cm.

Angle $GEF = \theta$ radians and the shaded region has an area of $A\,$cm^2.

  a Show that $A = 36\theta + 18\sin 2\theta$.

  b Given that θ is increasing at a rate of 0.05 radians per second, find the rate of

  change of A when $\theta = \dfrac{\pi}{6}$ radians.

17 CHALLENGE QUESTION

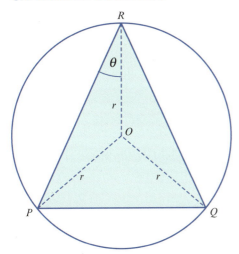

The diagram shows an isosceles triangle PQR inscribed in a circle, centre O, radius $r\,$cm.

$PR = QR$ and angle $ORP = \theta$ radians.

Triangle PQR has an area of $A\,$cm^2.

  a Show that $A = r^2 \sin 2\theta + r^2 \sin 2\theta \cos 2\theta$.

  b Find the value of θ for which A has a stationary value and determine the
nature of this stationary value.

SUMMARY

Exponential functions		
$\dfrac{d}{dx}(e^x) = e^x$	$\dfrac{d}{dx}\left[e^{ax+b}\right] = a e^{ax+b}$	$\dfrac{d}{dx}\left[e^{f(x)}\right] = f'(x) \times e^{f(x)}$
Logarithmic functions		
$\dfrac{d}{dx}(\ln x) = \dfrac{1}{x}$	$\dfrac{d}{dx}\left[\ln(ax+b)\right] = \dfrac{a}{ax+b}$	$\dfrac{d}{dx}\left[\ln(f(x))\right] = \dfrac{f'(x)}{f(x)}$
Trigonometric functions		
$\dfrac{d}{dx}(\sin x) = \cos x$	$\dfrac{d}{dx}\left[\sin(ax+b)\right] = a\cos(ax+b)$	
$\dfrac{d}{dx}(\cos x) = -\sin x$	$\dfrac{d}{dx}\left[\cos(ax+b)\right] = -a\sin(ax+b)$	
$\dfrac{d}{dx}(\tan x) = \sec^2 x$	$\dfrac{d}{dx}\left[\tan(ax+b)\right] = a\sec^2(ax+b)$	

Past paper questions

Worked example

a Given that $y = \dfrac{e^{3x}}{4x^2 + 1}$, find $\dfrac{dy}{dx}$. [3]

b Variables x, y and t are such that $y = 4\cos\left(x + \dfrac{\pi}{3}\right) + 2\sqrt{3}\sin\left(x + \dfrac{\pi}{3}\right)$ and $\dfrac{dy}{dt} = 10$.

    i Find the value of $\dfrac{dy}{dx}$ when $x = \dfrac{\pi}{2}$. [3]

    ii Find the value of $\dfrac{dx}{dt}$ when $x = \dfrac{\pi}{2}$. [2]

Cambridge IGCSE Additional Mathematics 0606 Paper 11 Q10 Jun 2017

Answers

a $y - \dfrac{e^{3x}}{4x^2 + 1}$ use the quotient rule

$$\frac{dy}{dx} = \frac{(4x^2 + 1)3e^{3x} - e^{3x}8x}{(4x^2 + 1)^2}$$

$$\frac{dy}{dx} = \frac{3e^{3x}(4x^2 + 1) - 8x e^{3x}}{(4x^2 + 1)^2}$$

b i $y = 4\cos\left(x + \dfrac{\pi}{3}\right) + 2\sqrt{3}\sin\left(x + \dfrac{\pi}{3}\right)$

$\dfrac{dy}{dx} = -4\sin\left(x + \dfrac{\pi}{3}\right) + 2\sqrt{3}\cos\left(x + \dfrac{\pi}{3}\right)$

When $x = \dfrac{\pi}{2}, \dfrac{dy}{dx} = -4\sin\left(\dfrac{5\pi}{6}\right) + 2\sqrt{3}\cos\left(\dfrac{5\pi}{6}\right)$

$= -4 \times \dfrac{1}{2} - 2\sqrt{3} \times \dfrac{\sqrt{3}}{2}$

$= -2 - 3$

$= -5$

ii Using the chain rule: $\dfrac{dy}{dx} = \dfrac{dy}{dt} \times \dfrac{dt}{dx}$

$-5 = 10 \times \dfrac{dt}{dx}$

$-\dfrac{1}{2} = \dfrac{dt}{dx}$

$\dfrac{dx}{dt} = -2$

1 a Find the equation of the tangent to the curve $y = x^3 - \ln x$ at the point on the curve where $x = 1$. [4]

b Show that this tangent bisects the line joining the points $(-2, 16)$ and $(12, 2)$. [2]

Cambridge IGCSE Additional Mathematics 0606 Paper 11 Q5i,ii Nov 2014

2 Find $\dfrac{dy}{dx}$ when

a $y = \cos 2x \sin\left(\dfrac{x}{3}\right)$, [4]

b $y = \dfrac{\tan x}{1 + \ln x}$. [4]

Cambridge IGCSE Additional Mathematics 0606 Paper 21 Q10 Jun 2014

3 i Given that $y = \dfrac{\tan 2x}{x}$, find $\dfrac{dy}{dx}$. [3]

ii Hence find the equation of the normal to the curve $y = \dfrac{\tan 2x}{x}$ at the point where $x = \dfrac{\pi}{8}$. [3]

Cambridge IGCSE Additional Mathematics 0606 Paper 12 Q6 Mar 2015

4 The point A, where $x = 0$, lies on the curve $y = \dfrac{\ln(4x^2 + 3)}{x - 1}$. The normal to the curve at A meets the x-axis at the point B.

i Find the equation of this normal. [7]

ii Find the area of the triangle AOB, where O is the origin. [2]

Cambridge IGCSE Additional Mathematics 0606 Paper 11 Q7 Jun 2015

5 Variables x and y are such that $y = (x - 3)\ln(2x^2 + 1)$.

i Find the value of $\dfrac{dy}{dx}$ when $x = 2$. [4]

ii Hence find the approximate change in y when x changes from 2 to 2.03. [2]

Cambridge IGCSE Additional Mathematics 0606 Paper 11 Q5 Nov 2015

6

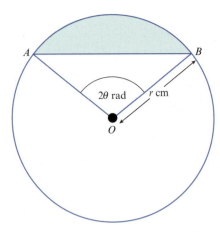

The diagram shows a circle, centre O, radius r cm. The points A and B lie on the circle such that angle $AOB = 2\theta$ radians.

i Find, in terms of r and θ, an expression for the length of the chord AB. [1]

ii Given that the perimeter of the shaded region is 20 cm, show that $r = \dfrac{10}{\theta + \sin \theta}$. [2]

iii Given that r and θ can vary, find the value of $\dfrac{\mathrm{d}r}{\mathrm{d}\theta}$ when $\theta = \dfrac{\pi}{6}$. [4]

iv Given that r is increasing at the rate of $15\,\mathrm{cm\,s^{-1}}$, find the corresponding rate of change of θ when $\theta = \dfrac{\pi}{6}$. [3]

Cambridge IGCSE Additional Mathematics 0606 Paper 11 Q11 Jun 2016

7 Variables x and y are such that $y = \dfrac{\mathrm{e}^{3x} \sin x}{x^2}$.

Use differentiation to find the approximate change in y as x increases from 0.5 to 0.5 + h, where h is small. [6]

Cambridge IGCSE Additional Mathematics 0606 Paper 22 Q9 Mar 2020

8 The tangent to the curve $y = \ln(3x^2 - 4) - \dfrac{x^3}{6}$, at the point where $x = 2$, meets the y-axis at the point P.
Find the exact coordinates of P. [6]

Cambridge IGCSE Additional Mathematics 0606 Paper 12 Q4 Mar 2020

9 Variables x and y are related by the equation $y = \dfrac{\ln x}{\mathrm{e}^x}$.

i Show that $\dfrac{\mathrm{d}y}{\mathrm{d}x} = \dfrac{1 - x\ln x}{x\mathrm{e}^x}$. [4]

ii Hence find the approximate change in y as x increases from 2 to 2 + h, where h is small. [2]

Cambridge IGCSE Additional Mathematics 0606 Paper 22 Q2 Mar 2019

10 A curve has equation $y = 4 + 5\sin 3x$.

i Find $\dfrac{\mathrm{d}y}{\mathrm{d}x}$. [2]

ii Hence find the equation of the tangent to the curve $y = 4 + 5\sin 3x$ at the point where $x = \dfrac{\pi}{3}$. [3]

Cambridge IGCSE Additional Mathematics 0606 Paper 12 Q2 Mar 2018

11 Differentiate with respect to x.

   i $4x \tan x$,   [2]

   ii $\dfrac{e^{3x+1}}{x^2 - 1}$.   [3]

Cambridge IGCSE Additional Mathematics 0606 Paper 21 Q7 Jun 2018

12 Two variables x and y are such that $y = \dfrac{\ln x}{x^3}$ for $x > 0$.

   i   Show that $\dfrac{dy}{dx} = \dfrac{1 - 3\ln x}{x^4}$.   [3]

   ii Hence find the approximate change in y as x increases from e to e $+ h$, where h is small.   [2]

Cambridge IGCSE Additional Mathematics 0606 Paper 21 Q2 Jun 2019

13 **a**   Find the x-coordinates of the stationary points of the curve $y = e^{3x}(2x + 3)^6$.   [6]

    b   A curve has equation $y = f(x)$ and has exactly two stationary points. Given that

        $f''(x) = 4x - 7$, $f'(0.5) = 0$, and $f'(3) = 0$, use the second derivative test to determine the nature of each of the stationary points of this curve.   [2]

Cambridge IGCSE Additional Mathematics 0606 Paper 21 Q12a & b Jun 2020

14 Find the equation of the tangent to the curve $y = \dfrac{\ln(3x^2 - 1)}{x + 2}$ at the point where $x = 1$.

   Give your answer in the form $y = mx + c$, where m and c are constants correct to 3 decimal places.   [6]

Cambridge IGCSE Additional Mathematics 0606 Paper 11 Q5 Jun 2020

15 Variables x and y are such that $y = e^{\frac{x}{2}} + x \cos 2x$, where x is in radians.

   Use differentiation to find the approximate change in y as x increases from 1 to $1 + h$, where h is small.   [6]

Cambridge IGCSE Additional Mathematics 0606 Paper 21 Q6 Jun 2021

> Chapter 15

Calculus – Integration

THIS SECTION WILL SHOW YOU HOW TO:

- use integration as the reverse process of differentiation
- integrate sums of terms in powers of x, including $\dfrac{1}{x}$ and $\dfrac{1}{ax+b}$
- integrate functions of the form $(ax+b)^n$, e^{ax+b}, $\sin(ax+b)$, $\cos(ax+b)$, $\sec^2(ax+b)$
- evaluate definite integrals and apply integration to the evaluation of plane areas.

Before you start…

Where it comes from	What you should be able to do	Check your skills
Cambridge IGCSE/O Level Mathematics	Substitute values for x and y into equations of the form $y = f(x) + c$ and solve to find c.	**1 a** Given that the line $y = 3x + c$ passes through the point $(2, -1)$, find the value of c. **b** Given that the curve $y = x^2 + 3x + c$ passes through the point $(-1, -3)$, find the value of c.
Cambridge IGCSE/O Level Mathematics	Find the x-coordinates of the point where a curve crosses the x-axis.	**2** Find the x-coordinates of the points where the curve crosses the x-axis. **a** $y = 2x^2 + x - 15$ **b** $y = 5x\sqrt{x} - 2x$
Chapter 12	Differentiate constant multiples, sums and differences of expressions containing terms of the form ax^n.	**3** Find $\dfrac{dy}{dx}$. **a** $y = 5x^4 - 5x + 3$ **b** $y = 6x\sqrt{x} + 2x^5 - 1$

15.1 Differentiation reversed

In Chapter 12, you learned about the process of obtaining $\dfrac{dy}{dx}$ when you know y.

This process was called differentiation.

You learned the rule for differentiating power functions: if $y = x^n$, then $\dfrac{dy}{dx} = nx^{n-1}$.

Applying this rule to functions of the form $y = x^2 + c$ you obtain:

$$\left.\begin{array}{l} y = x^2 + 5.3 \\ y = x^2 + 2 \\ y = x^2 \\ y = x^2 - 3 \end{array}\right\} \quad \frac{dy}{dx} = 2x$$

In this chapter you will learn about the reverse process of obtaining y when you know $\dfrac{dy}{dx}$. This reverse process is called **integration**.

KEY WORDS

integration

indefinite integral

definite integral

Note:

There are an infinite number of functions that when differentiated give $2x$.

CLASS DISCUSSION

Find $\dfrac{dy}{dx}$ for each of the following functions.

$$y = \frac{1}{4}x^4 + 3 \qquad y = \frac{1}{7}x^7 - 0.8 \qquad y = \frac{1}{3}x^3 + 4 \qquad y = \frac{1}{-2}x^{-2} + 1 \qquad y = \frac{1}{\left(\frac{1}{2}\right)}x^{\frac{1}{2}} - 7$$

Discuss your results with your classmates and try to find a rule for obtaining y if $\dfrac{dy}{dx} = x^n$. Describe your rule in words.

From the class discussion you should have concluded that:

If $\dfrac{dy}{dx} = x^n$ then $y = \dfrac{1}{n+1}x^{n+1} + c$, where c is an arbitrary constant and $n \neq -1$.

It is easier to remember this rule as:

'increase the power n by 1 to get the new power, then divide by the new power'.

WORKED EXAMPLE 1

Find y in terms of x for each of the following.

a $\dfrac{dy}{dx} = x^4$ **b** $\dfrac{dy}{dx} = \sqrt{x}$ **c** $\dfrac{dy}{dx} = \dfrac{1}{x^3}$

Answers

a $\dfrac{dy}{dx} = x^4$

$y = \dfrac{1}{4+1}x^{4+1} + c$

$= \dfrac{1}{5}x^5 + c$

b $\dfrac{dy}{dx} = x^{\frac{1}{2}}$

$y = \dfrac{1}{\frac{1}{2}+1}x^{\frac{1}{2}+1} + c$

$= \dfrac{2}{3}x^{\frac{3}{2}} + c$

$= \dfrac{2}{3}\sqrt{x^3} + c$

c $\dfrac{dy}{dx} = x^{-3}$

$y = \dfrac{1}{-3+1}x^{-3+1} + c$

$= \dfrac{1}{-2}x^{-2} + c$

$= \dfrac{1}{2x^2} + c$

WORKED EXAMPLE 2

Find y in terms of x for each of the following.

a $\dfrac{dy}{dx} = 6x^2 - \dfrac{5}{x^2} + 4x$ **b** $\dfrac{dy}{dx} = 8x^3 - \dfrac{4}{3x^5} - 2$ **c** $\dfrac{dy}{dx} = \dfrac{(x-2)(x+5)}{\sqrt{x}}$

Answers

a $\dfrac{dy}{dx} = 6x^2 - 5x^{-2} + 4x^1$ write in index form ready for integration

$y = \dfrac{6}{3}x^3 - \dfrac{5}{(-1)}x^{-1} + \dfrac{4}{2}x^2 + c$

$= 2x^3 + 5x^{-1} + 2x^2 + c$

$= 2x^3 + \dfrac{5}{x} + 2x^2 + c$

CONTINUED

b $\dfrac{dy}{dx} = 8x^3 - \dfrac{4}{3}x^{-5} - 2x^0$ write in index form ready for integration

$\quad\quad y = \dfrac{8}{4}x^4 - \dfrac{4}{3(-4)}x^{-4} - \dfrac{2}{1}x^1 + c$

$\quad\quad\quad = 2x^4 + \dfrac{1}{3}x^{-4} - 2x + c$

$\quad\quad\quad = 2x^4 + \dfrac{1}{3x^4} - 2x + c$

c $\dfrac{dy}{dx} = \dfrac{x^2 + 3x - 10}{\sqrt{x}}$ write in index form ready for integration

$\quad\quad\quad = x^{\frac{3}{2}} + 3x^{\frac{1}{2}} - 10x^{-\frac{1}{2}}$

$\quad\quad y = \dfrac{1}{\left(\frac{5}{2}\right)}x^{\frac{5}{2}} + \dfrac{3}{\left(\frac{3}{2}\right)}x^{\frac{3}{2}} - \dfrac{10}{\left(\frac{1}{2}\right)}x^{\frac{1}{2}} + c$

$\quad\quad\quad = \dfrac{2}{5}x^{\frac{5}{2}} + 2x^{\frac{3}{2}} - 20x^{\frac{1}{2}} + c$

WORKED EXAMPLE 3

A curve is such that $\dfrac{dy}{dx} = (1 - x)(3x - 2)$ and $(2, 8)$ is a point on the curve.

Find the equation of the curve.

Answers

$\dfrac{dy}{dx} = (1 - x)(3x - 2)$ expand brackets

$\quad\quad = -3x^2 + 5x - 2$

$\quad\quad = -3x^2 + 5x^1 - 2x^0$ write in index form ready for integration

$y = -x^3 + \dfrac{5}{2}x^2 - 2x + c$

When $x = 2$, $y = 8$

$\quad\quad 8 = -(2)^3 + \dfrac{5}{2}(2)^2 - 2(2) + c$

$\quad\quad 8 = -8 + 10 - 4 + c$

$\quad\quad c = 10$

The equation of the curve is $y = -x^3 + \dfrac{5}{2}x^2 - 2x + 10$.

Exercise 15.1

1 Find y in terms of x for each of the following.

   a $\dfrac{dy}{dx} = 12x^4$ **b** $\dfrac{dy}{dx} = 5x^8$ **c** $\dfrac{dy}{dx} = 7x^3$

   d $\dfrac{dy}{dx} = \dfrac{4}{x^3}$ **e** $\dfrac{dy}{dx} = \dfrac{1}{2x^2}$ **f** $\dfrac{dy}{dx} = \dfrac{3}{\sqrt{x}}$

2 Find y in terms of x for each of the following.

   a $\dfrac{dy}{dx} = 7x^6 + 2x^4 + 3$ **b** $\dfrac{dy}{dx} = 2x^5 - 3x^3 + 5x$

   c $\dfrac{dy}{dx} = \dfrac{3}{x^4} - \dfrac{15}{x^2} + x$ **d** $\dfrac{dy}{dx} = \dfrac{18}{x^{10}} + \dfrac{6}{x^7} - 2$

3 Find y in terms of x for each of the following.

   a $\dfrac{dy}{dx} = 3x(x - 2)$ **b** $\dfrac{dy}{dx} = x^2(4x^2 - 3)$

   c $\dfrac{dy}{dx} = (x + 2\sqrt{x})^2$ **d** $\dfrac{dy}{dx} = x(x - 3)(x + 4)$

   e $\dfrac{dy}{dx} = \dfrac{x^5 - 3x}{2x^3}$ **f** $\dfrac{dy}{dx} = \dfrac{(2x - 3)(x - 1)}{x^4}$

   g $\dfrac{dy}{dx} = \dfrac{x^5 - 4x^2 + 1}{2x^2}$ **h** $\dfrac{dy}{dx} = \dfrac{(3x + 5)(x - 2)}{\sqrt{x}}$

4 A curve is such that $\dfrac{dy}{dx} = 3x^2 - 4x + 1$.

Given that the curve passes through the point $(0, 5)$ find the equation of the curve.

5 A curve is such that $\dfrac{dy}{dx} = 6x(x - 1)$.

Given that the curve passes through the point $(1, -5)$ find the equation of the curve.

6 A curve is such that $\dfrac{dy}{dx} = \dfrac{2x^3 + 6}{x^2}$.

Given that the curve passes through the point $(-1, 10)$, find the equation of the curve.

7 A curve is such that $\dfrac{dy}{dx} = \dfrac{(2 - \sqrt{x})^2}{\sqrt{x}}$.

Given that the curve passes through the point $(9, 14)$, find the equation of the curve.

8   A curve is such that $\dfrac{dy}{dx} = kx^2 - 2x$ where k is a constant.

Given that the curve passes through the points $(1, 6)$ and $(-2, -15)$ find the equation of the curve.

9   A curve is such that $\dfrac{d^2y}{dx^2} = 12x - 12$.

The gradient of the curve at the point $(2, 9)$ is 8.
a   Express y in terms of x.
b   Show that the gradient of the curve is never less than 2.

10   A curve is such that $\dfrac{dy}{dx} = kx - 5$ where k is a constant.

The gradient of the normal to the curve at the point $(2, -1)$ is $-\dfrac{1}{3}$.

Find the equation of the curve.

15.2 Indefinite integrals

The symbol $\int$ is used to denote integration.

When you need to integrate x^2, for example, you write

$$\int x^2 \, dx = \frac{1}{3}x^3 + c.$$

$\int x^2 \, dx$ is called the **indefinite integral** of x^2 with respect to x.

It is called 'indefinite' because it has infinitely many solutions.

Using this notation, the rule for integrating powers of x can be written as:

$$\int x^n \, dx = \frac{1}{n+1}x^{n+1} + c, \text{ where } c \text{ is a constant and } n \neq -1$$

This section provides you with practice at using this new notation together with the following rules:

$$\int k\,\mathrm{f}(x)\,dx = k\int \mathrm{f}(x)\,dx, \text{ where } k \text{ is a constant}$$

$$\int [\mathrm{f}(x) \pm \mathrm{g}(x)]\,dx = \int \mathrm{f}(x)\,dx \pm \int \mathrm{g}(x)\,dx$$

WORKED EXAMPLE 4

Find:　**a** $\displaystyle\int x^2(10x^2 - 8x + 3)\,\mathrm{d}x$　　**b** $\displaystyle\int \frac{x^4 - 2}{x\sqrt{x}}\,\mathrm{d}x$

Answers

a $\displaystyle\int x^2(10x^2 - 8x + 3)\,\mathrm{d}x = \int (10x^4 - 8x^3 + 3x^2)\,\mathrm{d}x$

$$= \frac{10x^5}{5} - \frac{8x^4}{4} + \frac{3x^3}{3} + c$$

$$= 2x^5 - 2x^4 + x^3 + c$$

b $\displaystyle\int \frac{x^4 - 2}{x\sqrt{x}}\,\mathrm{d}x = \int \frac{x^4 - 2}{x^{\frac{3}{2}}}\,\mathrm{d}x$

$$= \int \left(x^{\frac{5}{2}} - 2x^{-\frac{3}{2}} \right)\,\mathrm{d}x$$

$$= \frac{1}{\left(\frac{7}{2}\right)}x^{\frac{7}{2}} - \frac{2}{\left(-\frac{1}{2}\right)}x^{-\frac{1}{2}} + c$$

$$= \frac{2}{7}x^{\frac{7}{2}} + 4x^{-\frac{1}{2}} + c$$

$$= \frac{2}{7}x^3\sqrt{x} + \frac{4}{\sqrt{x}} + c$$

Exercise 15.2

1 Find each of the following.

　a $\displaystyle\int 4x^7\,\mathrm{d}x$　　　　**b** $\displaystyle\int 12x^5\,\mathrm{d}x$　　　　**c** $\displaystyle\int 2x^{-3}\,\mathrm{d}x$

　d $\displaystyle\int \frac{4}{x^2}\,\mathrm{d}x$　　　　**e** $\displaystyle\int \frac{3}{\sqrt{x}}\,\mathrm{d}x$　　　　**f** $\displaystyle\int \frac{6}{x^2\sqrt{x}}\,\mathrm{d}x$

2 Find each of the following.

　a $\displaystyle\int (x+2)(x+5)\,\mathrm{d}x$　　**b** $\displaystyle\int (x-1)(2x+3)\,\mathrm{d}x$　　**c** $\displaystyle\int (x-5)^2\,\mathrm{d}x$

　d $\displaystyle\int (\sqrt{x}+3)^2\,\mathrm{d}x$　　**e** $\displaystyle\int x(x-1)^2\,\mathrm{d}x$　　**f** $\displaystyle\int \sqrt[3]{x}(x-4)\,\mathrm{d}x$

3 Find each of the following.

　a $\displaystyle\int \frac{x^2-5}{x^2}\,\mathrm{d}x$　　**b** $\displaystyle\int \frac{x^4-8}{2x^3}\,\mathrm{d}x$　　**c** $\displaystyle\int \frac{(x+1)^2}{3x^4}\,\mathrm{d}x$

　d $\displaystyle\int \frac{4x^2-3\sqrt{x}}{x}\,\mathrm{d}x$　　**e** $\displaystyle\int \frac{x^5+5}{x^3\sqrt{x}}\,\mathrm{d}x$　　**f** $\displaystyle\int \left(\sqrt{x}+\frac{3}{x^2\sqrt{x}}\right)^2\,\mathrm{d}x$

15.3 Integration of functions of the form $(ax + b)^n$

In Chapter 12 you learned that: $\dfrac{d}{dx}\left[\dfrac{1}{2 \times 8}(2x + 5)^8\right] = (2x + 5)^7$

Hence $\displaystyle\int (2x + 5)^7\,dx = \dfrac{1}{2 \times 8}(2x + 5)^8 + c.$

This leads to the general rule:

$$\int (ax + b)^n\,dx = \dfrac{1}{a(n + 1)}(ax + b)^{n+1} + c,\ n \neq -1 \text{ and } a \neq 0$$

WORKED EXAMPLE 5

Find

a $\displaystyle\int (3x - 8)^5\,dx$ **b** $\displaystyle\int \dfrac{8}{(4x + 1)^3}\,dx$ **c** $\displaystyle\int \dfrac{12}{\sqrt{2x - 7}}\,dx$

Answers

a $\displaystyle\int (3x - 8)^5\,dx = \dfrac{1}{3(5 + 1)}(3x - 8)^{5+1} + c$

$\qquad = \dfrac{1}{18}(3x - 8)^6 + c$

b $\displaystyle\int \dfrac{8}{(4x + 1)^3}\,dx = 8\int (4x + 1)^{-3}\,dx$

$\qquad = \dfrac{8}{4(-3 + 1)}(4x + 1)^{-3+1} + c$

$\qquad = -(4x + 1)^{-2} + c$

$\qquad = -\dfrac{1}{(4x + 1)^2} + c$

c $\displaystyle\int \dfrac{12}{\sqrt{2x - 7}}\,dx = 12\int (2x - 7)^{-\frac{1}{2}}\,dx$

$\qquad = \dfrac{12}{2\left(-\dfrac{1}{2} + 1\right)}(2x - 7)^{-\frac{1}{2}+1} + c$

$\qquad = 12\sqrt{2x - 7} + c$

Exercise 15.3

1   Find:

a $\displaystyle\int (x + 2)^9\,dx$ **b** $\displaystyle\int (2x - 5)^6\,dx$ **c** $\displaystyle\int 2(3x + 2)^9\,dx$ **d** $\displaystyle\int 3(2 - 3x)^4\,dx$ **e** $\displaystyle\int (7x + 2)^{\frac{1}{3}}\,dx$

f $\displaystyle\int \sqrt{(3x - 1)^3}\,dx$ **g** $\displaystyle\int \dfrac{6}{\sqrt{x + 1}}\,dx$ **h** $\displaystyle\int \left(\dfrac{2}{5x + 3}\right)^3\,dx$ **i** $\displaystyle\int \dfrac{3}{2(3 - 2x)^4}\,dx$

2 A curve is such that $\dfrac{dy}{dx} = (4x+1)^4$.

Given that the curve passes through the point $(0, -1.95)$, find the equation of the curve.

3 A curve is such that $\dfrac{dy}{dx} = \sqrt{2x+1}$.

Given that the curve passes through the point $(4, 11)$, find the equation of the curve.

4 A curve is such that $\dfrac{dy}{dx} = \dfrac{1}{\sqrt{10-x}}$.

Given that the curve passes through the point $(6, 1)$, find the equation of the curve.

5 A curve is such that $\dfrac{dy}{dx} = k(2x-3)^3$, where k is a constant.

The gradient of the normal to the curve at the point $(2, 2)$ is $-\dfrac{1}{8}$.

Find the equation of the curve.

6 A curve is such that $\dfrac{dy}{dx} = 2(kx-1)^5$, where k is a constant.

Given that the curve passes through the points $(0, 1)$ and $(1, 8)$, find the equation of the curve.

15.4 Integration of exponential functions

In Chapter 15, you learned the following rules for differentiating exponential functions:

$$\dfrac{d}{dx}(e^x) = e^x \qquad \dfrac{d}{dx}(e^{ax+b}) = ae^{ax+b}$$

Since integration is the reverse process of differentiation, the rules for integrating exponential functions are:

$$\int e^x\,dx = e^x + c \qquad \int e^{ax+b}\,dx = \dfrac{1}{a}e^{ax+b} + c$$

WORKED EXAMPLE 6

Find

a $\displaystyle\int e^{2x}\,dx$ **b** $\displaystyle\int e^{-7x}\,dx$ **c** $\displaystyle\int e^{6x-5}\,dx$.

Answers

a $\displaystyle\int e^{2x}\,dx = \dfrac{1}{2}e^{2x} + c$

b $\displaystyle\int e^{-7x}\,dx = \dfrac{1}{-7}e^{-7x} + c$

$\qquad = -\dfrac{1}{7}e^{-7x} + c$

c $\displaystyle\int e^{6x-5}\,dx = \dfrac{1}{6}e^{6x-5} + c$

Exercise 15.4

1 Find

a $\displaystyle\int e^{5x}\,dx$ b $\displaystyle\int e^{9x}\,dx$ c $\displaystyle\int e^{\frac{1}{2}x}\,dx$

d $\displaystyle\int e^{-2x}\,dx$ e $\displaystyle\int 4e^{x}\,dx$ f $\displaystyle\int 2e^{4x}\,dx$

g $\displaystyle\int e^{7x+4}\,dx$ h $\displaystyle\int e^{5-2x}\,dx$ i $\displaystyle\int \frac{1}{3}e^{6x-1}\,dx$

2 Find

a $\displaystyle\int e^{x}(5-e^{2x})\,dx$ b $\displaystyle\int (e^{2x}+1)^{2}\,dx$ c $\displaystyle\int (3e^{x}+e^{-x})^{2}\,dx$

d $\displaystyle\int \frac{e^{2x}+4}{e^{x}}\,dx$ e $\displaystyle\int \frac{5e^{3x}-e^{2x}}{2e^{x}}\,dx$ f $\displaystyle\int \frac{(e^{4x}-2e^{x})^{2}}{e^{3x}}\,dx$

3 Find

a $\displaystyle\int \left(2e^{x}+\frac{1}{\sqrt{x}}\right)\,dx$ b $\displaystyle\int (x^{2}-3e^{2x+1})\,dx$ c $\displaystyle\int \frac{3x^{2}e^{2x}-4e^{x}}{12x^{2}e^{x}}\,dx$

4 A curve is such that $\dfrac{dy}{dx}=2e^{2x}+e^{-x}$.

Given that the curve passes through the point $(0, 4)$, find the equation of the curve.

5 A curve is such that $\dfrac{dy}{dx}=ke^{2-x}+4x$, where k is a constant.

At the point $(2, 10)$ the gradient of the curve is 1.

a Find the value of k.

b Find the equation of the curve.

6 A curve is such that $\dfrac{d^{2}y}{dx^{2}}=8e^{-2x}$.

Given that $\dfrac{dy}{dx}=2$ when $x=0$ and that the curve passes through the point $\left(1,\dfrac{2}{e^{2}}\right)$,

find the equation of the curve.

7 The point $P\left(\dfrac{3}{2},5\right)$ lies on the curve for which $\dfrac{dy}{dx}=2e^{3-2x}$.

The point $Q(1, k)$ also lies on the curve.

a Find the value of k.

The normals to the curve at the points P and Q intersect at the point R.

b Find the coordinates of R.

15.5 Integration of sine, cosine and $\sec^2$ functions

In Chapter 14, you learned how to differentiate sine and cosine functions:

$$\frac{d}{dx}(\sin x) = \cos x \qquad \frac{d}{dx}[\sin(ax+b)] = a\cos(ax+b)$$

$$\frac{d}{dx}(\cos x) = -\sin x \qquad \frac{d}{dx}[\cos(ax+b)] = -a\sin(ax+b)$$

$$\frac{d}{dx}(\tan x) = \sec^2 x \qquad \frac{d}{dx}[\tan(ax+b)] = a\sec^2(ax+b)$$

Since integration is the reverse process of differentiation, the rules for integrating sine and cosine functions are:

$$\int \cos x \, dx = \sin x + c \qquad \int [\cos(ax+b)] \, dx = \frac{1}{a}\sin(ax+b) + c$$

$$\int \sin x \, dx = -\cos x + c \qquad \int [\sin(ax+b)] \, dx = -\frac{1}{a}\cos(ax+b) + c$$

$$\int \sec^2 x \, dx = \tan x + c \qquad \int [\sec^2(ax+b)] \, dx = \frac{1}{a}\tan(ax+b) + c$$

Note:

It is important to remember that the formulae for differentiating and integrating these trigonometric functions only apply when x is measured in radians.

WORKED EXAMPLE 7

Find

a $\displaystyle\int \sin 2x \, dx$ **b** $\displaystyle\int \cos 5x \, dx$ **c** $\displaystyle\int 3\sec^2\left(\frac{x}{2}\right) dx$

Answers

a $\displaystyle\int \sin 2x \, dx = -\frac{1}{2}\cos 2x + c$

b $\displaystyle\int \cos 5x \, dx = \frac{1}{5}\sin 5x + c$

c $\displaystyle\int 3\sec^2\left(\frac{x}{2}\right) dx = 3\int \sec^2\left(\frac{x}{2}\right) dx$

$$= 3 \times \left(\tan\frac{x}{2}\right) + c$$

$$= 6\tan\left(\frac{x}{2}\right) + c$$

WORKED EXAMPLE 8

Find

a $\int (3\cos 2x + 5\sin 3x)\,dx$

b $\int \left[x^2 + 2\cos (5x - 1) \right]\,dx$

Answers

a $\int (3\cos 2x + 5\sin 3x)\,dx = 3\int \cos 2x\,dx + 5\int \sin 3x\,dx$

$$= 3 \times \left(\frac{1}{2}\sin 2x \right) + 5 \times \left(-\frac{1}{3}\cos 3x \right) + c$$

$$= \frac{3}{2}\sin 2x - \frac{5}{3}\cos 3x + c$$

b $\int \left[x^2 + 2\cos (5x - 1) \right]\,dx = \int x^2\,dx + 2\int \cos (5x - 1)\,dx$

$$= \frac{1}{3}x^3 + 2 \times \left[\frac{1}{5}\sin (5x - 1) \right] + c$$

$$= \frac{1}{3}x^3 + \frac{2}{5}\sin (5x - 1) + c$$

Exercise 15.5

1   Find

a $\int \sin 4x\,dx$

b $\int \cos 2x\,dx$

c $\int \sin \frac{x}{3}\,dx$

d $\int \sec^2 (5x)\,dx$

e $\int 2\cos 2x\,dx$

f $\int 6\sin 3x\,dx$

g $\int 2\sec^2 (3x)\,dx$

h $\int 3\cos (2x + 1)\,dx$

i $\int 5\sin (2 - 3x)\,dx$

j $\int 2\cos (2x - 7)\,dx$

k $\int 4\sin (1 - 5x)\,dx$

l $\int 4\sec^2 (1 - 2x)\,dx$

2   Find

a $\int (1 - \sin x)\,dx$

b $\int (\sqrt{x} - 2\cos 3x)\,dx$

c $\int \left(3\cos 2x - \pi \sin \frac{5x}{2} \right)\,dx$

d $\int \left(\frac{1}{x^2} - \cos \frac{3x}{2} \right)\,dx$

e $\int (e^{2x} - 5\sin 2x)\,dx$

f $\int \left(\frac{2}{\sqrt{x}} + \sin \frac{x}{2} \right)\,dx$

3   A curve is such that $\dfrac{dy}{dx} = \cos x - \sin x$.

Given that the curve passes through the point $\left(\dfrac{\pi}{2}, 3 \right)$, find the equation of the curve.

4   A curve is such that $\dfrac{dy}{dx} = 1 - 4\cos 2x$.

Given that the curve passes through the point $\left(\dfrac{\pi}{4}, 1 \right)$, find the equation of the curve.

5   A curve is such that $\dfrac{dy}{dx} = 4x - 6\sin 2x$.

Given that the curve passes through the point $(0, -2)$, find the equation of the curve.

6 A curve is such that $\dfrac{d^2y}{dx^2} = 45\cos 3x + 2\sin x$.

Given that $\dfrac{dy}{dx} = -2$ when $x = 0$ and that the curve passes through the point

$(\pi, -1)$, find the equation of the curve.

7 A curve is such that $\dfrac{dy}{dx} = k\cos 3x - 4$, where k is a constant.

At the point $(\pi, 2)$, the gradient of the curve is -10.

a Find the value of k.

b Find the equation of the curve.

8 The point $\left(\dfrac{\pi}{2}, 5\right)$ lies on the curve for which $\dfrac{dy}{dx} = 4\sin\left(2x - \dfrac{\pi}{2}\right)$.

a Find the equation of the curve.

b Find the equation of the normal to the curve at the point where $x = \dfrac{\pi}{3}$.

9 The point $P\left(\dfrac{\pi}{3}, 3\right)$ lies on the curve for which $\dfrac{dy}{dx} = 3\cos\left(3x - \dfrac{\pi}{2}\right)$.

The point $Q\left(\dfrac{\pi}{2}, k\right)$ also lies on the curve.

a Find the value of k.

The tangents to the curve at the points P and Q intersect at the point R.

b Find the coordinates of R.

10 The point $T\left(\dfrac{\pi}{4}, -\dfrac{4}{3}\right)$ lies on the curve for which $\dfrac{dy}{dx} = 4\sec^2(3x - \pi)$.

The point $V\left(\dfrac{\pi}{3}, k\right)$ also lies on the curve.

a Find the value of k.

The tangents to the curve at the points T and V intersect at the point W.

b Find the coordinates of W.

15.6 Integration of functions of the form $\dfrac{1}{x}$ and $\dfrac{1}{ax+b}$

In Chapter 14, you learned the following rules for differentiating logarithmic functions:

$$\dfrac{d}{dx}(\ln x) = \dfrac{1}{x}, \ x > 0 \qquad \dfrac{d}{dx}[\ln(ax+b)] = \dfrac{a}{ax+b}, \ ax+b > 0$$

It is important to remember that $\ln x$ is defined only for $x > 0$.

Since integration is the reverse process of differentiation, the rules for integration are:

$$\int \dfrac{1}{x}\,dx = \ln x + c, \ x > 0 \qquad \int \dfrac{1}{ax+b}\,dx = \dfrac{1}{a}\ln(ax+b) + c, \ ax+b > 0$$

WORKED EXAMPLE 9

Find each of these integrals and state the values of x for which the integral is valid.

a $\int \dfrac{5}{x}\,\mathrm{d}x$ **b** $\int \dfrac{6}{2x+5}\,\mathrm{d}x$ **c** $\int \dfrac{8}{3-4x}\,\mathrm{d}x$

Answers

a $\int \dfrac{5}{x}\,\mathrm{d}x = 5\int \dfrac{1}{x}\,\mathrm{d}x$

$\qquad = 5\ln x + c, \ x > 0$

b $\int \dfrac{6}{2x+5}\,\mathrm{d}x = 6\int \dfrac{1}{2x+5}\,\mathrm{d}x$

$\qquad\qquad = 6\left(\dfrac{1}{2}\right)\ln(2x+5) + c$ valid for $2x + 5 > 0$

$\qquad\qquad = 3\ln(2x+5) + c, \ x > -\dfrac{5}{2}$

c $\int \dfrac{8}{3-4x}\,\mathrm{d}x = 8\int \dfrac{1}{3-4x}\,\mathrm{d}x$

$\qquad\qquad = 8\left(\dfrac{1}{-4}\right)\ln(3-4x) + c$ valid for $3 - 4x > 0$

$\qquad\qquad = -2\ln(3-4x) + c, \ x < \dfrac{3}{4}$

CLASS DISCUSSION

Nicola is asked to find $\int x^{-1}\,\mathrm{d}x$.

She tries to use the formula $\int x^{n}\,\mathrm{d}x = \dfrac{1}{n+1}x^{n+1} + c$ to obtain her answer.

Nicola is also asked to find $\int (5x+2)^{-1}\,\mathrm{d}x$.

She tries to use the formula $\int (ax+b)^{n}\,\mathrm{d}x = \dfrac{1}{a(n+1)}(ax+b)^{n+1} + c$ to obtain her answer.

Discuss with your classmates why Nicola's methods do not work.

CLASS DISCUSSION

Raju and Sara are asked to find $\int \dfrac{1}{2(3x-1)}\,dx$.

Raju writes: $\displaystyle\int \frac{1}{2(3x-1)}\,dx = \int \frac{1}{6x-2}\,dx$

$$= \frac{1}{6}\ln(6x-2) + c$$

Sara writes: $\displaystyle\int \frac{1}{2(3x-1)}\,dx = \frac{1}{2}\int \frac{1}{3x-1}\,dx$

$$= \left(\frac{1}{2}\right)\left(\frac{1}{3}\right)\ln(3x-1) + c$$

$$= \frac{1}{6}\ln(3x-1) + c$$

Decide who is correct and discuss the reasons for your decision with your classmates.

Exercise 15.6

1 Find:

a $\displaystyle\int \frac{8}{x}\,dx$

b $\displaystyle\int \frac{5}{x}\,dx$

c $\displaystyle\int \frac{1}{2x}\,dx$

d $\displaystyle\int \frac{5}{3x}\,dx$

e $\displaystyle\int \frac{1}{3x+2}\,dx$

f $\displaystyle\int \frac{1}{1-8x}\,dx$

g $\displaystyle\int \frac{7}{2x-1}\,dx$

h $\displaystyle\int \frac{4}{2-3x}\,dx$

i $\displaystyle\int \frac{5}{2(5x-1)}\,dx$

2 Find:

a $\displaystyle\int \left(7x+1+\frac{2}{x}\right)dx$

b $\displaystyle\int \left(1+\frac{2}{x}\right)^2 dx$

c $\displaystyle\int \left(5-\frac{3}{x}\right)^2 dx$

d $\displaystyle\int \frac{x+1}{x}\,dx$

e $\displaystyle\int \frac{3-5x^3}{2x^4}\,dx$

f $\displaystyle\int \left(3x-\frac{2}{x^2}\right)^2 dx$

g $\displaystyle\int \frac{5x+2\sqrt{x}}{x^2}\,dx$

h $\displaystyle\int \left(\frac{x^3-\sqrt{x}}{x\sqrt{x}}\right)dx$

i $\displaystyle\int \frac{5xe^{4x}-2e^x}{10xe^x}\,dx$

3 A curve is such that $\dfrac{dy}{dx} = \dfrac{5}{2x-1}$ for $x > 0.5$.

Given that the curve passes through the point $(1, 3)$, find the equation of the curve.

4 A curve is such that $\dfrac{dy}{dx} = 2x + \dfrac{5}{x}$ for $x > 0$.

Given that the curve passes through the point (e, e^2), find the equation of the curve.

5 A curve is such that $\dfrac{dy}{dx} = \dfrac{1}{x+e}$ for $x > -e$.

Given that the curve passes through the point $(e, 2 + \ln 2)$, find the equation of the curve.

6   CHALLENGE QUESTION

The point $P(1, -2)$ lies on the curve for which $\dfrac{dy}{dx} = 3 - \dfrac{2}{x}$.

The point $Q(2, k)$ also lies on the curve.

a   Find the value of k.

The tangents to the curve at the points P and Q intersect at the point R.

b   Find the coordinates of R.

15.7 Further indefinite integration

This section uses the concept that integration is the reverse process of differentiation to help integrate complicated expressions.

$$\text{If } \frac{dy}{dx}[\text{F}(x)] = \text{f}(x), \text{ then } \int \text{f}(x)\,dx = \text{F}(x) + c$$

WORKED EXAMPLE 10

Show that $\dfrac{d}{dx}\left(\dfrac{x^2 + 1}{\sqrt{4x - 3}}\right) = \dfrac{2(3x^2 - 3x - 1)}{\sqrt{(4x - 3)^3}}$.

Hence find $\displaystyle\int \dfrac{3x^2 - 3x - 1}{\sqrt{(4x - 3)^3}}\,dx$.

Answers

Let $y = \dfrac{x^2 + 1}{\sqrt{4x - 3}}$

$\dfrac{dy}{dx} = \dfrac{(\sqrt{4x - 3})\,2x - (x^2 + 1)\left[\frac{1}{2}(4x - 3)^{-\frac{1}{2}}(4)\right]}{4x - 3}$ quotient rule

$= \dfrac{2x\sqrt{4x - 3} - \dfrac{2(x^2 + 1)}{\sqrt{4x - 3}}}{4x - 3}$ multiply numerator and denominator by $\sqrt{4x - 3}$

$= \dfrac{2x(4x - 3) - 2(x^2 + 1)}{(4x - 3)\sqrt{4x - 3}}$

$= \dfrac{2(4x^2 - 3x - x^2 - 1)}{\sqrt{(4x - 3)^3}}$

$= \dfrac{2(3x^2 - 3x - 1)}{\sqrt{(4x - 3)^3}}$

$\displaystyle\int \dfrac{3x^2 - 3x - 1}{\sqrt{(4x - 3)^3}}\,dx = \dfrac{1}{2}\int \dfrac{2(3x^2 - 3x - 1)}{\sqrt{(4x - 3)^3}}\,dx$

$\qquad\qquad = \dfrac{x^2 + 1}{2\sqrt{4x - 3}} + c$

WORKED EXAMPLE 11

Differentiate $x \sin x$ with respect to x.

Hence find $\int x \cos x \, dx$.

Answers

Let $y = x \sin x$

$$\frac{dy}{dx} = (x)(\cos x) + (\sin x)(1) \qquad \text{product rule}$$

$$= x \cos x + \sin x$$

Hence $\int (x \cos x + \sin x) \, dx = x \sin x$

$$\int x \cos x \, dx + \int \sin x \, dx = x \sin x$$

$$\int x \cos x \, dx = x \sin x - \int \sin x \, dx$$

$$\int x \cos x \, dx = x \sin x + \cos x + c$$

WORKED EXAMPLE 12

Differentiate $x^3 \sqrt{2x - 1}$ with respect to x.

Hence find $\int \dfrac{7x^3 - 3x^2 + 5}{\sqrt{2x - 1}} \, dx$.

Answers

Let $y = x^3 \sqrt{2x - 1}$

$$\frac{dy}{dx} = (x^3) \left[\frac{1}{2} \times 2 \times (2x - 1)^{-\frac{1}{2}} \right] + \sqrt{2x - 1} \times (3x^2) \qquad \text{product rule}$$

$$= \frac{x^3}{\sqrt{2x - 1}} + 3x^2 \sqrt{2x - 1}$$

$$= \frac{x^3 + 3x^2(2x - 1)}{\sqrt{2x - 1}}$$

$$= \frac{7x^3 - 3x^2}{\sqrt{2x - 1}}$$

$$\int \frac{7x^3 - 3x^2 + 5}{\sqrt{2x - 1}} \, dx = \int \frac{7x^3 - 3x^2}{\sqrt{2x - 1}} \, dx + \int \frac{5}{\sqrt{2x - 1}} \, dx$$

$$= \int \frac{7x^3 - 3x^2}{\sqrt{2x - 1}} \, dx + 5 \int (2x - 1)^{-\frac{1}{2}} \, dx$$

$$= x^3 \sqrt{2x - 1} + \frac{5}{2 \times \frac{1}{2}} (2x - 1)^{\frac{1}{2}} + c$$

$$= x^3 \sqrt{2x - 1} + 5 \sqrt{2x - 1} + c$$

$$= (x^3 + 5) \sqrt{2x - 1} + c$$

Exercise 15.7

1   a   Given that $y = \dfrac{x + 5}{\sqrt{2x - 1}}$, show that $\dfrac{dy}{dx} = \dfrac{x - 6}{\sqrt{(2x - 1)^3}}$.

   b   Hence find $\displaystyle\int \dfrac{x - 6}{\sqrt{(2x - 1)^3}}\, dx$.

2   a   Differentiate $(3x^2 - 1)^5$ with respect to x.

   b   Hence find $\displaystyle\int x(3x^2 - 1)^4\, dx$.

3   a   Differentiate $x \ln x$ with respect to x.

   b   Hence find $\displaystyle\int \ln x\, dx$.

4   a   Show that $\dfrac{d}{dx}\left(\dfrac{\ln x}{x}\right) = \dfrac{1 - \ln x}{x^2}$.

   b   Hence find $\displaystyle\int \left(\dfrac{\ln x}{x^3}\right) dx$.

5   a   Given that $y = x\sqrt{x^2 - 4}$, find $\dfrac{dy}{dx}$.

   b   Hence find $\displaystyle\int \dfrac{x^2 - 2}{\sqrt{x^2 - 4}}\, dx$.

6   a   Given that $y = 3(x + 1)\sqrt{x - 5}$, show that $\dfrac{dy}{dx} = \dfrac{9(x - 3)}{2\sqrt{x - 5}}$.

   b   Hence find $\displaystyle\int \dfrac{(x - 3)}{\sqrt{x - 5}}\, dx$.

7   a   Find $\dfrac{d}{dx}\left(x e^{2x} - \dfrac{e^{2x}}{2}\right)$.

   b   Hence find $\displaystyle\int x e^{2x}\, dx$.

8   a   Show that $\dfrac{d}{dx}\left(\dfrac{\sin x}{1 - \cos x}\right)$ can be written in the form $\dfrac{k}{\cos x - 1}$, and state the value of k.

   b   Hence find $\displaystyle\int \dfrac{5}{\cos x - 1}\, dx$.

9   a   Given that $y = (x + 8)\sqrt{x - 4}$, show that $\dfrac{dy}{dx} = \dfrac{kx}{\sqrt{x - 4}}$, and state the value of k.

   b   Hence find $\displaystyle\int \dfrac{x}{\sqrt{x - 4}}\, dx$.

10 a   Given that $y = \dfrac{1}{x^2 - 7}$, show that $\dfrac{dy}{dx} = \dfrac{kx}{(x^2 - 7)^2}$, and state the value of k.

   b   Hence find $\displaystyle\int \dfrac{4x}{(x^2 - 7)^2}\, dx$.

11 a   Find $\dfrac{d}{dx}(2x^3 \ln x)$.

   b   Hence find $\displaystyle\int x^2 \ln x\, dx$.

12 a   Differentiate $x \cos x$ with respect to x.

  b   Hence find $\int x \sin x \, \mathrm{d}x$.

13 a   Given that $y = \mathrm{e}^{2x}(\sin 2x + \cos 2x)$, show that $\dfrac{\mathrm{d}y}{\mathrm{d}x} = 4\mathrm{e}^{2x}\cos 2x$.

  b   Hence find $\int \mathrm{e}^{2x}\cos 2x \, \mathrm{d}x$.

14 CHALLENGE QUESTION

  a   Find $\dfrac{\mathrm{d}}{\mathrm{d}x}\left(x^2\sqrt{2x-7}\right)$.

  b   Hence find $\int \dfrac{5x^2 - 14x + 3}{\sqrt{2x-7}} \, \mathrm{d}x$

15.8 Definite integration

You have learned about indefinite integrals such as

$$\int x^2 \, \mathrm{d}x = \frac{1}{3}x^3 + c$$

where c is an arbitrary constant.

$\int x^2 \, \mathrm{d}x$ is called the indefinite integral of x^2 with respect to x.

It is called 'indefinite' because it has infinitely many solutions.

You can integrate a function between two defined limits.

The integral of the function x^2 with respect to x between the limits $x = 1$ and $x = 4$ is

written as: $\displaystyle\int_{1}^{4} x^2 \, \mathrm{d}x$

The method for evaluating this integral is

$$\int_{1}^{4} x^2 \, \mathrm{d}x = \left[\frac{1}{3}x^3 + c\right]_{1}^{4}$$

$$= \left(\frac{1}{3}\times 4^3 + c\right) - \left(\frac{1}{3}\times 1^3 + c\right)$$

$$= 21$$

Note that the c's cancel out, so the process can be simplified to:

$$\int_{1}^{4} x^2 \, \mathrm{d}x = \left[\frac{1}{3}x^3\right]_{1}^{4}$$

$$= \left(\frac{1}{3}\times 4^3\right) - \left(\frac{1}{3}\times 1^3\right)$$

$$= 21$$

$\displaystyle\int_{1}^{4} x^2 \, \mathrm{d}x$ is called the **definite integral** of x^2 with respect to x.

It is called 'definite' because there is only one solution.

Hence, the evaluation of a definite integral can be written as:

$$\int_a^b f(x)\,dx = [F(x)]_a^b = F(b) - F(a)$$

The following rules for definite integrals may also be used.

$$\int_a^b k\,f(x)\,dx = k\int_a^b f(x)\,dx, \text{ where } k \text{ is a constant}$$

$$\int_a^b [f(x) \pm g(x)]\,dx = \int_a^b f(x)\,dx \pm \int_a^b g(x)\,dx$$

WORKED EXAMPLE 13

Evaluate: **a** $\displaystyle\int_1^2 \frac{x^5 + 3}{x^2}\,dx$ **b** $\displaystyle\int_0^5 \sqrt{3x + 1}\,dx$ **c** $\displaystyle\int_{-1}^1 \frac{10}{(3 - 2x)^2}\,dx$

Answers

a $\displaystyle\int_1^2 \frac{x^5 + 3}{x^2}\,dx = \int_1^2 (x^3 + 3x^{-2})\,dx$

$$= \left[\frac{1}{4}x^4 + \frac{3}{-1}x^{-1}\right]_1^2$$

$$= \left(\frac{1}{4}(2)^4 - 3(2)^{-1}\right) - \left(\frac{1}{4}(1)^4 - 3(1)^{-1}\right)$$

$$= \left(4 - \frac{3}{2}\right) - \left(\frac{1}{4} - 3\right)$$

$$= 5\frac{1}{4}$$

b $\displaystyle\int_0^5 \sqrt{3x + 1}\,dx = \int_0^5 (3x + 1)^{\frac{1}{2}}\,dx$

$$= \left[\frac{1}{(3)\left(\frac{3}{2}\right)}(3x + 1)^{\frac{3}{2}}\right]_0^5$$

$$= \left[\frac{2}{9}(3x + 1)^{\frac{3}{2}}\right]_0^5$$

$$= \left(\frac{2}{9} \times 16^{\frac{3}{2}}\right) - \left(\frac{2}{9} \times 1^{\frac{3}{2}}\right)$$

$$= \left(\frac{128}{9}\right) - \left(\frac{2}{9}\right)$$

$$= 14$$

CONTINUED

c $\displaystyle\int_{-1}^{1}\frac{10}{(3-2x)^2}\,\mathrm{d}x = \int_{-1}^{1} 10(3-2x)^{-2}\,\mathrm{d}x$

$\displaystyle = \left[\frac{10}{(-2)(-1)}(3-2x)^{-1}\right]_{-1}^{1}$

$\displaystyle = \left[\frac{5}{3-2x}\right]_{-1}^{1}$

$\displaystyle = \left(\frac{5}{1}\right) - \left(\frac{5}{5}\right)$

$\displaystyle = 4$

WORKED EXAMPLE 14

Evaluate: **a** $\displaystyle\int_{1}^{2} 4e^{2x-3}\,\mathrm{d}x$ **b** $\displaystyle\int_{0}^{\frac{\pi}{4}}(3\cos 2x + 5)\,\mathrm{d}x$

Answers

a $\displaystyle\int_{1}^{2} 4e^{2x-3}\,\mathrm{d}x = \left[\frac{4}{2}e^{2x-3}\right]_{1}^{2}$

$\displaystyle = (2e^{1}) - (2e^{-1})$

$\displaystyle = 2e - \frac{2}{e}$

$\displaystyle = \frac{2e^2 - 2}{e}$

b $\displaystyle\int_{0}^{\frac{\pi}{4}}(5 + 3\cos 2x)\,\mathrm{d}x = \left[5x + \frac{3}{2}\sin 2x\right]_{0}^{\frac{\pi}{4}}$

$\displaystyle = \left(\frac{5\pi}{4} + \frac{3}{2}\sin\frac{\pi}{2}\right) - \left(0 + \frac{3}{2}\sin 0\right)$

$\displaystyle = \left(\frac{5\pi}{4} + \frac{3}{2}\right) - (0 + 0)$

$\displaystyle = \frac{5\pi + 6}{4}$

In Section 15.10 you will learn why the modulus signs are included in the following integration formulae:

$$\int\frac{1}{x}\,\mathrm{d}x = \ln|x| + c \qquad \int\frac{1}{ax+b}\,\mathrm{d}x = \frac{1}{a}\ln|ax+b| + c$$

The next two worked examples show how these formulae are used in definite integrals.

Note:
It is normal practice to include the modulus sign only when finding definite integrals.

WORKED EXAMPLE 15

Find the value of $\displaystyle\int_2^3 \frac{8}{2x+1}\,dx$.

Answers

$$\int_2^3 \frac{8}{2x+1}\,dx = \left[\frac{8}{2}\ln|2x+1|\right]_2^3 \qquad \text{substitute limits}$$

$$= (4\ln|7|) - (4\ln|5|) \qquad \text{simplify}$$

$$= 4(\ln 7 - \ln 5)$$

$$= 4\ln\frac{7}{5}$$

WORKED EXAMPLE 16

Find the value of $\displaystyle\int_3^5 \frac{9}{2-3x}\,dx$.

Answers

$$\int_3^5 \frac{9}{2-3x}\,dx = \left[\frac{9}{-3}\ln|2-3x|\right]_3^5 \qquad \text{substitute limits}$$

$$= (-3\ln|-13|) - (-3\ln|-7|) \qquad \text{simplify}$$

$$= -3\ln 13 + 3\ln 7$$

$$= 3(\ln 7 - \ln 13)$$

$$= 3\ln\frac{7}{13}$$

Exercise 15.8

1   Evaluate.

a $\displaystyle\int_1^2 7x^6\,dx$

b $\displaystyle\int_1^2 \frac{4}{x^3}\,dx$

c $\displaystyle\int_1^3 (5x+2)\,dx$

d $\displaystyle\int_0^3 (x^2+2)\,dx$

e $\displaystyle\int_{-1}^2 (5x^2-3x)\,dx$

f $\displaystyle\int_1^4 \left(2+\frac{2}{x^2}\right)dx$

g $\displaystyle\int_2^4 \left(x^2-3-\frac{1}{x^2}\right)dx$

h $\displaystyle\int_1^3 \left(\frac{2x^2-1}{x^5}\right)dx$

i $\displaystyle\int_{-3}^{-2} (2x-1)(3x-5)\,dx$

j $\displaystyle\int_1^4 \sqrt{x}\,(x+3)\,dx$

k $\displaystyle\int_1^2 \frac{(5-x)(2+x)}{x^4}\,dx$

l $\displaystyle\int_1^4 \left(2\sqrt{x}-\frac{2}{\sqrt{x}}\right)dx$

2 Evaluate.

   a $\displaystyle\int_{1}^{2}(2x+1)^3\,dx$ **b** $\displaystyle\int_{0}^{6}\sqrt{2x+4}\,dx$ **c** $\displaystyle\int_{0}^{3}\sqrt{(x+1)^3}\,dx$

   d $\displaystyle\int_{-1}^{2}\frac{12}{(x+4)^3}\,dx$ **e** $\displaystyle\int_{-1}^{1}\frac{2}{(3x+5)^2}\,dx$ **f** $\displaystyle\int_{-4}^{0}\frac{6}{\sqrt{4-3x}}\,dx$

3 Evaluate.

   a $\displaystyle\int_{0}^{1}e^{2x}\,dx$ **b** $\displaystyle\int_{0}^{\frac{1}{4}}e^{4x}\,dx$ **c** $\displaystyle\int_{0}^{2}5e^{-2x}\,dx$

   d $\displaystyle\int_{0}^{\frac{1}{3}}e^{1-3x}\,dx$ **e** $\displaystyle\int_{0}^{1}\frac{5}{e^{2x-1}}\,dx$ **f** $\displaystyle\int_{0}^{1}(e^x+1)^2\,dx$

   g $\displaystyle\int_{0}^{1}(e^x+e^{2x})^2\,dx$ **h** $\displaystyle\int_{0}^{1}\left(3e^x-\frac{2}{e^x}\right)^2\,dx$ **i** $\displaystyle\int_{0}^{2}\frac{3+8e^{2x}}{2e^x}\,dx$

4 Evaluate.

   a $\displaystyle\int_{0}^{\pi}\sin x\,dx$ **b** $\displaystyle\int_{0}^{\frac{\pi}{2}}(3+\cos 2x)\,dx$ **c** $\displaystyle\int_{0}^{\frac{\pi}{3}}\sin\left(2x-\frac{\pi}{6}\right)\,dx$

   d $\displaystyle\int_{\frac{\pi}{6}}^{\frac{\pi}{3}}(2\cos x-\sin 2x)\,dx$ **e** $\displaystyle\int_{0}^{\frac{\pi}{4}}(2x-\sin 2x)\,dx$ **f** $\displaystyle\int_{\frac{\pi}{4}}^{\frac{\pi}{2}}(\sin 3x-\cos 2x)\,dx$

5 Evaluate.

   a $\displaystyle\int_{1}^{5}\frac{2}{3x+1}\,dx$ **b** $\displaystyle\int_{-1}^{4}\frac{1}{2x+3}\,dx$ **c** $\displaystyle\int_{2}^{8}\frac{3}{2x-1}\,dx$

   d $\displaystyle\int_{-2}^{-1}\frac{5}{2x+1}\,dx$ **e** $\displaystyle\int_{1}^{4}\frac{2}{1-3x}\,dx$ **f** $\displaystyle\int_{-3}^{-2}\frac{4}{3-2x}\,dx$

6 Evaluate.

   a $\displaystyle\int_{2}^{4}\left(1+\frac{2}{3x-1}\right)dx$ **b** $\displaystyle\int_{1}^{3}\left(\frac{2}{x}-\frac{1}{2x+1}\right)dx$ **c** $\displaystyle\int_{0}^{1}\left(3+\frac{1}{5-2x}-2x\right)dx$

7 Given that $\displaystyle\int_{1}^{k}\frac{2}{3x-1}\,dx=\frac{2}{3}\ln 7$, find the value of k.

8 **a** Given that $\dfrac{4x}{2x+3}=2+\dfrac{A}{2x+3}$, find the value of the constant A.

   b Hence show that $\displaystyle\int_{0}^{1}\frac{4x}{2x+3}\,dx=2-3\ln\frac{5}{3}$.

9 **a** Find the quotient and remainder when $4x^2+4x$ is divided by $2x+1$.

   b Hence show that $\displaystyle\int_{0}^{1}\frac{4x^2+4x}{2x+1}\,dx=2-\frac{1}{2}\ln 3$.

10 CHALLENGE QUESTION

   Find the value of $\displaystyle\int_{1}^{2}\frac{6x^2-8x}{2x-1}\,dx$.

15.9 Further definite integration

This section uses the concept that integration is the reverse process of differentiation to help evaluate complicated definite integrals.

WORKED EXAMPLE 17

Given that $y = \dfrac{3x}{\sqrt{x^2 + 5}}$, find $\dfrac{dy}{dx}$.

Hence evaluate $\displaystyle\int_0^2 \dfrac{3}{\sqrt{(x^2 + 5)^3}}\,dx$.

Answers

$$y = \frac{3x}{\sqrt{x^2 + 5}}$$

$$\frac{dy}{dx} = \frac{(\sqrt{x^2 + 5})(3) - (3x)\left[\frac{1}{2}(x^2 + 5)^{-\frac{1}{2}}(2x)\right]}{x^2 + 5} \qquad \text{quotient rule}$$

$$= \frac{3\sqrt{x^2 + 5} - \dfrac{3x^2}{\sqrt{x^2 + 5}}}{x^2 + 5} \qquad \text{multiply numerator and denominator by } \sqrt{x^2 + 5}$$

$$= \frac{3(x^2 + 5) - 3x^2}{(x^2 + 5)\sqrt{x^2 + 5}}$$

$$= \frac{15}{\sqrt{(x^2 + 5)^3}}$$

$$\int_0^2 \frac{3}{\sqrt{(x^2 + 5)^3}}\,dx = \frac{1}{5}\int_0^2 \frac{15}{\sqrt{(x^2 + 5)^3}}\,dx$$

$$= \frac{1}{5}\left[\frac{3x}{\sqrt{x^2 + 5}}\right]_0^2$$

$$= \frac{1}{5}\left[\left(\frac{6}{\sqrt{4 + 5}}\right) - \left(\frac{0}{\sqrt{0 + 5}}\right)\right]$$

$$= \frac{2}{5}$$

WORKED EXAMPLE 18

Given that $y = x \cos 2x$, find $\dfrac{dy}{dx}$.

Hence evaluate $\displaystyle\int_0^{\frac{\pi}{6}} 2x \sin 2x \, dx$.

Answers

Let $y = x \cos 2x$

$\dfrac{dy}{dx} = (x) \times (-2\sin 2x) + (\cos 2x) \times (1)$ product rule

$\phantom{\dfrac{dy}{dx}} = \cos 2x - 2x \sin 2x$

$\displaystyle\int_0^{\frac{\pi}{6}} (\cos 2x - 2x \sin 2x) \, dx = [x \cos 2x]_0^{\frac{\pi}{6}}$

$\displaystyle\int_0^{\frac{\pi}{6}} \cos 2x \, dx - \int_0^{\frac{\pi}{6}} 2x \sin 2x \, dx = \left(\dfrac{\pi}{6} \times \cos \dfrac{\pi}{3}\right) - (0 \times \cos 0)$

$\left[\dfrac{1}{2} \sin 2x\right]_0^{\frac{\pi}{6}} - \displaystyle\int_0^{\frac{\pi}{6}} 2x \sin 2x \, dx = \dfrac{\pi}{12}$

$\displaystyle\int_0^{\frac{\pi}{6}} 2x \sin 2x \, dx = \left[\dfrac{1}{2} \sin 2x\right]_0^{\frac{\pi}{6}} - \dfrac{\pi}{12}$

$\phantom{\displaystyle\int_0^{\frac{\pi}{6}} 2x \sin 2x \, dx} = \left(\dfrac{1}{2} \sin \dfrac{\pi}{3}\right) - \left(\dfrac{1}{2} \sin 0\right) - \dfrac{\pi}{12}$

$\phantom{\displaystyle\int_0^{\frac{\pi}{6}} 2x \sin 2x \, dx} = \dfrac{\sqrt{3}}{4} - \dfrac{\pi}{12}$

$\phantom{\displaystyle\int_0^{\frac{\pi}{6}} 2x \sin 2x \, dx} = \dfrac{3\sqrt{3} - \pi}{12}$

Exercise 15.9

1　**a**　Given that $y = (x + 1)\sqrt{2x - 1}$, find $\dfrac{dy}{dx}$.

　　b　Hence evaluate $\displaystyle\int_1^5 \dfrac{x}{\sqrt{2x - 1}} \, dx$.

2　**a**　Given that $y = x\sqrt{3x^2 + 4}$, find $\dfrac{dy}{dx}$.

　　b　Hence evaluate $\displaystyle\int_0^2 \dfrac{3x^2 + 2}{\sqrt{3x^2 + 4}} \, dx$.

3　**a**　Given that $y = \dfrac{1}{x^2 + 5}$, find $\dfrac{dy}{dx}$

　　b　Hence evaluate $\displaystyle\int_1^2 \dfrac{4x}{(x^2 + 5)^2} \, dx$.

4 **a** Given that $y = \dfrac{x+2}{\sqrt{3x+4}}$, find $\dfrac{dy}{dx}$.

    b Hence evaluate $\displaystyle\int_0^4 \dfrac{6x+4}{\sqrt{(3x+4)^3}}\,dx$.

5 **a** Show that $\dfrac{d}{dx}\left(\dfrac{x}{\cos x}\right) = \dfrac{\cos x + x\sin x}{\cos^2 x}$.

    b Hence evaluate $\displaystyle\int_0^{\frac{\pi}{4}} \dfrac{\cos x + x\sin x}{5\cos^2 x}\,dx$.

6 **CHALLENGE QUESTION**

    a Find $\dfrac{d}{dx}(x\sin x)$.

    b Hence evaluate $\displaystyle\int_0^{\frac{\pi}{2}} x\cos x\,dx$.

7 **CHALLENGE QUESTION**

    a Find $\dfrac{d}{dx}(x^2\ln x)$.

    b Hence evaluate $\displaystyle\int_1^e 4x\ln x\,dx$.

8 **CHALLENGE QUESTION**

    a Given that $y = x\sin 3x$, find $\dfrac{dy}{dx}$.

    b Hence evaluate $\displaystyle\int_0^{\frac{\pi}{6}} x\cos 3x\,dx$.

> **REFLECTION**
>
> Without looking back in this chapter, make a list of all the integration formulae that you have learned.

15.10 Area under a curve

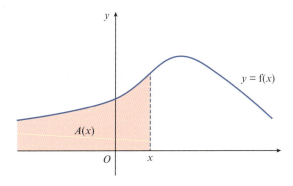

If you define the area under the curve $y = f(x)$ to the left of x as $A(x)$, then as x increases then $A(x)$ also increases.

Now consider a small increase in x, say δx, which results in a small increase, δA, in area.

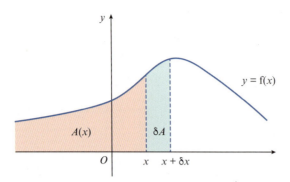

δA = area to the left of $(x + \delta x)$ − area to the left of x

$\delta A = A(x + \delta x) - A(x)$

Now consider the area δA which is approximately a rectangle.

$$\delta A \approx y\,\delta x$$

so $\dfrac{\delta A}{\delta x} \approx y$

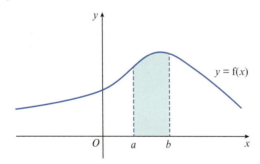

As $\delta x \to 0$, then $\dfrac{\delta A}{\delta x} \to \dfrac{\mathrm{d}A}{\mathrm{d}x}$ hence $\dfrac{\mathrm{d}A}{\mathrm{d}x} = y$.

If $\dfrac{\mathrm{d}A}{\mathrm{d}x} = y$, then $A = \displaystyle\int y\,\mathrm{d}x$.

Hence the area of the region bounded by the curve $y = \mathrm{f}(x)$, the lines $x = a$ and $x = b$ and the x-axis is given by the definite integral.

$$\text{Area} = \int_{a}^{b} \mathrm{f}(x)\,\mathrm{d}x, \quad \text{where } \mathrm{f}(x) \geqslant 0$$

WORKED EXAMPLE 19

Find the area of the shaded region.

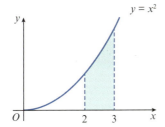

Answers

$$\text{Area} = \int_{2}^{3} x^2 \, dx$$

$$= \left[\frac{1}{3} x^3 \right]_{2}^{3}$$

$$= \left(\frac{27}{3} \right) - \left(\frac{8}{3} \right)$$

$$= 6\frac{1}{3} \text{ units}^2$$

WORKED EXAMPLE 20

Find the area of the shaded region.

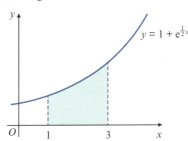

Answers

$$\text{Area} = \int_{1}^{3} \left(1 + e^{\frac{1}{2}x} \right) dx$$

$$= \left[x + \frac{1}{\left(\frac{1}{2} \right)} e^{\frac{1}{2}x} \right]_{1}^{3}$$

$$= \left[x + 2e^{\frac{1}{2}x} \right]_{1}^{3}$$

$$= \left(3 + 2e^{\frac{3}{2}} \right) - \left(1 + 2e^{\frac{1}{2}} \right)$$

$$= 2 + 2e^{\frac{3}{2}} - 2e^{\frac{1}{2}}$$

$$= 2[1 + \sqrt{e}\,(e - 1)]$$

$$\approx 7.67 \text{ units}^2$$

In the examples so far, the required area has been above the x-axis.

If the required area between $y = f(x)$ and the x-axis lies below the x-axis, then $\int_{a}^{b} f(x)\,dx$ will be a negative value.

Hence, for a region that lies below the x-axis, the area is given as $\left| \int_{a}^{b} f(x)\,dx \right|$.

WORKED EXAMPLE 21

Find the area of the shaded region.

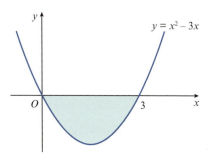

$y = x^2 - 3x$

Answers

$$\int_{0}^{3} (x^2 - 3x)\,dx = \left[\frac{1}{3}x^3 - \frac{3}{2}x^2 \right]_{0}^{3}$$

$$= \left(9 - \frac{27}{2} \right) - (0 - 0)$$

$$= -4.5$$

Area is 4.5 units^2.

The required region could consist of a section above the x-axis and a section below the x-axis.

If this happens you must evaluate each area separately.

This is illustrated in the following example.

WORKED EXAMPLE 22

Find the total area of the shaded regions.

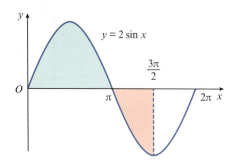

$y = 2\sin x$

CONTINUED

Answers

$$\int_0^\pi 2\sin x\,dx = [-2\cos x]_0^\pi$$
$$= (-2\cos\pi) - (-2\cos 0)$$
$$= (2) - (-2)$$
$$= 4$$

$$\int_\pi^{\frac{3\pi}{2}} 2\sin x\,dx = [-2\cos x]_\pi^{\frac{3\pi}{2}}$$
$$= \left(-2\cos\frac{3\pi}{2}\right) - (-2\cos\pi)$$
$$= (0) - (2)$$
$$= -2$$

Hence, the total area of the shaded regions = $4 + 2 = 6$ units^2.

In Section 15.8 you were given the formula $\displaystyle\int\frac{1}{x}\,dx = \ln|x| + c$

The use of the modulus symbols in this formula can be explained by considering the symmetry properties of the graph of $y = \dfrac{1}{x}$.

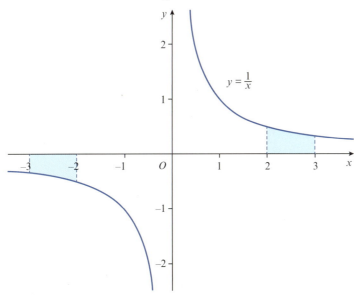

The shaded areas that represent the integrals $\displaystyle\int_2^3\frac{1}{x}\,dx$ and $\displaystyle\int_{-3}^{-2}\frac{1}{x}\,dx$ are equal in magnitude.

However, one of the areas is below the x-axis, which suggests that $\displaystyle\int_{-3}^{-2}\frac{1}{x}\,dx = -\int_2^3\frac{1}{x}\,dx$.

Evaluating $\displaystyle\int_2^3\frac{1}{x}\,dx$ gives:

$$\int_2^3\frac{1}{x}\,dx = [\ln x]_2^3 = \ln 3 - \ln 2 = \ln\frac{3}{2}$$

This implies that $\int_{-3}^{-2} \frac{1}{x}\,dx = -\ln\frac{3}{2} = \ln\frac{2}{3}$.

If you try using integration to find the value of $\int_{-3}^{-2} \frac{1}{x}\,dx$ you obtain

$$\int_{-3}^{-2} \frac{1}{x}\,dx = [\ln x]_{-3}^{-2} = \ln(-2) - \ln(-3) = \ln\frac{2}{3}$$

There is, however, a problem with this calculation in that $\ln x$ is only defined for $x > 0$ so $\ln(-2)$ and $\ln(-3)$ do not actually exist.

Hence for $x < 0$, we say that $\int \frac{1}{x}\,dx = \ln|x| + c$.

Exercise 15.10

1   Find the area of each shaded region.

a
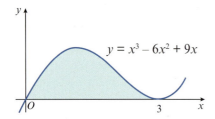
$y = x^3 - 6x^2 + 9x$

b
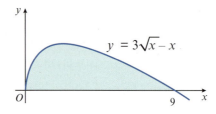
$y = 3\sqrt{x} - x$

c
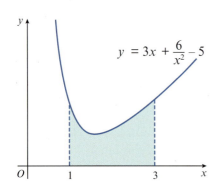
$y = 3x + \dfrac{6}{x^2} - 5$

d
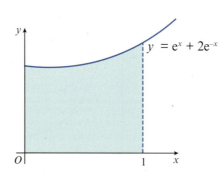
$y = e^x + 2e^{-x}$

e
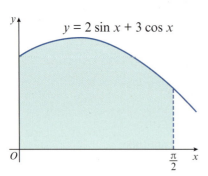
$y = 2\sin x + 3\cos x$

f
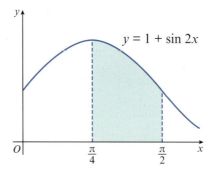
$y = 1 + \sin 2x$

2   Find the area of each shaded region.

a
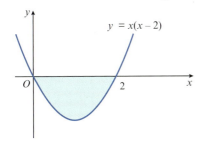
$y = x(x-2)$

b
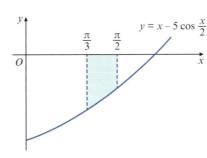
$y = x - 5\cos\dfrac{x}{2}$

c
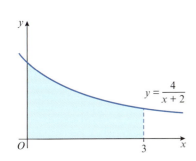
$y = \dfrac{4}{x+2}$

d
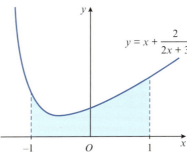
$y = x + \dfrac{2}{2x+3}$

3

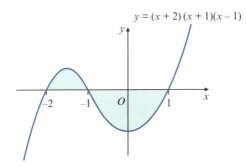
$y = (x+2)(x+1)(x-1)$

Find the total shaded region.

4   Sketch the following curves and find the area of the finite region or regions bounded by the curves and the x-axis.

a $y = x(x+1)$ **b** $y = (x+2)(3-x)$

c $y = x(x^2-4)$ **d** $y = x(x-2)(x+4)$

e $y = x(x-1)(x-5)$ **f** $y = x^2(4-x)$

5   Find the area enclosed by the curve $y = \dfrac{6}{\sqrt{x}}$, the x-axis and the lines $x = 4$ and $x = 9$.

6   **a**   Find the area of the region enclosed by the curve $y = \dfrac{12}{x^2}$, the x-axis and the lines $x = 1$ and $x = 4$.

     b   The line $x = p$ divides the region in **part a** into two equal parts. Find the value of p.

7   **a**   Show that $\dfrac{\mathrm{d}}{\mathrm{d}x}(x\mathrm{e}^x - \mathrm{e}^x) = x\mathrm{e}^x$.

  b   Use your result from **part a** to evaluate the area of the shaded region.

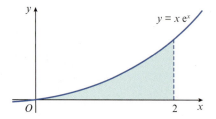

8   CHALLENGE QUESTION

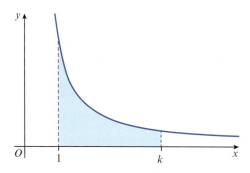

The diagram shows part of the curve $y = \dfrac{4}{2x - 1}$.

Given that the shaded region has area 8, find the exact value of k.

9   CHALLENGE QUESTION

  a   Show that $\dfrac{\mathrm{d}}{\mathrm{d}x}(x\ln x) = 1 + \ln x$.

  b   Use your result from **part a** to evaluate the area of the shaded region.

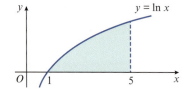

10   CHALLENGE QUESTION

  a   Show that $\dfrac{\mathrm{d}}{\mathrm{d}x}(x\cos x) = \cos x - x\sin x$.

  b   Use your result from **part a** to evaluate the area of the shaded region.

15.11 Area of regions bounded by a line and a curve

The following example shows a possible method for finding the area enclosed by a curve and a straight line.

WORKED EXAMPLE 23

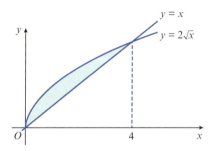

The curve $y = 2\sqrt{x}$ intersects the line $y = x$ at the point $(4, 4)$. Find the area of the shaded region bounded by the curve and the line.

Answers

Area = area under curve − area of triangle

$$= \int_0^4 2\sqrt{x}\, dx - \frac{1}{2} \times 4 \times 4$$

$$= \int_0^4 2x^{\frac{1}{2}}\, dx - 8$$

$$= \left[\frac{2}{\left(\frac{3}{2}\right)} x^{\frac{3}{2}} \right]_0^4 - 8$$

$$= \left(\frac{4}{3} \times 4^{\frac{3}{2}} \right) - \left(\frac{4}{3} \times 0^{\frac{3}{2}} \right) - 8$$

$$= 2\frac{2}{3} \text{ units}^2$$

There is an alternative method for finding the shaded area in the previous example.

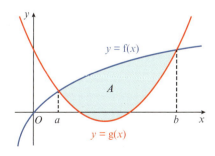

If two functions $f(x)$ and $g(x)$ intersect at $x = a$ and $x = b$, then the area, A, enclosed between the two curves is given by:

$$A = \int_a^b f(x)\,dx - \int_a^b g(x)\,dx$$

So, for the area enclosed by $y = 2\sqrt{x}$ and $y = x$:

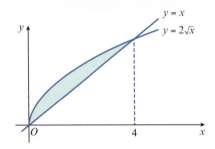

using $f(x) = 2\sqrt{x}$ and $g(x) = x$

$$\text{Area} = \int_0^4 f(x)\,dx - \int_0^4 g(x)\,dx$$

$$= \int_0^4 2\sqrt{x}\,dx - \int_0^4 x\,dx$$

$$= \int_0^4 (2\sqrt{x} - x)\,dx$$

$$= \left[\frac{2}{\left(\frac{3}{2}\right)} x^{\frac{3}{2}} - \frac{1}{2}x^2 \right]_0^4$$

$$= \left(\frac{4}{3} \times 4^{\frac{3}{2}} - \frac{1}{2} \times 4^2 \right) - \left(\frac{4}{3} \times 0^{\frac{3}{2}} - \frac{1}{2} \times 0^2 \right)$$

$$= 2\frac{2}{3} \text{ units}^2$$

This alternative method is the easiest method to use in the next example.

WORKED EXAMPLE 24

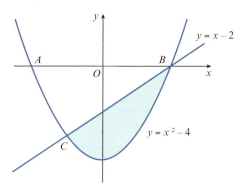

The curve $y = x^2 - 4$ intersects the x-axis at the points A and B and intersects the line $y = x - 2$ at the points B and C.

a   Find the coordinates of A, B and C.

b   Find the area of the shaded region bounded by the curve and the line.

Answers

a   When $y = 0$, $x^2 - 4 = 0$

$$x = \pm 2$$

A is the point $(-2, 0)$ and B is the point $(2, 0)$

For the intersection of a curve and a line:

$$x^2 - 4 = x - 2$$
$$x^2 - x - 2 = 0$$
$$(x + 1)(x - 2) = 0$$
$$x = -1 \text{ or } x = 2$$

When $x = -1$, $y = -3$

C is the point $(-1, -3)$

b   Area $= \displaystyle\int_{-1}^{2} (x - 2)\,dx - \int_{-1}^{2} (x^2 - 4)\,dx$

$$= \int_{-1}^{2} (x - 2 - x^2 + 4)\,dx = \int_{-1}^{2} (x + 2 - x^2)\,dx$$

$$= \left[\frac{1}{2}x^2 + 2x - \frac{1}{3}x^3 \right]_{-1}^{2}$$

$$= \left(2 + 4 - \frac{8}{3} \right) - \left(\frac{1}{2} - 2 + \frac{1}{3} \right)$$

$$= 4.5 \text{ units}^2$$

CLASS DISCUSSION

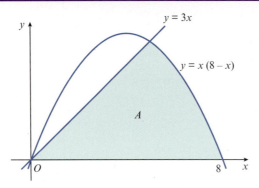

Discuss with your classmates how you could find the shaded area, A, enclosed by the curve $y = x(8 - x)$, the line $y = 3x$ and the x-axis.

Can you find more than one method?

Calculate the area using each of your different methods.

Discuss with your classmates which method you preferred.

Exercise 15.11

1

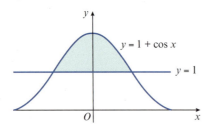

Find the area of the region enclosed by the curve $y = 1 + \cos x$ and the line $y = 1$.

2

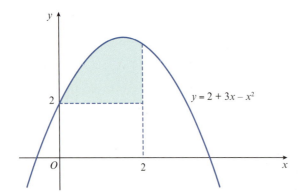

Find the area of the region bounded by the curve $y = 2 + 3x - x^2$, the line $x = 2$ and the line $y = 2$.

3

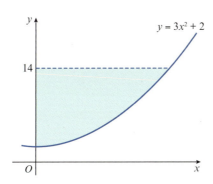

Find the area of the region bounded by the curve $y = 3x^2 + 2$, the line $y = 14$ and the y-axis.

4

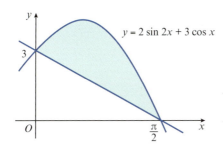

Find the area of the shaded region.

5   Sketch the following curves and lines and find the area enclosed between their graphs.

a $y = x^2 + 1$ and $y = 5$

b $y = x^2 - 2x + 3$ and $x + y = 9$

c $y = \sqrt{x}$ and $y = \dfrac{1}{2}x$

d $y = 4x - x^2$ and $2x + y = 0$

e $y = (x - 1)(x - 5)$ and $y = x - 1$

6   Sketch the following pairs of curves and find the area enclosed between their graphs for $x \geqslant 0$.

a $y = x^2$ and $y = x(2 - x)$

b $y = x^3$ and $y = 4x - 3x^2$

7

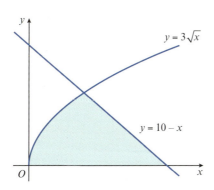

Find the shaded area enclosed by the curve $y = 3\sqrt{x}$, the line $y = 10 - x$ and the x-axis.

8

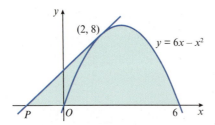

The tangent to the curve $y = 6x - x^2$ at the point $(2, 8)$ cuts the x-axis at the point P.

a Find the coordinates of P.

b Find the area of the shaded region.

9

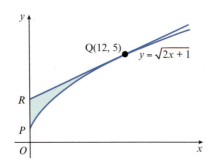

The curve $y = \sqrt{2x + 1}$ meets the y-axis at the point P.

The tangent at the point $Q(12, 5)$ to this curve meets the y-axis at the point R.

Find the area of the shaded region PQR.

10

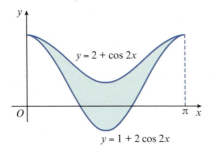

The diagram shows the graphs of $y = 2 + \cos 2x$ and $y = 1 + 2\cos 2x$ for $0 \leqslant x \leqslant \pi$.

Find the area of the shaded region.

11 CHALLENGE QUESTION

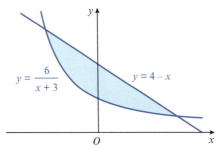

The diagram shows the graphs of $y = \dfrac{6}{x + 3}$ and $y = 4 - x$.

Find the area of the shaded region.

SUMMARY

Integration as the reverse of differentiation

If $\dfrac{d}{dx}[F(x)] = f(x)$, then $\displaystyle\int f(x)\,dx = F(x) + c$

Integration formulae

$\displaystyle\int x^n\,dx = \dfrac{1}{n+1}x^{n+1} + c$, where c is a constant and $n \neq -1$

$\displaystyle\int (ax+b)^n\,dx = \dfrac{1}{a(n+1)}(ax+b)^{n+1} + c,\, n \neq -1$ and $a \neq 0$

$\displaystyle\int e^x\,dx = e^x + c$ $\qquad\qquad$ $\displaystyle\int e^{ax+b}\,dx = \dfrac{1}{a}e^{ax+b} + c$

$\displaystyle\int \cos x\,dx = \sin x + c$ $\qquad\quad$ $\displaystyle\int [\cos(ax+b)]\,dx = \dfrac{1}{a}\sin(ax+b) + c$

$\displaystyle\int \sin x\,dx = -\cos x + c$ $\qquad$ $\displaystyle\int [\sin(ax+b)]\,dx = \dfrac{1}{a}\cos(ax+b) + c$

$\displaystyle\int \dfrac{1}{x}\,dx = \ln x + c,\, x > 0$ $\qquad$ $\displaystyle\int \dfrac{1}{ax+b}\,dx = \dfrac{1}{a}\ln(ax+b) + c,\, ax+b > 0$

$\displaystyle\int \dfrac{1}{x}\,dx = \ln|x| + c$ $\qquad\quad$ $\displaystyle\int \dfrac{1}{ax+b}\,dx = \dfrac{1}{a}\ln|ax+b| + c$

$\displaystyle\int \sec^2 x\,dx = \tan x + c$ $\qquad\;$ $\displaystyle\int [\sec^2(ax+b)]\,dx = \dfrac{1}{a}\tan(ax+b) + c$

Rules for indefinite integration

$\displaystyle\int k\,f(x)\,dx = k\int f(x)\,dx$, where k is a constant

$\displaystyle\int [f(x) \pm g(x)]\,dx = \int f(x)\,dx \pm \int g(x)\,dx$

Rules for definite integration

If $\displaystyle\int f(x)\,dx = F(x) + c$, then $\displaystyle\int_a^b f(x)\,dx = [F(x)]_a^b = F(b) - F(a)$.

$\displaystyle\int_a^b k\,f(x)\,dx = k\int_a^b f(x)\,dx$, where k is a constant

$\displaystyle\int_a^b [f(x) \pm g(x)]\,dx = \int_a^b f(x)\,dx \pm \int_a^b g(x)\,dx$

CONTINUED

Area under a curve

The area, A, bounded by the curve $y = f(x)$, the x-axis and the lines $x = a$ and $x = b$ is given by the formula

$A = \displaystyle\int_a^b f(x)\,dx$ if $f(x) \geqslant 0$.

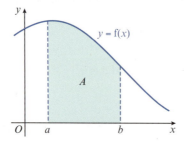

Area bounded by the graphs of two functions

If two functions $f(x)$ and $g(x)$ intersect at $x = a$ and $x = b$, then the area, A, enclosed between the two curves is given by the formula:

$A = \displaystyle\int_a^b f(x)\,dx - \int_a^b g(x)\,dx.$

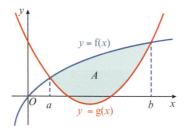

Past paper questions and practice questions

Worked example

A curve is such that when $x = 0$, both $y = -5$ and $\dfrac{dy}{dx} = 10$.

Given that $\dfrac{d^2y}{dx^2} = 4e^{2x} + 3$, find

i the equation of the curve, [7]

ii the equation of the normal to the curve at the point where $x = \dfrac{1}{4}$. [3]

Cambridge IGCSE Additional Mathematics 0606 Paper 12 Q10 Mar 2019

Answers

i $\dfrac{d^2y}{dx^2} = 4e^{2x} + 3$ integrate

$\dfrac{dy}{dx} = 2e^{2x} + 3x + c$

Using $\dfrac{dy}{dx} = 10$ when $x = 0$ gives

$10 = 2 + c$

$c = 8$

$\dfrac{dy}{dx} = 2e^{2x} + 3x + 8$ integrate

$y = e^{2x} + \dfrac{3}{2}x^2 + 8x + d$

Using $y = -5$ when $x = 0$ gives

$-5 = 1 + d$

$d = -6$

Equation of curve is $y = e^{2x} + \dfrac{3}{2}x^2 + 8x - 6$

ii When $x = \dfrac{1}{4}$, $y = e^{\frac{1}{2}} + \dfrac{3}{32} + 2 - 6 = -2.26$ to 3sf

$\dfrac{dy}{dx} = 2e^{\frac{1}{2}} + \dfrac{3}{4} + 8 = 12.0$ to 3sf

Hence gradient of normal $= -\dfrac{1}{12}$

Equation of normal is $y - (-2.26) = -\dfrac{1}{12}\left(x - \dfrac{1}{4}\right)$

$y + 2.26 = -\dfrac{1}{12}\left(x - \dfrac{1}{4}\right)$

1 a Given that $y = e^{x^2}$, find $\dfrac{dy}{dx}$.   [2]

b Use your answer to part **a** to find $\int x e^{x^2}\, dx$.   [2]

c Hence evaluate $\displaystyle\int_0^2 x e^{x^2}\, dx$.   [2]

Cambridge IGCSE Additional Mathematics 0606 Paper 11 Q5 Jun 2014

2 A curve is such that $\dfrac{dy}{dx} = 4x + \dfrac{1}{(x+1)^2}$ for $x > 0$. The curve passes through the point $\left(\dfrac{1}{2}, \dfrac{5}{6}\right)$.

a Find the equation of the curve.   [4]

b Find the equation of the normal to the curve at the point where $x = 1$.   [4]

Cambridge IGCSE Additional Mathematics 0606 Paper 11 Q7i,ii Jun 2014

3 Find the exact value of $\displaystyle\int_1^2 \left(5 + \dfrac{8}{4x-3}\right) dx$, giving your answer in the form $\ln(ae^b)$, where a and b are integers.   [5]

Practice question

4 a Find the value of the constant A such that $\dfrac{6x-5}{2x-3} = 3 + \dfrac{A}{2x-3}$.   [2]

b Hence show that $\displaystyle\int_3^5 \dfrac{6x-5}{2x-3}\, dx = 6 + 2\ln\dfrac{7}{3}$.   [5]

Practice question

5 **Do not use a calculator in this question.**

i Show that $\dfrac{\mathrm{d}}{\mathrm{d}x}\left(\dfrac{\mathrm{e}^{4x}}{4} - x\mathrm{e}^{4x}\right) = px\mathrm{e}^{4x}$, where p is an integer to be found. [4]

ii Hence find the exact value of $\displaystyle\int_{0}^{\ln 2} x\mathrm{e}^{4x}\,\mathrm{d}x$, giving your answer in the form $a\ln 2 + \dfrac{b}{c}$, where a, b and c are integers to be found. [4]

Cambridge IGCSE Additional Mathematics 0606 Paper 11 Q5 Jun 2016

6 **i** Find $\displaystyle\int\left(3x - x^{\frac{3}{2}}\right)\mathrm{d}x$. [2]

The diagram shows part of the curve $y = 3x - x^{\frac{3}{2}}$ and the lines $y = 3x$ and $2y = 27 - 3x$.

The curve and the line $y = 3x$ meet the x-axis at O and the curve and the line $2y = 27 - 3x$ meet the x-axis at A.

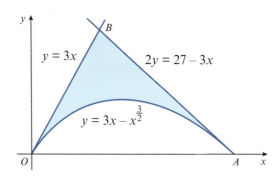

ii Find the coordinates of A. [1]

iii Verify that the coordinates of B are $(3, 9)$. [1]

iv Find the area of the shaded region. [4]

Cambridge IGCSE Additional Mathematics 0606 Paper 21 Q11 Jun 2016

7 The polynomial $\mathrm{p}(x) = 2x^3 - 3x^2 - x + 1$ has a factor $2x - 1$.

a Find $\mathrm{p}(x)$ in the form $(2x - 1)\mathrm{q}(x)$, where $\mathrm{q}(x)$ is a quadratic factor. [2]

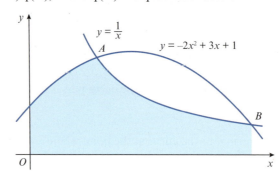

The diagram shows the graph of $y = \dfrac{1}{x}$ for $x > 0$, and the graph of $y = -2x^2 + 3x + 1$.

The curves intersect at the points A and B.

b Using your answer to **part a**, find the exact x-coordinate of A and of B. [4]

c Find the exact area of the shaded region. [6]

Cambridge IGCSE Additional Mathematics 0606 Paper 12 Q9 Mar 2021

8 **a** **i** Find $\int \dfrac{1}{(10x-1)^6}\,dx$. [2]

      ii Find $\int \dfrac{(2x^3+5)^2}{x}\,dx$. [3]

   b **i** Differentiate $y = \tan(3x+1)$ with respect to x. [2]

      ii Hence find $\displaystyle\int_{\frac{\pi}{12}}^{\frac{\pi}{10}} \left(\dfrac{\sec^2(3x+1)}{2} - \sin x \right) dx$. [4]

Cambridge IGCSE Additional Mathematics 0606 Paper 22 Q11 Mar 2021

9 Given that $\displaystyle\int_{1}^{a} \left(\dfrac{2}{2x+3} + \dfrac{3}{3x-1} - \dfrac{1}{x} \right) dx = \ln 2.4$ and that $a > 1$, find the value of a. [7]

Cambridge IGCSE Additional Mathematics 0606 Paper 12 Q11 Mar 2020

10

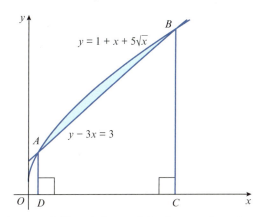

The diagram shows the curve $y = 1 + x + 5\sqrt{x}$ and the straight line $y - 3x = 3$.

The curve and line intersect at the points A and B. The lines BC and AD are perpendicular to the x-axis.

   i Using the substitution $u^2 = x$, or otherwise, find the coordinates A and B.
      You must show all your working. [6]

   ii Find the area of the shaded region, showing all your working. [6]

Cambridge IGCSE Additional Mathematics 0606 Paper 22 Q10 Mar 2019

11

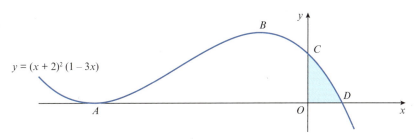

The diagram shows the graph of $y = (x+2)^2(1-3x)$. The curve has a minimum at the point A, a maximum at the point B and intersects the y-axis and the x-axis at the points C and D respectively.

   i Find the x-coordinate of A and of B. [5]

   ii Write down the coordinates of C and of D. [2]

   iii Showing all your working, find the area of the shaded region. [5]

Cambridge IGCSE Additional Mathematics 0606 Paper 12 Q10 Mar 2018

12 A curve is such that $\dfrac{d^2y}{dx^2} = 4\sin 2x$. The curve has a gradient of 5 at the point where $x = \dfrac{\pi}{2}$.

    i   Find an expression for the gradient of the curve at the point (x, y).   [4]

The curve passes through the point $P\left(\dfrac{\pi}{12}, -\dfrac{1}{2}\right)$.

    ii   Find the equation of the curve.   [4]

    iii   Find the equation of the normal to the curve at the point P, giving your answer in the form $y = mx + c$, where m and c are constants correct to 3 decimal places.   [3]

Cambridge IGCSE Additional Mathematics 0606 Paper 12 Q8 Mar 2017

13

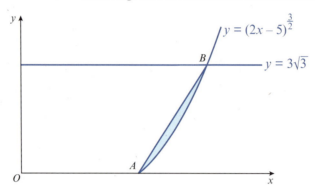

The diagram shows part of the curve $y = (2x - 5)^{\frac{3}{2}}$ and the line $y = 3\sqrt{3}$.

The curve meets the x-axis at the point A and the line $y = 3\sqrt{3}$ at the point B.

Find the area of the shaded region enclosed by the line AB and the curve, giving your answer in the form $\dfrac{p\sqrt{3}}{20}$, where p is an integer. You must show all your working.   [8]

Cambridge IGCSE Additional Mathematics 0606 Paper 12 Q10 Mar 2017

14 a   Find $\displaystyle\int e^{2x+1}\,dx$.   [2]

    b **i**   Given that $y = \dfrac{x}{\ln x}$, find $\dfrac{dy}{dx}$.   [3]

        ii   Hence find $\displaystyle\int\left(\dfrac{1}{\ln x} - \dfrac{1}{(\ln x)^2} + \dfrac{1}{x^2}\right)dx$.   [3]

Cambridge IGCSE Additional Mathematics 0606 Paper 22 Q9 Mar 2017

15 i   Find $\dfrac{d}{dx}\left(x(2x-1)^{\frac{3}{2}}\right)$.   [3]

    ii   Hence, show that $\displaystyle\int x(2x-1)^{\frac{1}{2}}\,dx = \dfrac{(2x-1)^{\frac{3}{2}}}{15}(px + q) + c$, where c is a constant of integration, and p and q are integers to be found.   [6]

    iii   Hence find $\displaystyle\int_{0.5}^{1} x(2x-1)^{\frac{1}{2}}\,dx$.   [2]

Cambridge IGCSE Additional Mathematics 0606 Paper 12 Q10 Mar 2016

16 Find the equation of the curve which passes through the point $(1, 7)$ and for which $\dfrac{dy}{dx} = \dfrac{9x^4 - 3}{x^2}$.   [4]

Cambridge IGCSE Additional Mathematics 0606 Paper 22 Q3 Mar 2016

17

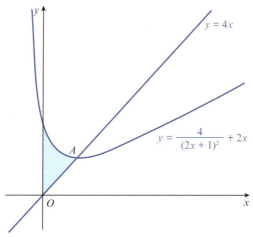

The diagram shows part of the curve $y = \dfrac{4}{(2x + 1)^2} + 2x$ and the line $y = 4x$.

i Find the coordinates of A, the stationary point of the curve. [5]

ii Verify that A is also the point of intersection of the curve $y = \dfrac{4}{(2x + 1)^2} + 2x$ and the line $y = 4x$. [1]

iii **Without using a calculator**, find the area of the shaded region enclosed by the line $y = 4x$, the curve and the y-axis. [6]

Cambridge IGCSE Additional Mathematics 0606 Paper 22 Q9 Mar 2015

〉 Chapter 15

Integration Matcher

The graphs of six functions and the graphs of their integrals have been mixed up below. Can you match them together?

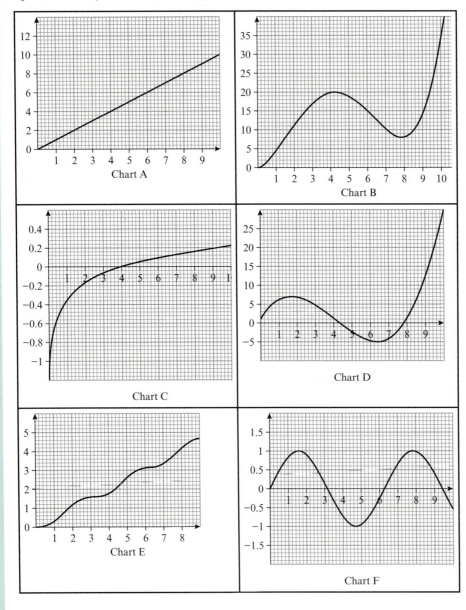

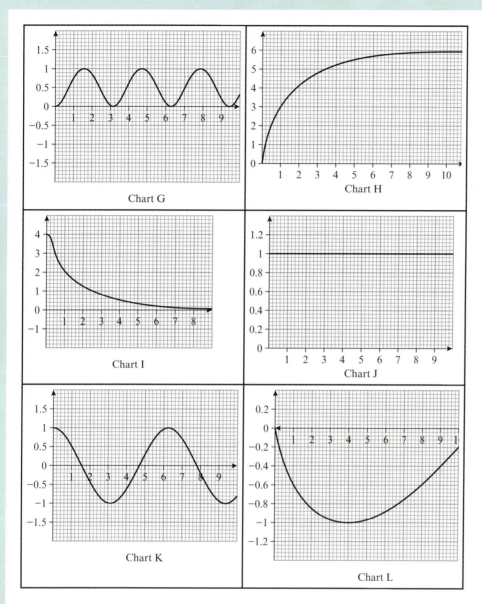

Chart G

Chart H

Chart I

Chart J

Chart K

Chart L

What strategies did you use to match them up?

How would you explain your strategies for matching the charts to a friend?

> Chapter 16

Kinematics

THIS SECTION WILL SHOW YOU HOW TO:

- apply differentiation and integration to kinematics problems that involve the displacement, velocity and acceleration of a particle moving in a straight line with variable or constant acceleration, and the use of x-t and v-t graphs.

PRE-REQUISITE KNOWLEDGE

Before you start…

Where it comes from	What you should be able to do	Check your skills
Cambridge IGCSE/O Level Mathematics	Calculate speed from a distance-time graph.	**1** The diagram shows a distance-time graph for a swimmer in a 100 m race. The swimming pool is 50 m long. 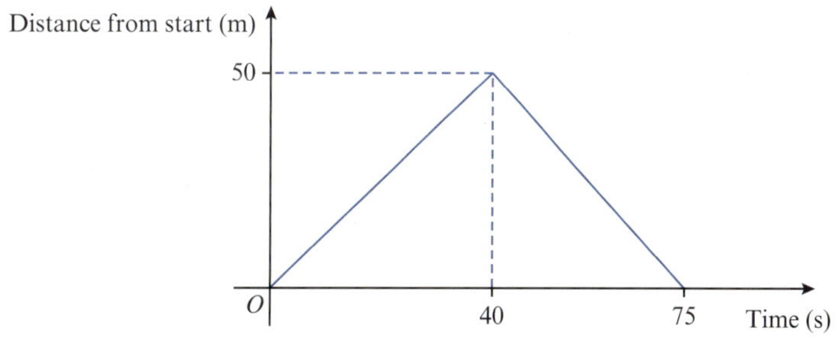 **a** Calculate the speed of the swimmer for the first length. **b** Calculate the speed of the swimmer for the second length.
Cambridge IGCSE/O Level Mathematics	Calculate acceleration and distance travelled from a speed-time graph.	**2** The diagram shows a speed-time graph for the journey of a car. 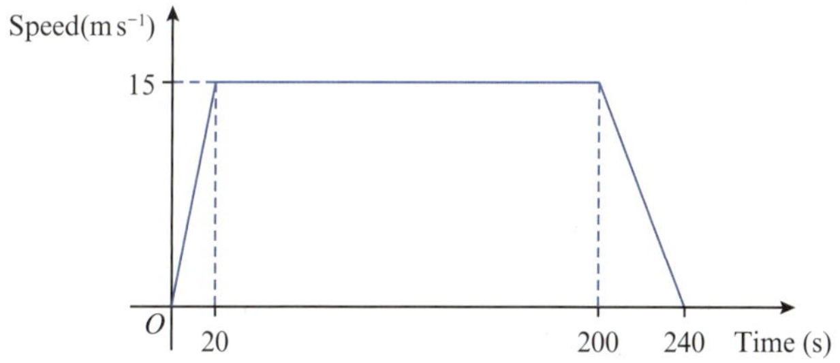 **a** Calculate the acceleration during the first 20 seconds. **b** Calculate the total distance travelled.

16.1 Applications of differentiation in kinematics

Displacement

Consider a particle, P, travelling along a straight line such that its **displacement**, s metres, from a fixed point O, t seconds after passing through O, is given by $s = 4t - t^2$.

The graph of s against t for $0 \leqslant t \leqslant 5$ is:

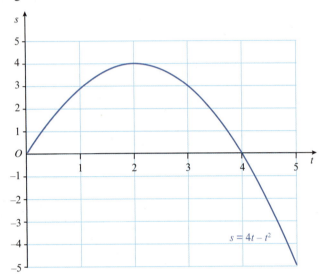

This can be represented on a motion diagram as:

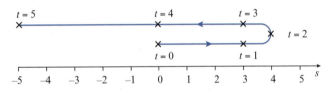

At time $t = 2$, the displacement of the particle from O is $4\,\text{m}$.

When $t = 2$ the particle stops instantaneously and reverses its direction of motion.

At time $t = 3$, the displacement is $3\,\text{m}$ and the total distance travelled is $4 + (4 - 3) = 5\,\text{m}$.

At time $t = 4$, the displacement is $0\,\text{m}$ and the total distance travelled is $4 + 4 = 8\,\text{m}$.

At time $t = 5$, the displacement is $-5\,\text{m}$ and the total distance travelled is $4 + 9 = 13\,\text{m}$.

Velocity and acceleration

If a particle moves in a straight line, with displacement function $s(t)$, then the rate of change of displacement with respect to time, $\dfrac{\mathrm{d}s}{\mathrm{d}t}$, is the **velocity**, v, of the particle at time t.

$$v = \frac{\mathrm{d}s}{\mathrm{d}t}$$

If the velocity function is $v(t)$, then the rate of change of velocity with respect to time, $\dfrac{dv}{dt}$, is the **acceleration**, a, of the particle at time t.

$$a = \frac{dv}{dt} = \frac{d^2 s}{dt^2}$$

CLASS DISCUSSION

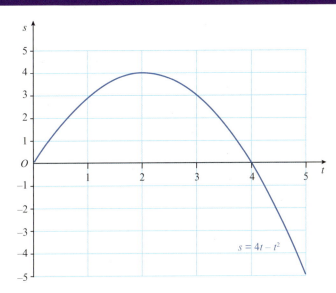

Earlier in this chapter you considered this displacement-time graph for the first 5 seconds of the motion of a particle P. The displacement of the particle, s metres, from a fixed point O, t seconds after passing through O, was given by $s = 4t - t^2$.

Working in small groups, complete the following tasks.

1   Find an expression, in terms of t, for the velocity, $v\,\mathrm{m\,s^{-1}}$, of this particle.

2   Draw the graph of v against t for $0 \leqslant t \leqslant 5$.

3   State the values of t for which v is

    a positive   **b** zero   **c** negative.

4   What can you say about the direction of motion of the particle when v is

    a positive   **b** zero   **c** negative?

5   Find the acceleration, $a\,\mathrm{m\,s^{-2}}$, of this particle and interpret your answer.

Now, discuss your conclusions with the whole class.

It is important that you are able to interpret the signs for displacements, velocities and accelerations.

These interpretations for a particle P relative to a fixed point O are summarised in the following tables.

Signs for displacement s

$s < 0$	$s = 0$	$s > 0$
P is to the left of O	P is at O	P is to the right of O

Signs for velocity v

$v < 0$	$v = 0$	$v > 0$
P is moving to the left	P is instantaneously at rest	P is moving to the right

Signs for acceleration a

$a < 0$	$a = 0$	$a > 0$
velocity is decreasing	velocity could be maximum or minimum or constant	velocity is increasing

WORKED EXAMPLE 1

A particle moves in a straight line so that, t seconds after passing a fixed point O, its displacement, s metres, from O is given by $s = t^3 - 12t^2 + 45t$.

a Find the velocity when $t = 2$.

b Find the acceleration when $t = 4$.

c Find the values of t when the particle is instantaneously at rest.

d Find the displacement of the particle from O when $t = 7$.

e Find the total distance travelled by the particle during the first 7 seconds.

Answers

a $s = t^3 - 12t^2 + 45t$

$v = \dfrac{ds}{dt} = 3t^2 - 24t + 45$

When $t = 2$, $v = 3(2)^2 - 24(2) + 45 = 9$

The velocity is $9\,\text{m s}^{-1}$ when $t = 2$.

b $v = 3t^2 - 24t + 45$

$a = \dfrac{dv}{dt} = 6t - 24$

When $t = 4$, $a = 6(4) - 24 = 0$

The acceleration is $0\,\text{m s}^{-2}$ when $t = 4$.

c The particle is at instantaneous rest when $v = 0$.

$3t^2 - 24t + 45 = 0$

$t^2 - 8t + 15 = 0$

$(t - 3)(t - 5) = 0$

$t = 3$ or $t = 5$

The particle is at instantaneous rest when $t = 3$ and $t = 5$.

CONTINUED

d   When $t = 7$, $s = (7)^3 - 12(7)^2 + 45(7) = 70$.

The particle is $70\,\text{m}$ from O when $t = 7$.

e   When $t = 0$, $s = 0$.
Critical values are when $t = 3$ and $t = 5$.
When $t = 3$, $s = (3)^3 - 12(3)^2 + 45(3) = 54$.
When $t = 5$, $s = (5)^3 - 12(5)^2 + 45(5) = 50$.

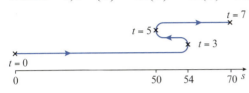

Total distance travelled $= 54 + (54 - 50) + (70 - 50) = 78\,\text{m}$.

WORKED EXAMPLE 2

A particle moves in a straight line so that the displacement, s metres, from a fixed point O, is given by $s = \ln(2t + 5)$, where t is the time in seconds after passing a point X on the line.

a   Find OX.

b   Find the velocity when $t = 10$.

c   Find the distance travelled during the third second.

d   Find the acceleration of the particle when $t = 2.5$.

Answers

a   When $t = 0$, $s = \ln[2(0) + 5] = \ln 5$ the particle is at X at time $t = 0$

$OX = \ln 5 \approx 1.61\,\text{m}$

b $s = \ln(2t + 5)$

$v = \dfrac{\text{d}s}{\text{d}t} = \dfrac{2}{2t + 5}$

When $t = 10$, $v = \dfrac{2}{2(10) + 5} = 0.08$

The velocity is $0.08\,\text{m}\,\text{s}^{-1}$.

CONTINUED

c Since $v > 0$ for all values of t, there is no change in the direction of motion of the particle.

The third second is from $t = 2$ to $t = 3$.

When $t = 2$, $s = \ln[2(2) + 5] = \ln 9$.

When $t = 3$, $s = \ln[2(3) + 5] = \ln 11$.

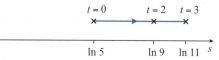

Distance travelled during the third second $= \ln 11 - \ln 9 = \ln \dfrac{11}{9} \approx 0.201\,\text{m}$.

d $v = \dfrac{2}{2t + 5} = 2(2t + 5)^{-1}$

$a = \dfrac{\mathrm{d}v}{\mathrm{d}t} = -2(2t + 5)^{-2} \times 2 = -\dfrac{4}{(2t + 5)^2}$

When $t = 2.5$, $a = -\dfrac{4}{[2(2.5) + 5]^2} = -0.04$

The acceleration is $-0.04\,\text{m s}^{-2}$.

WORKED EXAMPLE 3

A particle moves in a straight line such that its displacement, s metres, from a fixed point O on the line at time t seconds is given by $s = 20[\mathrm{e}^{-2t} - \mathrm{e}^{-3t}]$.

a Find the value of t when the particle is instantaneously at rest.

b Find the displacement of the particle from O when $t = 1$.

c Find the total distance travelled during the first second of its motion.

Answers

a $s = 20[\mathrm{e}^{-2t} - \mathrm{e}^{-3t}]$

$v = \dfrac{\mathrm{d}s}{\mathrm{d}t} = 20[-2\mathrm{e}^{-2t} + 3\mathrm{e}^{-3t}]$

$v = 0$ when $-2\mathrm{e}^{-2t} + 3\mathrm{e}^{-3t} = 0$

$$2\mathrm{e}^{-2t} = 3\mathrm{e}^{-3t}$$

$$\mathrm{e}^{t} = \frac{3}{2}$$

$$t = \ln\frac{3}{2}$$

It is instantaneously at rest when $t = \ln\dfrac{3}{2} = 0.405\,\text{s}$.

CONTINUED

b   When $t = 1$, $s = 20[e^{-2(1)} - e^{-3(1)}] \approx 1.711$.

Displacement from O is $1.71\,\text{m}$.

c   When $t = 0$, $s = 0$.

Critical value is when $t = \ln\dfrac{3}{2}$.

When $t = \ln\dfrac{3}{2}$, $s = 20\left[e^{-2\left(\ln\frac{3}{2}\right)} - e^{-3\left(\ln\frac{3}{2}\right)}\right] \approx 2.963$.

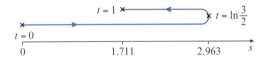

Total distance travelled $= 2.963 + (2.963 - 1.711) \approx 4.21\,\text{m}$.

WORKED EXAMPLE 4

A particle travels in a straight line so that, t seconds after passing through a fixed point O, its velocity $v\,\text{m}\,\text{s}^{-1}$, is given by $v = 5\cos\left(\dfrac{t}{2}\right)$.

a   Find the value of t when the particle first comes to instantaneous rest.

b   Find the acceleration of the particle when $t = \dfrac{\pi}{3}$.

Answers

a   The particle is at rest when $v = 0$.

$$5\cos\left(\frac{t}{2}\right) = 0$$

$$\cos\left(\frac{t}{2}\right) = 0$$

$$\frac{t}{2} = \frac{\pi}{2}$$

$$t = \pi$$

It is at instantaneous rest when $t = \pi$.

b $v = 5\cos\left(\dfrac{t}{2}\right)$

$a = \dfrac{\mathrm{d}v}{\mathrm{d}t} = -\dfrac{5}{2}\sin\left(\dfrac{t}{2}\right)$

When $t = \dfrac{\pi}{3}$, $a = -\dfrac{5}{3}\sin\left(\dfrac{\pi}{6}\right) = -1.25$

When $t = \dfrac{\pi}{3}$ the acceleration is $-1.25\,\text{m}\,\text{s}^{-2}$.

REFLECTION

Without looking back in this chapter, explain to a friend the difference between displacement and distance travelled.

Exercise 16.1

1   A particle, moving in a straight line, passes through a fixed point O.
    Its velocity $v\,\mathrm{m\,s}^{-1}$, t seconds after passing through O, is given by $v = \dfrac{50}{(3t + 2)^2}$.

    a   Find the velocity of the particle as it passes through O.

    b   Find the value of t when the velocity is $0.125\,\mathrm{m\,s}^{-1}$.

    c   Find the acceleration of the particle when $t = 1$.

2   A particle, moving in a straight line, passes through a fixed point O.
    Its velocity $v\,\mathrm{m\,s}^{-1}$, t seconds after passing through O, is given by $v = 6e^{2t} - 2t$.

    a   Find the initial velocity of the particle.

    b   Find the initial acceleration of the particle.

3   A particle starts from rest and moves in a straight line so that t seconds after
    passing through a fixed point O, its velocity $v\,\mathrm{m\,s}^{-1}$, is given by $v = 5(1 - e^{-t})$.

    a   Find the velocity of the particle when $t = \ln 100$.

    b   State the value which v approaches as t becomes very large.

    c   Find the acceleration of the particle when $v = 4$.

    d   Sketch the velocity-time graph for the motion of the particle.

4   A particle moves in a straight line such that its displacement, s metres, from a fixed
    point O on the line at time t seconds is given by $s = 9[\ln(3t + 2)]$.

    a   Find the value of t when the displacement of the particle from O is $36\,\mathrm{m}$.

    b   Find the velocity of the particle when $t = 1$.

    c   Show that the particle is decelerating for all values of t.

5   A particle moves in a straight line so that, t seconds after passing through a fixed
    point O, its displacement, s metres, from O is given by $s = \ln(1 + 2t)$.

    a   Find the value of t when the velocity of the particle is $0.4\,\mathrm{m\,s}^{-1}$.

    b   Find the distance travelled by the particle during the third second.

    c   Find the acceleration of the particle when $t = 1.5$.

6   A particle travels in a straight line so that, t seconds after passing through a fixed
    point O, its velocity $v\,\mathrm{m\,s}^{-1}$, is given by $v = 8\cos\left(\dfrac{t}{4}\right)$.

    a   Find the value of t when the velocity of the particle first equals $4\,\mathrm{m\,s}^{-1}$.

    b   Find the acceleration of the particle when $t = 5$.

    c   Sketch the acceleration-time graph for the motion of the particle.

7   A particle, moving in a straight line, passes through a fixed point O.
    Its velocity $v\,\mathrm{m\,s}^{-1}$, t seconds after passing through O, is given by $v = \cos 3t + \sin 3t$.

    a   Find the value of t when the particle is first instantaneously at rest.

    b   Find the acceleration of the particle when $t = \pi$.

8   A particle starts from rest and moves in a straight line so that, t seconds after
    leaving a fixed point O, its displacement, s metres, is given by $s = 2 - 2\cos 2t$.

    a   Find an expression for the velocity and the acceleration of the particle in terms of t.

    b   Find the time when the particle first comes to rest and its distance from O at this instant.

9   A particle moves in a straight line such that its displacement, s metres, from a fixed
    point O on the line at time t seconds is given by $s = 50(e^{-2t} - e^{-4t})$.

    a   Find the time when the particle is instantaneously at rest.

    b   Find the displacement of the particle from O when $t = 2$.

    c   Find the total distance travelled during the first 2 seconds of its motion.

10 A particle moves in a straight line so that its displacement from a fixed point O,
    is given by $s = 2t + 2\cos 2t$, where t is the time in seconds after the motion begins.

    a   Find the initial position of the particle.

    b   Find an expression for the velocity and the acceleration of the particle in terms of t.

    c   Find the time when the particle first comes to rest and its distance from O at this instant.

    d   Find the time when the acceleration of the particle is zero for the first time
    and its distance from O at this instant.

11 A particle, moving in a straight line, passes through a fixed point O.
    Its velocity $v\,\mathrm{m\,s}^{-1}$, t seconds after passing through O, is given by $v = k\cos 4t$,
    where k is a positive constant.

    a   Find the value of t when the particle is first instantaneously at rest.

    b   Find an expression for the acceleration, a, of the particle t seconds after
    passing through O.

    c   Given that the acceleration of the particle is $10\,\mathrm{m\,s}^{-2}$ when $t = \dfrac{7\pi}{24}$, find the value of k.

12 CHALLENGE QUESTION

A particle moves in a straight line such that its displacement, s metres, from a fixed
point O at a time t seconds, is given by

$s = 2t$ for $0 \leqslant t \leqslant 4$,

$s = 8 + 2\ln(t - 3)$ for $t > 4$.

    a   Find the initial velocity of the particle.

    b   Find the velocity of the particle when

        i $t = 2$ **ii** $t = 6$.

    c   Find the acceleration of the particle when

        i $t = 2$ **ii** $t = 6$.

    d   Sketch the displacement-time graph for the motion of the particle.

    e   Find the distance travelled by the particle in the 8th second.

13 CHALLENGE QUESTION

A particle moves in a straight line so that the displacement, s metres, from a fixed
point O, is given by $s = 2t^3 - 17t^2 + 40t - 2$, where t is the time in seconds after
passing a point X on the line.

    a   Find the distance OX.

    b   Find the value of t when the particle is first at rest.

    c   Find the values of t for which the velocity is positive.

    d   Find the values of t for which the velocity is negative.

16.2 Applications of integration in kinematics

In the last section you learned that when a particle moves in a straight line where the displacement from a fixed point O on the line is s, then:

$$v = \frac{ds}{dt} \quad \text{and} \quad a = \frac{dv}{dt} = \frac{d^2s}{dt^2}$$

Conversely, if a particle moves in a straight line where the acceleration of the particle is a, then:

$$v = \int a \, dt \quad \text{and} \quad s = \int v \, dt$$

In this section you will solve problems that involve both differentiation and integration.

The following diagram should help you remember when to differentiate and when to integrate.

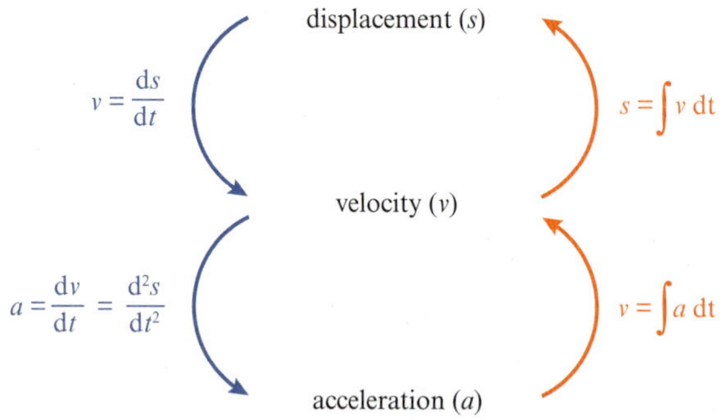

displacement (s)

$v = \dfrac{ds}{dt}$ $\qquad s = \int v \, dt$

velocity (v)

$a = \dfrac{dv}{dt} = \dfrac{d^2s}{dt^2}$ $\qquad v = \int a \, dt$

acceleration (a)

WORKED EXAMPLE 5

A particle moving in a straight line passes a fixed point O with velocity $8\,\text{m s}^{-1}$.
Its acceleration $a\,\text{m s}^{-2}$, t seconds after passing through O is given by $a = 2t + 1$.

a Find the velocity when $t = 3$.

b Find the displacement from O when $t = 3$.

Answers

a $a = 2t + 1$

$v = \int a \, dt$

$= \int (2t + 1) \, dt$

$= t^2 + t + c$

Using $v = 8$ when $t = 0$, gives $c = 8$

$v = t^2 + t + 8$

When $t = 3$, $v = (3)^2 + (3) + 8 = 20$

The particle's velocity when $t = 3$ is $20\,\text{m s}^{-1}$.

CONTINUED

b $v = t^2 + t + 8$

$$s = \int v\,dt$$

$$= \int (t^2 + t + 8)\,dt$$

$$= \frac{1}{3}t^3 + \frac{1}{2}t^2 + 8t + c$$

Using $s = 0$ when $t = 0$, gives $c = 0$

$$s = \frac{1}{3}t^3 + \frac{1}{2}t^2 + 8t$$

When $t = 3$, $s = \frac{1}{3}(3)^3 + \frac{1}{2}(3)^2 + 8(3) = 37.5$

Its displacement when $t = 3$ is $37.5\,\text{m}$.

WORKED EXAMPLE 6

A particle, moving in a straight line, passes through a fixed point O.
Its velocity $v\,\text{m}\,\text{s}^{-1}$, t seconds after passing through O, is given by $v = 6e^{3t} + 2t$.

a Find the acceleration of the particle when $t = 1$.

b Find an expression for the displacement of the particle from O.

c Find the total distance travelled by the particle in the first 2 seconds of its motion.

Answers

a $v = 6e^{3t} + 2t$

$$a = \frac{dv}{dt} = 18e^{3t} + 2$$

When $t = 1$, $a = 18e^{3(1)} + 2 \approx 364$

Its acceleration when $t = 1$ is $364\,\text{m}\,\text{s}^{-2}$.

b $v = 6e^{3t} + 2t$

$$s = \int v\,dt$$

$$= \int (6e^{3t} + 2t)\,dt$$

$$= 2e^{3t} + t^2 + c$$

Using $s = 0$ when $t = 0$, gives $c = -2$

The displacement, s metres, is given by $s = 2e^{3t} + t^2 - 2$.

CONTINUED

c   Since $v > 0$ for all values of t, there is no change in the direction of motion of the particle.

When $t = 0$, $s = 0$

When $t = 2$, $s = 2e^{3(2)} + (2)^2 - 2 = 2e^6 + 2$

Distance travelled during the first 2 seconds $= 2e^6 + 2 \approx 809\,\text{m}$.

Alternative method

Since there is no change in direction of motion:

$$s = \int v \, dt$$

$$= \int_0^2 (6e^{3t} + 2t) \, dt$$

$$= [2e^{3t} + t^2]_0^2$$

$$= (2e^6 + 4) - (2e^0 + 0)$$

$$= 2e^6 + 2$$

$$\approx 809\,\text{m}$$

> **TIP**
>
> This alternative method can be used only when there has been no change in direction of motion during the relevant time interval.

WORKED EXAMPLE 7

A particle starts from rest and moves in a straight line so that, t seconds after leaving a fixed point O, its velocity, $v\,\text{m s}^{-1}$, is given by $v = 4 + 8\cos 2t$.

a   Find the range of values of the velocity.

b   Find the range of values of the acceleration.

c   Find the value of t when the particle first comes to instantaneous rest.

d   Find the distance travelled during the time interval $0 \leqslant t \leqslant \dfrac{\pi}{2}$

Answers

a $v = 4 + 8\cos 2t$ and $-1 \leqslant \cos 2t \leqslant 1$

$v_{\min} = 4 + 8(-1) = -4$ and $v_{\max} = 4 + 8(1) = 12$

Hence, $-4 \leqslant v \leqslant 12$.

b $v = 4 + 8\cos 2t$

$a = \dfrac{dv}{dt} = -16\sin 2t$ and $-1 \leqslant \sin 2t \leqslant 1$

$a_{\min} = -16(1) = -16$ and $a_{\max} = -16(-1) = 16$

Hence, $-16 \leqslant a \leqslant 16$.

CONTINUED

c   When $v = 0$, $4 + 8\cos 2t = 0$

$$\cos 2t = -\frac{1}{2}$$

$$2t = \frac{2\pi}{3}$$

$$t = \frac{\pi}{3}$$

The particle first comes to rest when $t = \frac{\pi}{3}$

d $s = \int v \, dt$

$$= \int (4 + 8\cos 2t) \, dt$$

$$= 4t + 4\sin 2t + c$$

Using $s = 0$ when $t = 0$, gives $c = 0$

$$s = 4t + 4\sin 2t$$

Particle changes direction when $t = \frac{\pi}{3}$.

When $t = \frac{\pi}{3}$, $s = 4\left(\frac{\pi}{3}\right) + 4\sin\left(\frac{2\pi}{3}\right) \approx 7.6529$

When $t = \frac{\pi}{2}$, $s = 4\left(\frac{\pi}{2}\right) + 4\sin(\pi) = 2\pi \approx 6.2832$

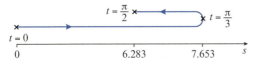

Total distance travelled $= 7.653 + (7.653 - 6.283) \approx 9.02 \, \text{m}$

Exercise 16.2

1   A particle, moving in a straight line, passes through a fixed point O.
Its velocity $v \, \text{m s}^{-1}$, t seconds after passing through O, is given by $v = 10t - t^2$.

   a   Find the velocity of the particle when the acceleration is $6 \, \text{m s}^{-2}$.

   b   Find the time taken before the particle returns to O.

   c   Find the distance travelled by the particle in the 2nd second.

   d   Find the distance travelled by the particle before it comes to instantaneous rest.

   e   Find the distance travelled by the particle in the first 12 seconds.

2   A particle, moving in a straight line, passes through a fixed point O.
Its velocity $v \, \text{m s}^{-1}$, t seconds after passing through O, is given by $v = \dfrac{32}{(t + 2)^2}$.

   a   Find the acceleration of the particle when $t = 2$.

   b   Find an expression for the displacement of the particle from O.

   c   Find the distance travelled by the particle in the 3rd second.

3 A particle, moving in a straight line, passes through a fixed point O.

Its velocity $v\,\text{m s}^{-1}$, t seconds after passing through O, is given by $v = 4e^{2t} + 2t$.

    a Find the acceleration of the particle when $t = 1$.

    b Find an expression for the displacement of the particle from O.

    c Find the total distance travelled by the particle in the first 2 seconds of its motion.

       Give your answer correct to the nearest metre.

4 A particle, moving in a straight line, passes through a fixed point O.

Its velocity $v\,\text{m s}^{-1}$, t seconds after passing through O, is given by $v = t + 2\cos\left(\dfrac{t}{3}\right)$.

Find the displacement of the particle from O when $t = \dfrac{3\pi}{2}$ and its acceleration at this instant.

5 A particle, moving in a straight line, passes through a fixed point O.

Its velocity $v\,\text{m s}^{-1}$, t seconds after passing through O, is given by $v = 4e^{2t} + 6e^{-3t}$.

    a Show that the velocity is never zero.

    b Find the acceleration when $t = \ln 2$.

    c Find, to the nearest metre, the displacement of the particle from O when $t = 2$.

6 A particle moves in a straight line, so that, t seconds after leaving a fixed point O, its velocity, $v\,\text{m s}^{-1}$, is given by $v = pt^2 + qt - 12$, where p and q are constants.

When $t = 2$, the acceleration of the particle is $18\,\text{m s}^{-2}$.

When $t = 4$, the displacement of the particle from O is $32\,\text{m}$.

Find the value of p and the value of q.

7 A particle moving in a straight line passes a fixed point O with velocity $10\,\text{m s}^{-1}$.

Its acceleration $a\,\text{m s}^{-2}$, t seconds after passing through O is given by $a = 3 - 2t$.

    a Find the value of t when the particle is instantaneously at rest.

    b Sketch the velocity-time graph for the motion of the particle.

    c Find the total distance travelled in the first 7 seconds of its motion.
       Give your answer correct to 3 sf.

8 A particle moving in a straight line passes a fixed point O with velocity $18\,\text{m s}^{-1}$.

Its acceleration $a\,\text{m s}^{-2}$, t seconds after passing through O is given by $a = 3t - 12$.

    a Find the values of t when the particle is instantaneously at rest.

    b Find the distance the particle travels in the 4th second.

    c Find the total distance travelled in the first 10 seconds of its motion.

9 A particle starts from rest and moves in a straight line so that, t seconds after leaving a fixed point O, its velocity, $v\,\text{m s}^{-1}$, is given by $v = 3 + 6\cos 2t$.

    a Find the range of values for the acceleration.

    b Find the distance travelled by the particle before it first comes to instantaneous rest.
       Give your answer correct to 3 sf.

10 A particle starts from rest at a fixed point O and moves in a straight line towards a point A.

The velocity, $v\,\text{m s}^{-1}$, of the particle, t seconds after leaving O, is given by $v = 8 - 8e^{-2t}$.

    a Find the acceleration of the particle when $t = \ln 5$.

    b Given that the particle reaches A when $t = 2$ find the distance OA.
       Give your answer correct to 3 sf.

11 A particle travels in a straight line so that, t seconds after passing through a fixed point O, its velocity, $v\,\mathrm{m\,s^{-1}}$, is given by $v = 2\cos\left(\dfrac{t}{2}\right) - 1$.

    a Find the value of t when the particle first comes to instantaneous rest at the point P.

    b Find the total distance travelled from $t = 0$ to $t = 2\pi$.

12 CHALLENGE QUESTION

A particle moves in a straight line so that t seconds after passing through a fixed point O, its acceleration, $a\,\mathrm{m\,s^{-2}}$, is given by $a = pt + q$, where p and q are constants. The particle passes through O with velocity $3\,\mathrm{m\,s^{-1}}$ and acceleration $-2\,\mathrm{m\,s^{-2}}$.

The particle first comes to instantaneous rest when $t = 2$.

    a Find the value of p and the value of q.

    b Find an expression, in terms of t, for the displacement of the particle.

    c Find the second value of t for which the particle is at instantaneous rest.

    d Find the distance travelled during the 4th second.

13 CHALLENGE QUESTION

A particle starts from a point O and moves in a straight line so that its displacement, $s\,\mathrm{cm}$, from O at time t seconds is given by $s = 2t\sin\dfrac{\pi t}{3}$.

    a Find expressions for the velocity, v, and the acceleration, a, of the particle at time t seconds.

    b Show that $18(vt - s) = t^2(9a + \pi^2 s)$.

SUMMARY

Relationships between displacement, velocity and acceleration

The relationships between displacement, velocity and acceleration are:

$$v = \frac{ds}{dt} \qquad s = \int v\,dt$$

$$a = \frac{dv}{dt} = \frac{d^2s}{dt^2} \qquad v = \int a\,dt$$

A particle is at instantaneous rest when $v = 0$.

Past paper questions

Worked example

A particle moves in a straight line such that its displacement, s metres, from a fixed point O at time t seconds, is given by $s = 4 + \cos 3t$, where $t \geqslant 0$. The particle is initially at rest.

i Find the exact value of t when the particle is next at rest. [2]

ii Find the distance travelled by the particle between $t = \dfrac{\pi}{4}$ and $t = \dfrac{\pi}{2}$ seconds. [3]

iii Find the greatest acceleration of the particle. [2]

Cambridge IGCSE Additional Mathematics 0606 Paper 21 Q10 Jun 2018

Answers

i $s = 4 + \cos 3t$ differentiate

$v = -3\sin 3t$

When $v = 0$, $-3\sin 3t = 0$

$\qquad\qquad \sin 3t = 0$

$\qquad\qquad 3t = 0, \pi, 2\pi, 3\pi, \ldots$

$\qquad\qquad t = 0, \dfrac{\pi}{3}, \dfrac{2\pi}{3}, \pi, \ldots$

Particle is next at rest when $t = \dfrac{\pi}{3}$.

ii The particle changes direction when $t = \dfrac{\pi}{3}$.

When $t = \dfrac{\pi}{4}$, $s = 4 + \cos\dfrac{3\pi}{4} = 4 - \dfrac{1}{\sqrt{2}} = 3.2929$

When $t = \dfrac{\pi}{3}$, $s = 4 + \cos\dfrac{3\pi}{3} = 3$

When $t = \dfrac{\pi}{2}$, $s = 4 + \cos\dfrac{3\pi}{2} = 4$

Distance travelled $= (3.2929 - 3) + (4 - 3)$

$\qquad\qquad = 0.2929 + 1$

$\qquad\qquad = 1.29\,\text{m to 3 sf}$

iii $v = -3\sin 3t$ differentiate

$a = -9\cos 3t$

Since $-1 \leqslant \cos 3t \leqslant 1$, maximum value of a is 9.

Hence greatest acceleration is $9\,\text{m s}^{-2}$.

1 A particle P is projected from the origin O so that it moves in a straight line. At time t seconds after projection, the velocity of the particle, $v\,\text{m s}^{-1}$, is given by $v = 2t^2 - 14t + 12$.

  i Find the time at which P first comes to instantaneous rest. [2]

  ii Find an expression for the displacement of P from O at time t seconds. [3]

  iii Find the acceleration of P when $t = 3$. [2]

Cambridge IGCSE Additional Mathematics 0606 Paper 21 Q6 Jun 2015

2 **a** A particle P moves in a straight line. Starting from rest, P moves with constant acceleration for 30 seconds after which it moves with constant velocity, $k\,\mathrm{m\,s}^{-1}$, for 90 seconds. P then moves with constant deceleration until it comes to rest; the magnitude of the deceleration is twice the magnitude of the initial acceleration.

      i Use the information to complete the velocity–time graph. [2]

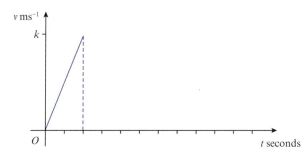

      ii Given that the particle travels 450 metres while it is accelerating, find the value of k and the acceleration of the particle. [4]

  b A body Q moves in a straight line such that, t seconds after passing a fixed point, its acceleration, $a\,\mathrm{m\,s}^{-2}$, is given by $a = 3t^2 + 6$. When $t = 0$, the velocity of the body is $5\,\mathrm{m\,s}^{-1}$.

      Find the velocity when $t = 3$. [5]

Cambridge IGCSE Additional Mathematics 0606 Paper 22 Q11 Mar 2015

3 A particle is moving in a straight line such that its velocity, $v\,\mathrm{m\,s}^{-1}$, t seconds after passing a fixed point O is
$v = \mathrm{e}^{2t} - 6\mathrm{e}^{-2t} - 1$.

  i Find an expression for the displacement, $s\,\mathrm{m}$, from O of the particle after t seconds. [3]

  ii Using the substitution $u = \mathrm{e}^{2t}$, or otherwise, find the time when the particle is at rest. [3]

  iii Find the acceleration at this time. [2]

Cambridge IGCSE Additional Mathematics 0606 Paper 21 Q10 Nov 2015

4 A particle P is projected from the origin O so that it moves in a straight line. At time t seconds after projection, the velocity of the particle, $v\,\mathrm{m\,s}^{-1}$, is given by $v = 9t^2 - 63t + 90$.

  i Show that P first comes to instantaneous rest when $t = 2$. [2]

  ii Find the acceleration of P when $t = 3.5$. [2]

  iii Find an expression for the displacement of P from O at time t seconds. [3]

  iv Find the distance travelled by P

      a in the first 2 seconds, [2]

      b in the first 3 seconds. [2]

Cambridge IGCSE Additional Mathematics 0606 Paper 22 Q12 Mar 2016

5 A particle P, moving in a straight line, passes through a fixed point O at time $t = 0\,\mathrm{s}$.
At time $t\,\mathrm{s}$ after leaving O, the displacement of the particle is $x\,\mathrm{m}$ and its velocity is $v\,\mathrm{m\,s}^{-1}$, where $v = 12\mathrm{e}^{2t} - 48t$, $t \geqslant 0$.

  i Find x in terms of t. [4]

  ii Find the value of t when the acceleration of P is zero. [3]

  iii Find the velocity of P when the acceleration is zero. [2]

Cambridge IGCSE Additional Mathematics 0606 Paper 12 Q8 Mar 2018

6 a

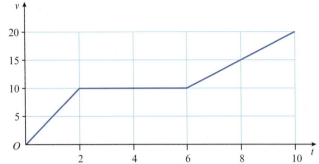

The diagram shows the velocity-time graph of a particle P moving in a straight line with velocity $v\,\mathrm{m\,s^{-1}}$ at time t seconds after leaving a fixed point.

   i   Write down the value of the acceleration of P when $t = 5$. [1]

   ii Find the distance travelled by the particle P between $t = 0$ and $t = 10$. [2]

b A particle Q moves such that its velocity, $v\,\mathrm{m\,s^{-1}}$, t seconds after leaving a fixed point, is given by $v = 3\sin 2t - 1$.

   i   Find the speed of Q when $t = \dfrac{7\pi}{12}$. [2]

   ii Find the least value of t for which the acceleration of Q is zero. [3]

Cambridge IGCSE Additional Mathematics 0606 Paper 12 Q8 Mar 2019

7 A particle P moves in a straight line such that, t seconds after passing through a fixed point O, its acceleration, $a\,\mathrm{m\,s^{-2}}$, is given by $a = -6$. When $t = 0$, the velocity of P is $18\,\mathrm{m\,s^{-1}}$.

a Find the time at which P comes to instantaneous rest. [3]

b Find the distance travelled by P in the 3rd second. [3]

Cambridge IGCSE Additional Mathematics 0606 Paper 22 Q12 Mar 2020

8 A particle P travels in a straight line such that, t seconds after passing through a fixed point O, its velocity, $v\,\mathrm{m\,s^{-1}}$, is given by

$$v = \frac{t}{2e} \qquad \text{for } 0 \leqslant t \leqslant 2,$$

$$v = e^{-\frac{t}{2}} \qquad \text{for } t > 2.$$

Given that, after leaving O, particle P is never at rest, find the distance it travels between $t = 1$ and $t = 3$. [6]

Cambridge IGCSE Additional Mathematics 0606 Paper 22 Q12 Mar 2021

9 A particle moves in a straight line so that, t seconds after passing through a fixed point O, its displacement, $s\,\mathrm{m}$, from O is given by

$$s = 1 + 3t - \cos 5t.$$

   i   Find the distance between the particle's first two positions of instantaneous rest. [7]

   ii Find the acceleration when $t = \pi$. [2]

Cambridge IGCSE Additional Mathematics 0606 Paper 21 Q12 Jun 2017

10

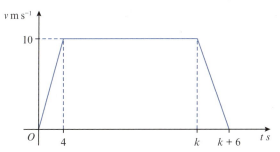

The velocity–time graph represents the motion of a particle travelling in a straight line.

i   Find the acceleration during the last 6 seconds of the motion.   [1]

ii   The particle travels with constant velocity for 23 seconds. Find the value of k.   [1]

iii Using your answer to **part ii**, find the total distance travelled by the particle.   [3]

Cambridge IGCSE Additional Mathematics 0606 Paper 21 Q5 Jun 2019

11 A particle travels in a straight line. As it passes through a fixed point O, the particle is travelling at a velocity of $3\,\text{m\,s}^{-1}$. The particle continues at this velocity for 60 seconds then decelerates at a constant rate for 15 seconds to a velocity of $1.6\,\text{m\,s}^{-1}$. The particle then decelerates again at a constant rate for 5 seconds to reach point A, where it stops.

a   Sketch the velocity-time graph for this journey on the axes below.   [3]

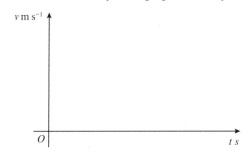

b   Find the distance between O and A.   [3]

c   Find the deceleration in the last 5 seconds.   [1]

Cambridge IGCSE Additional Mathematics 0606 Paper 21 Q9 Jun 2020

12 A particle is moving in a straight line such that t seconds after passing a fixed point O its displacement, s m, is given by $s = 3\sin 2t + 4\cos 2t - 4$.

i   Find expressions for the velocity and acceleration of the particle at time t.   [3]

ii   Find the first time when the particle is instantaneously at rest.   [3]

iii Find the acceleration of the particle at the time found in **part ii**.   [2]

Cambridge IGCSE Additional Mathematics 0606 Paper 21 Q5 Nov 2019

13 A particle moves in a straight line such that, t seconds after passing a fixed point O, its displacement, s, from O is s m, where $s = e^{2t} - 10e^{t} - 12t + 9$.

i   Find expressions for the velocity and acceleration at time t.   [3]

ii   Find the time when the particle is instantaneously at rest.   [3]

iii Find the acceleration at this time.   [2]

Cambridge IGCSE Additional Mathematics 0606 Paper 21 Q9 Nov 2020

> Answers

The questions and example answers that appear in this resource were written by the author. In examination, the way marks would be awarded to answers like these may be different.

Chapter 1

Pre-requisite Knowledge

1 14

2 $10 - 3x$

3 $f^{-1}(x) = \dfrac{x-5}{3}$

4 a

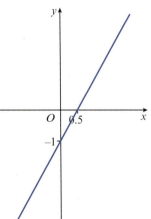

b

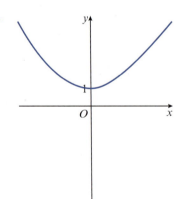

5 a $x = -1$

   b $x = 2$ or $x = -6$

Exercise 1.1

1 one-one

2 many-one

3 one-one

4 one-one

5 one-one

6 one-one

7 one-one

8 one-many

Exercise 1.2

1 1, 2, 3, 4, 5, 6 and 7

2 a $-7 \le f(x) \le 2$

   b $2 \le f(x) \le 17$

   c $-1 \le f(x) \le 9$

   d $0 \le f(x) \le 9$

   e $\dfrac{1}{8} \le f(x) \le 8$

   f $\dfrac{1}{5} \le f(x) \le 1$

3 $g(x) \ge 2$

4 $f(x) \ge -4$

5 $f(x) \ge 5$

6 $f(x) \ge -5$

7 $f(x) \le 10$

8 $f(x) \ge 3$

Exercise 1.3

1 9

2 51

3 675

4 1

5 67

6 **a** hk **b** kh

7 $x = -\dfrac{1}{3}$

8 $x = -\dfrac{2}{3}$ or $x = 4$

9 $x = \pm\dfrac{3}{4}$

10 $x = 2.5$

11 $x = 5$

12 $x = 3$ or $x = 4$

13 **a** fg **b** gf

    c g^2 **d** f^2

14 $x \in \mathbb{R}, x \neq -\dfrac{3}{2}, x \neq 0$

15 Domain: $x \in \mathbb{R}, x > 20$
Range: $fg(x) \in \mathbb{R}, fg(x) > -9$

16 **a** $x \in \mathbb{R}, x \neq 1$

    b $x \in \mathbb{R}, x \neq 0, x \neq 1$

17 **a** Domain: $x \in \mathbb{R}, x \geq 0$
     Range: $fg(x) \in \mathbb{R}, fg(x) \geq -6$

    b Domain: $x \in \mathbb{R}, x \geq 3$
     Range: $gf(x) \in \mathbb{R}, gf(x) \geq 0$

18 **a** **i** $f(x) \in \mathbb{R}, f(x) \geq -\dfrac{1}{8}$

      ii $g(x) \in \mathbb{R}, g(x) \neq 0$

    b **i** Domain: $x \in \mathbb{R}, x \neq 0$

        Range: $fg(x) \in \mathbb{R}, fg(x) \geq -\dfrac{1}{8}$

      ii Domain: $x \in \mathbb{R}, x \neq 0, x \neq \dfrac{1}{2}$

        Range: $gf(x) \in \mathbb{R}, gf(x) \leq -8$
        or $gf(x) > 0$

19 **a** **i** $f(x) \in \mathbb{R}, f(x) < 9$

      ii $g(x) \in \mathbb{R}, g(x) > 0$

    b $gf(x) = (2x + 2)^2$

    c Domain: $x \in \mathbb{R}, -1 < x < 3$
     Range: $gf(x) \in \mathbb{R}, 0 < gf(x) < 64$

20 **a** **i** $f(x) \in \mathbb{R}, f(x) > 5$

      ii $g(x) \in \mathbb{R}, g(x) > 47$

    b $gf(x) = 4x^2 - 4x - 1$

    c Domain: $x \in \mathbb{R}, x > 4$
     Range: $gf(x) \in \mathbb{R}, gf(x) > 47$

21 **a** **i** $f(x) \in \mathbb{R}, f(x) > 5$

      ii $g(x) \in \mathbb{R}, g(x) > 65$

    b $gf(x) = 9x^2 + 12x + 5$

    c Domain: $x \in \mathbb{R}, x > 2$
     Range: $gf(x) \in \mathbb{R}, gf(x) > 65$

Exercise 1.4

1 **a** $x = -2\dfrac{2}{3}, x = 4$

    b $x = -2, x = -7$

    c $x = 0.8, x = 1.6$

    d $x = -23, x = 25$

    e $x = -5, x = -2$

    f $x = -0.5, x = 7.5$

    g $x = 16, x = 24$

    h $x = -5, x = 3\dfrac{8}{9}$

    i $x = 1\dfrac{2}{3}, x = 5$

2 **a** $x = -4\dfrac{5}{6}, x = -1.9$

    b $x = -0.8, x = 0$

    c $x = -5.6, x = 4$

    d $x = 0.75, x = 3.5$

    e $x = 6.5$

    f $x = 3\dfrac{1}{3}$

3 **a** $x = -2, x = 2$

    b $x = -3, x = 3$

    c $x = -3, x = -1, x = 2$

    d $x = 0, x = 4, x = 6$

    e $x = -2, x = 1, x = 3$

    f $x = -2, x = -1, x = 0, x = 3$

    g $x = 0.5, x = 1$

    h $x = -1, x = 2$

    i $x = 0, x = 2, x = 6$

4 **a** $(-4, 0), (3, 7), (5, 9)$

    b $(0, 0), (1, 1), (2, 2)$

    c $(1, 3), (2.5, 7.5)$

Exercise 1.5

1 **a** $\vee$ shape, vertex at $(-1, 0)$, y-intercept 1

    b $\vee$ shape, vertex at $(1.5, 0)$, y-intercept 3

    c $\vee$ shape, vertex at $(5, 0)$, y-intercept 5

    d $\vee$ shape, vertex at $(-6, 0)$, y-intercept 3

    e $\vee$ shape, vertex at $(5, 0)$, y-intercept 10

    f $\vee$ shape, vertex at $(18, 0)$, y-intercept 6

2 a

x	−2	−1	0	1	2	3	4
y	7	6	5	4	3	4	5

b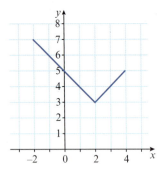

3 a ∨ shape, vertex at (0, 1)

  b ∨ shape, vertex at (0, −3)

  c ∧ shape, vertex at (0, 2)

  d ∨ shape, vertex at (3, 1), y-intercept 4

  e ∨ shape, vertex at (−3, −3), y-intercept 3

4 a $-3 \leqslant f(x) \leqslant 11$

  b $0 \leqslant g(x) \leqslant 11$

  c $-3 \leqslant h(x) \leqslant 5$

5 $-5 \leqslant f(x) < 5, 0 \leqslant g(x) < 5, -5 \leqslant h(x) < 3$

6 a ∨ shape, vertex at (−2, 0), y-intercept 4

  b straight line through (−5, 0), (0, 5)

  c $x = -3, x = 1$

7 a ∨ shape, vertex at (3, −3), y-intercept 3

  b $-3 \leqslant f(x) \leqslant 7$

  c $x = 0.5, x = 5.5$

8 a ∨ shape, vertex at $\left(\frac{4}{3}, 0\right)$, y-intercept 4

  b straight line through (−2, −4), (5, 10)

  c $x = 0.8, x = 4$

9 a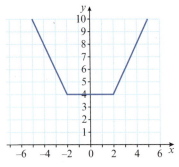

  b $x = -3, x = 3$

Exercise 1.6

1 $f^{-1}(x) = \sqrt{x + 7} - 5$

2 $f^{-1}(x) = \dfrac{6 - 2x}{x}$

3 $f^{-1}(x) = \dfrac{3 + \sqrt{x - 1}}{2}$

4 $f^{-1}(x) = (8 - x)^2 + 3$

5 $f^{-1}(x) = \dfrac{x + 3}{5}, g^{-1}(x) = \dfrac{2x - 7}{x}$

6 a $f^{-1}(x) = \sqrt{x + 5} - 2$

  b $x = 20$

7 a $f^{-1}(x) = 4 + \sqrt{x - 5}$

  b $x = 294$

8 a $g^{-1}(x) = \dfrac{x + 3}{x - 2}$

  b $x = 3.25$

9 a $f^{-1}(x) = 2(x - 2)$

  b no solution

10 $x = 0$

11 $x = -2, 3.5$

12 $f^{-1}(x) \geqslant 0$

13 $f^{-1}g^{-1}$

14 f, h

15 a $x \geqslant 2$

  b $g(x) \geqslant 0$

16 a $f^{-1}(x) = \dfrac{x + k}{3}, g^{-1}(x) = \dfrac{x + 14}{5 - x}$

  b $k = 13$

  c x

17 a $f^{-1}g$ **b** $g^{-1}f$

  c gf^{-1} **d** $f^{-1}g^{-1}$

Exercise 1.7

1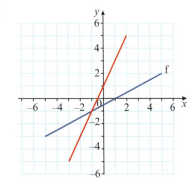

2

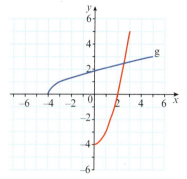

3

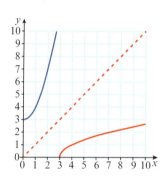

4

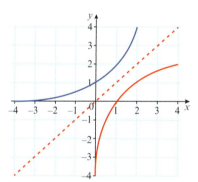

5

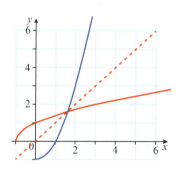

6

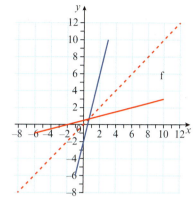

7 **a** f is a one-one function

b $f^{-1}(x) = \sqrt[3]{3-x} - 1$

c

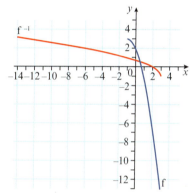

8 **a** $f^{-1}(x) = \dfrac{2x+7}{x-2}$

b The curve is symmetrical about the line $y = x$

Past paper questions

1 **a** $\dfrac{1}{3}$

b $f^{-1}(x) = (x+3)^2 + 1$

c $g^{-1}(x) = \dfrac{3x-2}{2x-1}$

2 **i** $2 - \sqrt{5} < f(x) \leqslant 2$

ii $f^{-1}(x) = (2-x)^2 - 5$

domain: $2 - \sqrt{5} < x \leqslant 2$

range: $-5 \leqslant f^{-1}(x) < 0$

iii $x = -4$

3 **i**

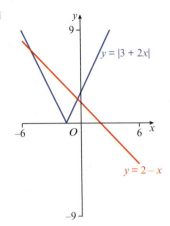

$y = |3 + 2x|$

$y = 2 - x$

ii $x = -\dfrac{1}{3}, x = -5$

4

	A	B	C	D
		✓		
			✓	✓
			✓	
	✓			

5 **i** $f^{-1}(x) = \dfrac{5x + 1}{2x}$

ii $x > 0$

iii $\dfrac{2x - 5}{-10x + 27}$

6 **i**

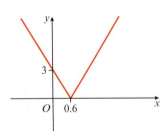

ii $x = \dfrac{1}{4}$ or $x = \dfrac{5}{6}$

7 **a** **i** $f(x) > 3$

ii $x = 0$ or $x = \dfrac{1}{2}$

b **i** $h^{-1}(x) = \dfrac{4x + 1}{x - 2}$, range $h^{-1}(x) \neq 4$

ii $h^2(x) = \dfrac{5x - 2}{17 - 2x}$

8 **i** $g(x) \geqslant -\dfrac{1}{2}$

ii $g(1) = 0$, domain of f is $x \geqslant 2$

iii $\dfrac{1}{2}x^2 - \dfrac{5}{2} + \dfrac{2}{x^2}$

iv $x \geqslant 2$

v Put into a standard quadratic form and then use the quadratic formula. Discard the negative square root.

9 **i**

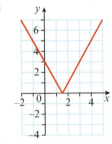

ii $x = 5$ or $x = -2$

Chapter 2

Pre-requisite Knowledge

1 **a** $x = \dfrac{5}{2}, y = -3$

b $x = 3, y = 5$

2 **a** $x = 2, y = -4$

b $x = 8, y = \dfrac{3}{2}$

3 **a** $x = -3, x = 2$

b $x = 2, x = 8$

c $x = -\dfrac{5}{2}, x = \dfrac{2}{3}$

4 **a** $2\left(x + \dfrac{7}{4}\right)^2 - \dfrac{25}{8}$

b $x = -3, x = -\dfrac{1}{2}$

5 $x = 1.22, x = 3.28$

Exercise 2.1

1 $x = 3, y = 9$ and $x = -2, y = 4$

2 $x = -1, y = -7$ and $x = 4, y = -2$

3 $x = -3, y = -4$ and $x = 4, y = 3$

4 $x = 1, y = 4$ and $x = -2, y = -2$

5 $x = 0.5, y = 0.5$ and $x = 0, y = 1$

6 $x = -1, y = -3$ and $x = 2, y = 1$

7 $x = 1.5, y = 4$ and $x = 2, y = 3$

8 $x = 3, y = 1$ and $x = 9, y = 7$

9 $x = 1.8, y = 2.6$ and $x = 1, y = 3$

10 $x = -1, y = -2$ and $x = 1, y = 2$

11 $x = 1, y = 2$ and $x = 2, y = 1$

12 $x = 1, y = 2$ and $x = 4, y = -4$

13 $x = 1, y = -\frac{1}{3}$ and $x = -\frac{1}{2}, y = \frac{1}{6}$

14 $x = 3, y = 1$ and $x = 1, y = 3$

15 $x = -1, y = -3$ and $x = 1, y = 3$

16 $x = 0, y = -0.5$ and $x = -1, y = -1$

17 $x = -1, y = 2$ and $x = 7\frac{1}{2}, y = -1.4$

18 $x = -1.5, y = -8$ and $x = 4, y = 3$

19 $(-0.2, 1.4)$ and $(1, -1)$

20 **a** $x + y = 11, xy = 21.25$

    b $x = 2.5, y = 8.5$ and $x = 8.5, y = 2.5$

21 17 cm and 23 cm

22 $6\sqrt{5}$ or 13.4 to 3 sf

23 $(0.5, 0)$

24 $5\sqrt{2}$ or 7.07 to 3 sf

25 $(2, 2)$

26 $y = -2x - 1$

Exercise 2.2

1 **a** min $(2.5, -12.25)$, axis crossing points $(-1, 0)$, $(6, 0)$, $(0, -6)$

    b min $(0.5, -20.25)$, axis crossing points $(-4, 0)$, $(5, 0)$, $(0, -20)$

    c min $(-2, -25)$, axis crossing points $(-7, 0)$, $(3, 0)$, $(0, -21)$

    d min $(-1.5, -30.25)$, axis crossing points $(-7, 0)$, $(4, 0)$, $(0, -28)$

    e min $(-2, -3)$, axis crossing points $(-2 - \sqrt{3}, 0)$, $(-2 + \sqrt{3}, 0)$, $(0, 1)$

    f max $(1, 16)$, axis crossing points $(-3, 0)$, $(5, 0)$, $(0, 15)$

2 **a** $(x - 4)^2 - 16$

    b $(x - 5)^2 - 25$

    c $(x - 2.5)^2 - 6.25$

    d $(x - 1.5)^2 - 2.25$

    e $(x - 2)^2 - 4$

    f $(x - 3.5)^2 - 12.25$

    g $(x - 4.5)^2 - 20.25$

    h $(x + 1.5)^2 - 2.25$

3 **a** $(x - 4)^2 - 1$

    b $(x - 5)^2 - 30$

    c $(x - 3)^2 - 7$

    d $(x - 1.5)^2 + 1.75$

    e $(x + 3)^2 - 4$

    f $(x + 3)^2 - 0$

    g $(x + 2)^2 - 21$

    h $(x + 2.5)^2 - 0.25$

4 **a** $2(x - 2)^2 - 5$

    b $2(x - 3)^2 - 17$

    c $3(x - 2)^2 - 7$

    d $2(x - 0.75)^2 + 0.875$

    e $2(x + 1)^2 - 1$

    f $2(x + 1.75)^2 - 9.125$

    g $2(x - 0.75)^2 + 3.875$

    h $3\left(x - \frac{1}{6}\right)^2 + 5\frac{11}{12}$

5 **a** $9 - (x - 3)^2$

    b $25 - (x - 5)^2$

    c $2.25 - (x - 1.5)^2$

    d $16 - (x - 4)^2$

6 **a** $6 - (x + 1)^2$

    b $12 - (x + 2)^2$

    c $16.25 - (x + 2.5)^2$

    d $9.25 - (x - 1.5)^2$

7 **a** $13.5 - 2(x + 1.5)^2$

    b $3 - 2(x + 1)^2$

    c $15 - 2(x - 2)^2$

    d $4\frac{1}{12} - 3\left(x - \frac{5}{6}\right)^2$

8 **a** $4\left(x + \frac{1}{4}\right)^2 + 4.75$

    b No, since $4.75 > 0$

9 **a** $2(x - 2)^2 - 7$

    b $(2, -7)$

10 **a** $-5.25, 0.5$

    b $x \geqslant 0.5$

11 **a** $11\frac{1}{8} - 2\left(x + 1\frac{3}{4}\right)^2$

    b $f(x) \leqslant 11\frac{1}{8}$

12 **a** $18.5 - 2(x - 1.5)^2$

    b $(1.5, 18.5)$

    c $\cap$ shaped curve, vertex $= (1.5, 18.5)$

13 **a** $13.25 - (x - 2.5)^2$

    b $(2.5, 13.25)$, maximum

    c $-7 \leqslant f(x) \leqslant 13.25$

    d No, it is not a one-one function.

14 **a** $2(x-2)^2 - 5$

  b $x \geqslant 2$

15 -0.75

16 **a** $5 - (x-2)^2$

  b $(2, 5)$ maximum

  c One-one function, $f^{-1}(x) = 2 + \sqrt{5-x} \geqslant 5$

Exercise 2.3

1 **a**

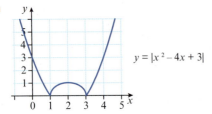

$y = |x^2 - 4x + 3|$

  b

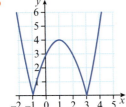

$y = |x^2 - 2x - 3|$

  c

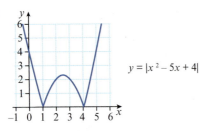

$y = |x^2 - 5x + 4|$

  d

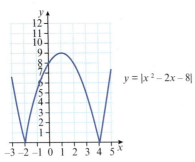

$y = |x^2 - 2x - 8|$

  e

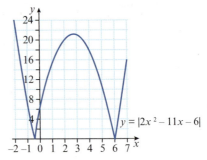

$y = |2x^2 - 11x - 6|$

  f

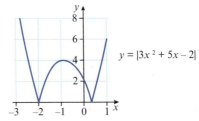

$y = |3x^2 + 5x - 2|$

2 **a** $f(x) = 5 - (x+2)^2$

  b $\cap$ shaped curve, vertex $= (-2, 5)$

  c

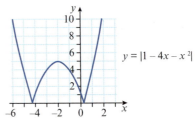

$y = |1 - 4x - x^2|$

3 **a** $f(x) = 2(x + 0.25)^2 - 3.125$

  b

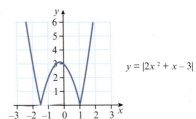

$y = |2x^2 + x - 3|$

4 **a** $(3, 16)$

  b

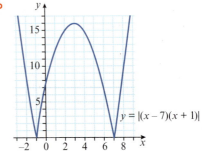

$y = |(x-7)(x+1)|$

  c $0 < k < 16$

5 a $(-3, 4)$

   b $k > 4$

6 a $(5.5, 6.25)$

   b $k = 6.25$

7 a $x = -4, x = 4$

   b $x = -2, x = 0, x = 2$

   c $x = -1, x = 2, x = 3, x = 6$

   d $x = -6, x = 4$

   e $x = \dfrac{5 - \sqrt{33}}{2}, x = 1, x = 4, x = \dfrac{5 + \sqrt{33}}{2}$

   f $x = -4, x = -2, x = -1, x = 1$

   g $x = -1 - \sqrt{10}, x = -1, x = -1 + \sqrt{10}$

   h $x = \dfrac{1 + \sqrt{7}}{2}, x = \dfrac{-1 + \sqrt{7}}{2}$

   i $x = 1, x = 3$

8 a $(-1, 0), (2, 3), (4, 5)$

   b $(-2, 1), (-1, 1.5), (2, 3), (5, 4.5)$

   c $(1, 2), (2, 4)$

Exercise 2.4

1 a $x < -3, x > 4$

   b $1 \leqslant x \leqslant 6$

   c $x \leqslant -7, x \geqslant 3$

   d $0 < x < 5$

   e $-0.5 < x < 4$

   f $-1 \leqslant x \leqslant 3$

   g $-1.5 < x < 5$

   h $-\infty < x < \infty$

   i $x = 3$

2 a $-7 < x < 2$

   b $x \leqslant -3, x \geqslant 2$

   c $4 \leqslant x \leqslant 5$

   d $x < -8, x > 6$

   e $-2.5 \leqslant x \leqslant 3$

   f $x < -1, x > -0.8$

3 a $-6 < x < 3$

   b $x < 5, x > 7$

   c $x \leqslant 0.5, x \geqslant 1$

   d $-3 < x < 2$

   e $x < -4, x > 1$

   f $-0.5 < x < 0.6$

4 a $3 < x < 5$

   b $1 < x < 6$

   c $0.5 < x < 1$

   d $3 < x < 5$

   e $x < -2, x \geqslant 3$

5 a $-5 < x < 3$

   b $0 < x < 2, 6 < x < 8$

   c $0 < x < 2, 4 < x < 6$

6 $-1\dfrac{1}{3} < x < 2$

Exercise 2.5

1 a two equal roots

   b two distinct roots

   c two distinct roots

   d no roots

   e two distinct roots

   f two equal roots

   g no roots

   h two distinct roots

2 $k = \pm 6$

3 $k < 0.5$

4 $k > \dfrac{1}{3}$

5 $0, -\dfrac{8}{9}$

6 $k > -1.5$

7 $k > 3\dfrac{2}{3}$

8 $k = -10, k = 14$

9 $k = 1, k = 4$

Exercise 2.6

1 $k = -3, k = 5$

2 $k = -7, k = -3$

3 $c = \pm 4$

4 $k < 1, k > 5$

5 a $k = \pm 10$

   b $(-2, 6), (2, -6)$

6 $k > -5$

7 $k \geqslant 0.75$

8 $-11 < m < 1$

9 $m = -2, m = -6$

Past paper questions

1 $3 < k < 4$

2 $-1 < x < 0$

3 **a** $2\left(x - \frac{1}{4}\right)^2 + \frac{47}{8}$

  b $\frac{47}{8}$ when $x = \frac{1}{4}$

4 $k < 2$

5 **a** $-5 \le x \le \frac{1}{4}$

  b **i** $(x + 4)^2 - 25$

   ii $25, -4$

   iii

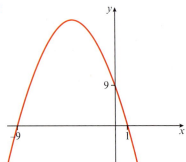

6 **i** $\left(-\frac{1}{3}, \frac{8}{3}\right), (1, 0)$

  ii $-3x + 6y = 7$

7 **a** $9\left(x - \frac{2}{3}\right)^2 + 1$

  b $\left(\frac{2}{3}, 1\right)$

8 **a** $(-2, -4), \left(\frac{4}{5}, 10\right)$

  b $\left(\frac{12}{5}, \frac{12}{5}\right)$

9 $x > 7$ or $x < -1$

10 $x = 4, y = 8$

  $x = -2, y = 2$

11 $k = 12, k = 4$

12 $k = \frac{1}{9}, k = 9$

13 midpoint $= (0.5, 10.5)$

14 **a** $2\left(x + \frac{5}{4}\right)^2 - \frac{49}{8}$

  b $\left(-\frac{5}{4}, -\frac{49}{8}\right)$

c

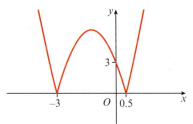

d $\frac{49}{8}$

15 **i** $\left(x - \frac{9}{2}\right)^2 - \frac{49}{4}$

  ii $\left(\frac{9}{2}, -\frac{49}{4}\right)$

  iii

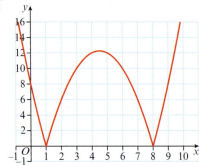

  iv $\frac{49}{4}$

Chapter 3

Pre-requisite Knowledge

1 $3x^3 - 8x^2 + 19x - 10$

2 $(2x - 3)(x + 4)$

3 **a** 122

  b 370 remainder 2

  c 384 remainder 18

Exercise 3.1

1 **a** $3x^4 + 2x^3 + 3x^2$

  b $9x^4 + 2x^3 + 7x^2 - 2$

  c $3x^4 - 4x^3 - 3$

  d $6x^7 + 3x^6 + 4x^5 + 5x^4 - 2x^3 + x^2 - 1$

2 **a** $8x^4 - 4x^3 + 2x^2 + 3x - 2$

  b $3x^4 + 8x^3 + 4x^2 - 3x - 2$

  c $3x^5 + 5x^4 - 3x^3 + 7x^2 + 8x - 20$

  d $3x^5 + 12x^4 + 13x^3 + 3x^2 - 4$

    e $x^4 - 10x^3 + 29x^2 - 20x + 4$

    f $27x^3 - 27x^2 + 9x - 1$

3 **a** $3x^2 + x - 7$

    b $2x^3 + 17x^2 + 21x - 4$

    c $2x^5 + 5x^4 - 8x^3 - x^2 - 19x$

4 **a** $x^3 + 7x^2 + x - 4$

    b $4x^4 - 4x^3 - 15x^2 + 8x + 16$

    c $8x^4 - 8x^3 - 32x^2 + 17x + 32$

    d $4x^4 - 4x^3 - 5x^2 + 3x - 2$

Exercise 3.2

1 **a** $x^2 + 2x - 48$ **b** $x^2 + x - 1$

    c $x^2 - 15x + 25$ **d** $x^2 + 2x + 1$

    e $x^2 + 4x - 5$ **f** $x^2 - 9$

2 **a** $3x^2 + 2x - 1$ **b** $2x^2 + 3x - 2$

    c $3x^2 - 5x - 10$ **d** $3x^2 + 4$

3 **a** $3x^2 - 4$ **b** $2x^2 - x + 5$

    c $x - 4$ **d** $x + 3$

4 **a** $x^3 - x^2 + x - 1$ **b** $x^2 + 2x + 4$

Exercise 3.3

2 **a** $a = 29$ **b** $a = 546$ **c** $a = 12$

3 $b = -2 - 2a$

4 **a** $a = 0, b = -19$

    b $a = 2, b = 38$

    c $a = -12, b = -7$

5 $a = -6$

6 $a = 13.5, b = 5.5$

7 **a** $p = 1, q = 9$

8 **b** $a = 0, 1, 3$

Exercise 3.4

1 **b** $(2x - 1)(x + 1)(x - 1)$

2 **a** $(x - 2)(x^2 + 4x + 5)$

    b $(x + 4)(x + 2)(x - 2)$

    c $x(2x + 3)(x - 6)$

    d $(x - 7)(x + 1)(x - 2)$

    e $(2x + 1)(x - 3)(x - 4)$

    f $(x - 2)(x + 3)(3x - 1)$

    g $(2x - 1)(2x + 1)(x - 2)$

    h $(2x - 1)(x - 3)(x + 5)$

3 **a** $x = -5, x = 1$ or $x = 7$

    b $x = 1, x = 2$ or $x = 3$

c $x = -4, x = -2$ or $x = \dfrac{1}{3}$

d $x = -4, x = 1$ or $x = 1.5$

e $x = -2, x = 0.5$ or $x = 3$

f $x = -4, x = -\dfrac{1}{2}$ or $x = 1$

g $x = -2, x = -1.5$ or $x = 0.5$

h $x = -4, x = 2.5$ or $x = 3$

4 **a** $x = 1$ or $x = -3 \pm \sqrt{7}$

    b $x = -3$ or $x = -\dfrac{5}{2} \pm \dfrac{1}{2}\sqrt{37}$

    c $x = 2$ or $x = -2 \pm \sqrt{3}$

    d $x = 1.5, x = 1, x = -4$

5 $x = -0.5$ or $x = -2 \pm \sqrt{13}$

6 $x = -5.54, x = -3$ or $x = 0.54$

7 **b** $x = 2$

8 **a** $x^3 - 4x^2 - 7x + 10$

    b $x^3 + 3x^2 - 18x - 40$

    c $x^3 + x^2 - 6x$

9 **a** $2x^3 - 11x^2 + 10x + 8$

    b $2x^3 - 7x^2 + 7x - 2$

    c $2x^3 - 9x^2 - 8x + 15$

10 $x^3 + x^2 - 7x - 3$

11 $2x^3 - 9x^2 + 6x - 1$

12 **b** $a = 2$ or $a = \dfrac{1 \pm \sqrt{33}}{8}$

Exercise 3.5

1 **a** 5 **b** -1

    c 76 **d** 2

2 **a** $a = 5$ **b** $b = 57$ **c** $c = 11$

3 $a = 4, b = 0$

4 $a = -6, b = -6$

5 **a** $a = -8, b = 15$

    b $x = 3, x = \dfrac{-1 + \sqrt{21}}{2}$ or $x = \dfrac{-1 - \sqrt{21}}{2}$

6 **a** $a = -9, b = 2$

    b 5

7 $b = 2$

8 $a = 6, b = -3$

9 **a** $b - 12 - 2a$ **b** $a = 5, b = 2$

10 **a** $k = 5$ **b** -72

11 **a** $a = -8, b = -5$ **b** -30

12 $k = 32$

13 **b** $a = 3, a = \dfrac{-7 + \sqrt{13}}{6}$ or $a = \dfrac{-7 - \sqrt{13}}{6}$

14 a $k = 3$

b -5

15 $a = 2, b = -5, c = 7$

16 b $k = -3$ or $k = 6$

Past paper questions

1 ii $(2x + 1)(x - 3)(x + 4)$

$x = -4, x = -\frac{1}{2}$ or $x = 3$

2 i $a = 3, b = 2$

ii $-\frac{35}{27}$

3 i -12

ii $k = -2$

iii discriminant $= -95$

4 a 0

b $(2x - 1)(x - 4)(x + 3)$

$x = -3, x = \frac{1}{2}$ or $x = 4$

5 ii $(2x - 1)(3x^2 + 5x - 2)$

$c = 3, d = 5, e = -2$

iii $(2x - 1)(3x - 1)(x + 2)$

6 ii $(x - 1)^2(x - 2)(x + 2)$

7 $a = 10, b = -12$

8 a $a = 6, b = -24$

b $(2x + 1)(3x^2 - 12)$

c $x = -\frac{1}{2}, x = \pm 2$

Chapter 4

Pre-requisite Knowledge

1 a $x = -\frac{4}{3}$ or $x = 2$

b $x = 3$

2 a $x = -3$ or $x = 2$

b $x = 1$

3 $x = -7$ or $x = 5$

Exercise 4.1

1 a $x = 1$ or $x = \frac{1}{3}$

b $x = -\frac{1}{2}$

c $x = -1$ or $x = \frac{7}{3}$

d $x = -1$ or $x = 0$

e $x = -\frac{1}{3}$ or $x = \frac{3}{5}$

f $x = -\frac{6}{5}$ or $x = -\frac{2}{7}$

g $x = -\frac{3}{5}$ or $x = 7$

h $x = \frac{7}{4}$

i $x = -\frac{14}{3}$ or $x = -\frac{6}{7}$

2 $x = \frac{13}{2}$ or $y = \frac{3}{2}$

3 $x = -\frac{7}{3}$ or $x = -\frac{5}{3}$

4 a $x = \pm 2$ or $x = \pm 4$

c $f(x) \geqslant -1$

5 $x = -2$ or or $x = \frac{10}{3}$

6 $x = 0, y = 5$ or $x = 2, y = 3$

7 $x = 4, y = -\frac{5}{2}$

Exercise 4.2

1 $4 < x < 8$

2 a

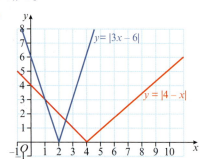

b $x \leqslant 1$ or $x \geqslant \frac{5}{2}$

3 a $x < -1$ or $x > 4$

b $-1 \leqslant x \leqslant \frac{13}{5}$

c $2 < x < \frac{10}{3}$

d $x < 2$ or $x > 5$

e $x < -3$ or $x > \frac{7}{3}$

f $-1 \leqslant x \leqslant 6$

4 a $\frac{4}{3} \leqslant x \leqslant 2$

b $x > \frac{2}{3}$

c $x \geqslant \frac{1}{4}$

5 **a** $x \le -1$ or $x \ge \frac{1}{5}$

b $x > -\frac{1}{2}$

c $\frac{1}{2} < x < 1$

d $x < -1$ or $x > -\frac{3}{5}$

e $-1 \le x \le 3$

f $-3 < x < 1$

6 **a** $x > \frac{3}{2}$

b $x \le -\frac{3}{2}$

c $x \le -2$ or $x \ge -\frac{3}{2}$

d $-6 \le x \le 0$

e $-14 < x < 2$

f $x \le -\frac{1}{2}$ or $x \ge 3$

7 **a** $-7 < x < 1$

b $\frac{2}{5} < x < 4$

c $x \le -2$ or $x \ge \frac{1}{4}$

8 $x \ge \frac{k}{2}$

9 $x < \frac{k}{5}$ or $x > \frac{7k}{3}$

10 $-\frac{4}{3} \le x \le \frac{4}{3}$

Exercise 4.3

1 $A(-1, 0)$, $B(2, 0)$, $C(3, 0)$, $D(0, 6)$

2 **a** ∿ shaped curve, axis intercepts
$(-3, 0)$, $(2, 0)$, $(4, 0)$, $(0, 24)$

b ∿ shaped curve, axis intercepts
$(-2, 0)$, $(-1, 0)$, $(3, 0)$, $(0, 6)$

c ∿ shaped curve, axis intercepts
$(-2, 0)$, $\left(-\frac{1}{2}, 0\right)$, $(2, 0)$, $(0, -4)$

d ∿ shaped curve, axis intercepts
$(-2, 0)$, $(1, 0)$, $\left(\frac{3}{2}, 0\right)$, $(0, -6)$

3 $A\left(\frac{7}{2}, 0\right)$, $B(0, 14)$

4 **a** ∿ shaped curve, axis intercepts
$(-2, 0)$ and $(0, 0)$ where $(0, 0)$ is a minimum point

b ∿ shaped curve, axis intercepts $\left(\frac{5}{2}, 0\right)$
and $(0, 0)$ where $(0, 0)$ is a minimum point

c ∿ shaped curve, axis intercepts
$(-1, 0)$, $(2, 0)$, $(0, -2)$ where $(-1, 0)$ is a maximum point

d ∿ shaped curve, axis intercepts
$(2, 0)$, $\left(\frac{10}{3}, 0\right)$, $(0, 40)$ where $(2, 0)$ is a
minimum point

5 **a**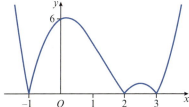

b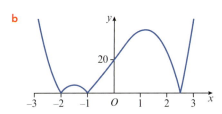

c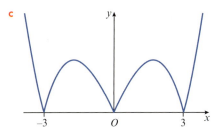

d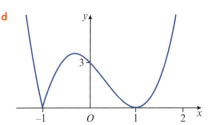

6 **a** ∿ shaped curve, axis intercepts
$(-3, 0)$, $(0, 0)$, $(3, 0)$, $(0, 9)$

b ∿ shaped curve, axis intercepts
$(-3, 0)$, $(-2, 0)$, $(1, 0)$, $(0, -6)$

c ∿ shaped curve, axis intercepts
$(-4, 0)$, $\left(\frac{1}{2}, 0\right)$, $(3, 0)$, $(0, 12)$

d ∿ shaped curve, axis intercepts
$(-3, 0)$, $\left(-\frac{5}{2}, 0\right)$, $(4, 0)$, $(0, -60)$

7 **a**

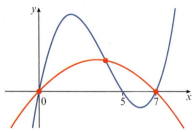

   b (0, 0), (4, 12), (7, 0)

8 **a**

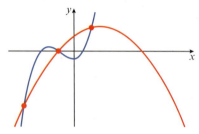

   blue curve axis intercepts: (−2, 0), (−1, 0),

   $\left(\dfrac{1}{2}, 0\right)$, (0, −2)

   red curve axis intercepts: (−1, 0), (4, 0), (0, 4)

   b (−3, −14), (−1, 0), (1, 6)

9 $a = 1$, $b = 2$ and $k = -2$

10 $a = -1$, $b = 1$, $c = 2$, $k = \pm 3$

Exercise 4.4

1 **a** $x \leqslant -1$ or $0 \leqslant x \leqslant 2$

   b $x \geqslant 2.15$

   c $x \leqslant -1.4$ or $1 \leqslant x \leqslant 1.4$

2 **a** $x \leqslant 2$

   b $x \geqslant -2$

   c $-1.9 \leqslant x \leqslant 0.35$ or $x \geqslant 1.5$

3 **a** $x \geqslant 2.5$

   b $-1 \leqslant x \leqslant 1$ or $x \geqslant 2$

   c $x \leqslant -0.8$ or $0.55 \leqslant x \leqslant 2.32$

Exercise 4.5

1 **a** $x = \pm 2$, $x = \pm 1$

   b $x = \pm\sqrt{2}$

   c $x = \pm 4$, $x = \pm 2$

   d $x = \pm\sqrt{2}$

   e $x = \pm\sqrt{7}$

   f $x = \pm 3$

   g $x = \pm\sqrt{2}$, $x = \pm\dfrac{\sqrt{3}}{2}$

   h $x = \pm\sqrt{\dfrac{2}{3}}$

   i $x = -1$, $x = 2$

2 **a** $x = \pm 0.356$, $x = \pm 2.81$

   b $x = \pm 2.32$

   c $x = \pm 1.16$

   d $x = -1.11$, $x = 1.42$

   e $x = 1.26$, $x = -0.693$

   f $x = \pm 1.40$

3 **a** $x = 4$, $x = 25$

   b $x = 16$

   c $x = 9$

   d $x = 25$

   e $x = \dfrac{9}{16}$, $x = \dfrac{9}{4}$

   f $x = \dfrac{25}{9}$

   g $x = \dfrac{1}{4}$, $x = 16$

   h $x = \dfrac{1}{9}$, $x = 25$

   i $x = \dfrac{1}{4}$, $x = 16$

4 $x = \dfrac{27}{8}$ or $x = 8$

5 **a** $5\sqrt{x} = x + 4$

   b (1, 1), (16, 4)

6 **a** $x = 1$ or $x = 2$

   b $x = 0$ or $x = 2$

   c $x = -1$ or $x = 2$

   d $x = -1$ or $x = 2$

   e $x = -2$ or $x = 3$

   f $x = -1$

7 $x = -2$

8 $x = 1$ or $x = 2$

9 $x = -4$, $x = -3$, $x = 0$ or $x = 1$

Past paper questions and practice questions

1 $x = \dfrac{8}{5}$ or $x = 2$

2 $x < -3$ or $x > 4$

3 $\dfrac{4}{5} < x < 2$

4 $\dfrac{1}{2} < x < 1$

5 $x \geqslant -\dfrac{1}{2}$

6 $-6 < x < -\dfrac{2}{3}$

7 $x > -\dfrac{k}{2}$

8 **a** $x = -1$ or $x = 27$

   b $y = -1$ or $y = 3$

9 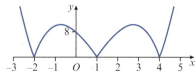 shaped curve, axis intercepts $(0, 0)$,

   $\left(\dfrac{3}{2}, 0\right)$, $(4, 0)$

10 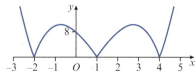 shaped curve, axis intercepts

   $(-1, 0)$, $\left(\dfrac{1}{2}, 0\right)$, $(3, 0)$, $(0, 6)$

11 $x < -0.91$ or $0.77 < x < 2.14$

12 **a** 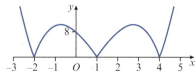 shaped curve, axis intercepts $(-2, 0)$, $(1, 0)$,
   $(4, 0)$, $(0, 8)$

   b
   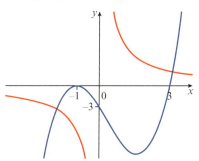

13 **a** $x(x - 2)(x + 3)$

   b 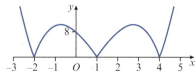 shaped curve, axis intercepts
   $(-3, 0)$, $(0, 0)$, $(2, 0)$

14 **a**
   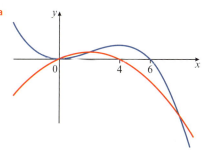

   b 2

15 **a** $(x + 4)(2x - 1)(x - 3)$

   b 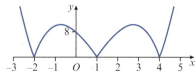 shaped curve, axis intercepts
   $(-4, 0)$, $\left(\dfrac{1}{2}, 0\right)$, $(3, 0)$, $(0, 12)$

16 **a**
   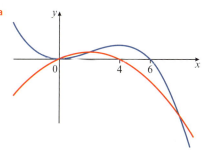

   b $(0, 0)$, $(2, 16)$, $(8, -128)$

17 $x < -2.5, x > 3$

18 **a**
   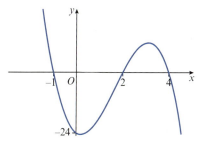

   b $x < -1$ or $2 < x < 4$

19 **a** $y = -\dfrac{1}{2}(x + 2)(x + 1)(x - 5)$

   b $-2 \leqslant x \leqslant -1$ or $x \geqslant 5$

20 $p(x) = \pm 3(x + 2)(x + 1)(x - 4)$

21 **a** $a = 2, b = 1, c = -1$

   b $x \leqslant -1.45$ or $-0.4 \leqslant x \leqslant 0.85$

Chapter 5

Pre-requisite Knowledge

1
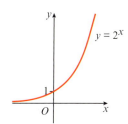

2
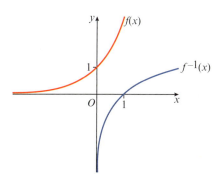

3 **a** 6   **b** 2   **c** 3

Exercise 5.1

1 **a** $\lg 1000 = 3$ **b** $\lg 100 = 2$

   c $\lg 1\,000\,000 = 6$ **d** $x = \lg 2$

   e $x = \lg 15$ **f** $x = \lg 0.06$

2 a 1.88
  b 2.48

  c 2.86
  d 1.19

  e -1.70
  f -2.30

3 a $10^5 = 100\,000$
  b $10^1 = 10$

  c $10^{-3} = \dfrac{1}{1000}$
  d $x = 10^{7.5}$

  e $x = 10^{1.7}$
  f $x = 10^{-0.8}$

4 a $126\,000$
  b 1450

  c 145
  d 0.501

  e 0.0316
  f 0.00145

5 a 4
  b -2

  c 0.5
  d $\dfrac{1}{3}$

  e 1.5
  f 2.5

Exercise 5.2

1 a $\log_4 64 = 3$
  b $\log_2 32 = 5$

  c $\log_5 125 = 3$
  d $\log_6 36 = 2$

  e $\log_2\left(\dfrac{1}{32}\right) = -5$
  f $\log_3\left(\dfrac{1}{81}\right) = -4$

  g $\log_a b = 2$
  h $\log_x 4 = y$

  i $\log_a c = b$

2 a $2^2 = 4$
  b $2^6 = 64$

  c $5^0 = 1$
  d $3^2 = 9$

  e $36^{\frac{1}{2}} = 6$
  f $8^{\frac{1}{3}} = 2$

  g $x^0 = 1$
  h $x^y = 8$

  i $a^c = b$

3 a 16
  b 9

  c 625
  d $\sqrt{3}$

  e 12
  f 3

  g 17
  h 4

  i -41

4 a 2
  b 4

  c 3
  d 2

  e 5
  f 3.5

  g 2.5
  h -1.5

  i 0.5
  j -0.5

  k $\dfrac{1}{3}$
  l -0.5

5 a 2
  b $\dfrac{1}{3}$

  c 1.5
  d -2

  e -6
  f 3.5

  g $\dfrac{2}{3}$
  h $\dfrac{7}{6}$

6 a 8
  b 625

Exercise 5.3

1 a $\log_2 15$
  b $\log_3 6$

  c $\log_5 64$
  d $\log_7 2$

  e $\log_3 20$
  f $\log_7\left(\dfrac{1}{2}\right)$

  g $\log_4 12$
  h $\lg\left(\dfrac{1}{20}\right)$

  i $\log_4 6.4$

2 a 3
  b 2

  c 1
  d 1

  e 1.5
  f -2

3 a $\log_5 36$
  b $\log_2 \dfrac{4}{3}$

4 a $2^4, 2^{-2}$
  b -2

5 a 2
  b 3

  c -3
  d -1

6 a 5^u
  b $u - 2$

  c $1 + \dfrac{1}{2}u$
  d $\dfrac{3}{2}u - 3$

7 a $1 + x$
  b $2 - x$

  c $x + 2y$
  d 4^{x+y}

8 a -8
  b -5.5

  c 13
  d 34

9 a 8
  b 20

  c 14
  d 0

Exercise 5.4

1 a $x = 5$
  b $x = 7.5$

  c $x = 25$
  d $x = 77$

2 a $x = 12$
  b $x = 8$

  c $x = 10$
  d $x = 3$

  e $x = 0.7$
  f $x = 7.5$

3 a $x = 2, 8$
  b $x = 5$

  c $x = 6$
  d $x = 4$

  e $x = 3$
  f $x = \dfrac{1}{4}, x = \dfrac{3}{2}$

  g $x = 5$
  h $x = 20$

4 a $x = 16$
  b $x = 4$

  c $x = \dfrac{2}{3}$
  d $x = \sqrt{3}$

5 a $x = 5, x = 25$

  b $x = \dfrac{1}{125}, x = 3125$

  c $x = \dfrac{1}{125}, x = 15\,625$

  d $x = \dfrac{1}{512}, x = 16$

6 **a** $x = 4, y = 16$

   b $x = 20, y = 10$

   c $x = 4, y = 12$

   d $x = 40, y = 16$

   e $a = 6.25, b = 2.5$

   f $x = 1.25, y = 2.5$

7 **b** $\lg x = 8, \lg y = 2$

Exercise 5.5

1 **a** $x = 6.13$ **b** $x = 2.73$

   c $x = 0.861$ **d** $x = 2.41$

   e $x = 2.65$ **f** $x = 1.66$

   g $x = 6.90$ **h** $x = 5.19$

   i $x = 1.15$ **j** $x = -13.4$

   k $x = 0.641$ **l** $x = 0.262$

2 **b** $x = 10$

   c $x = 3.32$

3 **a** $x = 0.415$ **b** $x = 2.42$

   c $x = 2.46$ **d** $x = 1.63$

   e $x = 1.03$

4 $x = -0.751, x = 1.38$

5 **a** $x = 0, 1.46$ **b** $x = 1.40$

   c $x = 2.32$ **d** $x = 0.683$

6 $x = 0.683, x = 1.21$

7 **a** $x = 1.58$

   b $x = 0.257, x = 0.829$

   c $x = 0.631, x = 1.26$

   d $x = 0.792, x = 0.161$

8 **a** $x = 2.32$

   b $x = 1.16$

   c $x = 0.631, x = 1.26$

   d $x = 0.431, x = 1.43$

9 **a** $x = 0.6$ **b** $x = -0.189$

10 **a** $x = 1.26$ **b** $x = 2.81$

   c $x = \pm 1.89$ **d** $x = \pm 2.81$

11 $x < \log_2 3$

Exercise 5.6

1 **a** 3.32   **b** 3.18

   c 1.29   **d** -3.08

2 **a** $\dfrac{1}{u}$ **b** $\dfrac{2}{u}$

   c $\dfrac{1}{2u}$ **d** $\dfrac{3}{2u}$

3 **a** $\dfrac{1}{x}$ **b** $1 + x$

   c $2x$ **d** $4 + 2x$

4 **a** 4   **b** 0.4

5 $\dfrac{1}{3}$

6 **a** $x = 23$ **b** $x = 23$

7 **a** $\dfrac{1}{2}\log_2 x$ **b** $x = 256$

8 **a** $x = 16$ **b** $x = 9$

   c $x = 1.59$ **d** $x = 1.87$

9 **a** $\dfrac{1}{\log_3 x}$ **b** $x = 3, x = 9$

10 **a** $x = \dfrac{1}{27}, x = 27$ **b** $x = 5$

   c $x = \dfrac{1}{256}, x = 4$ **d** $x = 16, x = 64$

   e $x = \dfrac{1}{2}, x = 512$ **f** $x = 25$

11 **a** $\dfrac{1}{2}\log_2 x$ **b** $\dfrac{1}{3}\log_2 y$

   c $x = 32, y = 2$

12 $x = 6.75, y = 13.5$

Exercise 5.7

1 **a** 7.39   **b** 4.48

   c 1.22   **d** 0.0498

2 **a** 1.39   **b** 0.742

   c -0.357 **d** -0.942

3 **a** 5   **b** 8

   c 6   **d** -2

4 **a** $x = 7$ **b** $x = 2.5$

   c $x = 6$ **d** $x = 0.05$

5 **a** $x = 4.25$ **b** $x = 1.67$

   c $x = 1.77$ **d** $x = 1.30$

6 **a** $x = \ln 7$ **b** $x = \ln 3$

   c $x = \dfrac{5 + \ln 3}{2}$ **d** $x = \dfrac{1 + \ln 8}{3}$

7 **a** $x = 20.1$ **b** $x = 0.135$

   c $x = 1100$ **d** $x = 12.5$

8 **a** $x = 3.49$ **b** $x = -2.15$

   c $x = 0.262$

9 **a** $x = 3 + e^2$ **b** $x = \dfrac{1 + \ln 7}{2}$

c $x = 2\ln 2$ **d** $x = \dfrac{\ln 2}{2}$

e $x = 2\ln 2, \ln 5$ **f** $x = \ln 2, \ln 3$

10 a $x = 1.79$

   b $x = 0, 1.39$

   c $x = -3.69, x = 4.38$

11 a $x = \dfrac{1}{e^2}, y = \dfrac{1}{e}$

   b $x = \ln 0.6, y = \ln 0.12$

12 $x = 0.822$

13 $x = 0, x = 0.4$

14 $x = \sqrt{2}, x = -\sqrt{2}, x = \dfrac{\ln 5}{2}$

Exercise 5.8

1 a $409\,600$ **b** 16.61

2 a $43\,000$ **b** 2097

3 a 0.0462 **b** 724

4 a 500 **b** 82.6

   c 2.31

5 a $\$250\,000$ **b** 0.112

   c 6.19 years

6 a $\dfrac{4}{3}$

   b $\dfrac{81}{80}$, A_0 represents the area of the patch at the start of the measurements

   c 6.73 days

Exercise 5.10

1 a asymptote: $y = -4$, y-intercept: $(0, -2)$ x-intercept: $(\ln 2, 0)$

   b asymptote: $y = 6$, y-intercept: $(0, 9)$

   c asymptote: $y = 2$, y-intercept: $(0, 7)$

   d asymptote: $y = 6$, y-intercept: $(0, 8)$

   e asymptote: $y = -1$, y-intercept: $(0, 2)$, x-intercept: $(\ln 3, 0)$

   f asymptote: $y = 4$, y-intercept: $(0, 2)$, x-intercept: $(-\ln 2, 0)$

   g asymptote: $y = 1$, y-intercept: $(0, 5)$

   h asymptote: $y = 8$, y-intercept: $(0, 10)$

   i asymptote: $y = 2$, y-intercept: $(0, 1)$, x-intercept: $\left(\dfrac{1}{4}\ln 2, 0\right)$

2 a asymptote: $x = -2$, x-intercept: $(-1.5, 0)$, y-intercept: $(0, \ln 4)$

   b asymptote: $x = 2$, x-intercept: $\left(2\dfrac{1}{3}, 0\right)$

c asymptote: $x = 4$, x-intercept: $(3.5, 0)$, y-intercept: $(0, \ln 8)$

d asymptote: $x = -1$, x-intercept: $(-0.5, 0)$, y-intercept: $(0, 2\ln 2)$

e asymptote: $x = 2$, x-intercept: $(2.5, 0)$

f asymptote: $x = 1.5$, x-intercept: $\left(1\dfrac{2}{3}, 0\right)$

Exercise 5.11

1 a $f^{-1}(x) = \ln(x - 4), x > 4$

   b $f^{-1}(x) = \ln(x + 2), x > -2$

   c $f^{-1}(x) = \ln\left(\dfrac{x+1}{5}\right), x > -1$

   d $f^{-1}(x) = \ln\left(\dfrac{x-1}{3}\right), x > 1$

   e $f^{-1}(x) = \ln\left(\dfrac{x-3}{5}\right), x > 3$

   f $f^{-1}(x) = -\dfrac{1}{3}\ln\left(\dfrac{x-5}{4}\right), x > 5$

   g $f^{-1}(x) = \ln(2 - x), x < 2$

   h $f^{-1}(x) = -\dfrac{1}{2}\ln\left(\dfrac{5-x}{2}\right), x < 5$

2 a $f^{-1}(x) = e^x - 1$

   b $f^{-1}(x) = e^x + 3$

   c $f^{-1}(x) = e^{0.5x} - 2$

   d $f^{-1}(x) = \dfrac{1}{2}(e^{0.5x} - 1)$

   e $f^{-1}(x) = \dfrac{1}{2}(e^{\frac{1}{3}x} + 5)$

   f $f^{-1}(x) = \dfrac{1}{3}(e^{-0.2x} + 1)$

3 a $f(x) > 1$

   b $f^{-1}(x) = \dfrac{1}{2}\ln(x - 1)$

   c $x > 1$

   d x

4 a i $5x$ **ii** $x + \ln 5$

   b $x = \sqrt{5}$

5 a i x^3 **ii** $3x$

   b $\dfrac{1}{2}\ln 2$

6 a $(2x + 1)^2$ **b** $x = \ln 2$

Past paper questions

1 a $5y^2 - 7y + 2 = 0$

   b $x = 0, x = -0.569$

2 $x = 5.8, y = 2.2$

3 **a** $\frac{1}{3}\log_3 x$ **b** $y = 125a$

4 **i** -2

   ii $-n$

   iii $y = 5$

   iv $x = 10$

5 **a** **i**

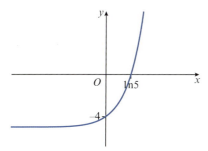

   ii $k \leqslant -5$

   b $\frac{5}{2}\log_a 2$

   c $x = 36$

6 **a** $p = 5, q = 2$ **b** $x = 4$

7 **a** **i** $x = -0.99(012\ldots)$ **ii** $y = 4.5$

   b $\log_2\left(\dfrac{p}{q}\right)$

8 $x = 12$

9 **a** $\log_3 pq$ **b** $a = 5, a = \sqrt[3]{5}$

10 **a** $x = 6.19$ **b** $y = -0.281$

11 $x = \dfrac{\sqrt[3]{e}}{5}$ or $x = \dfrac{1}{5e}$

12 **a** **i** both $x + 3 > 0$ and $2x - 1 > 0$,

       hence $x > \dfrac{1}{2}$

   ii $y = -3 + \sqrt{6}$

   b $2 + 3\log_a 9$

Chapter 6

Pre-requisite Knowledge

1 **a** 10   **b** $\dfrac{3}{4}$ **c** $(0, 2)$

2 **a** $-\dfrac{1}{2}$ **b** $y = -\dfrac{1}{2}x + \dfrac{3}{2}$

Exercise 6.1

1 **a** 3   **b** 4   **c** 10

   d 13   **e** 5   **f** 25

   g $\sqrt{74}$ **h** $\sqrt{13}$ **i** $12\sqrt{2}$

2 **a** $2\sqrt{5}, 3\sqrt{5}, \sqrt{65}$; right-angled

   b $2\sqrt{13}, 10, 4\sqrt{10}$; not right-angled

   c $4\sqrt{5}, 3\sqrt{5}, 5\sqrt{5}$; right-angled

4 5 and -5

5 1 and 3

6 **a** $(6, 4)$ **b** $(6.5, 7)$

   c $(3, 8)$ **d** $(0.5, 1.5)$

   e $(-4.5, -2.5)$ **f** $(3a, b)$

7 $a = 18, b = -8$

8 **a** $(0, 4.5)$ **b** $(1, 1)$

9 $k = 3$

10 $A(-5, 2), B(9, 4), C(-3, 6)$

Exercise 6.2

1 **a** -2 **b** -3 **c** 0

   d $3\dfrac{1}{3}$ **e** $\dfrac{1}{3}$ **f** $-\dfrac{3}{4}$

2 **a** $-\dfrac{1}{3}$ **b** 2   **c** $-\dfrac{5}{2}$

   d $-\dfrac{4}{5}$ **e** $\dfrac{2}{5}$

3 **a** $\dfrac{2}{3}$ **b** $-\dfrac{3}{2}$

4 $(-3, -1)$

6 **a** $-\dfrac{3}{5}, -\dfrac{1}{2}$

   b not collinear

7 $k = 5$

8 $k = 1$ or $k = 2$

9 $C(11, 0)$

Exercise 6.3

1 **a** $y = 3x - 13$ **b** $y = -4x + 7$

   c $y = -\dfrac{1}{2}x + 1$

2 **a** $2y = 5x - 11$ **b** $3x + 2y = 9$

   c $x + 2y = 1$

3 **a** $y = 2x - 10$ **b** $x + 2y = -8$

   c $2y = 3x - 15$ **d** $x + 4y = 0$

4   **a** $4x - 5y = -17$ **b** $(0, 3.4)$

    c 3.4 units^2

5   **a** $y = -2x$ **b** $3x + 4y = 2$

    c $5x + 7y = -26$

6   **a** $P\left(-1\frac{1}{2}, 0\right), Q\left(0, 2\frac{1}{4}\right)$

    b $\dfrac{3\sqrt{13}}{4}$

    c 1.6875 units^2

7   **a** $(2, 3)$ **b** $3y = 2x + 5$

8   **a** $(5, 6)$ **b** $k = 8$

9   **a** $y = -2x + 13$ **b** $(6, 1)$

    c $6\sqrt{5}, 3\sqrt{5}$ **d** 45 units^2

10   **a**   **i** $2x + 3y = 14$ **ii** $y = \frac{1}{2}x$

    b $(4, 2)$

Exercise 6.4

1   **a** 27.5 units^2 **b** 22 units^2

2   **a** 54.5 units^2 **b** 76 units^2

3   **a** $k = -9$ **b** 50 units^2

4   **a** $(-1, 1.5), (2, -4.5)$

    c 22.5 units^2

5   **a** $(4, 5), (0, -3)$ **b** 20 units^2

6   **a** $(0, -7)$ **b** 60 units^2

7   **a** $(7.5, 9)$ **b** 40.75 units^2

8   **a** $(2, 2), (4, -2), (0, 6)$ **b** 40 units^2

9   **a** $(5.5, 1)$ **b** $(6, 7)$ **c** 116 units^2

Exercise 6.5

1   **a** $y = ax^2 + b, Y = y, X = x^2, m = a, c = b$

    b $xy = ax^2 + b, Y = xy, X = x^2, m = a, c = b$

    c $\dfrac{y}{x} = ax - b, Y = \dfrac{y}{x}, X = x, m = a, c = -b$

    d $x = -b\dfrac{x}{y} + a, Y = x, X = \dfrac{x}{y}, m = -b, c = a$

    e $y\sqrt{x} = ax + b, Y = y\sqrt{x}, X = x, m = a, c = b$

    f $x^2y = bx^2 + a, Y = x^2y, X = x^2, m = b, c = a$

    g $\dfrac{x}{y} = ax + b, Y = \dfrac{x}{y}, X = x, m = a, c = b$

    h $\dfrac{\sqrt{x}}{y} = ax - b, Y = \dfrac{\sqrt{x}}{y}, X = x, m = a, c = -b$

2   **a** $\lg y = ax + b, Y = \lg y, X = x, m = a, c = b$

    b $\ln y = ax - b, Y = \ln y, X = x, m = a, c = -b$

    c $\lg y = b \lg x + \lg a, Y = \lg y, X = \lg x, m = b, c = \lg a$

    d $\lg y = x \lg b + \lg a, Y = \lg y, X = x, m = \lg b, c = \lg a$

    e $\ln y = -\dfrac{a}{b}\ln x + \dfrac{2}{b}, Y = \ln y, X = \ln x, m = -\dfrac{a}{b}, c = \dfrac{2}{b}$

    f $\lg x = -y \lg a + \lg b, Y = \lg x, X = y, m = -\lg a, c = \lg b$

    g $x^2 = -by + \ln a, Y = x^2, X = y, m = -b, c = \ln a$

    h $\ln y = bx + \ln a, Y = \ln y, X = x, m = b, c = \ln a$

Exercise 6.6

1   **a** $y = 2x$ **b** $y = \frac{1}{5}x^3 + 3$

    c $y = \sqrt{x} + 1$ **d** $y = -\frac{1}{2}x^4 + \frac{11}{2}$

    e $y = -2 \times 2^x + 7$ **f** $y = -\frac{5}{4}\ln x + \frac{27}{4}$

2   **a**   **i** $y = \dfrac{1}{x^2 + 1}$

      ii $y = \dfrac{1}{5}$

    b   **i** $y = 2x^2 + 3x$

      ii $y = 14$

    c   **i** $y = \dfrac{8}{x} - 1$

      ii $y = 3$

    d   **i** $y = (13 - 2x^2)^2$

      ii $y = 25$

    e   **i** $y = \dfrac{5}{4}x^2 - x - \dfrac{9}{2}$

      ii $y = -\dfrac{3}{2}$

    f   **i** $y = x^{0.5} + x^{-0.5}$

      ii $y = \dfrac{3\sqrt{2}}{2}$

3 $y = -2x^5 + 16x^2$

4   **a** $y^2 = 3(2^x) + 25$

    b $x = 5$

5   **a** $y = -2x^2 + 8x$

    b $x = 2.5, y = 7.5$

6   **a** $e^y = x^2 + 1$

    b $y = \ln(x^2 + 1)$

7   **a** $\lg y = \dfrac{3}{2}x - 7$

    b $y = 10^{-7} \times 10^{\frac{3}{2}x}$

8   **a** $y = 100x^{\frac{3}{2}}$

    b $x = 0.64$

9   **a** $\ln y = 3 \ln x - 1$

    b $y = \dfrac{x^3}{e}$

10   **a** $\ln y = 12.7$

    b $a = e^{12.7}, b = -2$

Exercise 6.7

1 **a**

x	0.5	1.0	1.5	2.0	2.5
xy	0.5	3	5.51	8	10.5

  c $y = 5 - \dfrac{2}{x}$

  d $x = 0.8,\ y = 2.5$

2 **a**

$\dfrac{1}{x}$	10	5	3.33	2.5	2
$\dfrac{1}{y}$	9.01	6.49	5.68	5.29	5

  c $y = \dfrac{2x}{1 + 8x}$

  d $x = 0.222$ (3 s.f.)

3 **b** $y = 0.8x + \dfrac{12}{x}$

  c $x = 0.6,\ y = 20.48$ or $x = -0.6,\ y = -20.48$

4 **b** $m_0 = 50,\ k = 0.02$

  c $m = 29.1$

5 **b** $k = 3,\ n = 0.7$

6 **b** $a = 5,\ b = 1.6$

7 **b** $a = 1.8,\ n = 0.5$

8 **b** $a = 0.2,\ b = 3$

  c 4.56

9 **b** $a = -1.5,\ b = 1.8$

  c 8.61

10 **b** $a = 3.6,\ b = -0.1$

  c gradient = 3.6, intercept = 0.1

11 **b** $y = e^3 \times x^{-0.8}$

  c gradient = -0.8, intercept = 3

Past paper questions

1 **a** $y = 7$

  b $3x + 4y = 31$

  c 12.5 units^2

2 **b** allow -1.4 to -1.6

  c allow 13 to 16

3 **i** $2x + 3y = 14$

  ii $y + 2 = \dfrac{3}{2}(x - 10)$

  iii $\sqrt{65}$

4 $x + 3y = 2$

5 **i** $(0, 3.5)$ **ii** $(3, 5)$

  iii $2x + y = 11$ **iv** $(0, 11)$

6 **a** $a = 4$ **b** $(-9, 16)$

7 **i** $A = -2,\ B = 3$ **ii** $y = 0.1$

  iii $x = 0.876$

8 $y = \left(\dfrac{10}{x} + 4\right)^4$

9 **i** $\ln y = \ln A + x \ln b$

  ii $\ln y = 1.4x + 2.2,\ A = 9,\ b = 4$

  iii $y = 400$

Chapter 7

Pre-requisite Knowledge

1 **a** -1 **b** 1

2 $(2, 6)$

3 **a** $\dfrac{4}{5}$ **b** $(0, -8)$ **c** $(10, 0)$

4 **a** $(x - 3)^2 - 6$

  b $x = 3 - \sqrt{6}$ or $x = 3 + \sqrt{6}$

Exercise 7.1

1 **a** centre $(0, 0)$, radius 5

  b centre $(0, 0)$, radius 9

  c centre $(0, 0)$, radius $4\sqrt{3}$

  d centre $(0, 0)$, radius $4\sqrt{2}$

  e centre $(0, 3)$, radius 1

  f centre $(5, 0)$, radius 3

  g centre $(-7, 0)$, radius $3\sqrt{2}$

  h centre $(-1, 2)$, radius 10

  i centre $(2, -4)$, radius 3

  j centre $(-3, -4)$, radius $3\sqrt{3}$

2 **a** $x^2 + y^2 = 16$

  b $(x - 3)^2 + (y + 1)^2 = 36$

  c $(x - 2)^2 + (y + 5)^2 = 20$

  d $\left(x + \dfrac{1}{2}\right)^2 + \left(y + \dfrac{3}{2}\right)^2 = 16$

3 **a** centre $(2, -5)$, radius 2

  b centre $(-7, 4)$, radius 5

  c centre $(-2, -6)$, radius 6

  d centre $(1, 8)$, radius $4\sqrt{2}$

  e centre $\left(\dfrac{1}{2}, 2\right)$, radius 2

  f centre $\left(-3, \dfrac{3}{2}\right)$, radius $2\sqrt{2}$

4 $(x - 3)^2 + (y - 2)^2 = 25$

5 $(x - 10)^2 + (y - 1)^2 = 41$

6

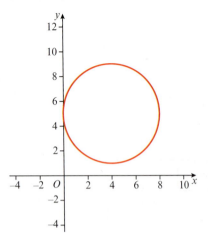

7 $(x + 3)^2 + (y - 2)^2 = 9$

8 $(x + 4)^2 + (y - 7)^2 = 75$

centre = $(-4, 7)$, radius $5\sqrt{3}$

10 $(x - 2)^2 + (y + 4)^2 = 13$

11 $(x - 2)^2 + (y - 1)^2 = 17$

12 $(x - 4)^2 + (y + 3)^2 = 26$ and
$(x + 2)^2 + (y - 1)^2 = 26$

13 $(x - 5)^2 + (y + 2)^2 = 25$ and
$(x - 5)^2 + (y + 10)^2 = 25$

14 $4x + 5y = 42$

15 $(x - 3)^2 + (y - 1)^2 = 25$

16 **b** $(x - 6)^2 + y^2 = 65$

17 **a** $(x - 8)^2 + (y - 5)^2 = 20$

b $(x + 1)^2 + (y + 1)^2 = 37$

c $x^2 + (y - 5)^2 = 25$

d $\left(x - \dfrac{7}{2}\right)^2 + \left(y - \dfrac{1}{2}\right)^2 = \dfrac{25}{2}$

e $(x + 3)^2 + (y - 1)^2 = 25$

18 One possible set of equations is:
$x^2 + (y - 1)^2 = 1$, $(x + 2.2)^2 + (y - 1)^2 = 1$,
$(x - 1.1)^2 + y^2 = 1$, $(x + 1.1)^2 + y^2 = 1$

19 **a** $1 + \sqrt{2}$

b One possible set of equations is:

$(x + 1)^2 + (y + 1)^2 = 1$,
$(x + 1)^2 + (y - 1)^2 = 1$

$(x - 1)^2 + (y + 1)^2 = 1$,
$(x - 1)^2 + (y - 1)^2 = 1$

$x^2 + y^2 = \left(1 + \sqrt{2}\right)^2$

20 One possible set of equations is:

$(x + 1)^2 + (y - 1)^2 = 1$,

$(x - 1)^2 + (y - 1)^2 = 1$

$x^2 + (y - 1 - \sqrt{3})^2 = 1$,

$x^2 + \left(y - 1 - \dfrac{\sqrt{3}}{3}\right)^2 = \left(1 + \dfrac{2\sqrt{3}}{3}\right)^2$

Exercise 7.2

1 **a** $(0, 4), (4, 4)$

b $(-2, -3), (1, 0)$

c $(-3, -3), (-1, 1)$

d $(-2, -3), (1.4, 3.8)$

2 **a** intersect

b intersect

c touch

d do not intersect

e do not intersect

3 $6\sqrt{2}$

4 $(-6, 5)$

5 Touch internally, $(-7, 0)$

6 **a** $4\sqrt{5}$ **b** $2x + y = 30$

8 **b** $4x + 3y = 49$

9 **a** $(3, 0), (1, 4)$

b $y = \dfrac{1}{2}x + 1$

c $(6 - 2\sqrt{5}, 4 - \sqrt{5}), (6 + 2\sqrt{5}, 4 + \sqrt{5})$

d $10\sqrt{5}$

10 **a** $(3, 2)$

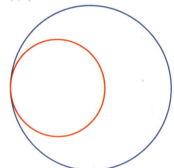

b (0, 1)

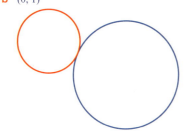

11 $-\dfrac{2}{3} < m < \dfrac{3}{2}$

12 $(x - 18)^2 + (y - 5)^2 = 25$ and $(x - 8)^2 + (y - 10)^2 = 100$

Practice questions

1 **a** $(-3, 8)$

   b $y = 6x - 11$

   c $x = 2 - \sqrt{2}, x = 2 + \sqrt{2}$

2 **a** $10x - 4y = 41$

   b $p = 3.5$

   c $x^2 + y^2 - 7x + 3y = 0$

3 **a** $y = 2x + 10$

   b $(x - 9)^2 + (y - 3)^2 = 145$

4 $(x - 5)^2 + (y + 4)^2 = 50$

5 **a** $2\sqrt{13}$

   b $(x - 3)^2 + (y - 4)^2 = 52$

   c $2x + 3y - 44 = 0$

6 **a** centre = (2, 5), radius = $5\sqrt{2}$

   b $y = 5 - \sqrt{46}, y = 5 + \sqrt{46}$

7 (4, 3)

Chapter 8

Pre-requisite Knowledge

1 **a** $\dfrac{4}{3}\pi$ cm   **b** $\dfrac{10}{3}\pi$ cm²

Exercise 8.1

1 **a** $\dfrac{\pi}{18}$ **b** $\dfrac{\pi}{9}$ **c** $\dfrac{2\pi}{9}$

   d $\dfrac{5\pi}{18}$ **e** $\dfrac{\pi}{12}$ **f** $\dfrac{2\pi}{3}$

   g $\dfrac{3\pi}{4}$ **h** $\dfrac{5\pi}{4}$ **i** 2π

   j 4π **k** $\dfrac{4\pi}{9}$ **l** $\dfrac{5\pi}{3}$

   m $\dfrac{\pi}{20}$ **n** $\dfrac{5\pi}{12}$ **o** $\dfrac{7\pi}{6}$

2 **a** 90°   **b** 30°   **c** 15°

   d 20°   **e** 120°   **f** 144°

   g 126°   **h** 75°   **i** 27°

   j 162°   **k** 216°   **l** 540°

   m 315°   **n** 480°   **o** 810°

3 **a** 0.559   **b** 0.960   **c** 1.47

   d 2.15   **e** 4.31

4 **a** 74.5°   **b** 143.2°   **c** 58.4°

   d 104.9°   **e** 33.2°

5 **a**

Degrees	0	45	90	135	180
Radians	0	$\dfrac{\pi}{4}$	$\dfrac{\pi}{2}$	$\dfrac{3\pi}{4}$	π

Degrees	225	270	315	360
Radians	$\dfrac{5\pi}{4}$	$\dfrac{3\pi}{2}$	$\dfrac{7\pi}{4}$	2π

   b

Degrees	0	30	60	90	120
Radians	0	$\dfrac{\pi}{6}$	$\dfrac{\pi}{3}$	$\dfrac{\pi}{2}$	$\dfrac{2\pi}{3}$

Degrees	150	180	210	240	270
Radians	$\dfrac{5\pi}{6}$	π	$\dfrac{7\pi}{6}$	$\dfrac{4\pi}{3}$	$\dfrac{3\pi}{2}$

Degrees	300	330	360
Radians	$\dfrac{5\pi}{3}$	$\dfrac{11\pi}{6}$	2π

6 **a** 0.964   **b** 1.03   **c** 0.932

   d 1   **e** 0.5   **f** 1

7 12.79°

Exercise 8.2

1 **a** $\dfrac{3\pi}{2}$ cm   **b** 2π cm

   c $\dfrac{15\pi}{4}$ cm   **d** 15π cm

2 **a** 9.6 cm   **b** 2 cm

3 **a** 1.25 rad   **b** 1.5 rad

4 **a** 12.4 cm   **b** 32 cm   **c** 31 cm

5 **a** 10 cm   **b** 1.85 rad   **c** 38.5 cm

6 **a** 23 cm   **b** 18.3 cm   **c** 41.3 cm

7 **a** 13.6 cm   **b** 21.1 cm   **c** 34.7 cm

Exercise 8.3

1 a $6\pi \,\text{cm}^2$ b $\dfrac{135}{2}\pi \,\text{cm}^2$

   c $35\pi \,\text{cm}^2$ d $\dfrac{135}{4}\pi \,\text{cm}^2$

2 a $10.4\,\text{cm}^2$ b $4.332\,\text{cm}^2$

3 a 1.11 rad   b 1.22 rad

4 a 0.8 rad   b $40\,\text{cm}^2$

5 $r(75 - r)$

6 a $9\sqrt{2}\,\text{cm}$ b $\dfrac{\pi}{2}$ c $\dfrac{81}{2}\pi \,\text{cm}^2$

7 a 1.24 rad   b $89.3\,\text{cm}^2$ c $121\,\text{cm}^2$

8 a 1 rad   b $49.8\,\text{cm}^2$ c $17.8\,\text{cm}^2$

9 a $24.3\,\text{cm}^2$ b $37.7\,\text{cm}^2$ c $13.4\,\text{cm}^2$

10 b $54\pi - 36\sqrt{2}\,\text{cm}^2$

11 a $\dfrac{3}{7}$ rad   b $18\,\text{cm}^2$

12 a $4.39\,\text{cm}$ b $2.40\,\text{cm}$

    c $15.5\,\text{cm}$ d $15.0\,\text{cm}^2$

13 $34.4\,\text{cm}$

14 a $45.5\,\text{cm}^2$ b $57.1\,\text{cm}^2$

    c $80.8\,\text{cm}^2$ d $21.7\,\text{cm}^2$

15 $14.6\,\text{cm}^2$

Past paper questions

1 a $0.4x^2$

   b 19.8 or $19.9\,\text{cm}$

   c 24.95 to $25\,\text{cm}^2$

2 i All sides are equal to the radii of the circles, which are also equal

   ii $\dfrac{2\pi}{3}$

   iii 58.2 or $58.3\,\text{cm}$

   iv $148\,\text{cm}^2$

3 i $19.3\,\text{cm}$ ii $79.1\,\text{cm}^2$ iii $57.5\,\text{cm}$

4 i $\dfrac{\pi}{4}$

   iii $11.6\,\text{cm}$

   iv $1.08 \leqslant \text{Area} \leqslant 1.11$

5 a 0.927 rad   b $33.3\,\text{cm}$ c $71.9\,\text{cm}^2$

6 a $35.6\,\text{cm}$ b $68.0\,\text{cm}^2$

7 a $30\pi \,\text{cm}^2$ b $10.9\,\text{cm}$

8 i $1.91\,\text{cm}^2$ ii $12.0\,\text{cm}$

Chapter 9

Pre-requisite Knowledge

1 a $r^2 - 4$ b $\dfrac{r^2 - 4}{r}$

   c $\dfrac{2}{r}$ d $\dfrac{r^2 - 4}{2}$

2 a i $\dfrac{\pi}{6}$ ii $\dfrac{3\pi}{4}$ iii 3π

   b i $60°$ ii $210°$ iii $80°$

3 a $0, -6$ b $-5, \dfrac{1}{2}$

Exercise 9.1

1 a $\dfrac{2\sqrt{13}}{13}$ b $\dfrac{3\sqrt{13}}{13}$ c $\dfrac{4}{13}$

   d 1 e $\dfrac{7 + \sqrt{13}}{9}$

2 a $\dfrac{\sqrt{23}}{5}$

   b $\dfrac{\sqrt{46}}{23}$

   c $\dfrac{23}{25}$

   d $\dfrac{\sqrt{2} + \sqrt{23}}{5}$

   e $\dfrac{23\sqrt{2} - 2\sqrt{23}}{10}$

3 a $\dfrac{4\sqrt{3}}{7}$ b $4\sqrt{3}$ c $\dfrac{4\sqrt{3}}{7}$

   d 1 e $\dfrac{7 - 196\sqrt{3}}{48}$

4 a $\dfrac{1}{2}$ b 3 c $\dfrac{2}{3}$

   d $\dfrac{\sqrt{3} + \sqrt{2}}{2}$ e $\dfrac{3\sqrt{2} - 3}{2}$ f $\dfrac{2}{7}$

5 a $\dfrac{\sqrt{2}}{4}$ b $\dfrac{1}{2}$

   c $\dfrac{\sqrt{6}}{3}$ d $\dfrac{-6 + 10\sqrt{3}}{3}$

   e $2 - \sqrt{2}$ f $2\sqrt{3} - \sqrt{6}$

Exercise 9.2

1 **a**

b

c

d

e

f

g

h

i

j

2 **a** second   **b** fourth

    c third   **d** third

    e third   **f** first

    g fourth   **h** third

    i first   **j** first

Exercise 9.3

1 **a** $-\sin 40°$ **b** $\cos 35°$

    c $-\tan 40°$ **d** $\cos 25°$

    e $\tan 60°$ **f** $\sin \dfrac{\pi}{5}$

    g $-\tan \dfrac{\pi}{4}$ **h** $\cos \dfrac{\pi}{6}$

    i $-\tan \dfrac{\pi}{3}$ **j** $\sin \dfrac{\pi}{4}$

2 **a** $-\dfrac{\sqrt{21}}{2}$ **b** $-\dfrac{\sqrt{21}}{5}$

3 **a** $\dfrac{\sqrt{3}}{2}$ **b** $-\dfrac{1}{2}$

4 **a** $-\dfrac{12}{13}$ **b** $-\dfrac{5}{12}$

5 **a** $-\dfrac{2\sqrt{13}}{13}$ **b** $-\dfrac{3\sqrt{13}}{13}$

6 **a** $-\dfrac{4}{5}$ **b** $-\dfrac{3}{5}$

    c $-\dfrac{\sqrt{6}}{3}$ **d** $\sqrt{2}$

7 **a** $\dfrac{5}{13}$ **b** $-\dfrac{12}{5}$

    c $-\dfrac{4}{5}$ **d** $-\dfrac{4}{3}$

Exercise 9.4

1 **a** **i** amplitude = 7, period = 360°, (0, 7), (180, −7), (360, 7)

        ii amplitude = 2, period = 180°, (45, 2), (135, −2), (225, 2), (315, −2)

iii   amplitude = 2, period = 120°, (0, 2), (60, −2), (120, 2), (180, −2), (240, 2), (300, −2), (360, 2)

iv   amplitude = 3, period = 720°, (180, 3)

v   amplitude = 4, period = 360°, (0, 5), (180, −3), (360, 5)

vi   amplitude = 5, period = 180°, (45, 3), (135, −7), (225, 3), (315, −7)

2 a i   amplitude = 4, period = 2π, $\left(\dfrac{\pi}{2}, 4\right)$, $\left(\dfrac{3\pi}{2}, -4\right)$

    ii   amplitude = 1, period = $\dfrac{2\pi}{3}$, (0, 1), $\left(\dfrac{\pi}{3}, -1\right)$, $\left(\dfrac{2\pi}{3}, 1\right)$, $(\pi, -1)$, $\left(\dfrac{4\pi}{3}, 1\right)$, $\left(\dfrac{5\pi}{3}, -1\right)$, $(2\pi, 1)$

    iii   amplitude = 2, period = $\dfrac{2\pi}{3}$, $\left(\dfrac{\pi}{6}, 2\right)$, $\left(\dfrac{\pi}{2}, -2\right)$, $\left(\dfrac{5\pi}{6}, 2\right)$, $\left(\dfrac{7\pi}{6}, -2\right)$, $\left(\dfrac{3\pi}{2}, 2\right)$, $\left(\dfrac{11\pi}{6}, -2\right)$

    iv   amplitude = 3, period = 4π, (0, 3), $(2\pi, -3)$

    v   amplitude = 1, period = π, $\left(\dfrac{\pi}{4}, 4\right)$, $\left(\dfrac{3\pi}{4}, 2\right)$, $\left(\dfrac{5\pi}{4}, 4\right)$, $\left(\dfrac{7\pi}{4}, 2\right)$

    vi   amplitude = 4, period = π, (0, 3), $\left(\dfrac{\pi}{2}, -5\right)$, $(\pi, 3)$, $\left(\dfrac{3\pi}{2}, -5\right)$, $(2\pi, 3)$

3 $a = 2, b = 4, c = 1$

4 $a = 3, b = 2, c = 4$

5 $a = 3, b = 2, c = 3$

6 a i   period = 90°, $x = 45°$, $x = 135°$, $x = 225°$, $x = 315°$

    ii   period = 360°, $x = 180°$

    iii   period = 60°, $x = 30°$, $x = 90°$, $x = 150°$, $x = 210°$, $x = 270°$, $x = 330°$

7 a i   period = $\dfrac{\pi}{4}$, $x = \dfrac{\pi}{8}$, $x = \dfrac{3\pi}{8}$, $x = \dfrac{5\pi}{8}$, $x = \dfrac{7\pi}{8}$, $x = \dfrac{9\pi}{8}$, $x = \dfrac{11\pi}{8}$, $x = \dfrac{13\pi}{8}$, $x = \dfrac{15\pi}{8}$

    ii   period = $\dfrac{\pi}{3}$, $x = \dfrac{\pi}{6}$, $x = \dfrac{\pi}{2}$, $x = \dfrac{5\pi}{6}$, $x = \dfrac{7\pi}{6}$, $x = \dfrac{3\pi}{2}$, $x = \dfrac{11\pi}{6}$

    iii   period = $\dfrac{\pi}{2}$, $x = \dfrac{\pi}{4}$, $x = \dfrac{3\pi}{4}$, $x = \dfrac{5\pi}{4}$, $x = \dfrac{7\pi}{4}$

8 $A = 1, B = 1, C = 3$

9 $a = 9, b = 4, c = 6$

10 a $A = 2, B = 5$

    b   3

    c

11 a $A = 1, B = 3, C = 4$

    b

12 b   2

13 b   2

14 b   4

Exercise 9.5

1 a $f(x) \geqslant 0$

    b $0 \leqslant f(x) \leqslant 1$ **c** $0 \leqslant f(x) \leqslant 3$

    d $0 \leqslant f(x) \leqslant 1$ **e** $0 \leqslant f(x) \leqslant 2$

    f $0 \leqslant f(x) \leqslant 2$ **g** $1 \leqslant f(x) \leqslant 3$

    h $0 \leqslant f(x) \leqslant 6$ **i** $0 \leqslant f(x) \leqslant 7$

2   **c**   4

3   **c**   2

4   **c**   3

5   **b**   2

6   **b**   5

7   **b**   2

8   **b**   4

9 $1 < k < 5$

10 $a = 1, b = 4, c = 2$

Exercise 9.6

1 a   17.5°, 162.5°   **b**   78.5°, 281.5°

    c   63.4°, 243.4°   **d**   216.9°, 323.1°

    e   125.5°, 305.5°   **f**   233.1°, 306.9°

    g   48.6°, 131.4°   **h**   120°, 240°

2 **a** $\dfrac{\pi}{3}, \dfrac{5\pi}{3}$

   c no solutions

   e 3.99, 5.44

   g 0.253, 2.89

   b 0.197, 3.34

   d 1.89, 5.03

   f 2.15, 4.13

   h 3.55, 5.87

3 **a** 26.6°, 63.4°

   c 31.7°, 121.7°

   e 18.4°, 161.6°

   g 80.8°, 170.8°

   b 63.4°, 116.6°

   d 108.4°, 161.6°

   f 98.3°, 171.7°

   h 20.9°, 69.1°

4 **a** 150°, 270°

   c 225°, 315°

   e 1.82, 2.75

   b 95.7°, 129.3°

   d $\dfrac{\pi}{2}, \dfrac{7\pi}{6}$

   f $\dfrac{\pi}{6}, \dfrac{7\pi}{6}$

5 **a** 14.0°, 194.0°

   b 126.9°, 306.9°

   c 31.0°, 211.0°

   d 25.7°, 115.7°, 205.7°, 295.7°

6 $x = 0.648$ or $x = 2.22$

7 **a** $x = 0°, 30°, 180°, 210°, 360°$

   b $x = 0°, 38.7°, 180°, 218.7°, 360°$

   c $x = 70.5°, 90°, 270°, 289.5°$

   d $x = 0°, 135°, 180°, 315°, 360°$

   e $x = 11.5°, 90°, 168.5°, 270°$

   f $x = 0°, 45°, 180°, 225°, 360°$

8 **a** $x = 30°, 150°, 210°, 330°$

   b $x = 31.0°, 149.0°, 211.0°, 329.0°$

9 **a** $x = 45°, 108.4°, 225°, 288.4°$

   b $x = 30°, 150°, 270°$

   c $x = 0°, 109.5°, 250.5°, 360°$

   d $x = 60°, 180°, 300°$

   e $x = 0°, 180°, 199.5°, 340.5°, 360°$

   f $x = 19.5°, 160.5°, 210°, 330°$

   g $x = 19.5°, 160.5°, 270°$

   h $x = 30°, 150°, 270°$

   i $x = 19.5°, 160.5°$

10 $x = 0.848$ rad

Exercise 9.8

1 **a** $x = 73.3°, 253.3°$

   b $x = 75.5°, 284.5°$

   c $x = 210°, 330°$

   d $x = 53.1°, 306.9°$

2 **a** $x = 0.201, 2.94$

   b $x = 0.896, 4.04$

   c $x = 1.82, 4.46$

   d $x = 2.55, 5.70$

3 **a** $x = 25.7°, 154.3°$

   b $x = 5.8°, 84.2°$

   c $x = 67.5°, 157.5°$

   d $x = 112.8°, 157.2°$

4 **a** $x = 100.5°, 319.5°$

   b $x = 73.3°, 151.7°$

   c $x = 2.56, 5.70$

   d $x = 1.28, 2.00$

5 **a** $x = 60°, 120°, 240°, 300°$

   b $x = 56.3°, 123.7°, 236.3°, 303.7°$

   c $x = 106.3°, 253.7°$

6 **a** $x = 41.4°, 180°, 318.6°$

   b $x = 113.6°, 246.4°,$

   c $x = 71.6°, 153.4°, 251.6°, 333.4°$

   d $x = 19.5°, 160.5°$

   e $x = 48.2°, 311.8°$

   f $x = 18.4°, 30°, 150°, 198.4°$

   g $x = 60°, 300°$

   h $x = 23.6°, 30°, 150°, 156.4°$

Exercise 9.9

5 26

6 **a** $-2 + 7\sin^2 x$

   b $-2 \leqslant \mathrm{f}(x) \leqslant 5$

7 **a** $7 - (\cos \vartheta - 2)^2$

   b maximum $= 6$, minimum $= -2$

Past paper questions

1 **b** **i** 4   **ii** 60°

2 **a** $x = 10°, 30°, 50°, 90°$

   b $y = \dfrac{\pi}{6}, \dfrac{5\pi}{6}$

3 **b** 54.7°, 125.3°, 234.7°, 305.3°

4 **a** 30°, 150°, 210°, 330°

   b 23.5°, 60°, 96.5°, 143.5°, 180°

   c $\dfrac{2\pi}{3}, \dfrac{5\pi}{3}$

6 **i** $x = 64.3°, 154.3°$

   ii $y = 90°, 194.5°, 345.5°$

   iii $z = \dfrac{5\pi}{12}, \dfrac{13\pi}{12}$

7 **a** $a = 10, b = 6, c = 4$

   b

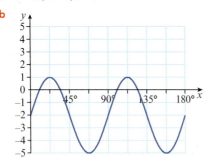

8 **i** $a = 2, b = 1$

   ii $x = \dfrac{\pi}{3}, x = \dfrac{\pi}{4}$

9 **ii** $x = 51.3°$ or $x = -128.7°$

10 **i**

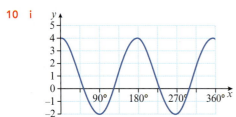

   ii $(90°, -2)$

11 **a** $x = 30°, 60°, 120°, 150°$

   b $y = \dfrac{5\pi}{12}, \dfrac{17\pi}{12}$

12 **a** $x = 0°, 45°, 135°, 180°$

   b **ii** $\theta = \pm\dfrac{\pi}{9}, \pm\dfrac{5\pi}{9}$

13 **a** **ii** $233.1°$

   b $\phi = 0.132, \phi = 1.18$

Chapter 10

Pre-requisite Knowledge

1 **a** 5040   **b** 12

   c 840   **d** 336

   e 6   **f** 60

   g 20   **h** 4200

2 **a** $2!$ **b** $6!$

   c $\dfrac{5!}{2!}$ **d** $\dfrac{17!}{13!}$

   e $\dfrac{10!}{7!3!}$ **f** $\dfrac{12!}{7!4!}$

3 **a** $\dfrac{n!}{(n-4)!}$ **b** $\dfrac{n!}{(n-6)!}$

   c $\dfrac{n!}{(n-3)!5!}$ **d** $\dfrac{n!}{(n-5)!3!}$

Exercise 10.2

1 **a** 24   **b** 5040

2 **a** 120   **b** 40 320   **c** 3 628 800

3 **a** 24

   b **i** 6   **ii** 6

4 **a** 5040   **b** 576

5 **a** 120   **b** 72

   c 72   **d** 42

6 **a** 12   **b** 36   **c** 24

7 **a** 720   **b** 120   **c** 48

8 **a** 600   **b** 312

9 4320

Exercise 10.3

1 **a** 6720   **b** 360

   c 6 652 800   **d** 5040

2 360

3 15 120

4 336

5 480

6 18

7 **a** 1680   **b** 840

   c 630   **d** 330

8 **a** 60   **b** 300

9 720

10 **a** $n = 6$ **b** $n = 9$ **c** $n = 7$

Exercise 10.4

1 **a** 5   **b** 20   **c** 1

   d 70   **e** 1   **f** 35

2 $\dfrac{8!}{3!5!} = \dfrac{8!}{5!3!}$

3 **a** 120   **b** 21   **c** 364

4 10

5 56

6 700

7 **a** 210   **b** 84

8 **a** 30 045 015   **b** 142 506

9 67 200

10	a	56	b	30	c	16

11	a	1287	b	756

12 1709

13	a	462	b	350

14	a	45	b	45

15	a	252	b	126	c	66

16 2000

17 $n = 14$

Past paper questions

1 a 28

   b i 420   ii 240

2 a i 84   ii 15 120

   b 240

3 a permutation, because the order matters

   b i 55   ii 420   iii 70

4 a i 28   ii 20 160   iii 720

   b 203

5 a i 60 480   ii 144   iii 1680

   b i 2100   ii 420

6 a i 40 320   ii 720   iii 5040

   b i 35   ii 1   iii 15

7 a 870

   b i 2002   ii 1092

8 i 18   ii 24   iii 48

9 i 3 991 680   ii 1 330 560   iii 1152

10 i 120   ii 720   iii 2520

11 a i 360   ii 60

   b i 1287   ii 1266   iii 45

Chapter 11

Pre-requisite Knowledge

1 a $4x^2 - 20x + 25$

   b $1 + x - x^2 + 15x^3$

2 a $16x^{12}$

   b $-243x^{20}$

3 a $3n - 8$

   b $9 - 2n$

Exercise 11.1

1 1, 6, 15, 20, 15, 6, 1

   1, 7, 21, 35, 35, 21, 7, 1

2 a $1 + 3x + 3x^2 + x^3$

   b $1 - 4x + 6x^2 - 4x^3 + x^4$

   c $p^4 + 4p^3q + 6p^2q^2 + 4pq^3 + q^4$

   d $8 + 12x + 6x^2 + x^3$

   e $x^5 + 5x^4y + 10x^3y^2 + 10x^2y^3 + 5xy^4 + y^5$

   f $y^3 + 12y^2 + 48y + 64$

   g $a^3 - 3a^2b + 3ab^2 - b^3$

   h $16x^4 + 32x^3y + 24x^2y^2 + 8xy^3 + y^4$

   i $x^3 - 6x^2y + 12xy^2 - 8y^3$

   j $81x^4 - 432x^3 + 864x^2 - 768x + 256$

   k $2 + x^2 + \dfrac{4}{x^2}$

   l $\dfrac{1}{8x^3} - \dfrac{3}{8x^5} + \dfrac{3}{8x^7} - \dfrac{1}{8x^9}$

3 a 16   b 10   c −12

   d 8   e 40   f 160

   g 5760   h $-\dfrac{3}{2}$

4 $A = 2048, B = 1280, C = 40$

5 $1 + 14x + 78x^2 + 216x^3 + 297x^4 + 162x^5$

6 $a = 8$

7 a $81 + 108x + 54x^2 + 12x^3 + x^4$

   b $376 + 168\sqrt{5}$

8 a $1 + 5x + 10x^2 + 10x^3 + 5x^4 + x^5$

   b i $76 + 44\sqrt{3}$ ii $76 - 44\sqrt{3}$

   c 152

9 a $16 - 32x^2 + 24x^4 - 8x^6 + x^8$

   b 64

10 90

11 $\dfrac{3}{4}$

12 a $32 + 80y + 80y^2$ b 400

13 $\dfrac{1}{5}$

14 a $a = 24, b = 128$ b $80\sqrt{2}$

15 a $y^3 - 3y$ b $y^5 - 5y^3 + 5y$

Exercise 11.2

1 a ${}^3C_0\ {}^3C_1\ {}^3C_2\ {}^3C_3$

   b ${}^4C_0\ {}^4C_1\ {}^4C_2\ {}^4C_3\ {}^4C_4$

   c ${}^5C_0\ {}^5C_1\ {}^5C_2\ {}^5C_3\ {}^5C_4\ {}^5C_5$

2 a $1 + 4x + 6x^2 + 4x^3 + x^4$

   b $1 - 5x + 10x^2 - 10x^3 + 5x^4 - x^5$

   c $1 + 8x + 24x^2 + 32x^3 + 16x^4$

   d $27 + 27x + 9x^2 + x^3$

   e $x^4 + 4x^3y + 6x^2y^2 + 4xy^3 + y^4$

   f $32 - 80x + 80x^2 - 40x^3 + 10x^4 - x^5$

   g $a^4 - 8a^3b + 24a^2b^2 - 32ab^3 + 16b^4$

   h $16x^4 + 96x^3y + 216x^2y^2 + 216xy^3 + 81y^4$

   i $\dfrac{1}{16}x^4 - \dfrac{3}{2}x^3 + \dfrac{27}{2}x^2 - 54x + 81$

   j $1 - \dfrac{x}{2} + \dfrac{x^2}{10} - \dfrac{x^3}{100} + \dfrac{x^4}{2000} - \dfrac{x^5}{100\,000}$

   k $x^5 - 15x^3 + 90x - \dfrac{270}{x} + \dfrac{405}{x^3} - \dfrac{243}{x^5}$

   l $x^{12} + 3x^8 + \dfrac{15}{4}x^4 + \dfrac{5}{2} + \dfrac{15}{16x^4} + \dfrac{3}{16x^8} + \dfrac{1}{64x^{12}}$

3 a $40x^3$

   b $175\,000x^3$

   c $160x^3$

   d $720x^3$

   e $-20x^3$

   f $-5376x^3$

   g $-9\,450\,000x^3$

   h $-954\,204\,160\,000x^3$

4 a $1 + 10x + 45x^2$

   b $1 + 16x + 112x^2$

   c $1 - 21x + 189x^2$

   d $729 + 2916x + 4860x^2$

   e $19\,683 - 59\,049x + 78\,732x^2$

   f $256 + 512x + 448x^2$

   g $1\,953\,125 - 3\,515\,625x^2 + 2\,812\,500x^4$

   h $1\,048\,576x^{10} - 13\,107\,200x^9y$
      $+ 73\,728\,000x^8y^2$

5 a $1 + 12x + 60x^2 + 160x^3$

   b 140

6 a $1 + \dfrac{13}{2}x + \dfrac{39}{2}x^2 + \dfrac{143}{4}x^3$

   b $\dfrac{377}{4}$

7 a $1 - 30x + 405x^2 - 3240x^3$

   b -4860

8 a $1 + 14x + 84x^2$

   b 47

9 a $1 + 7x + 21x^2 + 35x^3$

   b -7

10 -945

11 $\dfrac{21}{2}$

12 $a = \dfrac{3}{n - 2}$

Exercise 11.3

1 $a + 4d, a + 13d$

2 a 765   b -1310

   c 907.5   d $-2420x$

3 a 35, 3185   b $15, -1365$

4 9

5 a 5, 26   b 1287

6 -1875

7 2985

8 40

9 12 450

10 45

11 $2188

12 a $11, -3$ b 37

13 6, 8

14 $-5, -6$

15 $\dfrac{1}{12}(8n + 1)$

16 8°

17 a $a = 6d$ b $10a$

19 a $6 - 5\cos^2 x$

20 b 901

Exercise 11.4

1 a not geometric   b $-4, 16\,384$

   c $\dfrac{1}{3}, \dfrac{1}{27}$ d not geometric

   e not geometric   f $-1, 5$

2 ar^8, ar^{19}

3 $-\dfrac{2}{3}, 243$

4 ± 16.2

5 $\dfrac{3}{2}, 8$

6 $2, \dfrac{5}{64}, 40$

7 27

8 -12

9 22nd

10 19th

11 **a** 1020 **b** $1093\frac{1}{3}$

 c -3280 **d** -4166.656

12 14

13 **a** $10\left(\frac{4}{5}\right)^n$

 b 67.232 m

14 $k = -20$ or $k = 40$

15 50 minutes 19 seconds

18 $\dfrac{2(10^{n+1} - 10 - 9n)}{27}$

Exercise 11.5

1 **a** 4.5 **b** $\frac{2}{3}$

 c 10 **d** -97.2

2 50

3 $-\frac{1}{5}$, 250

4 $2\frac{7}{9}$

5 **a** $0.\dot{4}\dot{2} = \dfrac{42}{100} + \dfrac{42}{10\,000} + \dfrac{42}{1\,000\,000} + \ldots$

6 $-\frac{2}{3}$, $-93\frac{1}{3}$

7 0.5, 13

8 **a** -0.75, 128 **b** $73\frac{1}{7}$

9 **a** 105 **b** 437.5

10 **a** $\frac{3}{10}$, 60 **b** $85\frac{5}{7}$

11 **a** 60 **b** 375

12 162

13 $a = 50$, $r = \frac{1}{5}$

14 $-\dfrac{\pi}{6} < x < \dfrac{\pi}{6}$

15 84 m

16 **a** $3x, 4x, \dfrac{16}{3}x, \dfrac{64}{9}x$
  common ratio > 1

17 **a** 3π **b** $\dfrac{9\pi}{8}$

Exercise 11.6

1 **a** 108 **b** 166.25

2 **a** 125 **b** 31

3 **a** 1.5 **b** 364.5, 78

4 12.5, 63.5

5 **a** $\frac{3}{4}$

 b 84.375, 162.5

 c 800

6 **a** 0.75 **b** $\frac{5}{4}$, $n = 10$

7 19

8 **a** $x = -3$, 3rd term $= 33$ or $x = 8$, 3rd term $= 88$

 b $-\frac{2}{5}$, $-7\frac{1}{7}$

Past paper questions and practice questions

1 **i** $729 + 2916x + 4860x^2$

 ii 6804

2 **i** $64 + 192x^2 + 240x^4 + 160x^6$

 ii 1072

3 **a** **i** $a^4 + 4a^3b + 6a^2b^2 + 4ab^3 + b^4$

  ii $\dfrac{24}{25}$

 b $n = 6$

4 **i** $16 + 32ax + 24a^2x^2 + 8a^3x^3 + a^4x^4$

5 **i** $256x^8 - 64x^6 + 7x^4$

 ii 135

6 **i** $a = 256$, $b = \frac{1}{4}$, $c = 112$

 ii -2064

7 **i** $729 - 162x + 15x^2$

 ii -2856

8 $a = 3$, $b = -18$, $c = -6$

9 **a** -2.24 **b** 25

10 **a** 24 **b** 162

11 **a** $d = 2a$ **b** $99a$

12 **a** $a = 38$, $d = -2.5$ **b** $n = 25$

13 **a** $-\frac{3}{4}$ **b** 768 **c** $438\frac{6}{7}$

14 **a** 41 **b** $r = \frac{1}{4}$, $S = 10\frac{2}{3}$

15 **a** $a = -8$, $d = \frac{4}{3}$ **b** 80

16 **a** $\frac{1}{5}$ **b** 18

17 **a** **i** $r = -\frac{3}{5}$ and $a = 25$, or $r = \frac{3}{2}$ and $a = 4$

  ii $15\frac{5}{8}$

 b 119 382

18 **a** -16, 24

 b **i** $r = \dfrac{p}{3}$ **ii** $\dfrac{243p}{3 - p}$ **iii** $p = \dfrac{3}{4}$

19 a 25 terms

   b i $\dfrac{3}{4}$ **ii** 18

20 a i $-\dfrac{2}{3}$ **ii** 6.35

   b i $\log_x 3$ **ii** $\dfrac{n(n+1)}{2}\log_x 3$

      iii 78   **iv** 27

21 a ii 970

   b i 7776

      ii The sum to infinity does not exist as the common ratio is greater than 1.

Chapter 12

Pre-requisite Knowledge

1 a $2x^{\frac{3}{2}}$ **b** $7x^{\frac{2}{3}}$

   c $\dfrac{1}{3}x^{\frac{1}{2}}$ **d** $\dfrac{1}{2}x^{-2}$

2 a $3(x+1)^{-2}$ **b** $5(2x-3)^{-4}$

3 $\dfrac{4}{3}$

4 $y = 3x - 16$

Exercise 12.1

1 a $4x^3$ **b** $9x^8$

   c $-3x^{-4}$ **d** $-6x^{-7}$

   e $-x^{-2}$ **f** $-5x^{-6}$

   g $\dfrac{1}{2}x^{-\frac{1}{2}}$ **h** $\dfrac{5}{2}x^{\frac{3}{2}}$

   i $-\dfrac{1}{5}x^{-\frac{6}{5}}$ **j** $\dfrac{1}{3}x^{-\frac{2}{3}}$

   k $\dfrac{2}{3}x^{-\frac{1}{3}}$ **l** $-\dfrac{1}{2}x^{-\frac{3}{2}}$

   m 1   **n** $\dfrac{3}{2}x^{\frac{1}{2}}$

   o $\dfrac{5}{3}x^{\frac{2}{3}}$ **p** $6x^5$

   q $3x^2$ **r** $2x$

   s $\dfrac{1}{2}x^{-\frac{1}{2}}$ **t** $-\dfrac{3}{2}x^{-\frac{5}{2}}$

2 a $6x^2 - 5$

   b $40x^4 - 6x$

   c $-6x^2 + 4$

   d $6x - 2x^{-2} + 2x^{-3}$

   e $2 + x^{-2} + \dfrac{1}{2}x^{-\frac{3}{2}}$

   f $\dfrac{1}{2}x^{-\frac{1}{2}} - \dfrac{5}{2}x^{-\frac{3}{2}}$

   g $1 + 3x^{-2}$

   h $5 + \dfrac{1}{2}x^{-\frac{3}{2}}$

   i $\dfrac{3}{2}x^{\frac{1}{2}} - \dfrac{1}{2}x^{-\frac{1}{2}} + \dfrac{1}{2}x^{-\frac{3}{2}}$

   j $15x^2 + 10x$

   k $-2x^{-2} + 10x^{-3}$

   l $1 + 2x^{-2}$

   m $18x + 6$

   n $-6x^2 + 6x^5$

   o $12x + 5$

3 a 6   **b** 4   **c** -2

   d 0   **e** -0.2 **f** $\dfrac{2}{9}$

4 $(2, 5)$

5 0.25

6 5

7 $-11, 11$

8 $a = 3, b = -4$

9 $a = -2, b = 5$

10 $\left(1, 2\dfrac{5}{6}\right), \left(4, 4\dfrac{1}{3}\right)$

11 a $(-3, 2), (0, 5), (9, 14)$

   b $13, -8, 37$

12 a $12x^2 + 6x - 6$

   b $x \leqslant -1$ and $x \geqslant 0.5$

13 a $3x^2 + 2x - 16$

   b $-\dfrac{8}{3} \leqslant x \leqslant 2$

14 $\dfrac{dy}{dx} = 5[(x^2 - 2)^2 + (x + 5)^2] > 0$

Exercise 12.2

1 a $9(x + 2)^8$

   b $21(3x - 1)^6$

   c $-30(1 - 5x)^5$

   d $2\left(\dfrac{1}{2}x - 7\right)^3$

   e $4(2x + 1)^5$

   f $12(x - 4)^5$

   g $-30(5 - x)^4$

   h $8(2x + 5)^7$

   i $8x(x^2 + 2)^3$

   j $-28x(1 - 2x^2)^6$

k $5(2x - 3)(x^2 - 3x)^4$

l $8\left(x - \dfrac{1}{x^2}\right)\left(x^2 + \dfrac{2}{x}\right)^3$

2 a $-\dfrac{1}{(x + 4)^2}$ **b** $-\dfrac{6}{(2x - 1)^2}$

c $\dfrac{15}{(2 - 3x)^2}$ **d** $-\dfrac{64x}{(2x^2 - 5)^2}$

e $-\dfrac{8(x - 1)}{(x^2 - 2x)^2}$ **f** $-\dfrac{5}{(x - 1)^6}$

g $-\dfrac{30}{(5x - 1)^4}$ **h** $-\dfrac{6}{(3x - 2)^5}$

3 a $\dfrac{1}{2\sqrt{x + 2}}$ **b** $\dfrac{5}{2\sqrt{5x - 1}}$

c $\dfrac{2x}{\sqrt{2x^2 - 3}}$ **d** $\dfrac{3x^2 + 2}{2\sqrt{x^3 + 2x}}$

e $-\dfrac{2}{3(3 - 2x)^{\frac{2}{3}}}$ **f** $4/\sqrt{2x - 1}$

g $-\dfrac{3}{2(3x - 1)^{\frac{1}{2}}}$ **h** $\dfrac{5}{(2 - 5x)^{\frac{4}{3}}}$

4 8

5 2

6 0.75, −3

7 (3, 2)

8 $a = 8, b = 3$

Exercise 12.3

1 a $2x + 4$

b $12x + 10$

c $(x + 2)^2(4x + 2)$

d $x(x - 1)^2(5x - 2)$

e $\dfrac{3x - 10}{2\sqrt{x - 5}}$

f $\dfrac{3x + 2}{2\sqrt{x}}$

g $\dfrac{5x^2 + 12x}{2\sqrt{x + 3}}$

h $\dfrac{(3 - x^2)^2(3 - 13x^2)}{2\sqrt{x}}$

i $2(3x^2 + x + 5)$

j $(4x + 9)(x - 3)^2$

k $2(x - 1)(x + 2)(2x + 1)$

l $14(x - 1)(x - 3)^3(2x + 1)^2$

2 9

3 85

4 49, 0

5 $x = -2, 0, 1.5$

6 $x = 1\dfrac{2}{3}$

Exercise 12.4

1 a $\dfrac{11}{(5 - x)^2}$ **b** $\dfrac{10}{(x + 4)^2}$

c $\dfrac{7}{(3x + 4)^2}$ **d** $-\dfrac{1}{(8x - 3)^2}$

e $\dfrac{x(5x - 4)}{(5x - 2)^2}$ **f** $-\dfrac{x^2 + 1}{(x^2 - 1)^2}$

g $-\dfrac{15}{(3x - 1)^2}$ **h** $-\dfrac{x^2 + 8x + 2}{(x^2 - 2)^2}$

2 −4

3 (0, 0), (1, 1)

4 9

5 a $\dfrac{1 - 2x}{2\sqrt{x}(2x + 1)^2}$

b $\dfrac{1 - x}{(1 - 2x)^{\frac{1}{2}}}$

c $\dfrac{x(x^2 + 4)}{(x^2 + 2)^{\frac{1}{2}}}$

d $\dfrac{-5(x - 3)}{2\sqrt{x}(x + 3)^2}$

6 4

7 (3, −2)

8 a (−2, −2.4), (0.4, 0), (2, 1.6)

b $-\dfrac{23}{25}, \dfrac{125}{29}, -\dfrac{7}{25}$

Exercise 12.5

1 a $y = 4x - 6$

b $y = -x - 2$

c $y = 16x - 10$

d $y = -\dfrac{1}{2}x + 3$

e $y = -3x - 3$

f $y = \dfrac{1}{4}x + 2\dfrac{1}{4}$

2 a $y = -\dfrac{1}{3}x - 4\dfrac{1}{3}$

b $y = -\dfrac{1}{8}x + 5\dfrac{1}{4}$

c $y = \dfrac{1}{4}x - 3\dfrac{3}{4}$

d $y = 2x + 7.5$

e $y = -0.1x - 3.8$

f $y = 4x - 22$

3 $y = 8x - 6, y = -\dfrac{1}{8}x + \dfrac{17}{8}$

4 $(0, 5.2)$

5 $y = \dfrac{9}{16}x - \dfrac{1}{2}, y = -\dfrac{16}{9}x - \dfrac{1}{2}$

6 $(2, -3)$

7 $y = 2x - 20$

8 **a** $y = 2x + 8$

    b $(1, 6)$

    c $y = -\dfrac{1}{2}x + \dfrac{13}{2}$

9 $(1, 5.25)$

10 **a** $(7, 4)$ **b** 12 units^2

11 **b** $y = -0.4x - 0.6$

12 216 units^2

13 22.5 units^2

Exercise 12.6

1 0.21

2 0.68

3 $-0.8p$

4 $2p$

5 $25p$

6 $\dfrac{11}{3}p$

7 $\dfrac{\pi}{20}$

8 **a** $y = \dfrac{180}{x^2}$

    c $-254p$, decrease

Exercise 12.7

1 0.15 units per second

2 0.125 units per second

3 -4 units per second

4 -0.04 units per second

5 0.5 units per second

6 $\dfrac{1}{150}$ units per second

7 -0.02 units per second

8 $0.025 \, \text{cm s}^{-1}$

9 $\dfrac{1}{96} \, \text{cm s}^{-1}$

10 $324 \, \text{cm}^3 \, \text{s}^{-1}$

11 $\dfrac{1}{480} \, \text{cm s}^{-1}$

12 $12\pi \, \text{cm}^3 \, \text{s}^{-1}$

13 **a** $\dfrac{4}{5\pi} \, \text{cm s}^{-1}$

    b $\dfrac{1}{5\pi} \, \text{cm s}^{-1}$

14 **a** $\dfrac{1}{7} \, \text{cm s}^{-1}$

    b $\dfrac{1}{12} \, \text{cm s}^{-1}$

Exercise 12.8

1 **a** 10

    b $12x + 6$

    c $-\dfrac{18}{x^4}$

    d $320(4x + 1)^3$

    e $-\dfrac{1}{(2x + 1)^{\frac{3}{2}}}$

    f $\dfrac{3}{(x + 3)^{\frac{5}{2}}}$

2 **a** $12(x - 4)(x - 2)$

    b $\dfrac{8x - 6}{x^4}$

    c $\dfrac{8}{(x - 3)^3}$

    d $\dfrac{2(x^3 + 6x^2 + 3x + 2)}{(x^2 - 1)^3}$

    e $\dfrac{50}{(x - 5)^3}$

    f $\dfrac{102}{(3x - 1)^3}$

3 **a** -3 **b** -9 **c** -8

4 **b** $-18, 18$

5

x	0	1	2	3	4	5
$\dfrac{dy}{dx}$	+	0	−	−	0	+
$\dfrac{d^2y}{dx^2}$	−	−		+	+	+

6 $x > 2$

Exercise 12.9

1 **a** $(6, -28)$ minimum

    b $(-2, 9)$ maximum

    c $(-2, 18)$ maximum, $(2, -14)$ minimum

    d $\left(-2\dfrac{2}{3}, 14\dfrac{22}{27}\right)$ maximum,

      $(2, -36)$ minimum

e $(-3, -18)$ minimum, $\left(\frac{1}{3}, \frac{14}{27}\right)$ maximum

f $\left(\frac{2}{3}, \frac{14}{27}\right)$ maximum, $(4, -18)$ minimum

2 a $(4, 4)$ minimum

b $(-1, 3)$ minimum

c $(4, 3)$ minimum

d $\left(-3, -\frac{1}{3}\right)$ minimum, $\left(3, \frac{1}{3}\right)$ maximum

e $(-2, -4)$ maximum, $(0, 0)$ minimum

f $(-4, -13)$ maximum, $(2, -1)$ minimum

3 $\frac{dy}{dx} = -\frac{3}{(x+1)^2}$, numerator of $\frac{dy}{dx}$ is never zero

4 $a = 3$

5 a $a = -3, b = 5$

b minimum

c $(-1, 7)$, maximum

6 a $a = 2, b = -4$

b minimum

7 a $a = 8, b = -4$

b maximum

8 a $a = -36, b = 4$

b $(-2, 48)$

c $(-2, 48) =$ maximum, $(3, -77) =$ minimum

d $(0.5, -14.5), -37.5$

Exercise 12.10

1 a $y = 8 - x$

b i $P = 8x - x^2$ **ii** 16

c i $S = x^2 + (8 - x)^2$ **ii** 32

2 b $A = 1250, x = 50$

3 a $y = \frac{288}{x^2}$

c $A = 432$, 12 cm by 6 cm by 8 cm

4 a $h = \frac{V}{4x^2}$

c $\frac{5\sqrt{6}}{3}$

5 a $\theta = \frac{60}{r} - 2$

c $30 - 2r, -2$

d 15

e 225, maximum

6 a $h = 3 - \frac{1}{2}r(\pi + 2)$

c $6 - 4r - \pi r, -4 - \pi$

d $\frac{6}{4 + \pi}$

e $\frac{18}{4 + \pi}$, maximum

7 a $BC = 4 - p^2$

c $\frac{2\sqrt{3}}{3}$

d $\frac{32\sqrt{3}}{9}$, maximum

8 a $h = \frac{250}{r^2}$

c $4\pi r - \frac{500\pi}{r^2}, 4\pi + \frac{1000\pi}{r^3}$

d 5

e 150π, minimum

9 a $h = \frac{144}{r} - \frac{3}{2}r$

c $\frac{12\sqrt{10}}{5}$

10 a $r = \frac{25 - 2x}{\pi}$

c $A = 87.5, x = 7.00$

11 a $r = 2\sqrt{5 - h^2}$

c $\frac{5\sqrt{3}}{3}$

d maximum

12 a $h = 24 - 2r$

c 512π

13 a $r = \sqrt{20h - h^2}$

c $13\frac{1}{3}$

d $\frac{32\,000}{81}\pi$, maximum

Past paper exercises

1 ii $A = 246\,\text{cm}^2$, minimum

2 a $1.25\,\text{cm}\,\text{s}^{-1}$ **b** $120\,\text{cm}^2\,\text{s}^{-1}$

3 $439\,\text{cm}^2$

4 i $\frac{dy}{dx} = \frac{2x^2 + 1}{(x^2 + 1)^{\frac{1}{2}}}$

ii $2x^2$ cannot be -1

5 i $P = 2r + \frac{72}{r}$

ii 24, minimum

6 i $0.159\,\text{cm}\,\text{s}^{-1}$

ii $40\,\text{cm}^2\,\text{s}^{-1}$

7 i $\frac{dy}{dx} = -\frac{5}{2}(x - 9)^{-\frac{3}{2}}$

ii $-0.3125\,h$

8 i $(1, 0.5)$ and $(-1, -0.5)$

ii $p = 2, q = -6$.
Maximum when $x = 1$, minimum when $x = -1$

9 $6x + 8y - 27 = 0$

10 i $y = \dfrac{5 - 2r - \pi r}{2}$

    iii $r = 0.70$, $A = 1.75$

11 i $h = \dfrac{4}{3}(6 - r)$

    iii $r = 4$, $V = \dfrac{128}{3}\pi$, maximum

12 $4x - 9y + 1 = 0$

13 $(0, 32)$, minimum

14 i $V = 5\sqrt{3}\,h^2\,\text{m}^3$

    ii a $0.115\,\text{m s}^{-1}$ **b** $0.4\,\text{m s}^{-1}$

15 i $\dfrac{dy}{dx} = \dfrac{5x^2 - 6x + 1}{(2x - 3)^{\frac{1}{2}}}$

    ii $x + 9y - 54 = 0$

16 $30\pi\,\text{cm}^2\,\text{s}^{-1}$

Chapter 13

Pre-requisite Knowledge

1 a

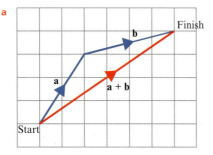

   b

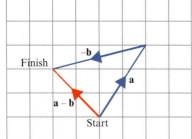

2 a $\begin{pmatrix} -6 \\ 1 \end{pmatrix}$ **b** $\begin{pmatrix} 4 \\ 5 \end{pmatrix}$

3 a $\begin{pmatrix} -2 \\ 4 \end{pmatrix}$ **b** $\begin{pmatrix} 3 \\ -6 \end{pmatrix}$

4 $\begin{pmatrix} -2 \\ 4 \end{pmatrix}\begin{pmatrix} -1 \\ 2 \end{pmatrix}\begin{pmatrix} 4 \\ -8 \end{pmatrix}$

5   10

6 a $\mathbf{q} - \mathbf{p}$ **b** $\dfrac{1}{2}(\mathbf{q} - \mathbf{p})$

   c $\dfrac{1}{2}(\mathbf{p} - \mathbf{q})$ **d** $\dfrac{1}{2}(\mathbf{p} + \mathbf{q})$

Exercise 13.1

1 a $\mathbf{i} - 3\mathbf{j}$ **b** $3\mathbf{i} - 2\mathbf{j}$

   c $4\mathbf{i} - \mathbf{j}$ **d** $2\mathbf{i}$

   e $\mathbf{i} + 3\mathbf{j}$ **f** $-2\mathbf{i} + \mathbf{j}$

   g $-2\mathbf{i}$ **h** $-3\mathbf{i} - 2\mathbf{j}$

   i $-\mathbf{i} - \mathbf{j}$

2 a 2   **b** 5

   c 13   **d** 10

   e 25   **f** 17

   g $4\sqrt{2}$ **h** $5\sqrt{5}$

3 $16\mathbf{i} + 12\mathbf{j}$

4 $36\mathbf{i} - 15\mathbf{j}$

5 a $\dfrac{1}{5}(3\mathbf{i} + 4\mathbf{j})$

   b $\dfrac{1}{13}(5\mathbf{i} + 12\mathbf{j})$

   c $-\dfrac{1}{5}(4\mathbf{i} + 3\mathbf{j})$

   d $\dfrac{1}{17}(8\mathbf{i} - 15\mathbf{j})$

   e $\dfrac{\sqrt{2}}{2}(\mathbf{i} + \mathbf{j})$

6 a $-4\mathbf{i} + 6\mathbf{j}$ **b** $14\mathbf{i} - 9\mathbf{j}$

   c $-26\mathbf{i} - 3\mathbf{j}$ **d** $-\mathbf{i} + 3\mathbf{j}$

7 a 15   **b** $\sqrt{461}$

8 $\lambda = 5$, $\mu = -3$

9 $\lambda = -2$, $\mu = 3$

Exercise 13.2

1 a $-\mathbf{i} - 3\mathbf{j}$ **b** $2\mathbf{i} - 10\mathbf{j}$

   c $3\mathbf{i} + \mathbf{j}$ **d** $-12\mathbf{i} + 3\mathbf{j}$

   e $\mathbf{i} + 7\mathbf{j}$ **f** $-6\mathbf{i} - \mathbf{j}$

2 a $\begin{pmatrix} 4 \\ 10 \end{pmatrix}$ **b** $\begin{pmatrix} -5 \\ 11 \end{pmatrix}$ **c** $\begin{pmatrix} 1 \\ 3 \end{pmatrix}$

3 a $-15\mathbf{i} + 20\mathbf{j}$ **b** $24\mathbf{i} + 10\mathbf{j}$

   c $39\mathbf{i} - 10\mathbf{j}$ **d** $\sqrt{1621}$

4 a $16\mathbf{i} + 12\mathbf{j}$ **b** $\dfrac{1}{5}(4\mathbf{i} + 3\mathbf{j})$ **c** $5\mathbf{i} + 2\mathbf{j}$

5 a $10\mathbf{i} + 24\mathbf{j}$ **b** 26

   c $\dfrac{1}{13}(5\mathbf{i} + 12\mathbf{j})$ **d** $3\mathbf{i} + 8\mathbf{j}$

6   7

7 a $\begin{pmatrix} 20 \\ -21 \end{pmatrix}$ **b** $\begin{pmatrix} 30 \\ -32 \end{pmatrix}$

8 a i 29   **ii** 30   **iii** 38.1

b $\begin{pmatrix} 22.5 \\ -1 \end{pmatrix}$

9 a $12\mathbf{i} + 9\mathbf{j}$ **b** $7\mathbf{i} + \mathbf{j}$

10 a $6\mathbf{i} - 8\mathbf{j}$ **b** $10.5\mathbf{i}$

11 $\begin{pmatrix} 9 \\ 7 \end{pmatrix}$

12 3

13 a 22 or -8
  b -9
  c 1

14 $6\mathbf{j}$

15 $\dfrac{20}{7}\mathbf{i}$

16 a i $2\sqrt{10}$ **ii** $\sqrt{130}$ **iii** $3\sqrt{10}$
  c $\lambda = \dfrac{7}{3}, \mu = \dfrac{2}{3}$

Exercise 13.3

1 a $(1-\lambda)\mathbf{a} + \lambda\mathbf{b}$
  b $\left(\dfrac{1}{2} - \dfrac{1}{2}\mu\right)\mathbf{a} + 3\mu\mathbf{b}$
  c $\lambda = \dfrac{3}{5}, \mu = \dfrac{1}{5}$

2 a i $2\mathbf{a} - \mathbf{b}$ **ii** $3\mathbf{a} + \mathbf{b}$
  b i $\lambda(5\mathbf{a} - \mathbf{b})$ **ii** $\mu(3\mathbf{a} + \mathbf{b})$
  c $\lambda = \dfrac{5}{8}, \mu = \dfrac{3}{8}$

3 a $\lambda(\mathbf{a} + 2\mathbf{b})$
  b $\mu\mathbf{a} + (3 - 3\mu)\mathbf{b}$
  c $\lambda = \dfrac{3}{5}, \mu = \dfrac{3}{5}$

4 a $(1-\lambda)\mathbf{a} + 2\lambda\mathbf{b}$
  b $\dfrac{5}{3}\mu\mathbf{a} + (1 - \mu)\mathbf{b}$
  c $\lambda = \dfrac{2}{7}, \mu = \dfrac{3}{7}$

5 a i $-\mathbf{a} + \mathbf{b}$ **ii** $\dfrac{1}{2}\mathbf{a} + \dfrac{1}{2}\mathbf{b}$
  b $\dfrac{1}{2}\lambda\mathbf{a} + \dfrac{1}{2}\lambda\mathbf{b}$
  c $\dfrac{3}{4}\mu\mathbf{a} + (1 - \mu)\mathbf{b}$
  d $\lambda = \dfrac{6}{7}, \mu = \dfrac{4}{7}$

6 a $\dfrac{3}{5}\lambda\mathbf{a} + \dfrac{2}{5}\lambda\mathbf{b}$

b $\dfrac{5}{7}\mu\mathbf{a} + (1 - \mu)\mathbf{b}$
  c $\lambda = \dfrac{25}{31}, \mu = \dfrac{21}{31}$

7 a i $\lambda\mathbf{a} - \mathbf{b}$ **ii** $-\mathbf{a} + \mu\mathbf{b}$
  b i $\dfrac{2}{5}\lambda\mathbf{a} + \dfrac{3}{5}\mathbf{b}$
   ii $\dfrac{1}{4}\mathbf{a} + \dfrac{3}{4}\mu\mathbf{b}$
   iii $\lambda = \dfrac{5}{8}, \mu = \dfrac{4}{5}$

8 a i $-9\mathbf{a} + 18\mathbf{b}$ **ii** $-5\mathbf{a} + 10\mathbf{b}$
  b $\overrightarrow{AC} = \dfrac{9}{5}\overrightarrow{AB}$, so AB and AC are parallel and A lies on both lines

9 a i $4\mathbf{a}$ **ii** $2\mathbf{b}$ **iii** $-2\mathbf{a} + 3\mathbf{b}$
  b i $6\mathbf{a} - 3\mathbf{b}$ **ii** $2\mathbf{a} - \mathbf{b}$ **iii** $4\mathbf{a} - 2\mathbf{b}$
  c $\overrightarrow{CE} = 3\overrightarrow{CD}$, so CE and CD are parallel and C lies on both lines
  d $1:2$

Exercise 13.4

1 a $(3.5\mathbf{i} + 9\mathbf{j})\,\text{m s}^{-1}$
  b $(30\mathbf{i} - 36\mathbf{j})\,\text{m}$
  c 12.5 hours

2 $(-22\mathbf{i} + 11.6\mathbf{j})\,\text{km h}^{-1}$

3 $125\mathbf{i}\,\text{km}$

4 a $(18\mathbf{i} + 18\mathbf{j})\,\text{km h}^{-1}$
  b $\left(10\mathbf{i} + 10\sqrt{3}\,\mathbf{j}\right)\text{km h}^{-1}$
  c $(-50\sqrt{3}\mathbf{i} - 50\mathbf{j})\,\text{m s}^{-1}$

5 a $20\,\text{m s}^{-1}$
  b i $(-68\mathbf{i} + 44\mathbf{j})\,\text{m}$
   ii $(-56\mathbf{i} + 28\mathbf{j})\,\text{m}$
   iii $(-44\mathbf{i} + 12\mathbf{j})\,\text{m}$
  c $\mathbf{r} = \begin{pmatrix} -80 \\ 60 \end{pmatrix} + t\begin{pmatrix} 12 \\ -16 \end{pmatrix}$

6 a $10\,\text{km h}^{-1}$
  b $(28\mathbf{i} + 14\mathbf{j})\,\text{km}$
  c $\mathbf{r} = \begin{pmatrix} 10 \\ 38 \end{pmatrix} + t\begin{pmatrix} 6 \\ -8 \end{pmatrix}$
  d 20 30 hours

7 a $(12\mathbf{i} - 12\mathbf{j})\,\text{km h}^{-1}$
  b i $(29\mathbf{i} - 12\mathbf{j})\,\text{km}$
   ii $(14\mathbf{i} + 3\mathbf{j})\,\text{km}$
  c $\mathbf{r} = \begin{pmatrix} 5 \\ 12 \end{pmatrix} + t\begin{pmatrix} 12 \\ -12 \end{pmatrix}$

8 **a** $(10\mathbf{i} + 6\mathbf{j})$ km

b $(5\mathbf{i} + 12\mathbf{j})$ km h^{-1}

c 13 km h^{-1}

d 52 km

9 **a** $(15\mathbf{i} + 20\mathbf{j})$ km

b $(8\mathbf{i} + 6\mathbf{j})$ km h^{-1}

c $(111\mathbf{i} + 92\mathbf{j})$ km

10 **a** $(50\mathbf{i} + 70\mathbf{j})$ km, $(40\mathbf{i} + 100\mathbf{j})$ km

b 31.6 km

11 **a** $(6\mathbf{i} + 8\mathbf{j})$ km, $21\mathbf{j}$ km

b 14.3 km

Past paper exercises

1 **i** $8\mathbf{i} - 15\mathbf{j}$

ii $\frac{1}{17}(8\mathbf{i} - 15\mathbf{j})$

iii $-53\mathbf{i}$

2 **a** **i** $5\sqrt{5}$

b $\frac{1}{5\sqrt{5}}(2\mathbf{i} + 11\mathbf{j})$

c $2\mathbf{i} + 1.5\mathbf{j}$

3 **i** 51 km h^{-1}

ii 40 minutes

4 **i** $\lambda\mathbf{b} - \mathbf{a}$

ii $\mu\mathbf{a} - \mathbf{b}$

iii $\frac{2}{3}\mathbf{a} + \frac{1}{3}\lambda\mathbf{b}$

iv $\frac{7}{8}\mu\mathbf{a} + \frac{1}{8}\mathbf{b}$

v $\lambda = \frac{3}{8}, \mu = \frac{16}{21}$

5 **a** **i** $\alpha = 2, \beta = 13$ **ii** $\frac{1}{5}(3\mathbf{i} - 4\mathbf{j})$

b $\overrightarrow{OC} = (1 - \lambda)\mathbf{a} + \lambda\mathbf{b}$ **c** $\mu = 3$

6 **a** $\begin{pmatrix} 48 \\ -90 \end{pmatrix}$

b $p = 2, q = 2$ or $p = 10, q = -38$

7 **i** $\mathbf{c} - \mathbf{a}$ **ii** $\frac{1}{3}\mathbf{a} + \frac{2}{3}\mathbf{c}$ **iii** $\frac{3}{5}\mathbf{b}$

iv $9\mathbf{b}$ **v** $-\frac{4}{9}\mathbf{a} + \frac{10}{9}\mathbf{c}$

8 **i** $4\mathbf{i} - 16\mathbf{j}$ **ii** $\frac{3\mathbf{i} + 8\mathbf{j}}{\sqrt{73}}$ **iii** $-\frac{\lambda}{1 + \lambda}(2\mathbf{i} + 12\mathbf{j})$

9 $\frac{1}{\sqrt{234}}\begin{pmatrix} 15 \\ -3 \end{pmatrix}$

10 **a** $\begin{pmatrix} -20 \\ 48 \end{pmatrix}$ km h^{-1} **b** $t\begin{pmatrix} -20 \\ 48 \end{pmatrix}$

c $\begin{pmatrix} 12 \\ 8 \end{pmatrix} + t\begin{pmatrix} -25 \\ 45 \end{pmatrix}$ **d** $\begin{pmatrix} 12 \\ 8 \end{pmatrix} + t\begin{pmatrix} -5 \\ -3 \end{pmatrix}$ km   **f** 2.15

11 **i** $\begin{pmatrix} 12 \\ 16 \end{pmatrix}$ m s^{-1} **ii** $\begin{pmatrix} 1 \\ 2 \end{pmatrix} + t\begin{pmatrix} 12 \\ 16 \end{pmatrix}$ **iii** $t = 4, \begin{pmatrix} 49 \\ 66 \end{pmatrix}$

12 **a** **i** $\frac{1}{2}(\mathbf{a} + \mathbf{c})$ **ii** $\frac{5}{4}(\mathbf{a} + \mathbf{c})$

b **i** $\mathbf{p} = -15\mathbf{i} + 36\mathbf{j}$ **ii** $\mathbf{q} = 30\mathbf{i} - 60\mathbf{j}$

iii $|\mathbf{q}| = 30\sqrt{5}, k = 30$

13 **a** $\begin{pmatrix} 30 \\ 10 \end{pmatrix} + t\begin{pmatrix} -8 \\ 6 \end{pmatrix}$ **b** 13 m s^{-1} **c** $100\sqrt{2}$ m

Chapter 14

Pre-requisite Knowledge

1 $6x + \frac{12}{x^4} + \frac{1}{\sqrt{x}}$

2 $8(2x - 3)^3$

3 **a** $2(x - 3)^3 (2x + 1)^4 (9x - 13)$

b $\frac{2x^2 + 2x + 10}{(2x + 1)^2}$

4 $y = \frac{x}{4} + 1$

5 $(2, -11)$ minimum, $(-2, 21)$ maximum

6 0.05

Exercise 14.1

1 **a** $7e^{7x}$ **b** $3e^{3x}$

c $15e^{5x}$ **d** $-8e^{-4x}$

e $-3e^{-\frac{x}{2}}$ **f** $3e^{3x + 1}$

g $2xe^{x^2 + 1}$ **h** $5 - \frac{3e^{\sqrt{x}}}{2\sqrt{x}}$

i $-3e^{-3x}$ **j** $-4e^{2x}$

k $\frac{e^x - e^{-x}}{2}$ **l** $10x(1 + e^{x^2})$

2 **a** $xe^x + e^x$ **b** $2x^2e^{2x} + 2xe^{2x}$

c $-3xe^{-x} + 3e^{-x}$ **d** $\sqrt{x}\,e^x + \frac{e^x}{2\sqrt{x}}$

e $\frac{xe^x - e^x}{x^2}$ **f** $\frac{e^{2x}(4x - 1)}{2x^{\frac{3}{2}}}$

g $-\frac{2e^x}{(e^x - 1)^2}$ **h** $2xe^{2x}$

i $\frac{e^x(x^2 + 2xe^x + 2x + 5)}{(e^x + 1)^2}$

3 **a** $5x + 8y = 10$ **b** $5x - 6y + 15\ln 5 = 0$

c $y = 3ex + 2x - 2e - 1$

4 $A\left(-\frac{1}{3}, 0\right)$

5 a $\left(-1, -\frac{1}{e}\right)$, minimum   b $y - e = -\frac{1}{2e}(x - 1)$

   c $\frac{(2e^2 + 1)^2}{4e}$

Exercise 14.2

1 a $\frac{1}{x}$ b $\frac{1}{x}$ c $\frac{2}{2x + 3}$

   d $\frac{2x}{x^2 - 1}$ e $\frac{6}{3x + 1}$ f $\frac{1}{2x + 4}$

   g $\frac{20}{5x - 2}$ h $2 - \frac{1}{x}$ i $\frac{3}{3x - 2}$

   j $\frac{1}{x \ln x}$ k $\frac{1}{(\sqrt{x} + 1)\sqrt{x}}$ l $\frac{2x^2 + 1}{x(x^2 + \ln x)}$

2 a $1 + \ln x$ b $2x + 4x \ln x$

   c $1 - \frac{1}{x} + \ln x$ d $10 + 5 \ln x^2$

   e $\frac{x}{\ln x} + 2x \ln(\ln x)$ f $\frac{1 - \ln 2x}{x^2}$

   g $-\frac{4}{x(\ln x)^2}$ h $\frac{2}{x^2(2x + 1)} - \frac{2\ln(2x + 1)}{x^3}$

   i $\frac{3x^2(2x + 3) - 2(x^3 - 1)\ln(x^3 - 1)}{(2x + 3)^2(x^3 - 1)}$

3   $2 + 4 \ln 6, \ 3 + 2 \ln 6$

4 a $\frac{3}{6x + 2}$ b $\frac{2}{5 - 2x}$

   c $\frac{5(x - 1)}{x(x - 5)}$ d $\frac{-3}{(2x + 1)(x - 1)}$

   e $\frac{x - 4}{x(2 - x)}$ f $\frac{x^2 + 4x + 2}{x(x + 1)(x + 2)}$

   g $\frac{2 - 6x - 2x^2}{(x - 5)(x + 1)(2x + 3)}$ h $\frac{-3x - 1}{(x + 3)(x - 1)}$

   i $\frac{-x^2 + 6x - 3}{x(x - 1)(x + 1)(2x - 3)}$

5 a $\frac{1}{x \ln 3}$ b $\frac{2}{x \ln 2}$ c $\frac{5}{(5x - 1)\ln 4}$

6 a $\frac{8x}{4x^2 - 1}$ b $\frac{15x^2 - 2}{5x^3 - 2x}$ c $\frac{2x - 1}{(x + 3)(x - 4)}$

7   0.2

Exercise 14.3

1 a $\cos x$ b $2 \cos x - 3 \sin x$

   c $-2 \sin x - \sec^2 x$ d $6 \cos 2x$

   e $20 \sec^2 5x$ f $-6 \sin 3x - 2 \cos 2x$

   g $3 \sec^2(3x + 2)$ h $2 \cos\left(2x + \frac{\pi}{3}\right)$

   i $-6 \sin\left(3x - \frac{\pi}{6}\right)$

2 a $3 \sin^2 x \cos x$ b $-30 \cos 3x \sin 3x$

   c $2 \sin x \cos x + 2 \sin x$ d $4(3 - \cos x)^3 \sin x$

   e $12 \sin^2\left(2x + \frac{\pi}{6}\right)\cos\left(2x + \frac{\pi}{6}\right)$

   f $-12 \sin x \cos^3 x$ $+8 \tan\left(2x - \frac{\pi}{4}\right)\sec^2\left(2x - \frac{\pi}{4}\right)$

3 a $x \cos x + \sin x$ b $-6 \sin 2x \sin 3x$ $+4 \cos 2x \cos 3x$

   c $x^2 \sec^2 x + 2x \tan x$ d $\frac{3}{2}x \tan^2\left(\frac{x}{2}\right)\sec^2\left(\frac{x}{2}\right)$ $+\tan^3\left(\frac{x}{2}\right)$

   e $15 \tan 3x \sec 3x$ f $\sec x + x \tan x \sec x$

   g $\frac{x \sec^2 x - \tan x}{x^2}$ h $\frac{1 + 2 \cos x}{(2 + \cos x)^2}$

   i $\frac{(3x - 1)\cos x - 3 \sin x}{(3x - 1)^2}$ j $-6 \cot 2x \operatorname{cosec}^3 2x$

   k $3 \operatorname{cosec} 2x - 6x \cot 2x \operatorname{cosec} 2x$ l $\frac{-2}{(\sin x - \cos x)^2}$

4 a $-\sin x \ e^{\cos x}$ b $-5 \sin 5x \ e^{\cos 5x}$

   c $\sec^2 x \ e^{\tan x}$ d $(\cos x - \sin x) e^{(\sin x + \cos x)}$

   e $e^x(\sin x + \cos x)$ f $\frac{1}{2}e^x\left(2 \cos\frac{1}{2}x - \sin\frac{1}{2}x\right)$

   g $2e^x \cos x$ h $xe^{\cos x}(2 - x \sin x)$

   i $\cot x$ j $x[2 \ln(\cos x) - x \tan x]$

   k $\frac{3 \cos 3x - 2 \sin 3x}{e^{2x - 1}}$ l $\frac{x \cos x + \sin x - x \sin x}{e^x}$

5 a -2 b $\frac{3\sqrt{2} - 8}{6}$

6 a $\tan x \sec x$ b $-\cot x \operatorname{cosec} x$

   c $-\operatorname{cosec}^2 x$

7 a $3 \cot 3x$ b $-2 \tan 2x$

8   $A = 3, B = -5$

9   $A = 4, B = -3$

Exercise 14.4

1   $4x - 24y = \pi$

2 a $2x + 2y = \pi$ b $Q\left(\frac{\pi}{2}, 0\right), R\left(0, \frac{\pi}{2}\right)$

   c $\frac{\pi^2}{8}$ units^2

3   $Q(1, 0)$

4 a $A\left(\frac{1}{2}\ln 5, 0\right), B(0, 4)$ b $C(-8, 0)$

5   $\frac{1}{2}e(2e^2 + 1)$ units^2

6 $\sqrt{2}\,p$

7 $\dfrac{2}{3}p$

8 $\dfrac{1}{4}p$

9 $\dfrac{9}{2}p$

10 $2\sqrt{3}\,p$

11 a $(-2, -2e^{-1})$ minimum

    b $(-1, e^{-2})$ maximum, $(0, 0)$ minimum

    c $(\ln 7, 9 - 7\ln 7)$ minimum

    d $(0, 4)$ minimum

    e $(-2, -4e^2)$ minimum, $(4, 8e^{-4})$ maximum

    f $\left(e^{-\frac{3}{2}}, -\dfrac{1}{2}e^{-1}\right)$, minimum

    g $\left(\sqrt{e}, \dfrac{1}{2}e^{-1}\right)$, maximum

    h $(-\sqrt{e-1},\, e^{-1})$, maximum; $(0, 0)$,

        minimum; $(\sqrt{e-1},\, e^{-1})$, maximum

12 a $(0.927, 5)$, maximum

    b $(1.86, 10)$, maximum

    c $(0, 5)$, maximum; $\left(\dfrac{\pi}{2}, -5\right)$, minimum

    d $\left(\dfrac{\pi}{4}, 3.10\right)$, minimum

    e $\left(\dfrac{\pi}{6}, \dfrac{3\sqrt{3}}{2}\right)$, maximum;

        $\left(\dfrac{5\pi}{6}, \dfrac{3\sqrt{3}}{2}\right)$, minimum

13 a $A = 2, B = 8$

    b $\left(\dfrac{1}{4}\ln 4, 8\right)$, minimum

14 $A\,(1, 0),\, B\,(e^{-1}, -e^{-1})$

15 a $P\,(0, 0),\, Q\,(-2, 4e^{-2})$ **b** $B\,(0, -2e),\, C\left(0, e + \dfrac{1}{3e}\right)$

    c $\dfrac{1}{2}\left(3e + \dfrac{1}{3e}\right)$ units²

16 b $2.7\,\text{cm}^2$ per second

17 b $\dfrac{\pi}{6}$, maximum

Past paper questions

1 a $y = 2x - 1$

2 a $\dfrac{1}{3}\cos 2x \cos\dfrac{x}{3} - 2\sin 2x \sin\dfrac{x}{3}$

    b $\dfrac{x(\sec^2 x)(1 + \ln x) - \tan x}{x(1 + \ln x)^2}$

3 i $\dfrac{2x\sec^2 2x - \tan 2x}{x^2}$

   ii $y - \dfrac{8}{\pi} = -\dfrac{\pi^2}{32(\pi - 2)}\left(x - \dfrac{\pi}{8}\right),$

       so $y = -0.27x + 2.65$

4 i $y + \ln 3 = \dfrac{1}{\ln 3}x$ **ii** $\dfrac{1}{2}(\ln 3)^3$

5 i $-\dfrac{8}{9} + \ln 9$ **ii** 0.0393

6 i $2r\sin\theta$ **iii** -17.8 **iv** -0.842

7 $7.14h$

8 $\left(0, \ln 8 - \dfrac{1}{3}\right)$

9 ii $-0.0261h$

10 i $15\cos 3x$ **ii** $y = -15x + 5\pi + 4$

11 i $4\tan x + 4x\sec^2 x$

   ii $\dfrac{(x^2 - 1)(3\,e^{3x+1}) - 2xe^{3x+1}}{(x^2 - 1)^2}$

12 ii $-\dfrac{2h}{e^4}$

13 a $x = -3.5, x = -1.5$

    b $x = 0.5$, maximum; $x = 3$, minimum

14 $y = 0.923x - 0.692$

15 $-1.41h$

Chapter 15

Pre-requisite Knowledge

1 a -7 **b** -1

2 a $2.5, -3$ **b** $0, \dfrac{4}{25}$

3 a $20x^3 - 5$ **b** $9\sqrt{x} + 10x^4$

Exercise 15.1

1 a $y = \dfrac{12}{5}x^5 + c$ **b** $y = \dfrac{5}{9}x^9 + c$

    c $y = \dfrac{7}{4}x^4 + c$ **d** $y = -\dfrac{2}{x^2} + c$

    e $y = -\dfrac{1}{2x} + c$ **f** $y = 6\sqrt{x} + c$

2 a $y = x^7 + \dfrac{2x^5}{5} + 3x + c$ **b** $y = \dfrac{x^6}{3} - \dfrac{3x^4}{4} + \dfrac{5x^2}{2} + c$

    c $y = -\dfrac{1}{x^3} + \dfrac{15}{x} + \dfrac{x^2}{2} + c$ **d** $y = -\dfrac{2}{x^9} - \dfrac{1}{x^6} - 2x + c$

3 **a** $y = x^3 - 3x^2 + c$ **b** $y = \dfrac{4}{5}x^5 - x^3 + c$

    c $y = \dfrac{8}{5}x^{\frac{5}{2}} + \dfrac{x^3}{3} + 2x^2 + c$ **d** $y = \dfrac{x^4}{4} + \dfrac{x^3}{3} - 6x^2 + c$

    e $y = \dfrac{x^3}{6} + \dfrac{3}{2x} + c$ **f** $y = -\dfrac{2}{x} + \dfrac{5}{2x^2} - \dfrac{1}{x^3} + c$

    g $y = \dfrac{x^4}{8} - 2x - \dfrac{1}{2x} + c$ **h** $y = \dfrac{6}{5}x^{\frac{5}{2}} - \dfrac{2x^{\frac{3}{2}}}{3} - 20\sqrt{x} + c$

4 $y = x^3 - 2x^2 + x + 5$

5 $y = 2x^3 - 3x^2 - 4$

6 $y = x^2 - \dfrac{6}{x} + 3$

7 $y = 8\sqrt{x} - 4x + \dfrac{2}{3}x^{\frac{3}{2}} + 8$

8 $y = 2x^3 - x^2 + 5$

9   **a** $y = 2x^3 - 6x^2 + 8x + 1$

10 $y = 2x^2 - 5x + 1$

Exercise 15.2

1 **a** $\dfrac{x^8}{2} + c$ **b** $2x^6 + c$ **c** $-\dfrac{1}{x^2} + c$

    d $-\dfrac{4}{x} + c$ **e** $6\sqrt{x} + c$ **f** $-\dfrac{4}{x\sqrt{x}} + c$

2 **a** $\dfrac{x^3}{3} + \dfrac{7x^2}{2} + 10x + c$ **b** $\dfrac{2x^3}{3} + \dfrac{x^2}{2} - 3x + c$

    c $\dfrac{x^3}{3} - 5x^2 + 25x + c$ **d** $4x^{\frac{3}{2}} + \dfrac{x^2}{2} + 9x + c$

    e $\dfrac{x^4}{4} - \dfrac{2x^3}{3} + \dfrac{x^2}{2} + c$ **f** $\dfrac{3x^{\frac{7}{3}}}{7} - 3x^{\frac{4}{3}} + c$

3 **a** $x + \dfrac{5}{x} + c$ **b** $\dfrac{x^2}{4} + \dfrac{2}{x^2} + c$

    c $-\dfrac{1}{3x} - \dfrac{1}{3x^2} - \dfrac{1}{9x^3} + c$ **d** $2x^2 - 6\sqrt{x} + c$

    e $\dfrac{2x^{\frac{5}{2}}}{5} - \dfrac{2}{x^{\frac{1}{2}}} + c$ **f** $\dfrac{x^2}{2} - \dfrac{6}{x} - \dfrac{9}{4x^4} + c$

Exercise 15.3

1 **a** $\dfrac{1}{10}(x + 2)^{10} + c$ **b** $\dfrac{1}{14}(2x - 5)^7 + c$

    c $\dfrac{1}{15}(3x + 2)^{10} + c$ **d** $-\dfrac{1}{5}(2 - 3x)^5 + c$

    e $\dfrac{3}{28}(7x + 2)^{\frac{4}{3}} + c$ **f** $\dfrac{2}{15}(3x - 1)^{\frac{5}{2}} + c$

    g $12\sqrt{x + 1} + c$ **h** $-\dfrac{4}{5(5x + 3)^2} + c$

    i $\dfrac{1}{4(3 - 2x)^3} + c$

2 $y = \dfrac{(4x + 1)^5}{20} - 2$

3 $y = \dfrac{\sqrt{(2x + 1)^3}}{3} + 2$

4 $y = 5 - 2\sqrt{10 - x}$

5 $y = (2x - 3)^4 + 1$

6 $y = \dfrac{(3x - 1)^6}{9} + 8$

Exercise 15.4

1 **a** $\dfrac{e^{5x}}{5} + c$ **b** $\dfrac{e^{9x}}{9} + c$ **c** $2e^{\frac{1}{2}x} + c$

    d $-\dfrac{e^{-2x}}{2} + c$ **e** $4e^x + c$ **f** $\dfrac{e^{4x}}{2} + c$

    g $\dfrac{e^{7x+4}}{7} + c$ **h** $-\dfrac{e^{5-2x}}{2} + c$ **i** $\dfrac{e^{6x-1}}{18} + c$

2 **a** $5e^x - \dfrac{e^{3x}}{3} + c$ **b** $\dfrac{e^{4x}}{4} + e^{2x} + x + c$

    c $\dfrac{9e^{2x}}{2} - \dfrac{e^{-2x}}{2} + 6x + c$ **d** $e^x - 4e^{-x} + c$

    e $\dfrac{5e^{2x}}{4} - \dfrac{e^x}{2} + c$ **f** $\dfrac{e^{5x}}{5} - 2e^{2x} - 4e^{-x} + c$

3 **a** $2e^x + 2\sqrt{x} + c$ **b** $\dfrac{1}{3}x^3 - \dfrac{3e^{2x+1}}{2} + c$

    c $\dfrac{e^x}{4} + \dfrac{1}{3x} + c$

4 $y = e^{2x} - e^{-x} + 4$

5   **a** $k = -7$ **b** $y = 7e^{2-x} + 2x^2 - 5$

6 $y = 2e^{-2x} + 6x - 6$

7   **a** $k = 6 - e$ **b** $\left(\dfrac{4e^2 - e - 2}{2e - 2}, \dfrac{24e - 4e^2 - 21}{4e - 4}\right)$

Exercise 15.5

1 **a** $-\dfrac{1}{4}\cos 4x + c$ **b** $\dfrac{1}{2}\sin 2x + c$

    c $-3\cos\dfrac{x}{3} + c$ **d** $\dfrac{1}{5}\tan 5x + c$

    e $\sin 2x + c$ **f** $-2\cos 3x + c$

    g $\dfrac{2}{3}\tan 3x + c$ **h** $\dfrac{3}{2}\sin(2x + 1) + c$

    i $\dfrac{5}{3}\cos(2 - 3x) + c$ **j** $\sin(2x - 7) + c$

    k $\dfrac{4}{5}\cos(1 - 5x) + c$ **l** $-2\tan(1 - 2x) + c$

2 **a** $x + \cos x + c$ **b** $\dfrac{2}{3}(x^{\frac{3}{2}} - \sin 3x) + c$

    c $\dfrac{3}{2}\sin 2x + \dfrac{2}{5}\pi\cos\dfrac{5x}{2} + c$ **d** $-\dfrac{1}{x} - \dfrac{2}{3}\sin\dfrac{3x}{2} + c$

    e $\dfrac{1}{2}(e^{2x} + 5\cos 2x) + c$ **f** $4\sqrt{x} - 2\cos\dfrac{x}{2} + c$

3 $y = \sin x + \cos x + 2$

4 $y = x - 2 \sin 2x + 3 - \dfrac{\pi}{4}$

5 $y = 2x^2 + 3 \cos 2x - 5$

6 $y = -5 \cos 3x - 2 \sin x - 6$

7 **a** $k = 6$ **b** $y = 2 \sin 3x - 4x + 2 + 4\pi$

8 **a** $y = 5 - 2\cos\left(2x - \dfrac{\pi}{2}\right)$ **b** $y = -\dfrac{1}{2}x + \dfrac{\pi}{6} + 5 - \sqrt{3}$

9 **a** $k = 2$ **b** $\left(\dfrac{3\pi - 2}{6}, 3\right)$

10 **a** $k = 0$

    b $W = \left(\dfrac{2 + \pi}{6}, \dfrac{2(\pi - 2)}{3}\right)$

    $\Bigg($Tangents to the curve are:

      $T : y = 8x - 2\pi - \dfrac{4}{3},\ V : y = 4x - \dfrac{4\pi}{3}\Bigg)$

Exercise 15.6

1 **a** $8 \ln x + c$ **b** $5 \ln x + c$ **c** $\dfrac{1}{2} \ln x + c$

    d $\dfrac{5}{3} \ln x + c$ **e** $\dfrac{1}{3} \ln (3x + 2) + c$

    f $-\dfrac{1}{8} \ln (1 - 8x) + c$ **g** $\dfrac{7}{2} \ln (2x - 1) + c$

    h $-\dfrac{4}{3} \ln (2 - 3x) + c$ **i** $\dfrac{1}{2} \ln (5x - 1) + c$

2 **a** $\dfrac{7x^2}{2} + x + 2\ln x + c$ **b** $x - \dfrac{4}{x} + 4\ln x + c$

    c $25x - \dfrac{9}{x} - 30\ln x + c$ **d** $x + \ln x + c$

    e $-\dfrac{1}{2x^3} - \dfrac{5\ln x}{2} + c$ **f** $3x^3 - \dfrac{4}{3x^3} - 12\ln x + c$

    g $5\ln x - \dfrac{4}{\sqrt{x}} + c$ **h** $\dfrac{2x^{\frac{5}{2}}}{5} - \ln x + c$

    i $\dfrac{e^{3x}}{6} - \dfrac{\ln x}{5} + c$

3 $y = \dfrac{5}{2} \ln (2x - 1) + 3$

4 $y = x^2 + 5\ln x - 5$

5 $y = 1 + \ln (x + e)$

6 **a** $1 - 2\ln 2$ **b** $(2\ln 2, 2\ln 2 - 3)$

Exercise 15.7

1 **b** $\dfrac{x + 5}{\sqrt{2x - 1}} + c$

2 **a** $30x(3x^2 - 1)^4$ **b** $\dfrac{1}{30}(3x^2 - 1)^5 + c$

3 **a** $\ln x + 1$ **b** $x\ln x - x + c$

4 **b** $-\dfrac{1}{4x^2} - \dfrac{\ln x}{2x^2} + c$

5 **a** $\dfrac{2(x^2 - 2)}{\sqrt{x^2 - 4}}$ **b** $\dfrac{1}{2}x\sqrt{x^2 - 4} + c$

6 **b** $\dfrac{2}{3}(x + 1)\sqrt{x - 5} + c$

7 **a** $2x\,e^{2x}$ **b** $\dfrac{1}{2}x e^{2x} - \dfrac{1}{4}e^{2x} + c$

8 **a** $k = 1$ **b** $\dfrac{5\sin x}{1 - \cos x} + c$

9 **a** $k = \dfrac{3}{2}$ **b** $\dfrac{2}{3}(x + 8)\sqrt{x - 4} + c$

10 **a** $k = -2$ **b** $-\dfrac{2}{x^2 - 7} + c$

11 **a** $2x^2 + 6x^2 \ln x$ **b** $\dfrac{1}{3}x^3 \ln x - \dfrac{1}{9}x^3 + c$

12 **a** $\cos x - x \sin x$ **b** $\sin x - x \cos x + c$

13 **b** $\dfrac{1}{4}e^{2x}(\sin 2x + \cos 2x) + c$

14 **a** $\dfrac{5x^2 - 14x}{\sqrt{2x - 7}}$ **b** $(x^2 + 3)\sqrt{2x - 7} + c$

Exercise 15.8

1 **a** 127 **b** 1.5 **c** 24

    d 15 **e** 10.5 **f** 7.5

    g $12\dfrac{5}{12}$ **h** $\dfrac{52}{81}$ **i** 75.5

    j 26.4 **k** $3\dfrac{13}{24}$ **l** $5\dfrac{1}{3}$

2 **a** 68 **b** $18\dfrac{2}{3}$ **c** 12.4

    d 0.5 **e** 0.25 **f** 8

3 **a** $\dfrac{1}{2}(e^2 - 1)$ **b** $\dfrac{1}{4}(e - 1)$

    c $\dfrac{5(e^4 - 1)}{2e^4}$ **d** $\dfrac{1}{3}(e - 1)$

    e $\dfrac{5(e^2 - 1)}{2e}$ **f** $\dfrac{1}{2}(e^2 + 4e - 3)$

    g $\dfrac{1}{12}(3e^4 + 8c^3 + 6e^7 - 17)$ **h** $\dfrac{9}{2}e^2 - \dfrac{2}{e^2} - \dfrac{29}{2}$

    i $4e^2 - \dfrac{3}{2e^2} - \dfrac{5}{2}$

4 **a** 2 **b** $\dfrac{3\pi}{2}$ **c** $\dfrac{\sqrt{3}}{4}$

    d $\sqrt{3} - \dfrac{3}{2}$ **e** $\dfrac{\pi^2 - 8}{16}$ **f** $\dfrac{1}{6}(3 - \sqrt{2})$

5 **a** $\dfrac{4}{3}\ln 2$ **b** $\dfrac{1}{2}\ln 11$ **c** $\dfrac{3}{2}\ln 5$

    d $-\dfrac{5}{2}\ln 3$ **e** $-\dfrac{2}{3}\ln\dfrac{11}{2}$ **f** $2\ln\dfrac{9}{7}$

6 **a** $2 + \dfrac{2}{3}\ln\dfrac{11}{5}$ **b** $\dfrac{1}{2}\left(\dfrac{5\ln 3}{\ln 7}\right)$ **c** $2 + \dfrac{1}{2}\ln\dfrac{5}{3}$

7 $k = 5$

8 **a** $A = -6$

9 **a** quotient $= 2x + 1$,
remainder $= -1$

10 $2 - \dfrac{5}{4}\ln 3$

Exercise 15.9

1 **a** $\dfrac{3x}{\sqrt{2x-1}}$ **b** $5\dfrac{1}{3}$

2 **a** $\dfrac{6x^2 + 4}{\sqrt{3x^2 + 4}}$ **b** 4

3 **a** $-\dfrac{2x}{(x^2+5)^2}$ **b** $\dfrac{1}{9}$

4 **a** $\dfrac{3x + 2}{2\sqrt{(3x+4)^3}}$ **b** 2

5 **b** $\dfrac{\pi\sqrt{2}}{20}$

6 **a** $\sin x + x\cos x$ **b** $\dfrac{1}{2}(\pi - 2)$

7 **a** $x + 2x\ln x$ **b** $1 + e^2$

8 **a** $\sin 3x + 3x\cos 3x$ **b** $\dfrac{1}{18}(\pi - 2)$

Exercise 15.10

1 **a** $6\dfrac{3}{4}$ **b** $13\dfrac{1}{2}$ **c** 6

    d $1 + e - \dfrac{2}{e}$ **e** 5 **f** $\dfrac{2 + \pi}{4}$

2 **a** $1\dfrac{1}{3}$ **b** $5\sqrt{2} - 5 - \dfrac{5\pi^2}{72}$

    c $4\ln\dfrac{5}{2}$ **d** $\ln 5$

3 $3\dfrac{1}{12}$

4 **a** $\dfrac{1}{6}$ **b** $20\dfrac{5}{6}$ **c** 8

    d $49\dfrac{1}{3}$ **e** $32\dfrac{3}{4}$ **f** $21\dfrac{1}{3}$

5 12

6 **a** 9 **b** 1.6

7 **b** $1 + e^2$

8 $k = \dfrac{1}{2}(e^4 + 1)$

9 **b** $5\ln 5 - 4$

10 **b** $\pi - 1$

Exercise 15.11

1 2

2 $3\dfrac{1}{3}$

3 16

4 $5 - \dfrac{3\pi}{4}$

5 **a** $10\dfrac{2}{3}$ **b** $20\dfrac{5}{6}$ **c** $1\dfrac{1}{3}$

    d 36 **e** $20\dfrac{5}{6}$

6 **a** $\dfrac{1}{3}$ **b** $\dfrac{3}{4}$

7 34

8 **a** $(-2, 0)$ **b** $42\dfrac{2}{3}$

9 $4\dfrac{4}{15}$

10 $\dfrac{5\pi}{3} - 2\sqrt{3}$

11 $\dfrac{35}{2} - 6\ln 6$

Past paper questions and practice questions

1 **a** $2xe^{x^2}$ **b** $\dfrac{1}{2}e^{x^2}$ **c** 26.8

2 **a** $y = 2x^2 - \dfrac{1}{x+1} + 1$ **b** $8x + 34y = 93$

3 $5 + 2\ln 5$

4 **a** $A = 4$

5 **i** $-4x\,e^{4x}$, so $p = -4$

    ii $4\ln 2 - \dfrac{15}{16}$, so $a = 4$, $b = -15$, $c = 16$

6 **i** $\dfrac{3}{2}x^2 - \dfrac{2}{5}x^{\frac{5}{2}} + c$ **ii** $(9, 0)$ **iv** 16.2

7 **a** $(2x - 1)(x^2 - x - 1)$

    b At A, $x = \dfrac{1}{2}$ and at B, $x = \dfrac{1 + \sqrt{5}}{2}$

    c $\ln(1 + \sqrt{5}) + \dfrac{19}{24}$

8 **a** **i** $-\dfrac{1}{50}(10x - 1)^{-5} + c$

      ii $\dfrac{2}{3}x^6 + \dfrac{20}{3}x^3 + 25\ln x + c$

    b **i** $3\sec^2(3x + 1)$ **ii** 0.322

9 $a = 3$

10 **i** $A(0.25, 3.75)$, $B(4, 15)$ **ii** $\dfrac{45}{16}$

11 i At A, $x = -2$ and at B, $x = -\dfrac{4}{9}$

   ii $C(0, 4)$, $D\left(\dfrac{1}{3}, 0\right)$ **iii** $\dfrac{241}{324}$

12 i $\dfrac{dy}{dx} = 3 - 2\cos 2x$ **ii** $y = 3x - \sin 2x - \dfrac{\pi}{4}$

   iii $y = -0.789x - 0.294$

13 $\dfrac{9\sqrt{3}}{20}$, so $p = 9$

14 a $\dfrac{1}{2}e^{2x+1} + c$

   b i $\dfrac{dy}{dx} = \dfrac{\ln x - 1}{(\ln x)^2}$ **ii** $\dfrac{x}{\ln x} - \dfrac{1}{x} + c$

15 i $3x(2x-1)^{\frac{1}{2}} + (2x-1)^{\frac{3}{2}}$ **ii** $p = 3$, $q = 1$ **iii** $\dfrac{4}{15}$

16 $y = 3x^3 + \dfrac{3}{x} + 1$

17 i $\left(\dfrac{1}{2}, 2\right)$ **iii** $\dfrac{3}{4}$

Chapter 16

Pre-requisite Knowledge

1 a $1.25\,\text{m s}^{-1}$ **b** $1\dfrac{3}{7}\,\text{m s}^{-1}$

2 a $0.75\,\text{m s}^{-2}$ **b** $1800\,\text{m}$

Exercise 16.1

1 a $12.5\,\text{m s}^{-1}$ **b** $t = 6$ **c** $-2.4\,\text{m s}^{-2}$

2 a $6\,\text{m s}^{-1}$ **b** $10\,\text{m s}^{-2}$

3 a $4.95\,\text{m s}^{-1}$ **b** 5 **c** $1\,\text{m s}^{-2}$

   d

4 a $t = \dfrac{1}{3}(e^4 - 2)$ **b** $5.4\,\text{m s}^{-1}$

5 a $t = 2$ **b** $\ln\left(\dfrac{7}{5}\right)\,\text{m}$ **c** $-0.25\,\text{m s}^{-2}$

6 a $t = \dfrac{4\pi}{3}$ **b** $-1.90\,\text{m s}^{-2}$

   c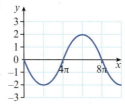

7 a $t = \dfrac{\pi}{4}$ **b** $-3\,\text{m s}^{-2}$

8 a $v = 4\sin 2t$, $a = 8\cos 2t$

   b $t = \dfrac{\pi}{2}$ seconds, $s = 4\,\text{m}$

9 a $t = \dfrac{1}{2}\ln 2$ **b** $0.899\,\text{m}$ **c** $24.1\,\text{m}$

10 a $2\,\text{m}$ away from O

   b $v = 2 - 4\sin 2t$, $a = -8\cos 2t$

   c $t = \dfrac{\pi}{12}$ seconds, $s = \dfrac{\pi}{6} + \sqrt{3}$

   d $t = \dfrac{\pi}{4}$ seconds, $s = \dfrac{\pi}{2}$

11 a $t = \dfrac{\pi}{8}$ seconds   **b** $a = -4k\sin 4t$ **c** $k = 5$

12 a $2\,\text{m s}^{-1}$ **b i** $2\,\text{m s}^{-1}$ **ii** $\dfrac{2}{3}\,\text{m s}^{-1}$

   c i $0\,\text{m s}^{-2}$ **ii** $-\dfrac{2}{9}\,\text{m s}^{-2}$

   d

   e $2\ln\dfrac{5}{4}\,\text{m}$

13 a $2\,\text{m}$ **b** $t = \dfrac{5}{3}$ seconds

   c $0 \le t < \dfrac{5}{3}$ and $t > 4$ **d** $\dfrac{5}{3} < t < 4$

Exercise 16.2

1 a $16\,\text{m s}^{-1}$ **b** 15 seconds   **c** $12\dfrac{2}{3}\,\text{m}$

   d $166\dfrac{2}{3}\,\text{m}$ **e** $189\dfrac{1}{3}\,\text{m}$

2 a $-1\,\text{m s}^{-2}$ **b** $s = -\dfrac{32}{t+2}$ **c** $1.6\,\text{m}$

3 **a** $2(4e^2 + 1)\,\text{m s}^{-2}$ **b** $s = 2e^{2t} + t^2 - 2$

c $111\,\text{m}$

4 $\left(\dfrac{9\pi^2}{8} + 6\right)\text{m}, \dfrac{1}{3}\,\text{m s}^{-2}$

5 **b** $29.75\,\text{m s}^{-2}$ **c** $109\,\text{m}$

6 $p = 6, q = -6$

7 **a** $t = 5$

b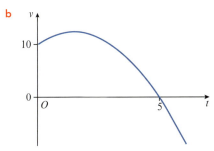

c $62.5\,\text{m}$

8 **a** $t = 2, t = 6$ **b** $5.5\,\text{m}$ **c** $112\,\text{m}$

9 **a** $-12 \leqslant a \leqslant 12$ **b** $5.74\,\text{m}$

10 **a** $0.64\,\text{m s}^{-2}$ **b** $12.1\,\text{m}$

11 **a** $t = \dfrac{2\pi}{3}$ **b** $\left(4\sqrt{3} + \dfrac{2\pi}{3}\right)\text{m}$

12 **a** $p = \dfrac{1}{2}, q = -2$ **b** $s = \dfrac{1}{12}t^3 - t^2 + 3t$

c $t = 6$ **d** $\dfrac{11}{12}\text{m}$

13 **a** $v = \dfrac{2\pi t}{3}\cos\dfrac{\pi t}{3} + 2\sin\dfrac{\pi t}{3},$

$a = -\dfrac{2\pi^2 t}{9}\sin\dfrac{\pi t}{3} + \dfrac{4\pi}{3}\cos\dfrac{\pi t}{3}$

Past paper questions

1 **i** $t = 1$ **ii** $\dfrac{2}{3}t^3 - 7t^2 + 12t$ **iii** $-2\,\text{m s}^{-2}$

2 **a** **i**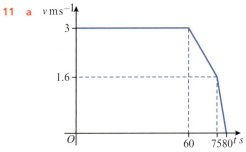

ii $k = 30, a = 1\,\text{m s}^{-2}$

b $50\,\text{m s}^{-1}$

3 **i** $\dfrac{1}{2}e^{2t} + 3e^{-2t} - t - 3.5$ **ii** $\dfrac{1}{2}\ln 3$ **iii** $10\,\text{m s}^{-2}$

4 **ii** $0\,\text{m s}^{-2}$ **iii** $3t^3 - \dfrac{63}{2}t^2 + 90t$

iv **a** $78\,\text{m}$ **b** $88.5\,\text{m}$

5 **i** $x = 6e^{2t} - 24t^2 - 6$ **ii** $\dfrac{1}{2}\ln 2$

iii $24 - 24\ln 2\,\text{m s}^{-1}$

6 **a** **i** $a = 0$ **ii** $110\,\text{m}$

b **i** $2.5\,\text{m s}^{-1}$ **ii** $t = \dfrac{\pi}{4}$

7 **a** $t = 3$ **b** $3\,\text{m}$

8 $0.565\,\text{m}$

9 **i** answer between $0.48\,\text{m}$ and $0.49\,\text{m}$

ii $-25\,\text{m s}^{-2}$

10 **i** $-\dfrac{5}{3}\,\text{m s}^{-2}$ **ii** $k = 27$ **iii** $280\,\text{m}$

11 **a**

b $218.5\,\text{m}$ **c** $0.32\,\text{m s}^{-2}$

12 **i** $v = 6\cos 2t - 8\sin 2t,$
$a = -12\sin 2t - 16\cos 2t$

ii $t = 0.322$ seconds **iii** $a = -20\,\text{m s}^{-2}$

13 **i** $v = 2e^{2t} - 10e^t - 12, a = 4e^{2t} - 10e^t$

ii $t = \ln 6$ seconds **iii** $84\,\text{m s}^{-2}$

> Index